D1619753

Edited by
Nazario Martin and
Jean-Francois Nierengarten

Supramolecular Chemistry of Fullerenes and Carbon Nanotubes

Related Titles

Klauk, H. (ed.)

Organic Electronics II

More Materials and Applications

2012

ISBN: 978-3-527-32647-1

Siebbeles, L. D. A., Grozema, F. C. (eds.)

Charge and Exciton Transport through Molecular Wires

2011

ISBN: 978-3-527-32501-6

Samori, P., Cacialli, F. (eds.)

Functional Supramolecular Architectures

for Organic Electronics and Nanotechnology

2011

ISBN: 978-3-527-32611-2

Guldi, D. M., Martín, N. (eds.)

Carbon Nanotubes and Related Structures

Synthesis, Characterization, Functionalization, and Applications

2010

ISBN: 978-3-527-32406-4

Martín, N., Giacalone, F. (eds.)

Fullerene Polymers

Synthesis, Properties and Applications

2009

ISBN: 978-3-527-32282-4

Cademartiri, L., Ozin, G. A.

Concepts of Nanochemistry

2009

ISBN: 978-3-527-32626-6

Steed, J. W., Atwood, J. L.

Supramolecular Chemistry

ISBN: 978-0-470-51234-0

Atwood, J. L., Steed, J. W. (eds.)

Organic Nanostructures

2008

ISBN: 978-3-527-31836-0

Diederich, F., Stang, P. J., Tykwinski, R. R. (eds.)

Modern Supramolecular Chemistry

Strategies for Macrocycle Synthesis

2008

ISBN: 978-3-527-31826-1

Edited by Nazario Martin
and Jean-Francois Nierengarten

Supramolecular Chemistry of Fullerenes and Carbon Nanotubes

WILEY-VCH Verlag GmbH & Co. KGaA

The Editors

Prof. Dr. Nazario Martín
Universidad Complutense de Madrid
Departamento de Química Orgánica
Ciudad Universitaria s/n
28040 Madrid
Spain

Prof. Jean-Franc. Nierengarten
Université de Strasbourg
UMR 7509
25 rue Becquerel
67087 Strasbourg Cedex 2
France

Library of Congress Card No.: applied for

British Library Cataloguing-in-Publication Data
A catalogue record for this book is available from the British Library.

Bibliographic information published by the Deutsche Nationalbibliothek
The Deutsche Nationalbibliothek lists this publication in the Deutsche Nationalbibliografie; detailed bibliographic data are available on the Internet at http://dnb.d-nb.de.

Print ISBN: 978-3-527-32789-8
ePDF ISBN: 978-3-527-65015-6
ePub ISBN: 978-3-527-65014-9
mobi ISBN: 978-3-527-65013-2
oBook ISBN: 978-3-527-65012-5

Typesetting Thomson Digital, Noida, India
Printing and Binding Markono Print Media Pte Ltd, Singapore
Cover Design Formgeber, Eppelheim

Printed in Singapore
Printed on acid-free paper

Contents

Preface

Three great scientists, Jean-Marie Lehn, Donald J. Cram, and Charles J. Pedersen, received the Nobel Prize in Chemistry in 1987 for the development and utilization of molecules with highly selective structure-specific interactions. Only 2 years before, in 1985, fullerenes were discovered and, in 1996, Sir Harold W. Kroto, Robert F. Curl, and the late Richard E. Smalley were awarded the Nobel Prize in Chemistry for the discovery of these new carbon allotropes. Soon after this seminal finding, the discovery of multi- and single-wall carbon nanotubes (CNTs) in 1991 and 1993, respectively, provided a new kind of carbon allotropes with cylindrical geometry that belong structurally to the family of fullerenes. Although both areas of chemistry (supramolecular chemistry and fullerenes) have been thoroughly studied in the past two decades and very significant advances in terms of basic knowledge and practical applications have independently been made, more recently they gave rise to a new interdisciplinary field in which the imagination of chemists has afforded unprecedented fullerene-based supramolecular architectures.

This year we celebrate the 25th anniversary of the awarding of Nobel to those scientists who brought to the attention of the scientific community the new concept of supramolecular chemistry. Therefore, it is an excellent opportunity to comment on the most important achievements and future goals in this emerging field of supramolecular chemistry of fullerenes and carbon nanotubes.

The huge number of publications during the past decade devoted to fullerenes and supamolecular chemistry attest the interest in this new avenue of chemistry stemming from both fields. The use of concepts and principles of supramolecular chemistry to fullerenes and carbon nanotubes has reached an outstanding position in its own right and we certainly believe that it constitutes a new interdisciplinary field with basic research interest and important potential applications in fields such as biomedical and materials sciences.

Therefore, in this timely book containing 14 chapters, we have gathered the most important developments authored by leading scientists actively engaged in supramolecular/fullerene research, thus giving a precise picture on the state of the art in this new hybrid field.

One of the major challenges nowadays in chemistry is the control of weak forces, on a molecular basis, which will eventually lead to the definition of the size and shape in relation to function of the resulting supramolecular ensembles. In this regard, the rigid structure and round and rod shape of fullerenes and CNTs, respectively, as well as their remarkable electronic properties result in rather unique scaffolds for the development of unprecedented carbon-based nanoarchitectures. For these purposes, the different types of weak forces, namely, hydrogen bonding, π–π stacking, coordination of metal cations, electrostatic interactions, and solvophobic forces have been used in the different chapters to construct new noncovalently bonded structures. A singular aspect in the construction of new architectures involving these weak forces is that, in contrast to covalently bonded structures, they are reversible and their binding energies can be tailored "at will" by means of the chemical environment and temperature.

The book starts with the first chapter devoted to the general introduction to the field of fullerenes and CNTs emphasizing the main achievements in covalent and macromolecular chemistry. The following chapters are devoted to the main supramolecular topics such as H-bonded fullerene assemblies, receptors for pristine fullerenes based on concave–convex π–π interactions, cooperative effects on the self-assembly of fullerene–donor ensembles, fullerene-containing catenanes and rotaxanes, biomimetic motifs toward the construction of artificial reaction centers, supramolecular chemistry of fullerene-containing micelles and gels, fullerene-containing supramolecular polymers and dendrimers, fullerene-containing thermotropic liquid crystals, organizing fullerenes on solid surfaces with STM, supramolecular chemistry of fullerenes and carbon nanotubes at interfaces: toward the application of supramolecular ensembles with fullerenes and CNTs, solar cells and transistors, and supramolecular chemistry of carbon nanotubes. The last chapter is dedicated to the experimental determination of association constants involving fullerenes.

The guest editors want to express their gratitude to the many authors who have participated in this venture for their efforts to bring out this outstanding and unique book in which, for the first time, the supramolecular chemistry of the important carbon allotropes, fullerenes and CNTs, are brought together for the benefit of the reader.

We hope that this book on this new interdisciplinary field be a stimulus for young researchers and would put a new heart into other colleagues to develop new chemical concepts and molecular architectures in which imagination would be the only limitation.

Nazario Martín and Jean-François Nierengarten, editors

List of Contributors

Davide Bonifazi
University of Namur (FUNDP)
Department of Chemistry
Rue Bruxelles 61
5000 Namur
Belgium

University of Trieste
Department of Chemical and Pharmaceutical Sciences
Piazzale Europa 1
34127 Trieste
Italy

Stéphane Campidelli
CEA, IRAMIS
Laboratoire d'Electronique Moléculaire
CEA Saclay
91191 Gif sur Yvette Cedex
France

Robert Deschenaux
Université de Neuchâtel
Institut de Chimie
Laboratoire de Chimie Macromoléculaire
Avenue de Bellevaux 51
2000 Neuchâtel
Switzerland

Bertrand Donnio
Université de Strasbourg
UMR 7504 CNRS
Institut de Physique et Chimie des Matériaux de Strasbourg
23 Rue du Loess, BP43
67034 Strasbourg Cedex 2
France

Arianna Filoramo
CEA, IRAMIS
Laboratoire d'Electronique Moléculaire
CEA Saclay
91191 Gif sur Yvette Cedex
France

José María Gallego
Instituto Madrileño de Estudios Avanzados en Nanociencia (IMDEA-Nanociencia)
Cantoblanco
28049 Madrid
Spain

Consejo Superior de Investigaciones Científicas
Instituto de Ciencia de Materiales de Madrid
Cantoblanco
28049 Madrid
Spain

Francesco Giacalone
Università di Palermo
Department of Organic Chemistry
"E. Paternò"
90128 Palermo
Italy

Davide Giust
University of Castilla la Mancha
Department of Organic
Inorganic and Biochemistry
Av. da José Cela 10
13071 Ciudad Real
Spain

Bruno Grimm
Friedrich-Alexander-Universitaet
Erlangen-Nuermberg
Interdisciplinary Center for Molecular
Materials (ICMM)
Department of Chemistry and
Pharmacy
Egerlandstr. 3
91058 Erlangen
Germany

Daniel Guillon
Université de Strasbourg
UMR 7504 CNRS
Institut de Physique et Chimie des
Matériaux de Strasbourg
23 Rue du Loess, BP43
67034 Strasbourg Cedex 2
France

Dirk M. Guldi
Friedrich-Alexander-Universitaet
Erlangen-Nuermberg
Interdisciplinary Center for Molecular
Materials (ICMM)
Department of Chemistry and
Pharmacy
Egerlandstr. 3
91058 Erlangen
Germany

Takeharu Haino
Hiroshima University
Graduate School of Science
Department of Chemistry
1-3-1 Kagamiyama
Higashi-Hiroshima City 739-8526
Japan

Mª Ángeles Herranz
Universidad Complutense de Madrid
Facultad de Ciencias Químicas
Departamento de Química Orgánica
28040 Madrid
Spain

Toshiaki Ikeda
Hiroshima University
Graduate School of Science
Department of Chemistry
1-3-1 Kagamiyama
Higashi-Hiroshima City 739-8526
Japan

Beatriz M. Illescas
Universidad Complutense de Madrid
Facultad de Ciencias Químicas
Departamento de Química Orgánica
28040 Madrid
Spain

Hiroshi Imahori
Kyoto University
Institute for Integrated Cell-Material
Sciences (iCeMS)
Kyotodaigaku-katsura
Nishikyo-ku
Kyoto 615-8501
Japan

Kyoto University
Graduate School of Engineering
Department of Molecular Engineering
Nishikyo-ku
Kyoto 615-8501
Japan

Bruno Jousselme
CEA, IRAMIS
Laboratoire de Chimie des Surfaces et Interfaces
CEA Saclay
91191 Gif sur Yvette Cedex
France

Takeshi Kawase
University of Hyogo
Graduate School of Engineering
Department of Materials Science and Chemistry
Shosha 2167
Himeji
Hyogo 671-2280
Japan

Adrian Kremer
University of Namur (FUNDP)
Department of Chemistry
Rue Bruxelles 61
5000 Namur
Belgium

Hongguang Li
Max Planck Institute of Colloids and Interfaces
Department of Interfaces
MPI-NIMS International Joint Laboratory
Am Mühlenberg 1
14476 Potsdam
Germany

Riccardo Marega
University of Namur (FUNDP)
Department of Chemistry
Rue Bruxelles 61
5000 Namur
Belgium

Nazario Martín
Universidad Complutense de Madrid
Facultad de Ciencias Químicas
Departamento de Química Orgánica
28040 Madrid
Spain

Universidad Autónoma de Madrid
Facultad de Ciencias
IMDEA-Nanociencia
Ciudad Universitaria de Cantoblanco
Módulo C-IX, 3ª Planta
28049 Madrid
Spain

Aurelio Mateo-Alonso
Albert-Ludwigs-Universität Freiburg
Freiburg Institute for Advanced Studies (FRIAS)
School of Soft Matter Research
Albertstrasse 19
79104 Freiburg im Breisgau
Germany

Albert-Ludwigs-Universität Freiburg
Institut für Organische Chemie und Biochemie
Albertstrasse 21
79104 Freiburg im Breisgau
Germany

Rodolfo Miranda
Universidad Autónoma de Madrid
Departamento de Física de la Materia Condensada
Cantoblanco
28049 Madrid
Spain

Instituto Madrileño de Estudios Avanzados en Nanociencia (IMDEA-Nanociencia)
Cantoblanco
28049 Madrid
Spain

Takashi Nakanishi
National Institute for Materials Science (NIMS)
Organic Materials Group
1-2-1 Sengen
Tsukuba 305-0047
Japan

Jean-François Nierengarten
Université Louis Pasteur et CNRS (UMR 7509)
Ecole Européenne de Chimie Polymères et Matériaux (ECPM)
Laboratoire de Chimie des Matériaux Moléculaires
25 rue Becquerel
67087 Strasbourg Cedex 2
France

Roberto Otero
Universidad Autónoma de Madrid
Departamento de Física de la Materia Condensada
Cantoblanco
28049 Madrid
Spain

Instituto Madrileño de Estudios Avanzados en Nanociencia (IMDEA-Nanociencia)
Cantoblanco
28049 Madrid
Spain

Emilio M. Pérez
IMDEA Nanociencia
Campus Universitario de Cantoblanco
Facultad de Ciencias Módulo C-IX, 3ª planta
28049 Madrid
Spain

Luis Sánchez
Universidad Complutense de Madrid
Facultad de Ciencias Químicas
Departamento de Química Orgánica
28040 Madrid
Spain

Sukumaran Santhosh Babu
National Institute for Materials Science (NIMS)
Organic Materials Group
1-2-1 Sengen
Tsukuba 305-0047
Japan

Max Planck Institute of Colloids and Interfaces
Department of Interfaces
MPI-NIMS International Joint Laboratory
Am Mühlenberg 1
14476 Potsdam
Germany

José Santos
Universidad Complutense de Madrid
Facultad de Ciencias Químicas
Departamento de Química Orgánica
28040 Madrid
Spain

Tomokazu Umeyama
Kyoto University
Graduate School of Engineering
Department of Molecular Engineering
Nishikyo-ku
Kyoto 615-8501
Japan

1
Carbon Nanostructures: Covalent and Macromolecular Chemistry

Francesco Giacalone, Mª Ángeles Herranz, and Nazario Martín

1.1
Introduction

The aim of this introductory chapter is to bring to the attention of the readers the achievements made in the chemistry of carbon nanostructures and, mostly, in the chemistry of fullerenes, carbon nanotubes (CNTs), and the most recent graphenes. Since the discovery of fullerenes in 1985 and their further preparation in multigram amounts, the chemistry and reactivity of these molecular carbon allotropes have been well established. Actually, this chemical reactivity has been used as a benchmark for further studies carried out in the coming carbon nanotubes (single and multiple wall) and graphenes. Assuming that the fundamental chemistry of fullerenes is known and basically corresponds to that of typical electron-deficient alkenes, we have mainly focused on the chemistry of fullerene-containing polymers. In this regard, the combination of the unique fullerenes with the highly versatile polymer chemistry has afforded a new and interdisciplinary field in which the resulting architectures are able to exhibit unprecedented properties. The basic knowledge of this important topic of macromolecular chemistry of fullerenes nicely complements the following chapters devoted to their supramolecular chemistry.

The chemistry of CNTs, on the other hand, is considerable less developed than that of fullerenes, and most of their studied reactions are generally based on those previously studied on fullerenes. Therefore, despite the recent reviews and books published on CNTs, we feel that an introductory chapter describing the most significant solubilization/derivatization covalent and noncovalent methods should be helpful and welcome by the readers, and particularly to those nonexperts in the field. This same objective has been pursued for the most recent and planar graphenes. The available literature on the chemistry of these one-atom thickness carbon allotropes is considerably less developed. Therefore, some useful chemical procedures reported so far for the functionalization and solubilization of graphenes – thus allowing its manipulation and application for the construction of devices – have also been included at the end of the chapter.

Supramolecular Chemistry of Fullerenes and Carbon Nanotubes, First Edition. Edited by Nazario Martin and Jean-Francois Nierengarten.

1.2 Fullerene-Containing Polymers

Since the achievement of [60]fullerenes in ponderable amounts [1], its combination with macromolecular chemistry provided an opportunity to generate new fullerene-containing polymers, with potential for a broad scope of real applications since it merges C_{60} properties with the ease and versatile processability and handling of polymers. This approach prompted chemists to design and develop synthetic strategies aimed to obtain even more complex and fascinating novel fullerene-based architectures with unprecedented properties that have been recently reviewed [2]. In fact, chemists are now able to tailor at will a polymeric backbone possessing C_{60} moieties in such way as to achieve peculiar properties of the final macromolecular material. In this way, block copolymer with well-defined donor–acceptor domains within the diffusion path of electron are created for solar devices [3], or water-soluble biocompatible and biodegradable polymers are designed in order to carry fullerene in circulation for photodynamic cancer therapy purposes [4]. These recent achievements are only the tip of the iceberg of a growing field in which almost all the materials display outstanding properties such as optical limiting [5], or photoinduced electron transfer [6] just to name a few. In addition, polyfullerenes have been successfully employed as active materials not only in electroluminescent devices [7] but also in nonvolatile flash devices [8], and in membranes both for gas separation [9] and for proton exchange fuel cells [10].

The many types of polymeric fullerene derivatives may be classified according to their structural features. As a criterion for the classification, polyfullerenes can be ordered as a function of their increasing chemical complexity and the difficulty to synthesize them, and other interesting classes of nondiscrete multifullerene-containing hybrid materials may also be included in this classification (Figure 1.1).

All-fullerene polymers are specifically those materials or structures that are constituted exclusively by fullerene units covalently linked to each other [11]. These "intrinsic polymers" are prepared simply by exposing pure fullerene to visible light [12], high pressure [13], electron beam [14], and plasma irradiation [15] without control or care for the final structure. Recently, it has been shown that unbound and bound states of C_{60} molecules can be reversibly controlled at the single-molecule level in ultrathin films of all-C_{60} polymers using a tip of a scanning tunneling microscope at RT, thus allowing single-molecule-level topochemical digital data [16].

Heteroatom-containing polymers have metals or elements other than carbon inserted in between two C_{60} moieties. In 1994, Forró discovered that the fulleride phases AC_{60} (A: K, Rb, Rb or Cs), undergo [2 + 2] cycloadditions producing polyfullerenes with alkali metals in the crystal voids [17]. For the organometallic polymers, the metal doping of C_{60} leads to the formation of the corresponding charge transfer polymer [18]. Several different metals have been employed in the polymerization with fullerenes, but among them palladium led to the most promising copolymers with outstanding properties [19]. In fact, the polymer $(C_{60}Pd_3)_n$ has been able to catalyze heterogeneous hydrogenation reactions of alkenes [20]. Very

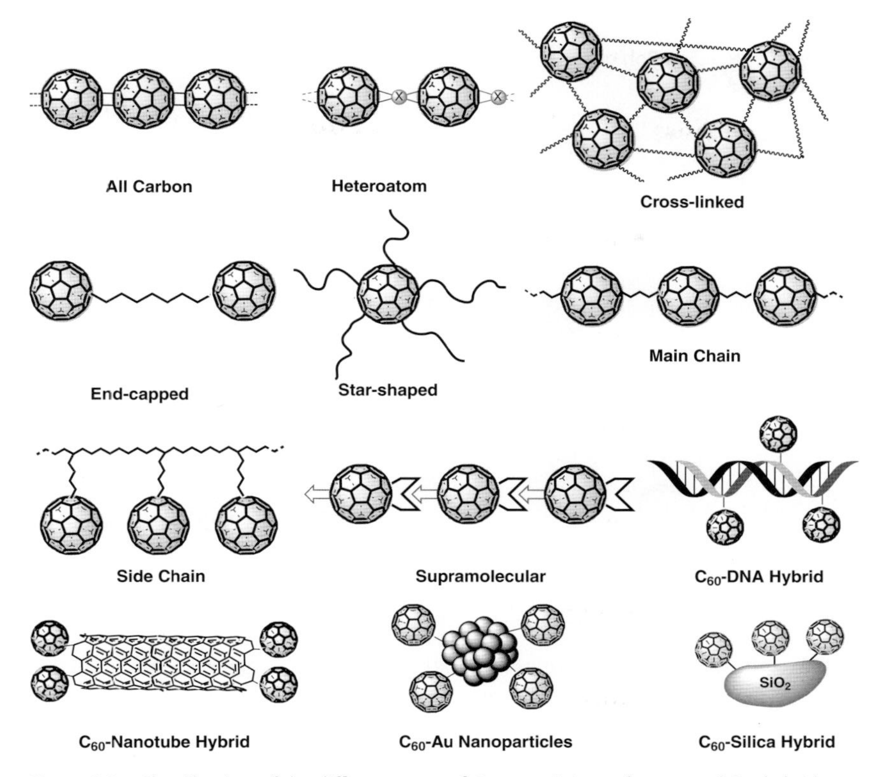

Figure 1.1 Classification of the different types of C_{60}-containing polymers and C_{60}-hybrid materials.

recently, 1D "zigzag" polymers have been achieved by mixing fullerene with the oxidizing superacid AsF_5 [21]. As a result, a polyfullerenium salt in the solid state has been prepared, with $C_{60}{}^{2+}$ units connected by an alternating sequence of four-membered carbon rings ([2 + 2] cycloaddition) and single C−C bonds stabilized by $ASF_6{}^-$ anions in the lattice.

Cross-linked C_{60} polymers are synthesized from random and quick reactions that proceed in three dimensions with the help of the fullerene topology. Nevertheless, some extent of control of the addition reactions to the 30 fullerene double bonds is required in order to avoid a drastic intractability of the final products. This class of polymers can be prepared by four main pathways (Figure 1.2):

a) C_{60} or a C_{60} derivative and a monomer are mixed together and allowed to react randomly.
b) A multisubstituted C_{60} derivative is homopolymerized in three dimensions.
c) A preformed polymer properly functionalized at the end termini is allowed to react with C_{60} or a multisubstituted C_{60} derivative.
d) Polymers endowed with pending reacting moieties are allowed to react with C_{60}.

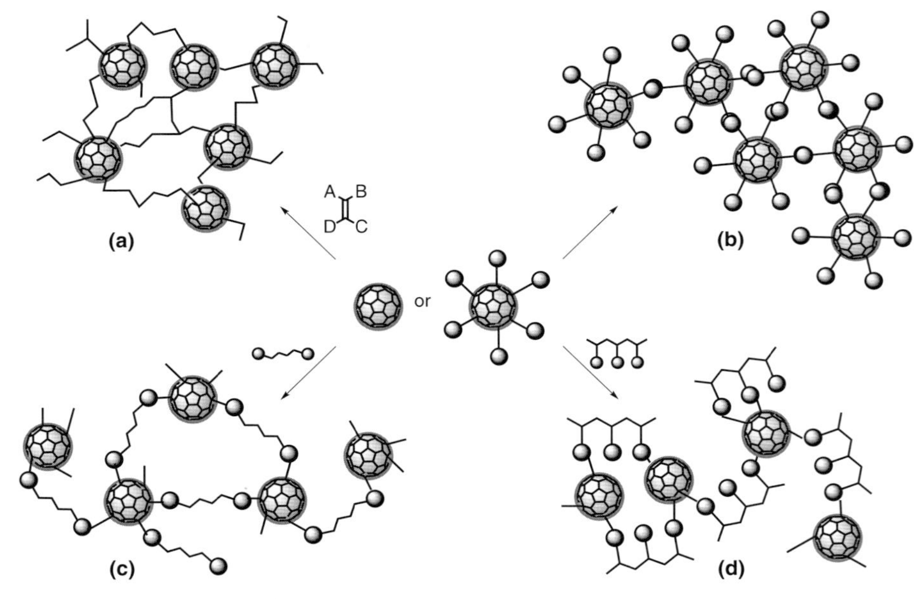

Figure 1.2 Strategies for the synthesis of C_{60} cross-linked polymers.

Although cross-linked polymers are rather intractable materials, even in this family of polyfullerenes there exist some interesting and useful examples: random fullerene-containing polystyrenes (PS) [22] and poly(methylmethacrylate) [23] showed outstanding nonlinear optics. Nevertheless, cross-linked fullerene-containing polyurethanes displayed third-order NLO response with one–two orders of enhancement in comparison with other C_{60} derivatives [24]. But the very "revenge" of this family against all the other processable C_{60} polymers stems from a novel approach described recently by Cheng and Hsu, which will create an organic photovoltaics revolution [25]. They generated *in situ* a robust cross-linked C_{60} polymer (**1**, Figure 1.3), which permits to sequentially deposit the active layer avoiding interfacial erosion. The inverted solar cell ITO/ZnO/**1**/P_3HT:PCBM/PEDOT:PSS/Ag showed an outstanding device with a characteristic power conversion efficiency (PCE) of 4.4% together with an improved cell lifetime with no need for encapsulation.

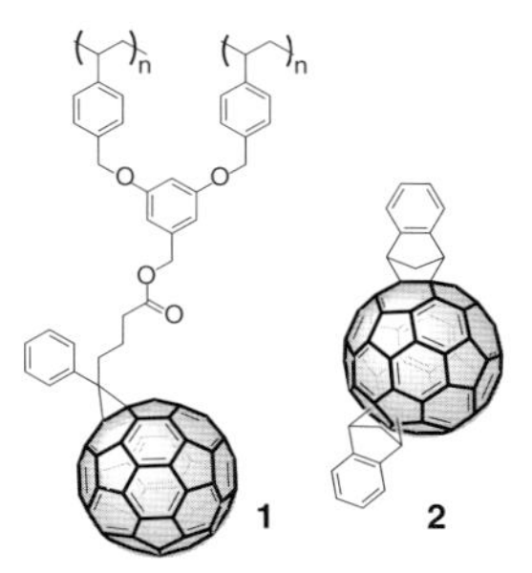

Figure 1.3 Cross-linked polymers used in solar cells.

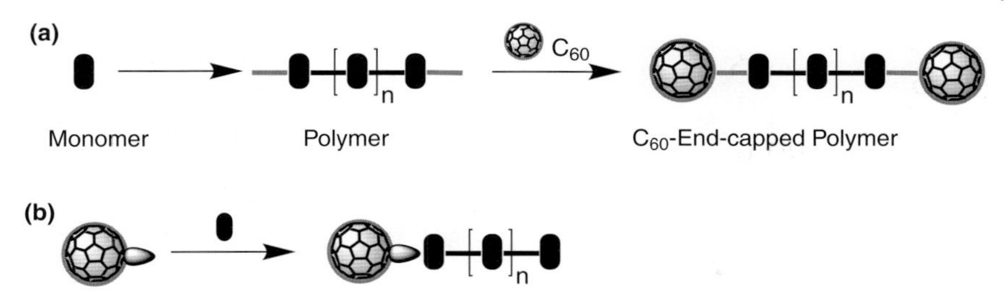

Figure 1.4 Strategies for the synthesis of C_{60} end-capped polymers.

Some months later, the same group reported the performance of a solar cell obtained employing fullerene bis-adduct **2** in a device with architecture ITO/ZnO/**1**/P3HT:2/PEDOT:PSS/Ag achieving the impressive value of 6.22% efficiency, which retains 87% of the magnitude of its original PCE value after being exposed to ambient conditions for 21 days [26].

End-capped polymers, also called telechelic polymer, represent a clear example in which the introduction of only one or two units of C_{60} in the terminal positions of the polymeric backbone may strongly influence both molecular and bulk behavior of well-established polymers. In fact, the presence of fullerene moieties at the end of the polymeric chain can modify the hydrophobicity of the parent polymer and, subsequently, its properties. Two different synthetic strategies have been followed in order to prepare these polymers (Figure 1.4):

a) "grafting to": the polymer is synthesized first, followed by incorporation of fullerenes.
b) "grafting from": the polymeric chain is grown from a C_{60} derivative as the starting material.

In the past years, water-soluble end-capped polymers are finding applications not only as radical scavengers [27], antitumor agents, in photodynamic cancer therapy (PDT) [28] but also as compatibilizer between the donor polymer and the fullerene acceptor in organic solar cells [29].

Star-shaped C_{60} polymers, also known as "flagellenes," are composed of 2- to 12-long and flexible polymer chains directly linked to a fullerene moiety resembling sea stars. So far, their syntheses have been carried out by following two different synthetic pathways, analogously to end-capped polymers (Figure 1.5):

a) The "graft-to" approach, in which preformed polymers are linked to the fullerene unit.
b) The "graft-from" approach, in which the polymeric chains grow from the surface of properly functionalized fullerene derivatives.

In the past years, water-soluble and biodegradable star-shaped C_{60}–poly(vinyl alcohol) [30] and star-shaped [31] poly(vinylpyrrolidone) proved to be effective as photosensitizers to produce singlet oxygen resulting in good candidates for PDT. Also, a star-shaped PEG-C_{60} showed good singlet oxygen generation and low toxicity

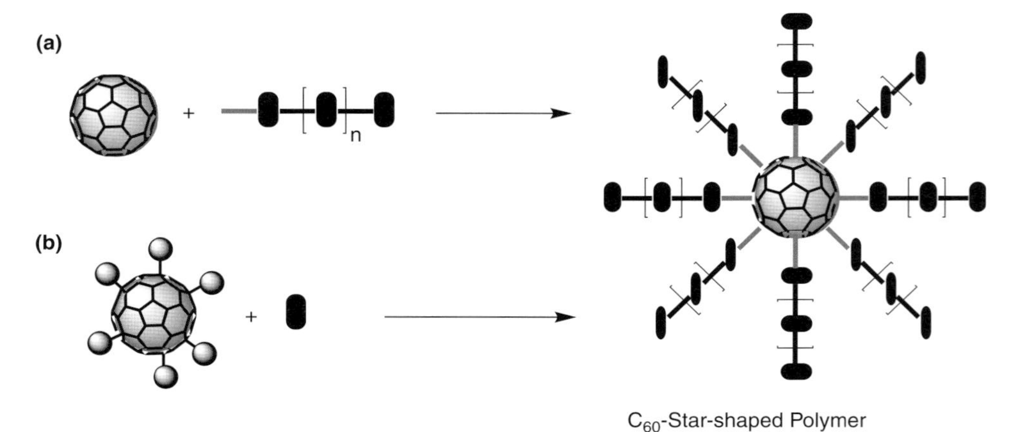

Figure 1.5 Strategies for the synthesis of C_{60} star-shaped polymers.

against human promonocytic THP-1 cells, and could find application in the treatment of multidrug-resistant pathogens [32].

In the *main-chain polymers,* also called in-chain or pearl necklace, C_{60} units are directly allocated in the polymer backbone forming a necklace-type structure. Unfortunately, double addition to the C_{60} sphere affords a complex regioisomeric mixture (up to eight isomers) besides the formation of cross-linking products by multiple additions. Main-chain polymers are prepared by following three different synthetic strategies (Figure 1.6):

a) A direct reaction between the C_{60} cage and a suitable symmetrically difunctionalized monomer;

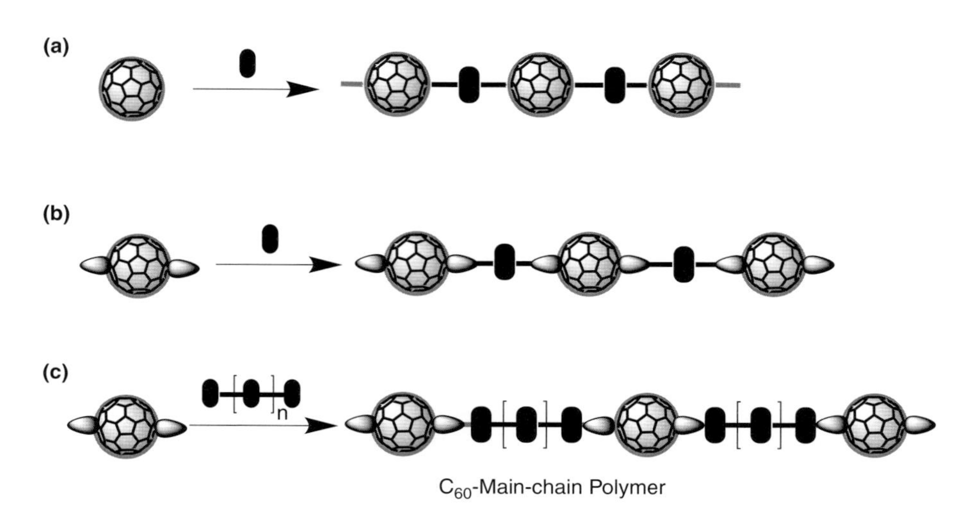

Figure 1.6 Strategies for the synthesis of C_{60} main-chain polymers.

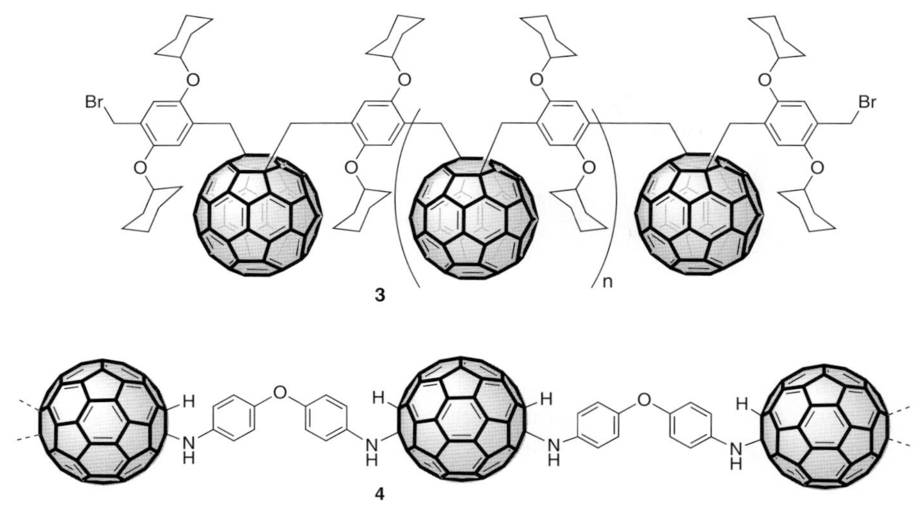

Figure 1.7 1,4-Connected in-chain polymer **3** and water-soluble polyrotaxane **4**.

b) A polycondensation between a fullerene bis-adduct (or a mixture) and a difunctionalized monomer;

c) Reaction between C_{60} or a fullerene bis-adduct and a linear polymer difunctionalized at both ends.

In 2009, 1,4-connected in-chain polymer **3** (Figure 1.7) worked well in nonoptimized photovoltaic devices showing the remarkable conversion of light efficiency of 1.6% [33]. On the other hand, water-soluble polyrotaxane **4** has been successfully employed as a highly efficient DNA-cleaving agent under visible light conditions (Figure 1.7) [34].

Side-chain polymers are the most “populated” family of C_{60} polymers, also called on-chain or “charm-bracelet,” with a wide range of potential applications. In this family are also included the “double-cable” polymers [35], in which the π-conjugated semiconducting polymer (p-type cable) with electron-donating characteristics is endowed with covalently connected electron-accepting fullerene units able to interact among themselves (n-type cable), thus forming a “double-cable” with *a priori* remarkable advantages for construction of photovoltaic devices.

Although in the literature a large number of examples have been reported, the synthetic strategies followed for the preparation of side-chain polymers can be summarized as only two different approaches (Figure 1.8):

a) Direct introduction of fullerene itself or a C_{60} derivative into a preformed polymer.

b) Synthesis of a C_{60} derivative that can be, in turn, directly homopolymerized or copolymerized together with other monomer(s). In the case of double-cable polymers, monomer electropolymerization is also possible.

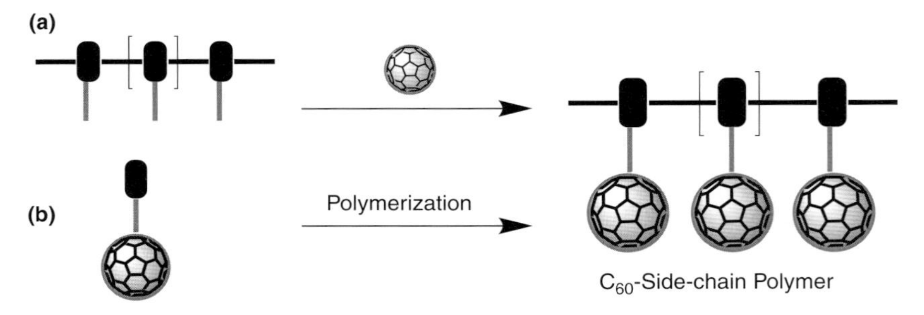

Figure 1.8 Strategies for the synthesis of side-chain C_{60} polymers.

In this family, the knowledge gathered during a century of studies on polymers has been used to bind C_{60} to all of the "classic" families of polymers such as polystyrenes [36], polyacrylates [37], polyethers [38], polycarbonates [39], polysiloxanes [40], polyvinylcarbazoles [41], and polysaccharides [42] in the search for improved processability and enhanced properties, with a wide range of potential applications. In the past years, much attention has been devoted to fullerene-containing side-chain block copolymers, especially for their use in solar cells, thanks to their natural tendency to self-assemble into periodic ordered nanostructures [43]. Moreover, they can be employed in diverse manners to control the final material morphology. A number of C_{60}-containing block copolymers have been prepared and their morphological organization [44], as well as their performances in solar cells [45], has been reported.

Supramolecular C_{60} polymers will be discussed in detail in Chapter 8. To date, four synthetic strategies have been followed in order to prepare C_{60} supramolecular polymers, as shown in Figure 1.9:

a) Systems obtained by interactions between functionalized polymers and C_{60} derivatives or fullerene itself;

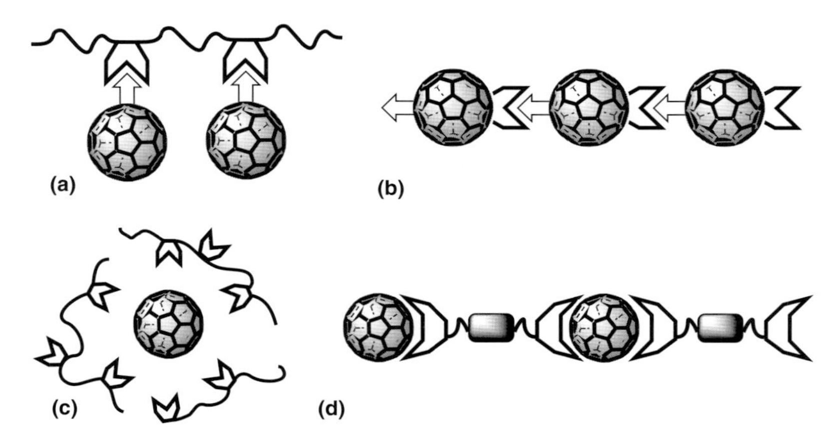

Figure 1.9 Different forms of C_{60} supramolecular polymers.

b) Assembly of self-complementary C_{60} derivatives;
c) Multisubstituted fullerene derivatives and complementary polymeric backbones;
d) Assemblies between ditopic concave guests and C_{60} by means of concave–convex complementary interactions.

Besides the above-mentioned well-established classes of fullerene polymers, a series of polydisperse nondiscrete hybrid materials with a huge potential for real applications is emerging.

Fullerene–DNA hybrids have been studied since 1994 [46], but recently are experiencing a new momentum due to their application as gene delivery agents [47]. Two different strategies have been followed in order to conjugate DNA with fullerenes:

1) The supramolecular complexation: since DNA may be regarded as an anionic polyelectrolyte, it can form complexes with positively charged fullerene derivatives.
2) The covalent binding of C_{60} at the end of a polynucleotide chain.

Carbon nanotube–fullerene hybrids were first discovered serendipitously by Luzzi in 1998 who found that several C_{60} units, by-products in the synthesis of single-walled carbon nanotubes (SWCNTs), were trapped inside the open-ended nanotubes, forming quasi-1D arrays called "peapods" [48]. Since then, these kinds of aggregates have been extensively studied and a series of endohedral and functionalized hollow fullerenes have been introduced in the inner space of nanotubes and have been investigated [49]. Besides the peapods, fullerene–CNT hybrids have been prepared both via supramolecular or covalent methods. To date five strategies have been disclosed for the combination of the two allotropes (Figure 1.10):

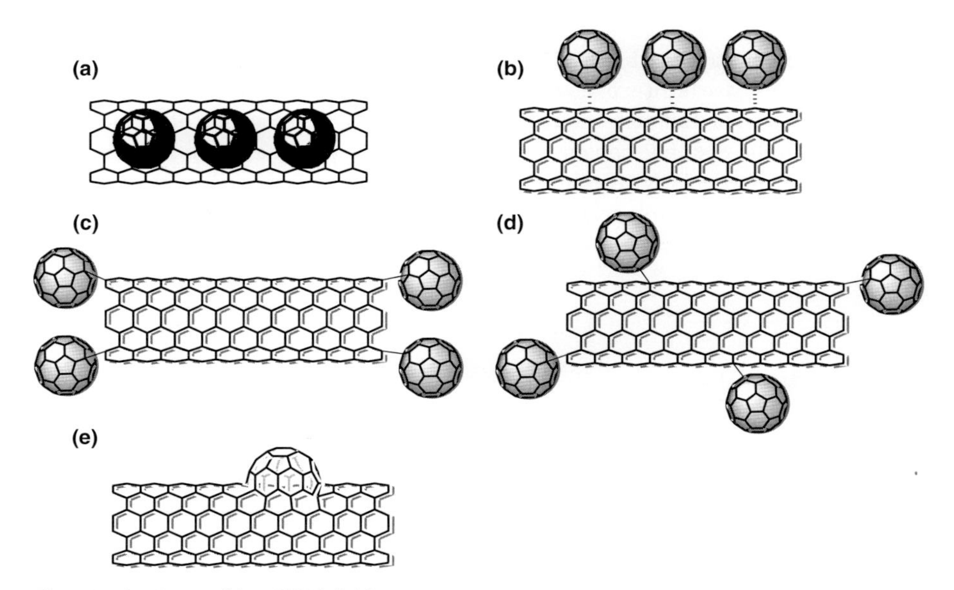

Figure 1.10 Types of C_{60}–CNT hybrids.

a) The supramolecular filling of the void space of CNTs with hollow fullerene or endohedral derivatives both pristine or functionalized;
b) Supramolecular interaction between C_{60}, or its derivatives, and the sidewall of CNTs;
c) The covalent derivatization of the external rims of CNTs;
d) The covalent random derivatization of the side-wall and rims of CNTs;
e) The covalent fusion of C_{60} onto the CNTs surface.

The external supramolecular linking has been achieved through π–π interactions using fullerene derivatives endowed with pyrene moieties [50], while the covalent linking has been carried out both at the ending rims [51] and at the wall level [52]. In 2007 were synthesized nanobuds, carbon nanostructures in which fullerene moieties are fused on the outer wall of the nanotube [53].

Other covalent hybrids between two different allotropes of carbon have also been reported such as a C_{60}-diamond hybrid prepared by deposition of evaporated fullerene onto the bare-diamond surface [54] or the synthesis and characterization of a graphene–fullerene hybrid [55].

Fullerene–gold nanoparticles (Au NP) were first prepared in 1998 [56]. Since then, several groups have been engaged in the synthesis of these hybrid systems in their search for new exotic properties. Their synthesis has been carried out employing three main approaches:

a) The supramolecular linking of C_{60} to Au NP;
b) The covalent linkage of fullerene or its derivatives on surface-functionalized Au NP;
c) Direct *in situ* electroreductive self-assembly between $C_{60}{}^{2-}$ and a gold precursor.

In the first approach, electrostatic [57] and complementary supramolecular interactions have been exploited [58], while in the second the nucleophilic addition of amines to a fullerene double bond has been the reaction of choice [59]. Alternatively, disulfide-containing fullerene derivatives have been successfully anchored onto gold NP surface [60]. Finally, the C_{60} dianion has been used to reduce tetrachloroauric acid obtaining three-dimensional fullerene bound gold nanoclusters [61].

1.3
Carbon Nanotubes

The extensive research carried out on CNTs during the past years has shown the unprecedented mechanical, electrical, and thermal properties that these macromolecules exhibit [62]. Of the wide range of carbon nanostructures available [63], CNTs stand out as truly unique materials that have caused a noteworthy impact on fields such as field-effect transistors (FETs), light-emitting diodes (LEDs), organic solar cells (OSCs), biochemical sensors, memory elements or additives in composite materials [64].

However, most CNT applications require manipulation, interaction, or bonding of CNTs to other materials, in solution or in high-viscosity matrices. These operations are often hindered by the chemical inertness of the CNT surface and by the large aggregates that the tubes form due to π–π stacking interactions. The chemical modification of the CNT surface is an approach generally used to overcome some of these problems [65].

In general, CNT functionalization can be divided into different categories [66]: the covalent attachment of functional entities onto the CNT scaffold that can take place at the termini of the tubes and/or at the sidewalls; the noncovalent interaction of organic and inorganic moieties with the carbon nanotube surface by the use of varying interaction forces such as van der Waals, charge transfer, and π–π interactions, and the special case of the endohedral filling of CNTs with atoms or small molecules (see previous section for the endohedral filling with fullerenes). In the following paragraphs, we will briefly introduce the recent advances in the covalent modification of CNTs toward the creation of multifunctional materials since the noncovalent modification of CNTs will be addressed in detail in chapters 11 to 13.

1.3.1 Defect Functionalization

Among the different surface functionalization techniques, oxidation of CNTs is probably the most widely studied. The treatment of the crude material under strong acidic and oxidative conditions, such as sonication in a mixture of concentrated HNO_3 and H_2SO_4, or heating in a mixture of H_2SO_4/H_2O_2, results in the formation of short opened tubes with oxygenated functions (carbonyl, carboxyl, hydroxyl, etc.) mostly at the edges of the CNTs [67]. This approach represents a popular pathway for further modification of the nanotubes since the acid functions can react with alcohols or amines to give ester or amide derivatives (Figure 1.11).

In one of the first examples, Haddon and coworkers [68] reported the functionalization of SWCNTs, with octadodecylamine (ODA) to produce the corresponding SWCNT-ODA amide **5**. The introduction of long hydrocarbon chains increases the solubility of the SWCNT-ODA in organic solvents, assisting its characterization and facilitating the obtainment of highly purified SWCNT-based materials that are suitable for physical property measurements. To increase the water solubility of SWCNTs, the same group described the covalent functionalization with poly(*m*-aminobenzene sulfonic acid) (PABS) (**6**) via an amide link [69]. The conductivity of the SWCNT–PABS copolymer was much higher than that of the parent PABS.

The solubilization of SWCNTs associated with biomolecules in both aqueous and organic solutions has been investigated at considerable length. The biomolecules employed in the functionalization of SWCNTs include simple saccharides and polysaccharides, peptides, proteins, lipids, enzymes, or DNA and RNA [62f]. For

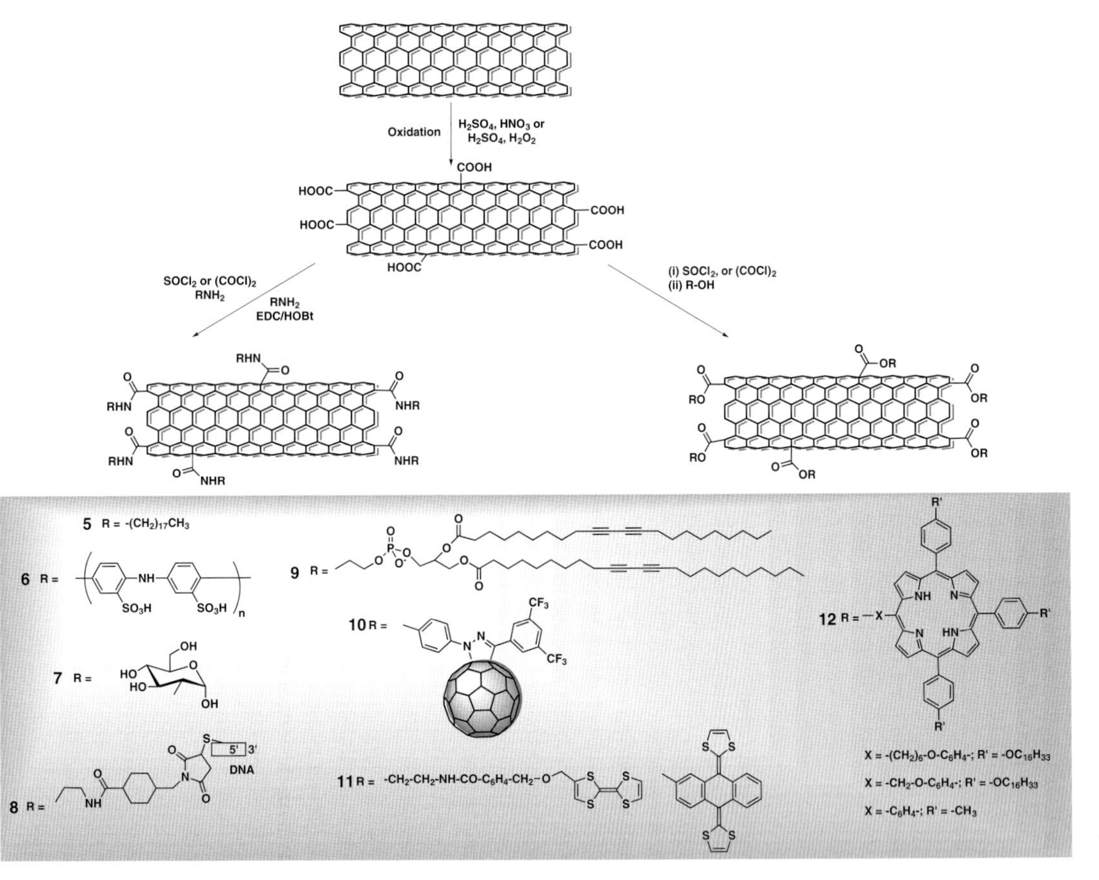

Figure 1.11 Schematic representation of the defect functionalization of CNTs and subsequent amidation and esterification reactions to obtain different materials.

example, considering the method developed by Pompeo and Resasco reported the coupling of glucosamine to acyl chloride-activated nanotubes (**7**) [70]. The solubility in water of the conjugates formed ranged from 0.1 to 0.3 mg/ml, depending on temperature. The preparation of covalent nanoconjugates of DNA and SWCNTs (**8**) was carried out by Baker *et al.* [71]. The high stability and accessibility in hybridization experiments indicates that DNA chains are chemically bound to the exterior surface of SWCNTs and not wrapped around or intercalated within the nanotubes. These DNA–SWCNT adducts hybridize selectively with complementary sequences, with only minimal interactions with mismatched sequences. Other biomolecules of interest are lipids, which have also been covalently linked to SWCNTs in the search for novel biosensor systems (**9** in Figure 1.11) [72].

The development of reliable and reproducible methodologies to integrate CNTs into functional structures such as donor–acceptor hybrids, able to transform sunlight into electrical or chemical energy, has emerged as an area of intensive research [73]. In this context, recently the synthesis of the first conjugated hybrid SWCNT-[60] fullerene materials (**10**) has been described, considering the reaction of amines with SWCNTs functionalized in their rims with acyl chlorides [51]. The authors decorated the SWCNTs with fragments of *N*-anilinopyrazolino[60]fullerene and with the aid of vibrational spectroscopy and high-resolution transmission electron microscopy (TEM) corroborated the existence of [60]fullerene units located at the rims of the SWCNTs. Considering a similar strategy, Martín and coworkers [74] described an efficient way to functionalize SWCNTs with the strong electron donor tetrathiafulvalene (TTF) or its π-extended analogue (exTTF) derivatives (**11**). The procedure implies the reaction of SWCNT–COOH with TTF or exTTF-based compounds, through esterification or amidation reactions. This work depicts the preparation of the first TTF-SWCNT donor–acceptor systems to evaluate the possible use of CNTs in solar energy conversion applications. The analysis of these compounds by different analytical, spectroscopic, and microscopic (TEM) techniques confirmed the presence of TTF or exTTF units connected to the SWCNTs. Photophysical studies by time-resolved spectroscopy revealed the presence of radical ion species ($TTF^{\bullet +}$, $exTTF^{\bullet +}$, and $SWCNT^{\bullet -}$) indicating the existence of an efficient photoinduced electron transfer (PET), a critical requirement to prepare photovoltaic devices. The covalent attachment of porphyrin units to SWCNTs (**12**) has also been investigated by several groups in an attempt to emulate the natural photosynthesis [75]. For example, Li *et al.* [75a] reported that quenching of the fluorescence of the porphyrin units was strongly dependent on the length of the spacer between the nanotube and the chromophore. In fact, a better interaction between the nanotube and the porphyrin was observed for longer and more flexible alkyl chains.

1.3.2
Sidewall Functionalization

As outlined in the previous section, a defect functionalization-based chemical attachment of organic moieties on CNTs will primarily proceed at the nanotube tips. As a consequence, the functionalization degree will be relatively low as the huge surface area of the CNT sidewall is hardly involved in the transformation sequence. Indeed, the direct chemical alteration of the CNT sidewall is a challenging task due to the close relationship of the CNT sp^2 carbon network with planar graphite and its low reactivity. In contrast to the esterification or amidation reactions of the acid groups of oxidized SWCNTs, the functionalization of the sidewalls of CNTs requires the use of highly reactive species.

Historically, based on the low reactivity of the CNT sidewall, direct sidewall addition of elemental fluorine was one of the first successful alterations of the sp^2 carbon framework [76]. During the past decade, a variety of pathways leading to fluorinated carbon nanotubes have been developed, and fluorinated nanotubes have become a widely used starting material for further chemical transformation steps, based on the nucleophilic substitution of fluorine atoms. For example, Barron and coworkers have shown that the reaction of fluorinated HiPco SWCNTs (**13**) with ω-amino acids leads to water-soluble SWCNT derivatives (**14**) [77]. In addition, the solubility in water is controlled by the length of the hydrocarbon chain of the amino acid and does strongly depend on the pH value of the solution. Kelly and coworkers have prepared thiol- and thiophene-functionalized SWCNTs (**14**) – which allow a subsequent coupling of the tubes to gold nanoparticles – via the reaction of bifunctional amines with fluorinated CNTs (**13**) [78].

Another useful methodology is the one introduced by Tour in 2001, where aryl radicals are generated from diazonium salts via one-electron electrochemical reduction [79] or considering a solvent-based thermally induced reaction with the *in situ* generation of the diazonium compound by reacting aniline derivatives with isoamyl nitrite (Figure 1.12, **15**) [80].

Cycloaddition reactions have attracted the interest of a great number of scientists and remarkable examples on the functionalization of the sidewalls of carbon nanotubes by means of carbene [81] or nitrene [82] additions, Bingel reaction [83], or Diels–Alder cycloadditions [84] have been reported. Although, probably the most commonly used sidewall derivatization reaction is the 1,3-dipolar cycloaddition of azomethine ylides, generated by the condensation of an α-amino acid and an aldehyde (Figure 1.13) [85].

Since this method was reported, a large variety of CNT derivatives have been described, and it is especially important to remark on the relevant development achieved in the field of drug design and discovery based on CNT derivatives [86]. Prato's methodology has allowed to synthesize water-soluble CNTs bearing pendant ammonium groups (NH_3^+-SWCNT) [87], which were tested in biocompatibility and cellular interaction studies, where different cells were cultured in the presence of these CNT derivatives. The ammonium-functionalized CNTs were observed by TEM microscopy and localized inside the cells [88]. In other appealing example, radioactive-labeled carbon nanotubes were intravenously injected in mice and found to be excreted in urine [89], showing that water-soluble CNTs can be well-tolerated *in vivo*, while exhibiting an exceptional capacity to cross cell membranes and localize into the cytoplasm.

Water-soluble CNTs have also been used to efficiently complex and translocate DNA inside cells in the pioneering studies exploring the ability of CNTs to deliver

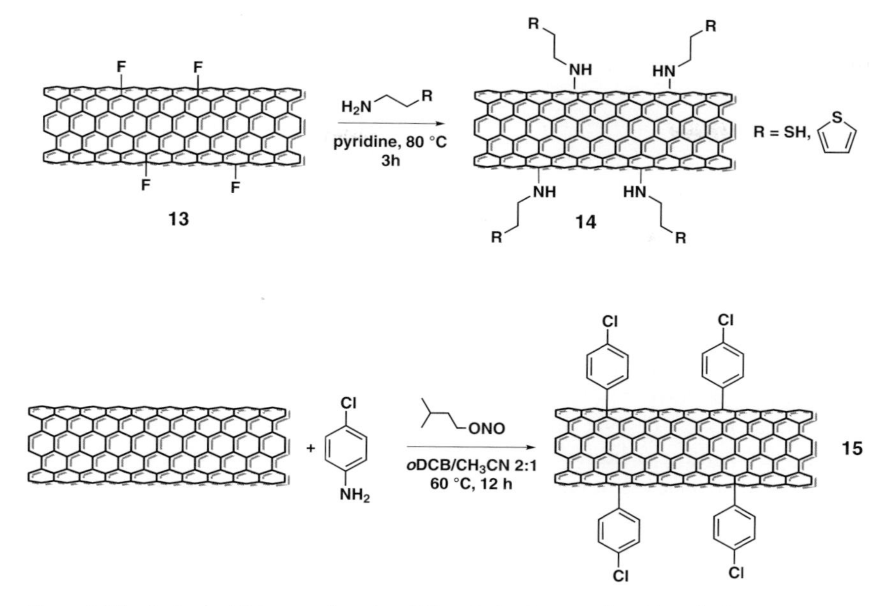

Figure 1.12 Radical additions and nucleophilic substitutions on the sidewall of CNTs.

genes [86]. Furthermore, CNTs were functionalized with antibiotics (Amphotericin B) by Prato's reaction showing a promising activity, even on strains usually resistant to Amphotericin B [90]. CNTs bearing methotrexate – a potent anticancer agent – have been synthesized as well by using this methodology. Preliminary results demonstrated that these CNT derivatives are, at least, as active as the methotrexate alone [91].

Although the chemical functionalization has brought about an important advance in the preparation and processing of CNTs, still the essential challenge is to control the degree of functionalization at the different levels that various applications require. In this sense, Hirsch and coworkers have recently reported a method that could be particularly useful for controlling the extent of chemical modification of CNTs (Figure 1.14).

The first step of the process is to covalently attach alkyl groups to the sidewalls of SWCNTs, following a reaction sequence that consists in the nucleophilic addition of

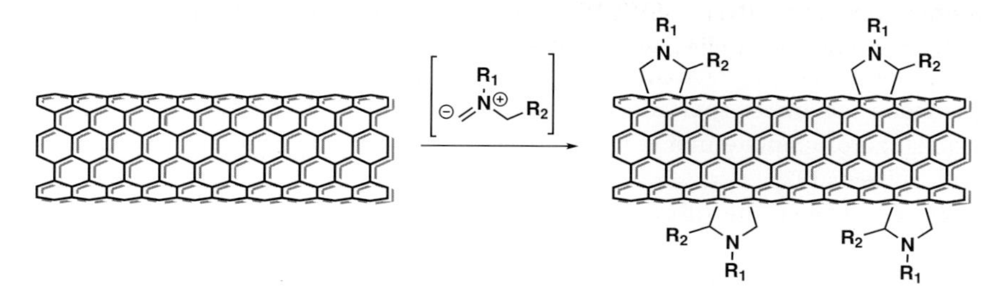

Figure 1.13 1,3-Dipolar cycloaddition of azomethine ylides.

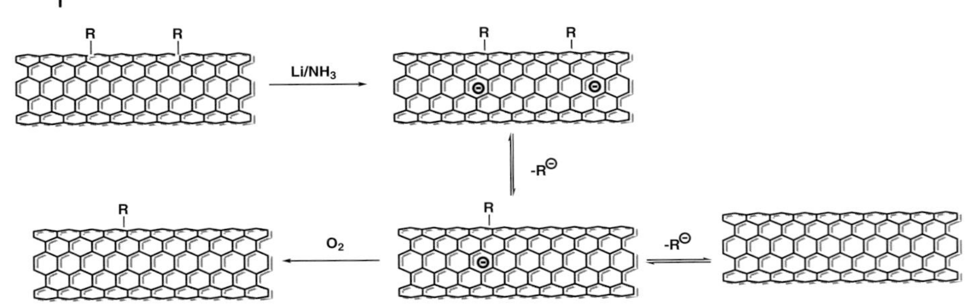

Figure 1.14 Reductive retrofunctionalization of SWCNTs (R = *n*Bu, *n*PrNH).

alkyllithium derivatives to the sidewalls of CNTs and subsequent reoxidation of the intermediate R_n–CNT^{n-} [92]. Subsequently, the highly functionalized SWCNTs were treated with sodium or lithium metals in liquid ammonia and the appended alkyl groups could be released, after donation of electrons from the metal to the SWCNTs. More interestingly, some of the alkyl groups could be replaced by *n*-propyl amino groups by simply adding *n*-propylamine to the reaction mixture [93]. In addition to controlling the functionalization degree, this protocol allows the attachment of two different chemical groups to the surface of SWCNTs.

1.4 Graphenes

Since the first report on the mechanical isolation of graphene from graphite, the interest in the physical properties and potential applications of the youngest member of the family of carbon nanostructures has led to an unprecedented increase in the number of publications on its synthesis, properties, and applications [94]. The revolution is going further, if possible, since in October 2010 the Nobel Prize for Physics was awarded to Andre Geim and Konstantin Novoselov for their work on these unique 2D macromolecules [95].

Graphenes offer important properties such as high elasticity, electric conductivity, or spin transport, and at the same time are free of some of the drawbacks that still hamper the implementation of SWCNTs in technological applications, such as the presence of metal catalyst impurities [96]. However, before one can chemically manipulate graphenes they have to be chemically exfoliated into individual or few-layer sheets and be stabilized. Typical methodologies for the production of graphene are direct sonication, chemical reduction, mechanical exfoliation, or exfoliation of graphite into individual sheets by stabilization through interactions with solvents [97].

Once obtained, the functionalization of graphenes is achieved by employing covalent or noncovalent methodologies similar to the ones used for CNTs, to make them tractable in different solvents and to decorate graphenes with different functional groups that can add new properties to the material.

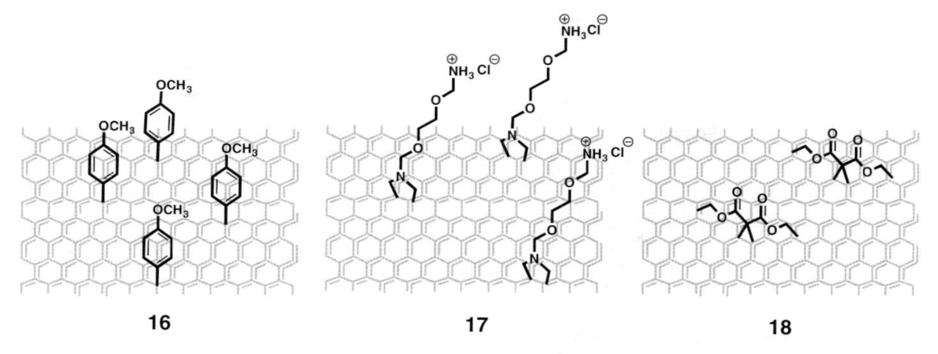

Figure 1.15 Covalently functionalized few-layer graphenes with diazonium salts (**16**), azomethyne ylides (**17**), or bromomalonates (**18**).

1.4.1 Covalent Functionalization

By using a covalent strategy, Haddon and coworkers [98] have reported an efficient method to obtain graphene sheets functionalized with long hydrocarbon chains (G-octadodecylamine, G-ODA). Oxidative treatment of microcrystalline graphite with nitric and sulfuric acids produces oxidized graphite (GO), decorated with carboxylic groups at the edges. The presence of these acidic functionalities provide a way to introduce long hydrocarbon chains that assist the solubilization of the graphene sheets in polar solvents, which is a fundamental issue in order to improve graphite processability.

In a similar approach, Tour and coworkers [99] have recently reported the functionalization of single graphene sheets using aryl diazonium salts (**16**, Figure 1.15). These graphene sheets were in turn produced via surfactant-wrapped chemically converted graphene nanosheets, and obtained from reduction of graphene oxide with hydrazine. The authors studied these new carbon nanomaterials by different spectroscopic and microscopic techniques, and the results proved an efficient functionalization, allowing the nanosheets to be dissolved in organic solvents.

In a recent work, graphenes produced by dispersion and exfoliation of graphite in *N*-methylpyrrolidone or in benzylamine have been successfully functionalized by using the 1,3-dipolar cycloaddition of azomethine ylides under thermal conditions (**17**, Figure 1.15) [100] or following a Bingel reaction under microwave irradiation (**18**, Figure 1.15) [101]. The graphenes incorporating pyrrolidine rings, which also bear an amino functional group, bind selectively to gold nanorods that served as contrast markers for the identification of the reactive sites and produce a hybrid material that can be useful for various applications.

1.4.2 Noncovalent Functionalization

Graphene can be functionalized using noncovalent methods such as wrapping with surfactants or ionic liquids [94a]. Water-soluble graphenes have been obtained with surfactants such as polyoxyethylene(40)nonylphenylether (Igepal CO-890), sodium dodecylsulfate (SDS) or cetyltrimethylammonium bromide (CTAB) [102]. The one-step electrochemical approach for the preparation of ionic liquid-functionalized graphite sheets has also been explored in the search for water-soluble graphenes, which in addition can be exfoliated into functionalized graphene nanosheets [103].

However, the graphene produced today is mostly going into composites that benefit from its mechanical, electronic, or optical properties [94]. *Mechanical* reinforcement can be achieved when graphene is dispersed with polyacrylonitrile nanofibers [104], or in poly(methyl methacrylate) (PMMA) [105]. Ramanathan *et al.* reported that an addition of approximately 1 wt% of graphene to PMMA leads to increases in the elastic modulus of 80% and in the tensile strength of about 20%. In a comparative study, these researchers demonstrated that among all the nanofiller materials considered, single-layer functionalized graphene provide the best results [106]. For *electronic* applications, graphene offers fewer problems with uncontrollable variety from batch to batch than does CNT synthesis and, it has been combined with different polymers in the search for materials with improved electronic properties. In this sense, Bai *et al.* modified graphenes with conjugated polymer sulfonated polyaniline (PANI) and investigated its electrochemical activity [107]. Pluronic copolymers (poly-(ethylene oxide)-block-poly(propylene oxide)-block-poly(ethylene oxide) were also employed as solubilizing agents for graphenes to form a supramolecular hydrogel [108]. The *optical* applications of graphene composites offer the advantage of a relatively easy processing condition of films [94d]. Recently, Rao *et al.* prepared graphenes dispersed in a polymer matrix that proved to be highly effective saturable absorbers – optical elements used in mode-locking of laser, based on their properties to exhibit optical loss at high optical intensity [109].

The interactions between graphene and a series of small aromatic molecules have also been investigated in detail. For example, the π–π interactions with pyrene derivatives (i.e., pyrene butanoic acid succidymidyl ester **19**, Figure 1.16) helped to stabilize graphene in aqueous solutions [110].

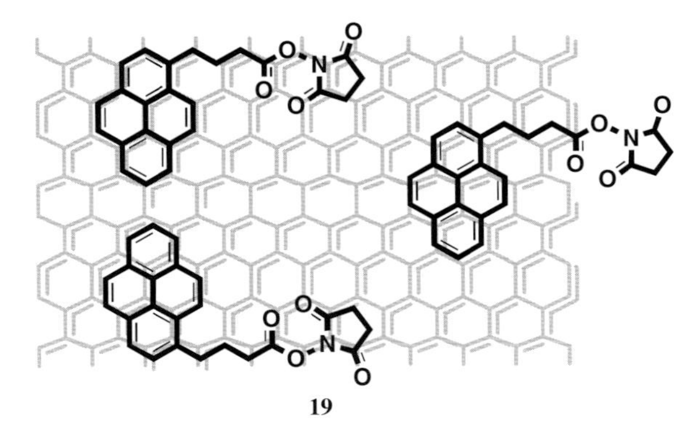

Figure 1.16 Noncovalent interactions between graphenes and the pyrene butanoic acid succidymidyl ester **19**.

In addition, different electroactive molecules have been found to interact with the graphene surface by π–π stacking interactions and form donor–acceptor complexes by using 1-pyrene-1-sulfonic acid sodium salts and the disodium salt of 3,4,9,10-perylenetetracarboxylic diimide bisbenzenesulfonic acid [111]. The negative charges in both molecules act as stabilizing species to maintain a strong static repulsion force between the negative-charged graphene sheets in solution, giving rise to durable monolayers. More interestingly, the noncovalent modification of graphene films with pyrene derivatives improves the hydrophilic characteristics, work function, and power conversion efficiency in photovoltaic cells [112]. The PCE of organic solar cells fabricated with the supramolecularly modified graphene increases from 0.21% for the unmodified films to 1.71% for anodes using the pyrene-modified graphene, which paves the way for the substitution of ITO in photovoltaic and electroluminescent devices with low-cost graphene films.

Molecular doping has turned out to be an important tool to tailor the electronic properties of graphene. The interaction of graphene with electron donors (aniline or tetrathiafulvalene, TTF) and electron acceptors (nitrobenzene or tetracyanoethylene, TCNE) caused marked changes in the electronic structure of graphene [113]. Thus, the Raman G band of few-layer graphene is softened in the presence of aniline or TTF, and strengthened with electron acceptor molecules such as nitrobenzene and TCNE. In addition, charge transfer bands are found in the visible region when TTF and TCNE interact with few-layer graphene [114]. The electrical conductivity of graphene also varies on interaction with both types of electroactive molecules: electron donor molecules decrease the conductivity of graphene, while electron acceptor molecules increase the conductivity. In addition, the magnitude of the interaction depends on the surface area of the graphene samples [115]. However, recent work has demonstrated that graphene can be electronically decoupled from the electroactive molecules adsorbed on it when grown on a metallic surface of Ir (111) [116]. Under such conditions, the self-organization of the molecules 7,7′,8,8′-tetracyano-*p*-quinodimethane, TCNQ, and 2,3,5,6-tetrafluoro-7,7′,8,8′-tetracyano-*p*-quinodimethane (F_4-TCNQ) is guided by the intermolecular interaction, attractive for TCNQ and repulsive for F_4-TCNQ [116].

So far, all the investigations on molecular doping have been carried out in the solid state and only recently, Hirsch and coworkers have reported the binding and electronic interaction of graphene with an organic dye molecule **20** (Figure 1.17) in homogeneous solution [117].

Although perylene bisimides exhibit strong fluorescence, no emission background is present in the Raman spectrum of a dispersion of turbostatic graphite with **20** in *N*-methylpyrrolidone, which indicates a pronounced quenching due to the electronic communication with graphene. The interaction between **20** and graphene was further corroborated by titration experiments, where the quantum yield of the graphene-titrated perylene solution was reduced by 65% compared to pristine perylene solutions [117]. These results demonstrated that the noncovalent functionalization of graphene in organic solvents could be a useful approach for the systematic tuning of the electronic properties of graphenes.

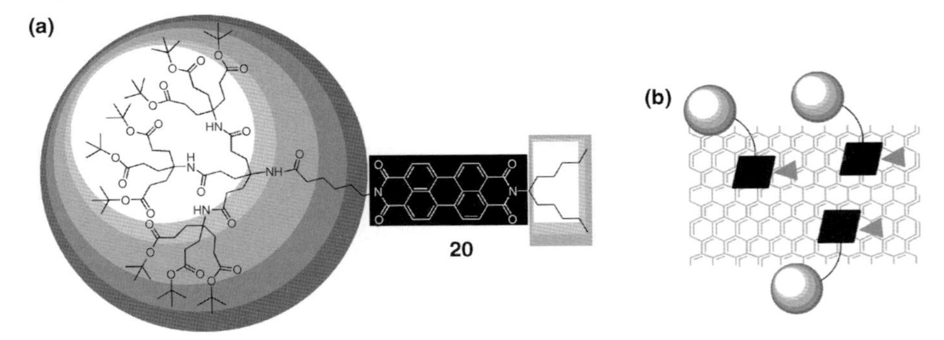

Figure 1.17 (a) Dendronized perylene bisimide **20** and (b) cartoon representing the noncovalent binding of **20** to graphene.

1.5
Summary and Conclusions

As stated at the outset, the goal of this chapter is to summarize the most important achievements in the covalent and macromolecular chemistry of fullerenes in order to have a better understanding of the chemistry of these molecular carbon allotropes, thus nicely complementing the following chapters devoted to supramolecular chemistry. Furthermore, both CNTs and graphenes have been considered in this chapter and, therefore, the most significant functionalization/solubilization methods known so far have also been presented.

Importantly, covalent functionalization is typically accompanied by a significant damage to the electronic structure of the carbon nanostructures. Therefore, the relevance of the supramolecular methods available so far is significantly emphasized in the following chapters of the book.

Needless to say, for a more comprehensive knowledge of the nowadays very rich chemistry of fullerenes, CNTs, and graphenes, the reader is encouraged to go to the many reviews and books available in the chemical literature mentioned in this book.

Acknowledgments

This work has been supported by the MICINN of Spain (CT2008-00795 and Consolider-Ingenio CSD2007-00010 on Molecular Nanoscience) and Comunidad de Madrid (MADRISOLAR-2, S2009/PPQ-1533).

References

1 Kratschmer, W., Lamb, L.D., Fostiropoulos, K., and Huffman, D.R. (1990) *Nature*, **347**, 354–358.
2 (a) Giacalone, F. and Martín, N. (2006) *Chem. Rev.*, **106**, 5136–5190; (b) Giacalone, F. and Martín, N. (2010) *Adv. Mater.*, **22**, 4220–4248; (c) Giacalone, F. and Martín, N. (2009) *Fullerene Polymers: Synthesis, Properties and Applications*,

Wiley-VCH Verlag GmbH, Weinheim, Germany.

3 Charvet, R., Acharya, S., Hill, J.P., Akada, M., Liao, M., Seki, S., Honsho, Y., Saeki, A., and Ariga, K. (2009) *J. Am. Chem. Soc.*, **131**, 18030–18031.

4 (a) Stoilova, O., Jérôme, C., Detrembleur, C., Mouithys-Mickalad, A., Manolova, N., Rashkov, I., and Jérôme, R. (2006) *Chem. Mater.*, **18**, 4917–4923; (b) Stoilova, O., Jérôme, C., Detrembleur, C., Mouithys-Mickalad, A., Manolova, N., Rashkov, I., and Jérôme, R. (2007) *Polymer*, **48**, 1835–1843.

5 Ma, B., Riggs, J.E., and Sun, Y.-P. (1998) *J. Phys. Chem. B*, **102**, 5999–6009.

6 Chen, Y., Huang, Z.-E., Cai, R.-F., Yu, B.-C., Ito, O., Zhang, J., Ma, W.-W., Zhong, C.-F., Zhao, L., Li, Y.-F., Zhu, L., Fujitsuka, M., and Watanabe, A. (1997) *J. Polym. Sci. B Polym. Phys.*, **35**, 1185–1190.

7 (a) Cloutet, E., Fillaut, J.-L., Gnanou, Y., and Astruc, D. (1994) *J. Chem. Soc. Chem. Commun.*, 2433–2434; (b) Fedurco, M., Costa, D.A., Balch, A.L., and Fawcett, W.R. (1995) *Angew. Chem. Int. Ed. Engl.*, **34**, 194–196; (c) Chen, X., Gholamkhass, B., Han, X., Vamvounis, G., and Holdcroft, S. (2007) *Macromol. Rapid Commun.*, **28**, 1792–1798; (d) Nanjo, M., Cyr, P.W., Liu, K., Sargent, E.H., and Manners, I. (2008) *Adv. Funct. Mater.*, **18**, 470–477.

8 Ling, Q.-D., Lim, S.-L., Song, Y., Zhu, C.-X., Chan, D.S.-H., Kang, E.-T., and Neoh, K.-G. (2007) *Langmuir*, **23**, 312–319.

9 (a) Sterescu, D.M., Bolhuis-Versteeg, L., van der Vegt, N.F.A., Stamatialis, D.F., and Wessling, M. (2004) *Macromol. Rapid Commun.*, **25**, 1674–1678; (b) Sterescu, D.M., Stamatialis, D.F., Mendes, E., Wibbenhorst, M., and Wessling, M. (2006) *Macromolecules*, **39**, 9234–9242; (c) Vinogradova, L.V., Polotskaya, G.A., Shevtsova, A.A., and Alentev, A.Y. (2009) *Polym. Sci. Ser. A*, **51**, 209–215.

10 Wang, H., DeSousa, R., Gasa, J., Tasaki, K., Stucky, G., Jousselme, B., and Wudl, F. (2007) *J. Membr. Sci.*, **289**, 277–283.

11 For detailed treatises on this topic: (a) Special issue on polymeric fullerenes (1997) *Appl. Phys. A Mater. Sci. Process.*, **64** (3); (b) Sundqvist, B. (1999) *Adv. Phys.*, **48**, 1–416; (c) Blank, V.D., Buga, S.G., Dubitsky, G.A., Serebryanaya, N.R., Yu Popov, M., and Sundqvist, B. (1998) *Carbon*, **36**, 319–343.

12 Rao, A.M., Zhou, P., Wang, K.-A., Hager, G.T., Holden, J.M., Wang, Y., Lee, W.-T., Bi, X.-X., Eklund, P.C., Cornett, D.S., Duncan, M.A., and Amster, I.J. (1993) *Science*, **259**, 955–957.

13 Iwasa, Y., Arima, T., Fleming, R.M., Siegrist, T., Zhou, O., Haddon, R.C., Rothberg, L.J., Lyons, K.B., Carter, H.L., Hebard, A.F., Jr., Tycko, R., Dabbagh, G., Krajewski, J.J., Thomas, G.A., and Yagi, T. (1994) *Science*, **264**, 1570–1572.

14 Rao, A.M., Eklund, P.C., Hodeau, J.L., Marques, L., and Nuñez-Regueiro, M. (1997) *Phys. Rev. B*, **55**, 4766–4773.

15 (a) Takahashi, N., Dock, H., Matsuzawa, N., and Ata, M. (1993) *J. Appl. Phys.*, **74**, 5790–5794; (b) Zou, Y.J., Zhang, X.W., Li, Y.L., Wang, B., Yan, H., Cui, J.Z., Liu, L.M., and Da, D.A. (2002) *J. Mater. Sci.*, **37**, 1043–1047.

16 Nakaya, M., Tsukamoto, S., Kuwahara, Y., Aono, M., and Nakayama, T. (2010) *Adv. Mater.*, **22**, 1622–1625.

17 (a) Stephens, P.W., Bortel, G., Faigel, G., Tegze, M., Janossy, A., Pekker, S., Oszlányi, G., and Forró, L. (1994) *Nature*, **370**, 636–639; (b) Pekker, S., Forró, L., Mihaly, L., and Janossy, A. (1994) *Solid State Commun.*, **90**, 349–351.

18 Rao, A.M., Eklund, P.C., Venkateswaran, U.D., Tucker, J., Duncan, M.A., Bendele, G.M., Stephens, P.W., Houdeau, J.-L., Marques, L., Nuñez-Regueiro, M., Bashkin, I.O., Ponyatovsky, E.G., and Morovsky, A.P. (1997) *Appl. Phys. A Mater. Sci. Process*, **64**, 231–2239.

19 Nagashima, H., Nakaoka, A., Saito, Y., Kato, M., Kawanishi, T., and Itoh, K. (1992) *J. Chem. Soc. Chem. Commun.*, 377–379.

20 Nagashima, H., Nahaoka, A., Tajima, S., Saito, Y., and Itoh, K. (1992) *Chem. Lett.*, 1361–1364.

21 Riccò, M., Pontiroli, D., Mazzani, M., Gianferrari, F., Pagliari, M., Goffredi, A., Brunelli, M., Zandomeneghi, G., Meier, B.H., and Shiroka, T. (2010) *J. Am. Chem. Soc.*, **132**, 2064–2068.

22 (a) Bunker, C.E., Lawson, G.E., and Sun, Y.P. (1995) *Macromolecules*, **28**, 3744–3746; (b) Kojima, Y., Matsuoka, T., Takahashi, H., and Karauchi, T. (1995) *Macromolecules*, **28**, 8868–8869; (c) Kojima, Y., Matsuoka, T., Takahashi, H., and Karauchi, T. (1997) *J. Mater. Sci. Lett.*, **16**, 2029–2031.

23 Lu, Z., Goh, S.H., Lee, S.Y., Sun, X., and Ji, W. (1999) *Polymer*, **40**, 2863–2867.

24 (a) Luang, L., Chen, Q., Sargent, E.H., and Wang, Z.Y. (2003) *J. Am. Chem. Soc.*, **125**, 13648–13649; (b) Chen, Q., Luang, L., Sargent, E.H., and Wang, Z.Y. (2003) *Appl. Phys Lett.*, **83**, 2115–2117; (c) Chen, Q., Luang, L., Wang, Z.Y., and Sargent, E.H. (2004) *Nano Lett.*, **9**, 1673–1678.

25 Hsieh, C.-H., Cheng, Y.-J., Li, P.-J., Chen, C.-H., Dubosc, M., Liang, R.-M., and Hsu, C.-S. (2010) *J. Am. Chem. Soc.*, **132**, 4887–4893.

26 Cheng, Y.-J., Hsieh, C.-H., He, Y., Hsu, C.-S., and Li, Y. (2010) *J. Am. Chem. Soc.*, **132**, 17381–17383.

27 (a) Kai, W., Hua, L., Dong, T., Pan, P., Zhu, B., and Inoue, Y. (2008) *Macromol. Chem. Phys.*, **209**, 104–111; (b) Zhou, G., Harruna, I.I., Zhou, W.L., Aicher, W.K., and Geckeler, K.E. (2007) *Chem. Eur. J.*, **13**, 569–573.

28 (a) Liu, J., Ohta, S., Sonoda, A., Yamada, M., Yamamoto, M., Nitta, N., Murata, K., and Tabata, Y. (2007) *J. Control. Release*, **117**, 104–110; (b) Liu, J. and Tabata, Y. (2011) *J. Drug. Target.*, **18**, 602–610.

29 Boudouris, B.W., Molins, F., Blank, D.A., Frisbie, C.D., and Hillmyer, M.C. (2009) *Macromolecules*, **42**, 4118–4126.

30 Detrembleur, C., Stoilova, O., Bryaskova, R., Debigne, A., Mouithys-Mickalad, A., and Jérôme, R. (2006) *Macromol. Rapid Commun.*, **27**, 498–504.

31 Hurtgen, M., Debuigne, A., Mouithys-Mickalad, A., Jérôme, R., Jérôme, C., and Detrembleur, C. (2010) *Chem. Asian J.*, **5**, 859–868.

32 Stoilova, O., Jérôme, C., Detrembleur, C., Mouithys-Mickalad, A., Manolova, N., Rashkov, I., and Jérôme, R. (2007) *Polymer*, **48**, 1835–1843.

33 (a) Hiorns, R.C., Cloutet, E., Ibarboure, E., Vignau, L., Lemaitre, N., Guillerez, S., Absalon, C., and Cramail, H. (2009) *Macromolecules*, **42**, 3549–3558; (b) Hiorns, R.C., Cloutet, E., Ibarboure, E., Khoukh, A., Bejbouji, H., Vignau, L., and Cramail, H. (2010) *Macromolecules*, **43**, 6033–6044.

34 Samal, S., Choi, B.-J., and Geckeler, K.E. (2001) *Macromol. Biosci.*, **1**, 329–331.

35 (a) Cravino, A. and Sariciftci, N.S. (2002) *J. Mater. Chem.*, **12**, 1931–2159; (b) Cravino, A. and Sariciftci, N.S. (2003) *Nat. Mater.*, **2**, 360–361; (c) Cravino, A. (2007) *Polymer Int.*, **56**, 943–956.

36 (a) For some selected example on side-chain polystyrene-C_{60} polymers: Liu, B., Bunker, C.E., and Sun, T.-P. (1996) *J. Chem. Soc. Chem. Commun.*, 1241–1242. (b) Chen, Y., Huang, Z.-E., Cai, R.-F., Kong, S.-Q., Chen, S., Shao, Q., Yan, X., Zhao, F., and Fu, D. (1996) *J. Polym. Sci. A Polym. Chem.*, **34**, 3297–3302; (c) Stalmach, U., de Boer, B., Videlot, C., van Hutten, P.F., and Hadziioannou, G. (2000) *J. Am. Chem. Soc.*, **122**, 5464–5472; (d) Cao, T., Wei, F., Yang, Y., Huang, L., Zhao, X., and Cao, W. (2002) *Langmuir*, **18**, 5186–5189.

37 For some selected example on side-chain polyacrylate-C_{60} polymers: (a) Zheng, J., Goh, S.H., and Lee, S.Y. (1997) *Polym. Bull.*, **39**, 79–84. (b) Zheng, J.W., Goh, S.H., and Lee, S.Y. (2000) *J. Appl. Polym. Sci.*, **75**, 1393–1396; (c) Wang, Z.Y., Kuang, L., Meng, X.S., and Gao, J.P. (1998) *Macromolecules*, **31**, 5556–5558; (d) Wang, C., Tao, Z., Yang, W., and Fu, S. (2001) *Macromol. Rapid Commun.*, **22**, 98–103.

38 (a) Goh, S.H., Zheng, J.W., and Lee, S.Y. (2000) *Polymer*, **41**, 8721–8724; (b) Gutiérrez-Nava, M., Masson, P., and Nierengarten, J.-F. (2003) *Tetrahedron Lett.*, **44**, 4487–4490.

39 For some selected examples on side-chain polyacarbonate-C_{60} polymers: (a) Tang, B.Z., Peng, H., Leung, S.M., Au, C.F., Poon, W.H., Chen, H., Wu, X.,

Fok, M.W., Yu, N.-T., Hiraoka, H., Song, C., Fu, J., Ge, W., Wong, G.K.L., Monde, T., Nemoto, F., and Su, K.C. (1998) *Macromolecules*, **31**, 103–108. (b) Wu, H., Li, F., Lin, Y., Cai, R.-F., Wu, H., Tong, R., and Qian, S. (2006) *Polym. Eng. Sci.*, **46**, 399–405; (c) Vitalini, D., Mineo, P., Iudicelli, V., Scamporrino, E., and Troina, G. (2000) *Macromolecules*, **33** 7300–7309.

40 (a) Kraus, A. and Müllen, K. (1999) *Macromolecules*, **32**, 4214–4219; (b) Li, Z. and Qin, J. (2003) *J. Appl. Polym. Sci.*, **89**, 2068–2071.

41 (a) Chen, Y., Huang, Z.-E., and Cai, R.-F. (1996) *J. Polym. Sci. B Polym. Phys.*, **34**, 631–640; (b) Gu, T., Chen, W.-X., and Xu, Z.-D. (1999) *Polym. Bull.*, **42**, 191–196.

42 (a) Okamura, H., Miyazono, K., Minoda, M., and Miyamoto, T. (1999) *Macromol. Rapid Commun.*, **20**, 41–45; (b) Ungurenasu, C. and Pienteala, M. (2007) *J. Polym. Sci. A Polym. Chem.*, **45**, 3124–3128.

43 Darling, S.B. (2009) *Energy Environ. Sci.*, **2**, 1266–1273.

44 (a) Chen, X., Gholamkhass, B., Han, X., Vamvounis, G., and Holdcroft, S. (2007) *Macromol. Rapid Commun.*, **28**, 1792–1797; (b) Richard, F., Brochon, C., Leclerc, N., Eckhardt, D., Heiser, T., and Hadziioannou, G. (2008) *Macromol. Rapid Commun.*, **29**, 885–891; (c) Lee, J.U., Cirpan, A., Emrik, T., Russell, T.P., and Jo, W.H. (2009) *J. Mater. Chem.*, **19**, 1483–1487; (d) Hu, Z., Zou, J., Deibel, C., Gesquiere, A.J., and Zhai, L. (2010) *Macromol. Chem. Phys.*, **211**, 2416–2424.

45 (a) Gholamkhass, B. and Holdcroft, S. (2010) *Chem. Mater.*, **22**, 5371–5376; (b) Gholamkhass, B., Peckham, T.J., and Holdcroft, S. (2010) *Polym. J.*, **1**, 708; (c) Yang, C., Lee, J.K., Heeger, A.J., and Wudl, F. (2009) *J. Mater. Chem.*, **19**, 5416–5423; (d) Dante, M., Yang, C., Walker, B., Wudl, F., and Nguyen, T.-Q. (2010) *Adv. Mater.*, **22**, 1835–1838.

46 Boutorine, A.S., Tokuyama, H., Takasugi, M., Isobe, H., Nakamura, E., and Hélène, C. (1994) *Angew. Chem. Int. Ed. Engl.*, **33**, 2462–2465.

47 (a) Isobe, H., Nakanishi, W., Tomita, N., Jinno, S., Okayama, H., and Nakamura, E. (2006) *Chem. Asian J.*, **1**, 167–175; (b) Maeda-Mamiya, R., Noiri, E., Isobe, H., Nakanishi, W., Okamoto, K., Doi, K., Sugaya, T., Izumi, T., Homma, T., and Nakamura, E. (2010) *Proc. Natl. Acad. Sci. USA*, **57**, 5339–5344.

48 Smith, B.W., Monthioux, M., and Luzzi, D.E. (1998) *Nature*, **396**, 323–324.

49 (a) Kitaura, R. and Shinohara, H. (2006) *Chem. Asian J.*, **1**, 646–655; (b) Britz, D.A. and Khlobystov, A.N. (2006) *Chem. Soc. Rev.*, **35**, 637–659.

50 Guldi, D.M., Menna, E., Maggini, M., Marcaccio, M., Paolucci, D., Paolucci, F., Campidelli, S., Prato, M., Rahman, G.M.A., and Schergna, S. (2006) *Chem. Eur. J.*, **12**, 3975–3983.

51 Delgado, J.L., de la Cruz, P., Urbina, A., López Navarrete, J.T., Casado, J., and Langa, F. (2007) *Carbon*, **45**, 2250–2252.

52 Giordani, S., Colomer, J.-F., Cattaruzza, F., Alfonsi, J., Meneghetti, M., Prato, M., and Bonifazi, D. (2009) *Carbon*, **47**, 578–588.

53 (a) Nasibulin, A.G., Pikhitsa, P.V., Jiang, H., Brown, D.P., Krasheninnikov, A.V., Anisimov, A.S., Queipo, P., Moisala, A., Gonzalez, D., Lientschnig, G., Hassanien, A., Shandakov, S.D., Lolli, G., Resasco, D.E., Choi, M., Tománek, D., and Kauppinen, E.I. (2007) *Nat. Nanotechnol.*, **2**, 156–161; (b) Nasibulin, A.G., Anisimov, A.S., Pikhitsa, P.V., Jiang, H., Brown, D.P., Choi, M., and Kauppinen, E.I. (2007) *Chem. Phys. Lett.*, **446**, 109–114; (c) He, M., Rikkinen, E., Zhu, Z., Tian, Y., Anisimov, A.S., Jiang, H., Nasibulin, A.G., Kauppinen, E.I., Niemelä, M., and Krause, A.O.I. (2010) *J. Phys. Chem. C* **114**, 13540–13545.

54 Ouyang, T., Loh, K.P., Qi, D., Wee, A.T.S., and Nesladek, M. (2008) *ChemPhysChem*, **9**, 1286–1293.

55 (a) Zhang, X., Huang, Y., Wang, Y., Ma, Y., Liu, Z., and Chen, Y. (2008) *Carbon*, **47**, 334–337; (b) Liu, Z.-B., Xu, Y.-F., Zhang, X.-Y., Zhang, X.-L., Chen, Y.-S., and Tian, J.-G. (2009) *J. Phys. Chem. B*, **113**, 9681–9686.

56 Brust, M., Kiely, C.J., Bethell, D., and Schiffrin, D.J. (1998) *J. Am. Chem. Soc.*, **120**, 12367–12367.
57 (a) Lim, I.-I.S., Ouyang, J., Luo, J., Wang, L., Zhou, S., and Zhong, C.-J. (2005) *Chem. Mater.*, **17**, 6528–6531; (b) Lim, I.-I.S., Pan, Y., Mott, D., Ouyang, J., Njoki, P.N., Luo, J., Zhou, S., and Zhong, C.-J. (2007) *Langmuir*, **23**, 10715–10724; (c) Yin, G., Xue, W., Chen, F., and Fan, X. (2009) *Coll. Surface A*, **340**, 121–125.
58 (a) Liu, J., Alvarez, J., Ong, W., and Kaifer, A.E. (2001) *Nano Lett.*, **1**, 57–60; (b) Liu, Y., Wang, H., Chen, Y., Ke, C.-F., and Liu, M. (2005) *J. Am. Chem. Soc.*, **127**, 657–666.
59 (a) Shon, Y.-S. and Choo, H. (2002) *Chem. Commun.*, 2560–2561; (b) Sudeep, P.K., Ipe, B.I., Thomas, K.G., George, M.V., Barazzouk, S., Hotchandani, S., and Kamat, P.V. (2002) *Nano Lett.*, **2**, 29–35; (c) Geng, M., Zhang, Y., Huang, Q., Zhang, B., Li, Q., Li, W., and Li, J. (2010) *Carbon*, **48**, 3570–3574.
60 Shih, S.-M., Su, W.-F., Lin, Y.-J., Wu, C.-S., and Chen, C.-D. (2002) *Langmuir*, **18**, 3332–3335.
61 Liu, W. and Gao, X. (2008) *Nanotechnology*, **19**, 405609.
62 For recent reviews, see (a) Peng, X. and Wong, S.S. (2009) *Adv. Mater.*, **21**, 625–642; (b) Singh, P., Campidelli, S., Giordani, S., Bonifazi, D., Bianco, A., and Prato, M. (2009) *Chem. Soc. Rev.*, **38**, 2214–2230; (c) Zhao, Y.-L. and Stoddart, J.F. (2009) *Acc. Chem. Res.*, **42**, 1161–1171; (d) Wang, H. (2009) *Curr. Opin. Colloid Interface Sci.*, **14**, 364–371; (e) Eder, D. (2010) *Chem. Rev.*, **110**, 1348–1385; (f) Karousis, N., Tagmatarchis, N., and Tasis, D. (2010) *Chem. Rev.*, **110**, 5366–5397.
63 Delgado, J.L., Herranz, M.A., and Martín, N. (2008) *J. Mater. Chem.*, **18**, 1417–1426.
64 Prato, M. (2010) *Nature*, **465**, 172–173.
65 (a) Reich, S., Thomsen, C., and Maultzsch, J. (2004) *Carbon Nanotubes: Basic Concepts and Physical Properties*, Wiley-VCH Verlag GmbH, Weinheim, Germany; (b) Popov, V.N. and Lambin, P. (2006) *Carbon Nanotubes*, Springer, Dordrecht; (c) Guldi, D.M. and Martín, N. (2010) *Carbon Nanotubes and Related Structures*, Wiley-VCH Verlag GmbH, Weinheim, Germany; (d) Akasaka, T., Wudl, F., and Nagase, S. (2010) *Chemistry of Nanocarbons*, John Wiley & Sons, Ltd., Chichester, United Kingdom.
66 (a) Hirsch, A. (2002) *Angew. Chem. Int. Ed.*, **41**, 1853–1859; (b) Hirsch, A. and Vostrowsky, O. (2005) *Top. Curr. Chem.*, **245**, 193–237.
67 Liu, J., Rinzler, A.G., Dai, H., Hafner, J.H., Bradley, R.K., Boul, P.J., Lu, A., Iverson, T., Shelimov, K., Huffman, C.B., Rodriguez-Marcias, F., Shon, Y.-S., Lee, T.R., Colbert, D.T., and Smalley, R.E. (1998) *Science*, **280**, 1253–1256.
68 Hamon, M.A., Chen, J., Hu, H., Chen, Y., Itkis, M.E., Rao, A.M., Eklund, P.C., and Haddon, R.C. (1999) *Adv. Mater.*, **11**, 834–840.
69 Zhao, B., Hu, H., and Haddon, R.C. (2004) *Adv. Funct. Mater.*, **14**, 71–76.
70 Pompeo, F. and Resasco, D.E. (2002) *Nano Lett.*, **2**, 369–373.
71 Baker, S.E., Cai, W., Lasseter, T.L., Weidkamp, K.P., and Hammers, R.J. (2002) *Nano Lett.*, **2**, 1413–1417.
72 He, P. and Urban, M.W. (2005) *Biomacromolecules*, **6**, 2455–2457.
73 (a) Guldi, D.M., Rahman, G.M.A., Sgobba, V., and Ehli, C. (2006) *Chem. Soc. Rev.*, **35**, 471–487; (b) Guldi, D.M. (2007) *Phys. Chem. Chem. Phys.*, 1400–1420; (c) Sgobba, V. and Guldi, D.M. (2009) *Chem. Soc. Rev.*, **38**, 165–184.
74 Herranz, M.A., Martín, N., Campidelli, S., Prato, M., Brehm, G., and Guldi, D.M. (2006) *Angew. Chem. Int. Ed.*, **45**, 4478–4482.
75 (a) Li, H., Martin, R.B., Harruff, B.A., Carino, R.A., Allard, L.F., and Sun, Y.-P. (2004) *Adv. Mater.*, **16**, 896–900; (b) Baskaran, D., Mays, J.W., Zhang, X.P., and Bratcher, M.S. (2005) *J. Am. Chem. Soc.*, **127**, 6916–6917.
76 (a) Van Lier, G., Ewels, C.P., Zuliani, F., De Vita, A., and Charlier, J.-C. (2005) *J. Phys. Chem. B*, **109**, 6153–6158; (b) Lee, Y.-S. (2007) *J. Fluorine Chem.*, **128**, 392–403.

77 Zeng, L., Zhang, L., and Barron, A.R. (2005) *Nano Lett.*, **5**, 2001–2004.

78 (a) Zhang, L., Zhang, J., Schmandt, N., Cratty, J., Khabashesku, V.N., Kelly, K.F., and Barron, A.R. (2005) *Chem. Commun.*, 5429–5431; (b) Zhang, J., Zhang, L., Khabashesku, V.N., Barron, A.R., and Kelly, K.F. (2008) *J. Phys. Chem. C*, **112**, 12321–12325.

79 Bahr, J.L., Yang, J., Kosynkin, D.V., Bronikowski, M.J., Smalley, R.E., and Tour, J.M. (2001) *J. Am. Chem. Soc.*, **123**, 6536–6542.

80 (a) Bahr, J.L. and Tour, J.M. (2001) *Chem. Mater.*, **13**, 3823–3824; (b) Dyke, A. and Tour, J.M. (2003) *Nano Lett.*, **3**, 1215–1218.

81 Holzinger, M., Vostrowsky, O., Hirsch, A., Hennrich, F., Kappes, M., Weiss, R., and Jellen, F. (2001) *Angew. Chem. Int. Ed.*, **40**, 4002–4005.

82 Pastine, S.J., Okawa, D., Kessler, B., Rolandi, M., Llorente, M., Zettl, A., and Frechet, J.M.J. (2008) *J. Am. Chem. Soc.*, **130**, 4238–4239.

83 Ashcroft, J.M., Hartman, K.B., Mackeyev, Y., Hofmann, C., Pheasant, S., Alemany, L.B., and Wilson, L. (2006) *Nanotechnology*, **17**, 5033–5037.

84 Delgado, J.L., de la Cruz, P., Langa, F., Urbina, A., Casado, J., and Lopez Navarrete, J.T. (2004) *Chem. Commun.*, 1734–1735.

85 Tagmatarchis, N. and Prato, M. (2004) *J. Mater. Chem.*, **14**, 437–439.

86 Prato, M., Kostarelos, K., and Bianco, A. (2008) *Acc. Chem. Res.*, **41**, 60–68.

87 Georgakilas, V., Tagmatarchis, N., Pantarotto, D., Bianco, A., Briand, J.-P., and Prato, M. (2002) *Chem. Commun.*, 3050–3051.

88 Pantarotto, D., Singh, R., McCarthy, D., Erhardt, M., Briand, J.-P., Prato, M., Kostarelos, K., and Bianco, A. (2004) *Angew. Chem. Int. Ed.*, **43**, 5242–5246.

89 Singh, R., Pantarotto, D., Lacerda, L., Pastorin, G., Klumpp, C., Prato, M., Bianco, A., and Kostarelos, K. (2006) *Proc. Natl. Acad. Sci. USA*, **103**, 3357–3362.

90 Wu, W., Wieckowski, S., Pastorin, G., Benincasa, M., Klumpp, C., Briand, J., Gennaro, R., Prato, M., and Bianco, A. (2005) *Angew. Chem. Int. Ed.*, **44**, 6358–6362.

91 Pastorin, G., Wu, W., Wieckowski, S., Briand, J.-P., Kostarelos, K., Prato, M., and Bianco, A. (2006) *Chem. Commun.*, 1182–1184.

92 Graupner, R., Abraham, J., Wunderlich, D., Vencelova, A., Lauffer, P., Roehrl, J., Hundhausen, M., Ley, L., and Hirsch, A. (2006) *J. Am. Chem. Soc.*, **128**, 6683–6689.

93 Syrgiannis, Z., Gebhardt, B., Dotzer, C., Hauke, F., Graupner, R., and Hirsch, A. (2010) *Angew. Chem. Int. Ed.*, **49**, 3322–3325.

94 For recent reviews, see (a) Rao, C.N.R., Sood, A.K., Subrahmanyam, K.S., and Govindaraj, A. (2009) *Angew. Chem. Int. Ed.*, **48**, 7752–7777; (b) Allen, M.J., Tung, V.C., and Kaner, R.B. (2010) *Chem. Rev.*, **110**, 132–145; (c) Dreyer, D.R., Park, S., Bielawski, C.W., and Rodney, S.R. (2010) *Chem. Soc. Rev.*, **39**, 228–240; (d) Ping Loh, K., Bao, Q., Ang, P.K., and Yang, J. (2010) *J. Mater. Chem.*, **20**, 2277–2289.

95 Van Noorden, R. (2011) *Nature*, **469**, 14–16.

96 Baughman, R.H., Zakhidov, A.A., and de Heer, W.A. (2002) *Science*, **297**, 787–792.

97 Hernandez, Y., Nicolosi, V., Lotya, M., Blighe, F.M., Sun, Z., De, S., McGovern, I.T., Holland, B., Byrne, M., Gun'ko, Y.K., Boland, J.J., Niraj, P., Duesberg, G., Krishnamurthy, S., Goodhue, R., Hutchison, J., Scardaci, V., Ferrari, A.C., and Coleman, J.M. (2008) *Nat. Nanotechnol.*, **3**, 563–568.

98 Niyogi, S., Bekyarova, E., Itkis, M.E., McWilliams, J.L., Hamon, M.A., and Haddon, R.C. (2006) *J. Am. Chem. Soc.*, **128**, 7720–7721.

99 Lomeda, R., Doyle, C.D., Kosynkin, D.V., Hwang, W.-F., and Tour, J.M. (2008) *J. Am. Chem. Soc.*, **130**, 16201–16206.

100 Quintana, M., Spyrou, K., Grelczak, M., Browne, W.R., Rudolf, P., and Prato, M. (2010) *ACS Nano*, **4**, 3527–3533.

101 Economopoulos, S.P., Rotas, G., Miyata, Y., Shinohara, H., and Tagmatarchis, N. (2010) *ACS Nano*, **4**, 7499–7507.

102 Subrahmanyam, K.S., Ghosh, A., Gomathi, A., Govindaraj, A., and Rao, C.N.R. (2009) *Nanosci. Nanotechnol. Lett.*, **1**, 28–31.

103 Liu, N., Luo, F., Wu, H., Liu, Y., Zhang, C., and Cheng, J. (2008) *Adv. Funct. Mater.*, **18**, 1518–1525.

104 Mack, J.J., Viculis, L.M., Luoh, A.A.R., Yang, G., Hahn, H.T., Ko, F.K., and Kaner, R.B. (2005) *Adv. Mater.*, **17**, 77–80.

105 Das, B., Prasad, K.E., Ramamurty, U., and Rao, C.N.R. (2009) *Nanotechnology*, **20**, 125705.

106 Ramanathan, T., Abdala, A.A., Stankovich, S., Dikin, D.A., Alonso, M.H., Piner, R.D., Adamson, D.H., Schniepp, H.C., Chen, X., Ruoff, R.S., Nguyen, S.T., Aksay, I.A., Prud-Homme, R.K., and Brinson, L.C. (2008) *Nat. Nanotechnol.*, **3**, 327–331.

107 Bai, H., Xu, Y., Zhao, L., Li, C., and Shi, G. (2009) *Chem. Commun.*, 1667–1669.

108 Zu, S.-Z. and Han, B.-H. (2009) *J. Phys. Chem. C*, **113**, 13651–13657.

109 Bao, Q., Zhang, H., Wang, Y., Ni, Z., Yan, Y., Shen, Z.X., Loh, K.P., and Tang, D.Y. (2009) *Adv. Funct. Mater.*, **19**, 3077–3083.

110 Xu, X.Y., Bai, H., Lu, G.W., Li, C., and Shi, G.Q. (2008) *J. Am. Chem. Soc.*, **130**, 5856–5857.

111 Su, Q., Pang, S., Alijani, V., Li, C., Feng, X., and Müllen, K. (2009) *Adv. Mater.*, **21**, 3191–3195.

112 Wang, Y., Chen, X., Zhong, Y., Zhu, F., and Loh, K.P. (2009) *Appl. Phys. Lett.*, **95**, 063302.

113 (a) Das, B., Voggu, R., Rout, C.S., and Rao, C.N.R. (2008) *Chem. Commun.*, 5155–5157; (b) Voggu, R., Das, B., Rout, C.S., and Rao, C.N.R. (2008) *J. Phys. Condens. Matter.*, **20**, 472204.

114 Manna, A.K. and Pati, S.K. (2009) *Chem. Asian J.*, **4**, 855–860.

115 Subrahmanyam, K.S., Voggu, R., Govindaraj, A., and Rao, C.N.R. (2009) *Chem. Phys. Lett.*, **472**, 96–98.

116 Barja, S., Garnica, M., Hinarejos, J.J., Vázquez de Parga, A.L., Martín, N., and Miranda, R. (2010) *Chem. Commun.*, **46**, 8198–8200.

117 Kozhemyakina, N.V., Englert, J.M., Yang, G., Spiecker, E., Schmidt, C.D., Hauke, F., and Hirsch, A. (2010) *Adv. Mater.*, **22**, 5483–5487.

2
Hydrogen-Bonded Fullerene Assemblies

José Santos, Beatriz M. Illescas, Luis Sánchez, and Nazario Martín

2.1
Introduction

The combination of covalent bonds and noncovalent interactions in *Mother Nature* creates an ample repertoire of complex structures such as DNA, amyloidal fibrils, tobacco mosaic virus, and so on, with a specific shape, size, and, more importantly, function. The noncovalent interactions at disposal of Nature are metal coordination, van der Waals forces, π–π interactions, electrostatic effects, and hydrogen bonding [1]. According to the bond strength, hydrogen bonds are highly selective and directional interactions, but they are also weak, with binding energies in the range of only 5 kcal/mol (Figure 2.1a) [2]. The strength of the hydrogen bonding interactions can be modulated by the incorporation of arrays of more than one H bond or by the synergy of two classes of noncovalent interactions (Figure 2.1b) [3].

A natural example of a complex ensemble organized by noncovalent interactions is the photosynthetic apparatus in which a highly organized array of electron donor (porphyrins) and electron acceptor (benzoquinones) moieties are able to transform sunlight into chemical energy. In this photosynthetic process, a cascade of energy (ET) and electron transfer (eT) events operate to convert CO_2 and H_2O into glucose and energy (ATP) [5]. Owing to the significance of this complex natural process, big efforts have been dedicated to the construction of chemical structures capable of reproducing it [6]. Efficient energy and electron transfer processes are also fundamental in the development of molecular-scale optoelectronics, photonics, or other areas included in nanoscience and nanotechnology [7]. A plethora of organic electroactive molecules have been extensively utilized in the preparation of models for the study of ET and eT processes and also for the fabrication of optoelectronic devices (solar cells, field-effect transistors, or light emitting diodes) [8]. Among the different electron acceptor molecules used for these purposes, [60]fullerene is probably the most studied system owing to several factors: (a) small reorganization energy, (b) high electron affinity, and (c) ability to transport negative charges [9]. Therefore, the combination of fullerenes with H bonding interactions should render

Supramolecular Chemistry of Fullerenes and Carbon Nanotubes, First Edition. Edited by Nazario Martin and Jean-Francois Nierengarten.

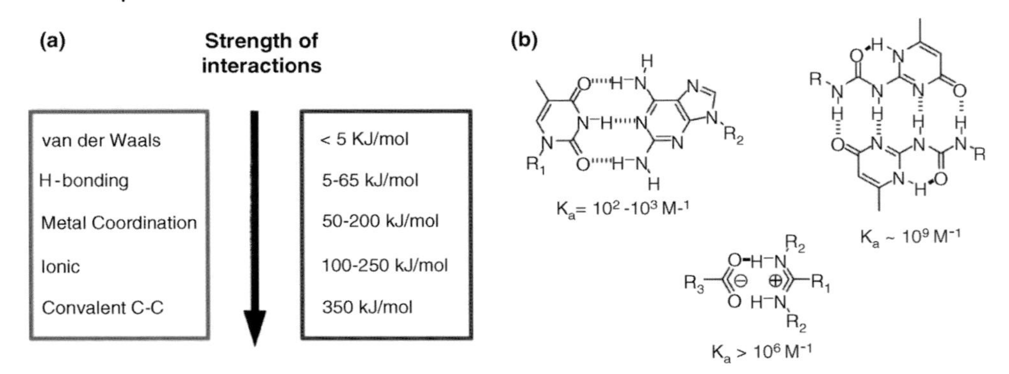

Figure 2.1 (a) Noncovalent interactions ordered according to their bond strength [4]. (b) H bonding motifs with high binding constant (K_a) [3].

a new family of noncovalent structures with unprecedented features relative to ET and/or eT processes of interest for the preparation of optoelectronic devices.

This chapter does not intend to be a comprehensive review of all the examples of H-bonded fullerenes reported so far, but is an update of our previously reported work published in 2005 [3] in which relevant examples of this class of noncovalent ensembles were highlighted. The examples now presented will be grouped in three different parts that comprise (1) H-bonded supramolecular structures endowed with fullerenes, (2) C_{60} electron donor supramolecular assemblies, and (3) applications.

2.2 Hydrogen-Bonded Fullerene-Based Supramolecular Structures

In this section, we highlight those examples of H-bonded fullerene ensembles that serve (a) as proof of principle for increasing the knowledge on basic aspects of hydrogen-bonded supramolecular structures or (b) as model systems to generate more complicated assemblies that exhibit a number of properties arising from the connection of C_{60} units through H bonds with some relevant moieties.

The starting point for the preparation of supramolecular structures in which fullerenes are attached by H bonds is 1999, when Diederich *et al.* reported on the formation of the rotaxane-like C_{60}-dimer **1** (Figure 2.2a) [10]. In this dimer, a Bingel-type C_{60}-cycloadduct endowed with an ammonium salt is threaded into a malonate-appended dibenzo-24-crown-8 fullerene derivative by means of the well-established N^+–H···O and C–H···O hydrogen bonds, resulting in a binding constant of 970 M^{-1} ($CHCl_3/CH_3CN$ 9 : 1). Very recently, a new family of interlocked architectures involving [60]fullerene has been reported in which a Leigh-type rotaxane [10] is connected to a C_{60} thread. The constitutive parts of all these kinds of rotaxanes are (1) a fullerene and a di-*tert*-butylphenyl stopper, (2) a glycylglycine template, (3) an alkyl chain, and (4) the macrocycle that can shuttle or pirouette (Figure 2.2b). In good

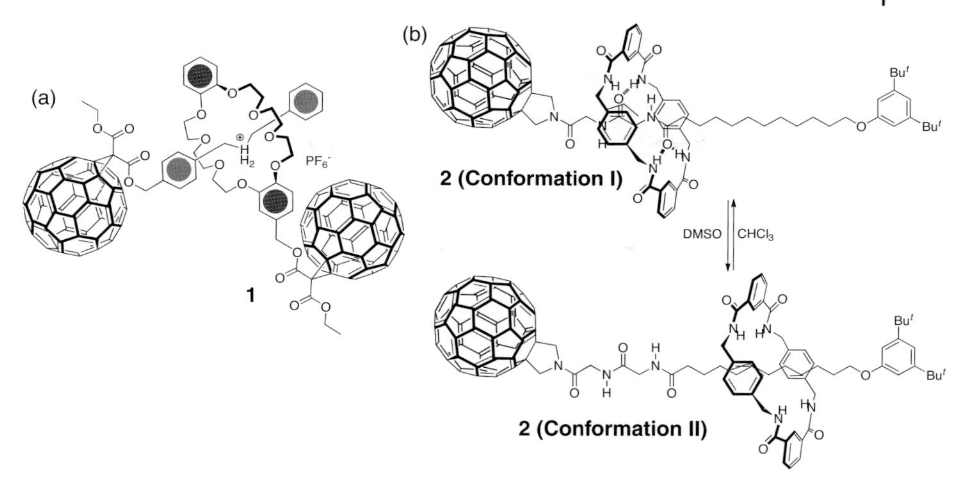

Figure 2.2 Chemical structures of the first C_{60}-based rotaxane (a) and Leigh-type rotaxane **2** showing the shuttling motion induced by the solvent.

solvents for H bonds, such as chloroform, the H bonds between the amide groups of the macrocycle and the carbonyl functionality of the glycylglycine unit locates the macrocycle in the proximity of the fullerene surface (conformation I). Addition of highly polar solvents, such as DMSO, disrupts the H bonds and shifts the macrocycle along the alkyl chain (conformation II). This shuttling effect can be monitored by ^{1}H NMR experiments [11].

Self-complementary 2-ureido-4-pyrimidinone (UP) is a structural H bonding motif in which a DDAA quadruple array of H bonds induces an extraordinary high binding constant due to the presence of attractive secondary interactions [12]. Two referable supramolecular dimers based on the UP motif were reported in 2001 (compounds **3a** and **3b** in Figure 2.3a and b) [13]. The photophysical studies carried out on dimer **3a** demonstrated that the strong electronic coupling between the two fullerene units is favored by the H bonding array. On the other hand, dimer **3b** takes the preliminary stage toward the consecution of more complex structures such as the supramolecular polymer (**4**) reported by Hummelen and coworkers in 2002 (Figure 2.3c) [14]. The dynamic character of these polymers was tested by concentration-dependent ^{1}H NMR experiments in $CDCl_3$ as solvent. Diagnostic resonances at $\delta = 12.87$, 11.63, 10.65, and 5.72 – ascribable to the four hydrogen bonds present in the UP units – were visible in concentrated solutions (100 mM). However, decreasing the concentration until 10 mM results in the appearance of multiple resonances in the same region of the ^{1}H NMR spectrum due to the presence of oligomeric species of different lengths. In addition, the cyclic voltammogram of polymer **4** exhibited four intense reduction waves corresponding to the C_{60} units and a small wave ascribable to the UP unit. The presence of only one set of reduction waves implies a negligible electronic communication between the constitutive C_{60} units of the supramolecular polymer.

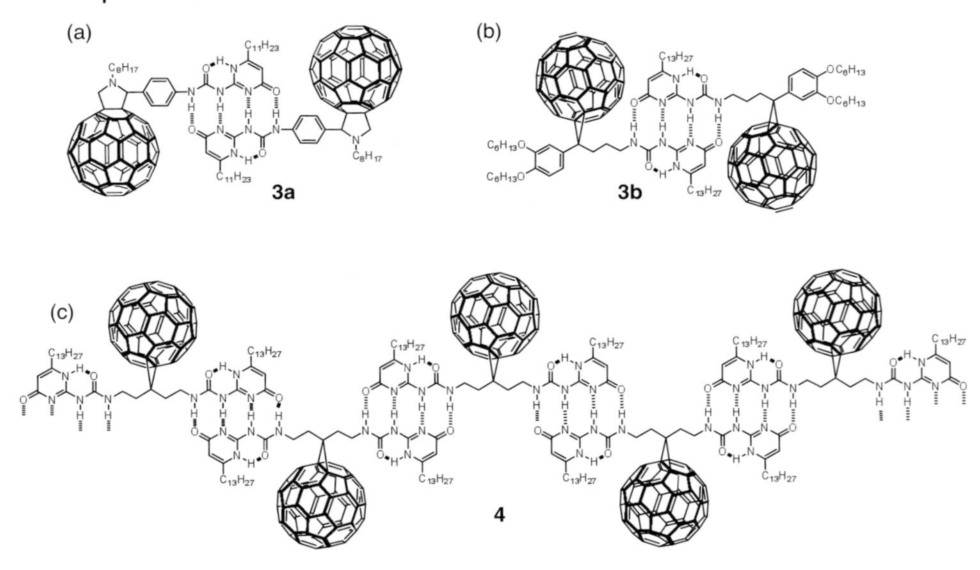

Figure 2.3 Chemical structures of dimers (a and b) and supramolecular polymer (c) in which the C_{60} units are attached through UP moieties.

Very recently, Granja and coworkers reported on the synthesis of a α,γ-octapeptide (**5**) with hydrophilic residues to increase the solubility in polar solvents enhancing its self-assembly in polar solvents as a consequence of the solvophobic effects (Figure 2.4) [15]. This octapeptide also possesses a lysine residue functionalized with a fullerene derivative. The self-assembly of the circular octapeptide **5** through H bonding between the N–H and the C=O functional groups of the peptide generates nanotubular structures. The attachment of the C_{60} units allows coiling of the self-assembled peptide nanotube with this redox-active molecule. The formation of such nanostructures has been visualized by atomic force microscopy (AFM) on mica as surface. AFM images of associated **5** show wire-like

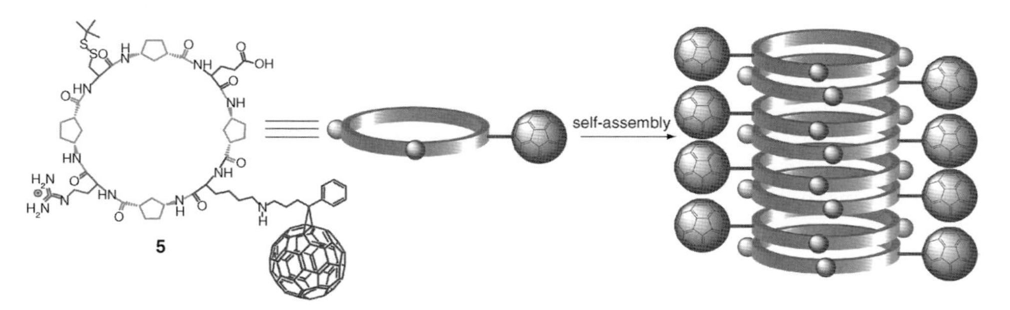

Figure 2.4 Chemical structure of the α,γ-octapeptide **5** and schematic illustration of its self-assembly forming nanotubes.

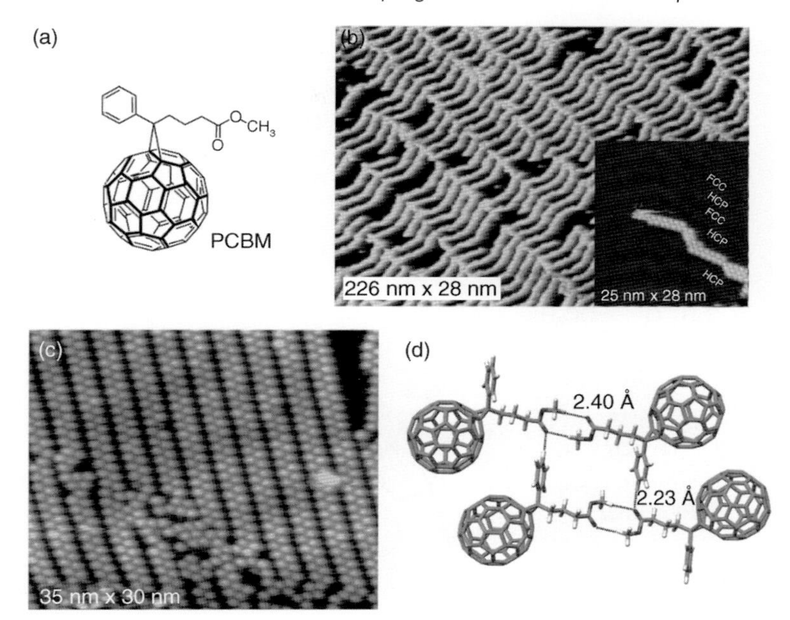

Figure 2.5 (a) Chemical structure of PCBM. STM images of the Au(111) after depositing 0.4 monolayer (b) (the inset shows a zoom of the zigzag structures) and 0.6 monolayer (c) of PCBM. (d) Optimized structure for a PCBM tetramer showing the C–H · · · O hydrogen bonds responsible for this organization.

structures several micrometers long with a height of ~3 nm, roughly corresponding to the diameter of the circular octapeptide with the dangling C_{60} units.

All the above collected examples of H-bonded supramolecular structures based on C_{60} exhibit polarized N–H · · · O hydrogen bonds as the structural motif responsible for the noncovalent binding. However, weaker C–H · · · X (X = O, N, . . .) H bonds have been demonstrated to play a relevant role in the organization of organic molecules onto surfaces [16]. We have recently visualized the self-organization of PCBM (phenyl-C_{61}-butyric acid ethyl ester, Figure 2.5a) – probably the most studied C_{60} derivative – onto a Au(111) surface by the operation of weak C–H · · · O hydrogen bonds and utilizing scanning tunneling microscopy (STM) at low temperatures under ultrahigh vacuum (UHV) conditions [17].

The nucleation of the PCBM molecules onto the Au(111) surface strongly depends on the coverage of the surface. Thus, at low coverage, PCBM molecules locate at the fcc areas of the herringbone reconstruction of the Au(111) [18] surface forming long isolated chains constituted by two arrays of PCBM units interacting by the π stacking of the fullerene cages (Figure 2.5b). These chains grow with the increasing number of PCBM molecules until covering completely the fcc areas and also interacting in the elbows of the reconstruction. The growth of the linear twin chains forms a bidimensional spiderweb-like network directed by the Au(111) surface (Figure 2.5b). This site selectivity in the adsorption of PCBM disappears when the coverage is higher than

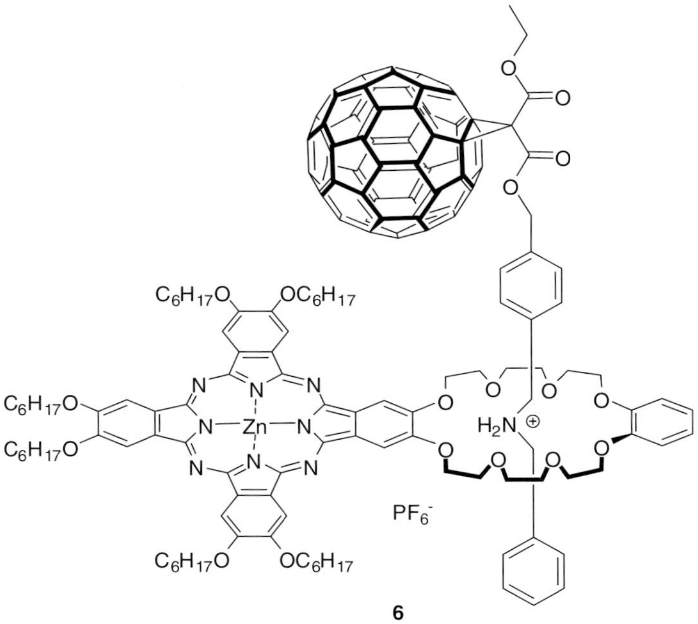

Figure 2.6 Chemical structure of the first H-bonded C_{60}–D dyad **6**.

the number of fcc areas. In this situation, a reorganization of the PCBM molecules is observed and a compact arrangement of twin chains appears (Figure 2.5c). This new organization has been justified by a complex network of weak C–H $\cdots$ O hydrogen bonds between the methyl ester and the phenyl groups of the tail of PCBM molecules with distances of 2.23 and 2.40 Å (Figure 2.5d). In addition, DFT theoretical calculations also demonstrate that the H bonding interactions are formed far away from the Au surface. This situation provokes a negligible contact of the tails of PCBM with the surface and explains the site-insensitive adsorption of PCBM onto Au(111) at high coverage [19].

2.3
Hydrogen-Bonded Fullerene-Based Donor–Acceptor Structures

Several C_{60}-based donor–acceptor conjugates using a hydrogen bonding strategy have been studied, showing that the electronic communication in these ensembles is, at least, as strong as that found in covalently linked systems [20]. This finding strengthens the concept of hydrogen bonding as a viable method for the construction of functional materials for various applications. The first H-bonded C_{60}-donor dyad was reported by Guldi *et al.* [21] and consisted in a pseudorotaxane-like complex between a zinc phthalocyanine as electron donor and a dibenzylammonium fullerene as the acceptor counterpart (Figure 2.6). An efficient intracomplex electron transfer

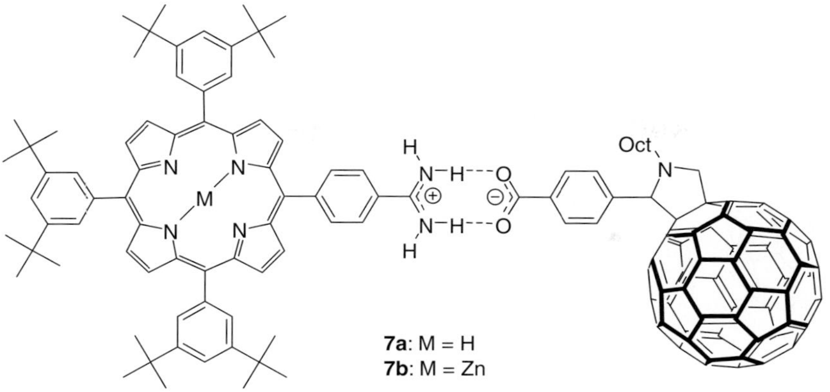

Figure 2.7 Supramolecular dyad based on the amidinium–carboxylate interaction between donor and acceptor units.

from the excited state of the Zn Pc moiety was proved and, importantly, the reported lifetime for the radical pair species (i.e., $C_{60}^{\bullet-}$ $ZnPc^{\bullet+}$) was in the range of microseconds.

Since this first promising example, many different supramolecular H-bonded ensembles interfacing C_{60} with a variety of electron donor moieties have been studied and they have been thoroughly reviewed [3]. Here, we will summarize outstanding and recent examples on this topic.

Owing to their similarity with natural electron–donor centers, porphyrins (P) play a major role for the preparation of supramolecular donor–acceptor dyads or triads. A set of noncovalently associated C_{60}–porphyrin ensembles (**7a–b**) was prepared by using a two-point amidinium carboxylate binding motif, which is particularly stable as a result of the synergy of hydrogen bonds and electrostatic interactions [20]. This amidinium carboxylate ion pairing diminishes other possible bonding modes, thus favoring the linearity of the donor–acceptor pair and ensuring an optimal pathway for the motion of the charges. The association constants for these pairs reach values up to $10^7\,M^{-1}$ in toluene or $10^5\,M^{-1}$ in THF. This strong binding gives rise to an exceptionally strong electronic coupling between both electroactive elements ($36\,cm^{-1}$ for **7b**), which in turn facilitates the formation of long-lived radical ion pairs with a lifetime of $\sim 1\,\mu s$ in THF (Figure 2.7).

Hirsch and coworkers employed a modular concept for the self-assembly of electron donor–acceptor complexes by introducing the Hamilton receptor/cyanuric acid binding motif in a series of porphyrin–fullerene systems (**8**·**9**, Figure 2.8) [22]. In particular, different fullerene malonates carrying a cyanuric acid moiety separated with either a propylene or a hexylene chain were prepared (**8**) and assembled with a porphyrin bearing the Hamilton receptor. Besides the free base porphyrin derivatives, zinc, with a tetragonal planar coordination motif, and tin, with an octahedral coordination pattern with two additional axial ligands, as central metals have also been studied. This methodology ensures a fine-tuning of the strength of the

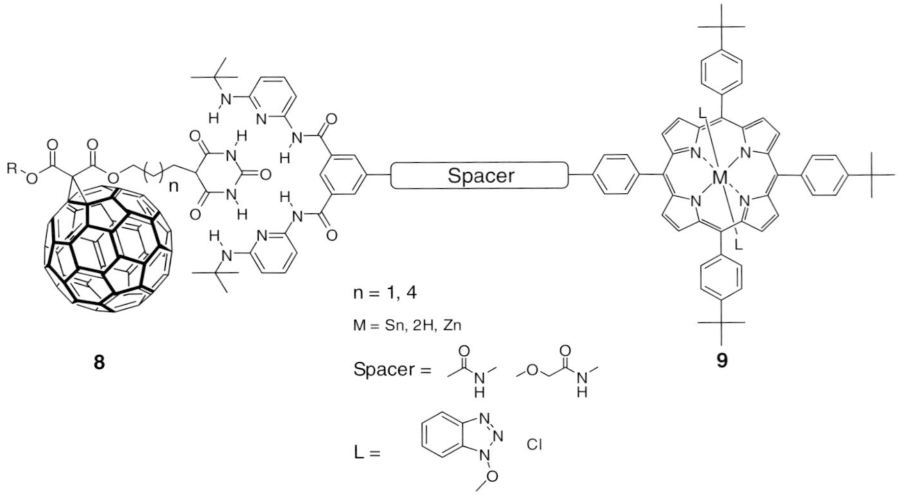

Figure 2.8 Hydrogen binding between Hamilton receptor/cyanuric acid of moieties **9** and **8**, respectively.

complexation and control of the electronic coupling to impact electron and energy transfer processes. Association constants K_a that range from $3.7 \times 10^3\,M^{-1}$ to $7.9 \times 10^5\,M^{-1}$ ($CHCl_3$) have been determined for these 1 : 1 complexes connected by six hydrogen bonds. Interestingly, the substitution of a hexylene alkyl spacer with a propylene spacer in **8** strengthens the formation of the nanohybrid. The differential absorption spectra of the different complexes revealed a fast charge separation from the photoexcited ZnP chromophores, giving rise to the $ZnP^{\bullet +} - C_{60}^{\bullet -}$ state with a rate constant of $4.3 \times 10^9\,s^{-1}$. In contrast, in the analogous SnP complexes, an exothermic energy transfer process, instead of electron transfer, is the main process of deactivation of the SnP excited state. This different behavior can be explained attending to the anodically shifted oxidation potential of SnP related to ZnP.

The same hydrogen bonding motif has been employed to form the electron donor–acceptor nanohybrids shown in Figure 2.9 [23]. The integration of conjugated spacers (i.e., *p*-phenylene ethynylene, *p*-phenylene vinylene, *p*-ethynylene, and fluorene) provides wire-like behavior and restricts the flexibility of the Hamilton receptor and cyanuric acid functionalities (Figure 2.9). The association constants were determined to be in the range of 10^4–$10^5\,M^{-1}$ (chloroform) from ^1H NMR and steady-state fluorescence assays. Transient absorption measurements confirm an electron transfer upon photoexcitation to yield the one-electron oxidized porphyrins and one-electron reduced fullerenes. The study of the dependence of the rate constants of charge separation and charge recombination on the donor–acceptor distances allowed to calculate the attenuation factor (β) for the first time in a hydrogen-bonding-mediated electron transfer. The obtained value of $0.11\,Å^{-1}$ is notably small and lies between the extremes of covalent *p*-phenylene ethynylene [24] and fluorene [25] systems.

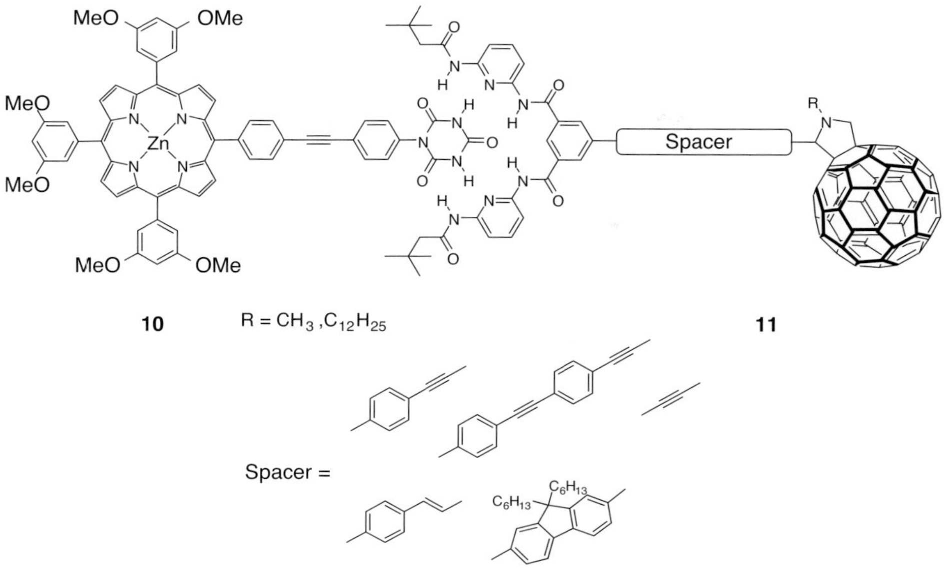

MeO
OMe
MeO
MeO
Zn
MeO
OMe
Spacer
10
R = CH_3 ,$C_{12}H_{25}$
11
Spacer =
C_6H_{13}
C_6H_{13}

Figure 2.9 Supramolecular wires based on fullerenes and porphyrins.

Using diacetylamidopyridine:uracil complementary hydrogen bonding motif, a bis (zinc porphyrin)-fullerene triad has been constructed (**12**, Figure 2.10) [26]. The geometry of the triad calculated from DFT-MO studies revealed the presence of the "three-point" hydrogen bonding and that one of the porphyrins was closer to the fullerene core than the other. Therefore, this triad offered the opportunity to compare the distance-dependent electron coupling and its effect on the electron transfer rate constants. The measured k_{CS} and k_{CR} values are different for the near-side and

12

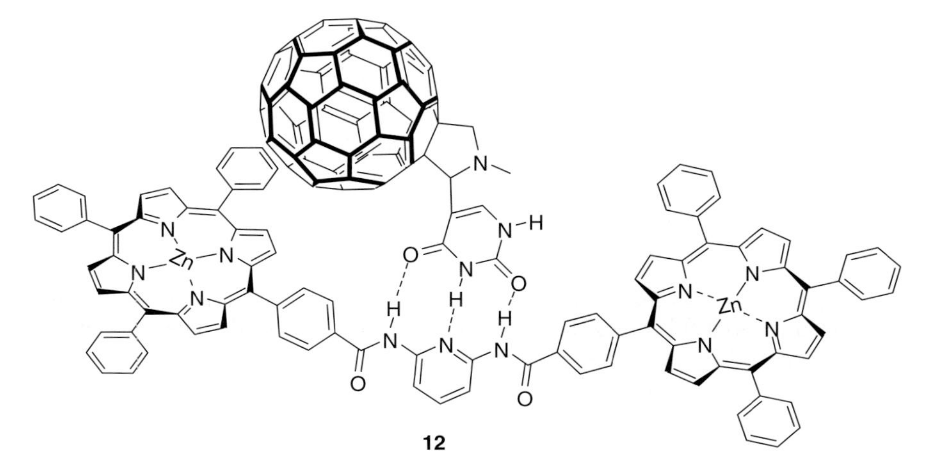

Figure 2.10 Molecular structure of the bis(zinc porphyrin)–fullerene supramolecular triad.

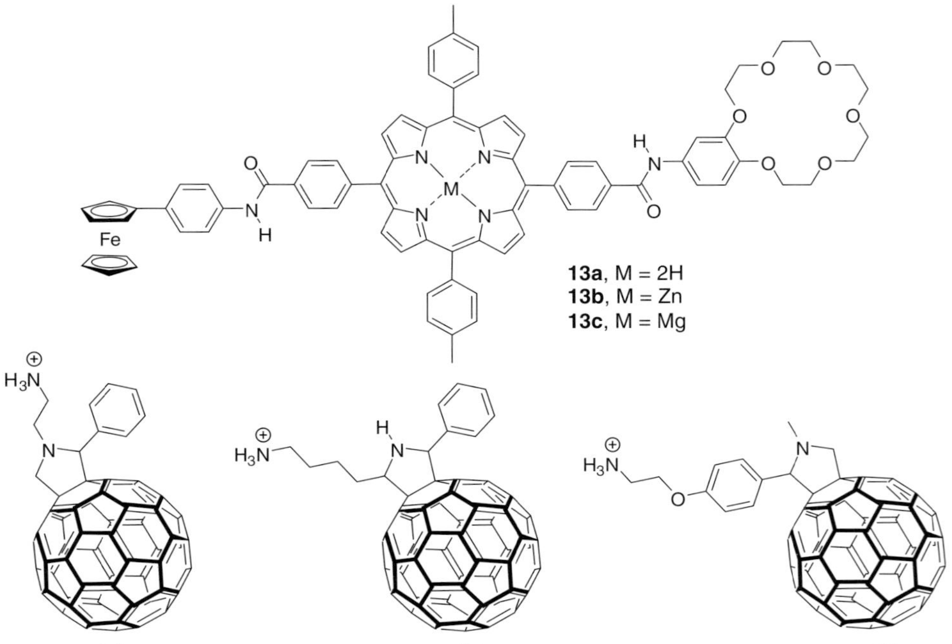

Figure 2.11 Structures of crown ether-appended porphyrin–ferrocene dyad and fullerene ammonium salts.

far-side located ZnP units with respect to the C_{60} entity, resulting in lifetimes for CS species of 45 ns and 172 ns, respectively. This triad constitutes an excellent example of the influence of the molecular geometry on the electron transfer rates, that is, on the kinetics of charge separation and charge recombination.

Several triads have also been constructed in which covalently linked ferrocene-porphyrin–crown ether compounds are self-assembled with different alkylammonium cation-functionalized fullerenes (Figure 2.11) [27]. These systems were designed to achieve stepwise electron transfer and hole shift to generate long-lived charge-separated states. Binding constants from 10^3 to $10^5\ M^{-1}$ (benzonitrile) were determined from fluorescence quenching experiments. Photoinduced charge transfer studies revealed that the charge recombination processes of the radical anion of the fullerene moiety take place in two steps, namely, a direct charge recombination from the porphyrin radical cation and a slower step involving distant charge recombination from the ferrocene cation moiety. The k_{CR} values for the latter route were found to be one–two orders of magnitude slower than the former route, clearly demonstrating charge stabilization in the supramolecular triads.

D'Souza *et al.* have successfully employed the so-called "two-point" binding strategy to form D–A systems with defined distance and orientation. In previous studies, two-point binding involving coordination of pyridine to zinc and crown ether–alkyl cation complexation resulted in stable porphyrin–fullerene conjugates

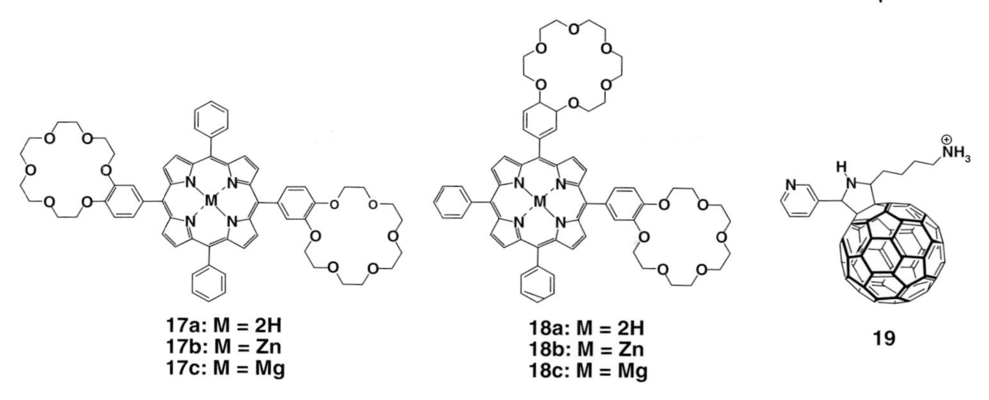

Figure 2.12 Structures of bis-benzo-18-crown-6-substituted porphyrins and pyridine-ammonium cation-derivatized fullerene.

where it was possible to monitor the photoinduced electron transfer processes in a polar solvent such as benzonitrile [28]. Extending these studies, they have reported on the supramolecular assembly of bis-fullerene–porphyrin triads by complexation of bis-benzo-18-crown-6-appended porhyrins (**17–18**) with alkyl ammonium-functionalized fullerenes **15** and **19** [29]. Different structures are formed depending on the nature of the fullerene derivative, as pyridine is able to coordinate axially to the zinc center of the porphyrin, and also depending on the metal of the macrocycle, as the presence of magnesium gives rise to six-coordinated complexes (Figure 2.12). In general, the obtained k_{CS} and Φ_{CS} values are higher for the present 1 : 2 porphyrin:fullerene conjugates than those of the earlier reported 1 : 1 porphyrin:fullerene supramolecular assemblies using the same binding methodologies, which implies that a higher number of acceptor entities improves the charge separation efficiency.

Different triads bearing porphyrin and fullerene have also been formed via this two-point binding strategy involving coordination and hydrogen bonding. In this approach, zinc porphyrin was functionalized with a pendant arm having either a carboxylic acid or an amide terminal group, and the fullerene derivative contained a pyridine coordinating ligand and a secondary donor moiety (Figure 2.13) [30]. The *K*-values obtained range from 1 to $10 \times 10^4\,M^{-1}$ (*o*-dichlorobenzene), which confirms the two-point bound triads. Theoretical calculations predict structures with a triangular disposition of the three entities and corroborate the existence of hydrogen bonds between the pyrrolidine and the carboxylic or amide groups. The photophysical study of these triads shows an efficient electron transfer from the excited singlet state of zinc porphyrin to the fullerene entity. In the case of triads having hydrogen bonding with the carboxylic acid group, a slower charge recombination and hence better charge stabilization was observed, which may be related to the larger *K*-values determined for these triads.

A biomimetic bacterial photosynthetic reaction center complex has been constructed in which the "special pair" donor, a cofacial porphyrin dimer, was formed via

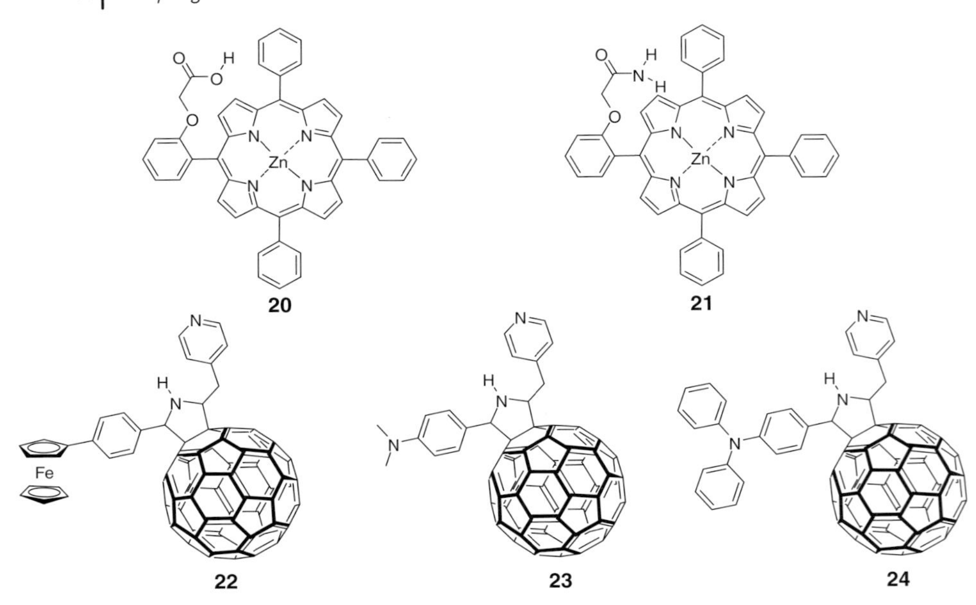

Figure 2.13 Structures of porphyrin and fullerene derivatives employed to construct supramolecular triads by the "two-point" binding strategy.

potassium ion induced dimerization of *meso*-(benzo-15-crown-5)porphyrinatozinc [31]. The dimer $K_4(ZnTCP)_2$ was subsequently self-assembled with functionalized fullerene **19** via axial coordination and crown ether–alkyl ammonium cation complexation, giving rise to a supramolecular tetrad, with defined geometry and orientation (see Scheme 2.1). Efficient charge separation from singlet excited zinc porphyrin dimer to the complexed fullerene within the supramolecular tetrad was observed, and the experimentally measured values of k_{CS} and k_{CR} ($\sim 10^9$ and $\sim 10^7\,s^{-1}$, respectively) reveal a slow reverse electron transfer.

The application of Watson–Crick hydrogen bonding paradigm to assemble porphyrins and fullerene had a remarkable impact on the lifetime of the photogenerated radical ion pair (Figure 2.14). Thus, the lifetime of the radical ion pair state for dyad **25**, constructed by means of a guanosine-cytidine scaffold, is 2.02 μs (CH_2Cl_2), substantially higher than those reported for related covalently linked C_{60}–ZnP dyads [32]. The use of phthalocyanine to form a noncovalent dyad based on the use of cytidine–guanosine hydrogen bonding interactions led to a highly stabilized complex **26**, for which a binding constant of $2.6 \times 10^6\,M^{-1}$ (CH_2Cl_2/toluene 4 : 1) was determined [33]. This value is several orders of magnitude larger than that obtained for the analogous porphyrin dyad ($5.1 \times 10^4\,M^{-1}$), which could be justified by additional π–π or charge transfer interactions in the Pc-based system. In this case, the radical ion pair state decays with a lifetime of 3.0 ns in toluene (versus 2.02 μs for **25** in dichloromethane). The pronounced coupling between the ZnPc and the C_{60} moieties, which is corroborated by the large association constant, can explain the shortening of the lifetime in ensemble **26**.

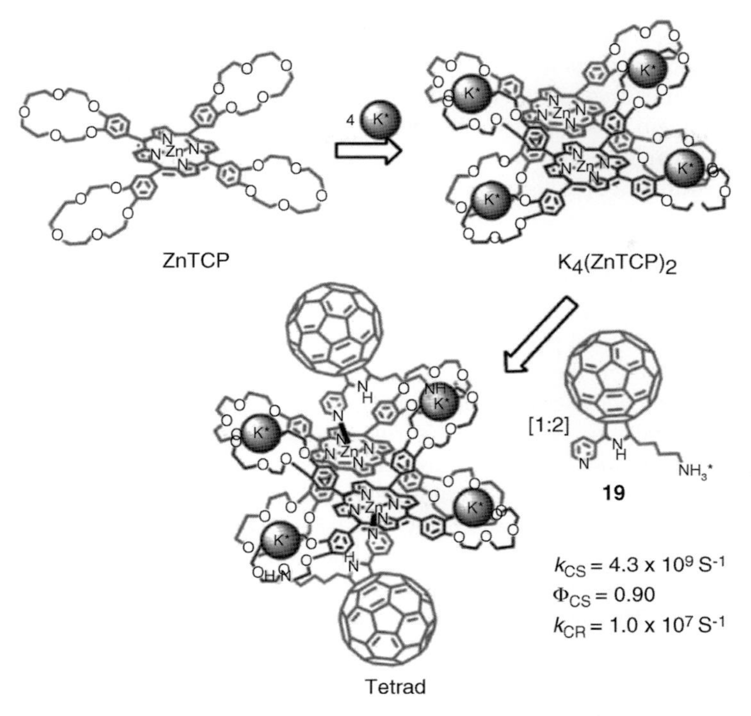

Scheme 2.1 Formation of the supramolecular porphyrin dimer–fullerene tetrad. (Copyright 2009 *Chemical Communications*).

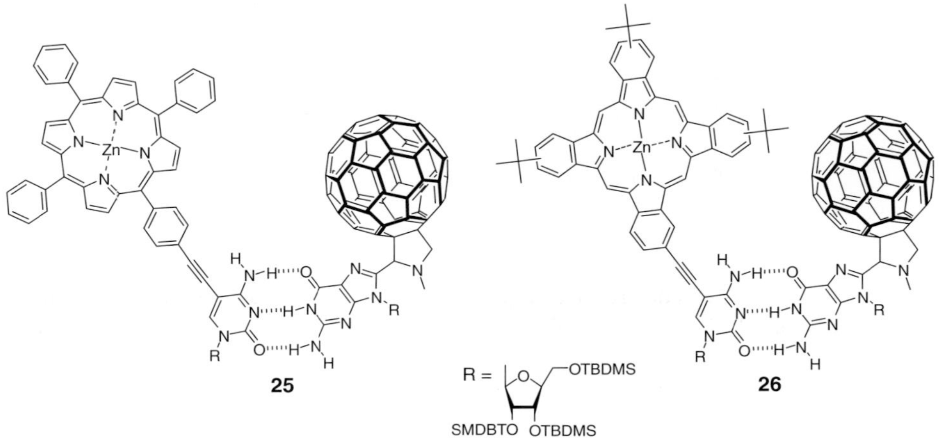

Figure 2.14 Watson–Crick H-bonded D–A dyads.

As stated above, porphyrin and related metallomacrocycles have constituted the main choice as electron donor moieties for the preparation of C_{60}–D dyads. However, other different donor fragments H bonding interfaced with C_{60} have also been studied. Among them, π-conjugated oligomer-C_{60} supramolecular ensembles have received much attention, particularly in an attempt to control the morphology of active layers in different optoelectronic devices.

Nierengarten and coworkers have prepared a series of C_{60}–oligophenylene vinylene (OPV) conjugates in which the inherent weakness of the ammonium–crown ether pair is overcome by the cooperative effect of π stacking interactions [34]. Recently, they have synthesized fullerodendrons with 1, 2, or 4 fullerene units and an ammonium unit at the focal point and studied their assembly with OPV receptors bearing one or two crown ether moieties (Figure 2.15) [35]. The resulting structures are multicomponent photoactive devices in which the emission of the central receptor is dramatically quenched by the peripheral fullerene units. On the basis of the photophysical studies of related covalent fullerene–OPV conjugates, the quenching of the photoexcited OPV within the supramolecular complexes is most probably ascribed to a photoinduced energy transfer [36]. Interestingly, a positive cooperative effect for the assembly of fullerodendrimers with the ditopic receptor has been observed, increasing the stability of the supramolecular ensembles with the size of the dendritic unit. This positive dendritic effect has been explained by the sum of

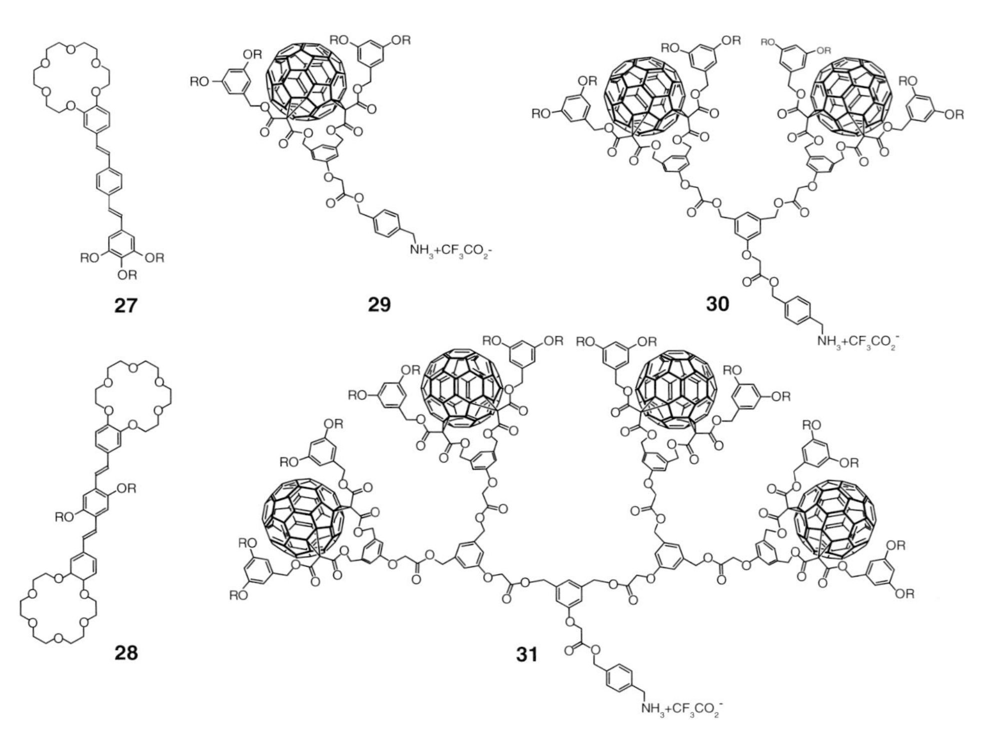

Figure 2.15 Molecular structures of fullerodendrons and oligomer crown ether receptors.

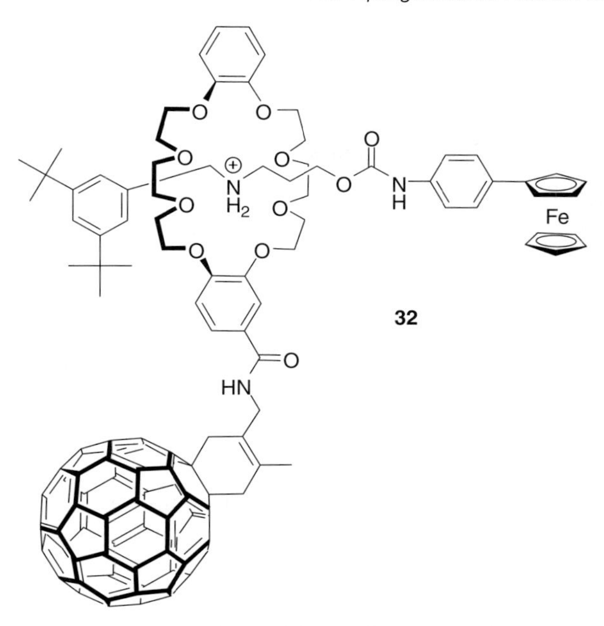

Figure 2.16 Supramolecular fullerene–ferrocene dyad.

secondary intramolecular interactions, such as π–π stacking and hydrophobic interactions.

Rotaxanes are interlocked compounds with unique structural features mainly related to the freedom and mobility of their components, which are linked by noncovalent linkage. Recently, rotaxanes containing fullerenes as electron acceptors and porphyrin as electron donors have been synthesized. In these systems, depending on the distance between the donor and the acceptor, through-space CS process takes place via the singlet and triplet excited states of the chromophores. This finding contrasts with covalent D–A dyads, in which CS usually proceeds via the excited singlet state with a through-bond mechanism. Takata and coworkers [37] have studied the photoinduced electron transfer in a rotaxane containing [60]fullerene as the electron acceptor and ferrocene as the electron donor unit (Figure 2.16). In rotaxane **32**, the charge-separated state ($C_{60}^{\bullet -}$–$Fc^{\bullet +}$) is formed via the $^1C_{60}{}^*$ and the $^3C_{60}{}^*$ states, with lifetimes of 20–33 ns in DMF and PhCN at room temperature. At low temperatures, the lifetime of the radical ion pair extends to 270 ns in DMF at −65 °C, due to the fluctuation of the distance between the components, which leaves the C_{60} and the Fc moieties at longer distance in the low-temperature region.

To study the possibility of controlling the reversible activation–deactivation of the electron transfer process, two different rotaxanes have been synthesized (Figure 2.17) [38]. In rotaxane **33**, the relative position of the ferrocene containing macrocycle is fixed, independently of the solvent used. In contrast, in rotaxane **34** the position of the macrocycle along the thread can be modified depending on the

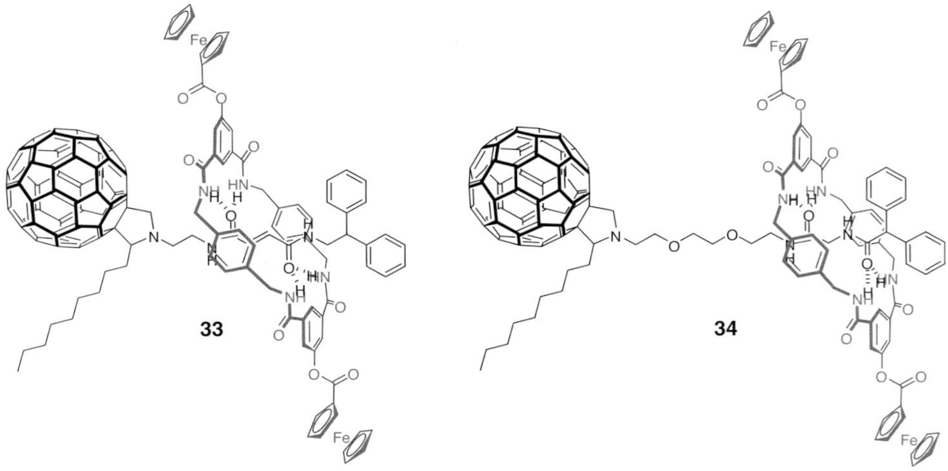

Figure 2.17 Structures of fullerene–ferrocene rotaxanes.

solvent employed. The lifetimes of the charge-separated states in CH_2Cl_2 are 8.7 ns for **33** and 26.2 ns for **34**, which are consistent with the larger separation between the macrocycle and the fullerene entity in **34**, and therefore the longer lifetimes. Addition of HFIP (hexafluoro-2-propanol) increases the hydrogen bond basicity of the solvent and shortens the lifetime of the charge-separated state of **34** to 13 ns as a consequence of a shorter relative separation between the Fc and the C_{60} moieties and a higher shuttling rate. The addition of HFIP strongly weakens the hydrogen bonds between the macrocycle and the peptide in **34**, promoting the formation of π stacking interactions between the macrocycle and the fullerene and therefore the shuttling along the thread. From this study, it can be concluded that submolecular translational motion can be employed to modulate the kinetics of the electron transfer process.

A structurally similar triad has been reported with the aim of constructing a supramolecular redox gradient and promoting an unidirectional cascade of two consecutive through-space charge transfer reactions between the three electroactive units [39]. The time-resolved absorption measurements demonstrate that, after excitation of the central Ru(CO)TPP unit, an electron transfer from the porphyrin to C_{60} is produced. A subsequent charge shift promotes the formation of the $C_{60}^{\bullet -}$–$Fc^{\bullet +}$ radical ion pair, and elongates the lifetime of $C_{60}^{\bullet -}$(Figure 2.18).

TTF (tetrathiafulvalene) and exTTF [9,10-di(1,3-dithiol-2-ylidene)-9,10-dihydroanthracene] donor moieties have demonstrated to be useful building blocks to form supramolecular H-bonded ensembles with C_{60} [40]. Thus, different exTTF-based secondary ammonium salts have been assembled to form a fullerene–crown ether derivative with K_a-values up to $1.4 \times 10^4\,M^{-1}$ (in CH_3CN:CH_2Cl_2 1 : 1) for triad **36** (Figure 2.19) [41].

In order to increase the stability of supramolecular ensembles based on ammonium–crown ether motifs, additional recognition motifs should be added [42]. In this

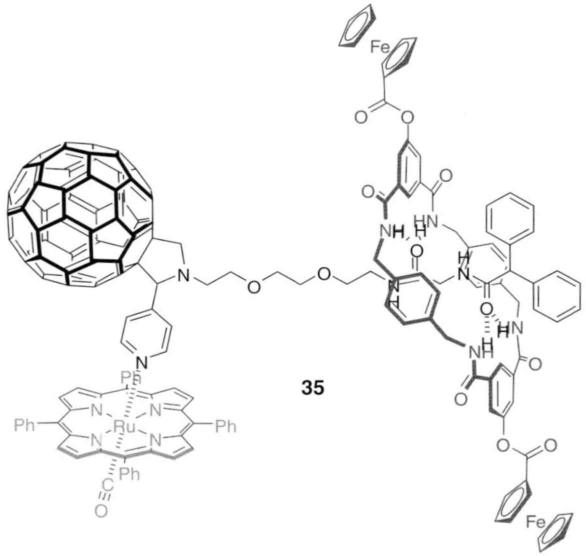

Figure 2.18 Fullerene–porphyrin–ferrocene supramolecular rotaxane-type triad.

sense, the concave aromatic face of exTTF has been demonstrated to act as a recognizing motif for fullerenes, and this noncovalent interaction should favor complexation between complementary ammonium salts and crown ethers [43]. Indeed, a dramatic increase in the association constant was found when the intramolecular interaction between the convex fullerene surface and the concave face of exTTF is allowed [44]. Thus, the binding constant obtained for the formation of supramolecular complex **37–38** is of $1.58 \times 10^6\ M^{-1}$ in chlorobenzene, around three orders of magnitude higher than the typical values observed for the ammonium salt–crown ether interaction (Scheme 2.2).

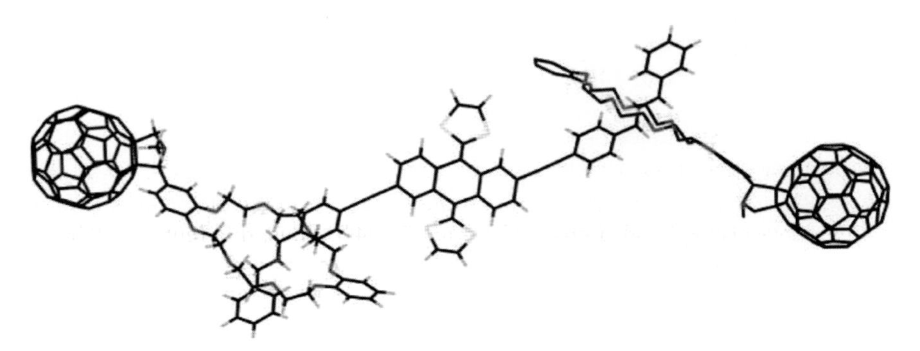

Figure 2.19 Structure of C_{60}–exTTF–C_{60} supramolecular triad **36**.

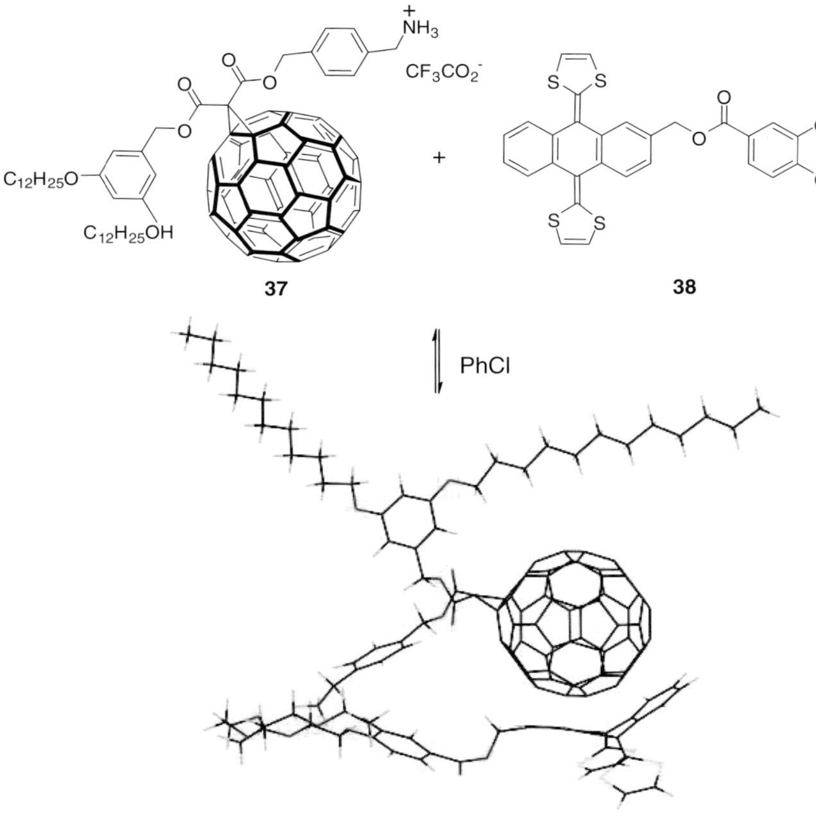

Scheme 2.2 Chemical structure of fullerene ammonium salt **37**, exTTF crown ether **38**, and molecular model showing its supramolecular complex.

The electrochemical study of complexation by cyclic voltammetry showed an anodic shift of ~100 mV in the oxidation potential of exTTF, justified by the strong donor–acceptor interaction. The photophysical study by time-resolved transient absorption spectroscopy showed the formation of a radical ion pair upon photoexcitation ($\tau = 9.3$ ps).

Bioinspired nanotubular cyclopeptidic heterodimers in which one cyclic peptide bears an electron acceptor fullerene moiety and the other an electron donor exTTF fragment have also been described (Figure 2.20) [45]. A remarkable association constant of $10^6\,M^{-1}$ has been determined for the equilibrium mixture of three species with different relative positions of the exTTF and C_{60} fragments. A photogenerated radical ion pair, with a lifetime of at least 1 μs was detected by steady-state and time-resolved spectroscopies. The extension of this heterodimer to form a nanotubular structure with alternating electron donors and acceptors suggests the possibility of designing materials for photonic and electronic applications.

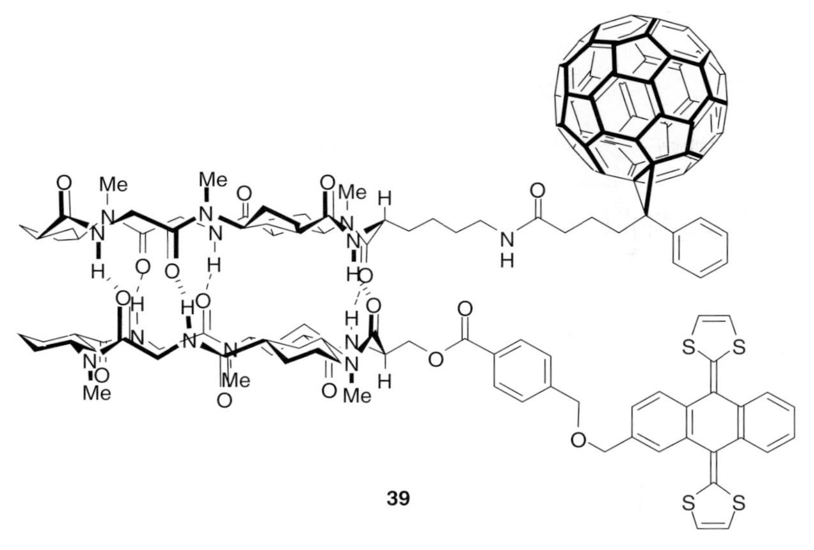

Figure 2.20 Structure of cyclopeptidic heterodimer **39**.

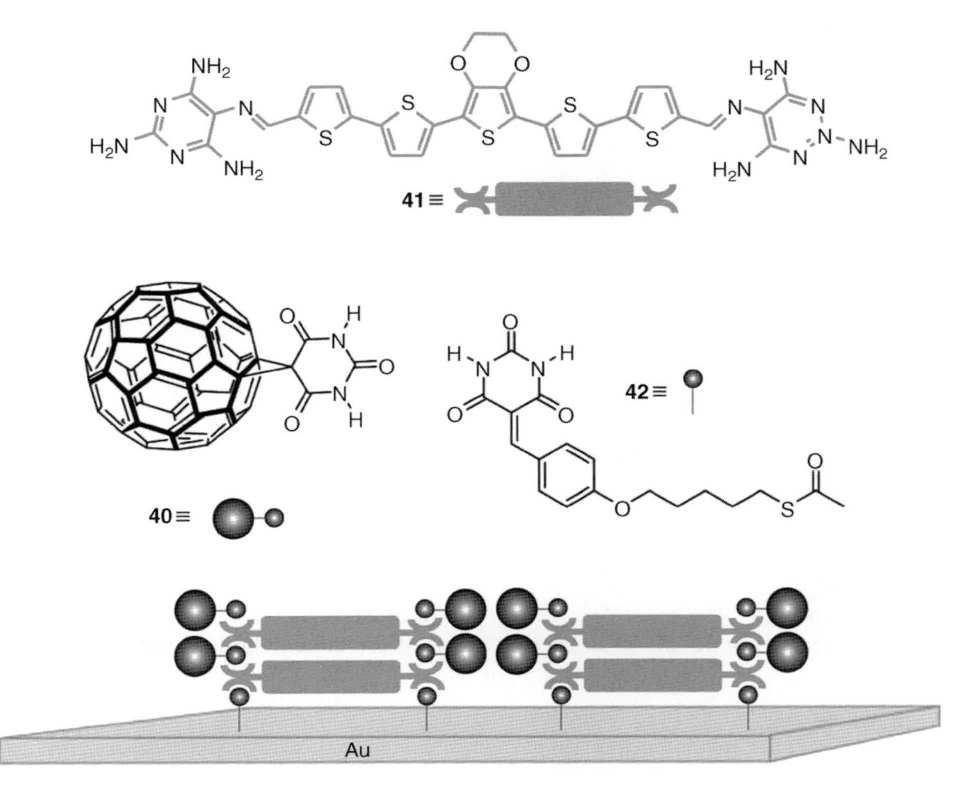

Figure 2.21 Linear ADA triads deposited over a barbiturate **42**-coated gold surface.

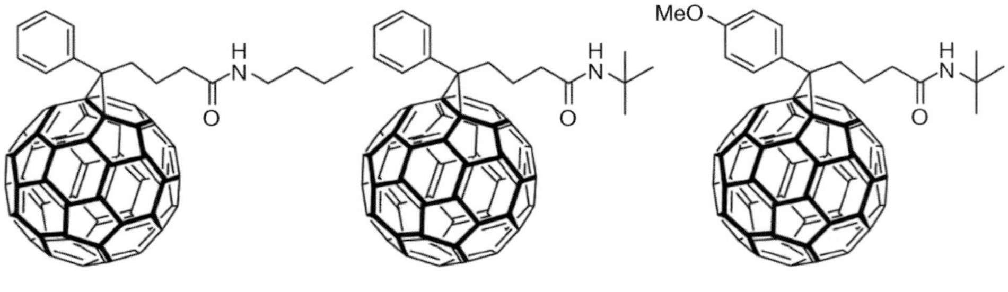

MeO

43 : PCB-*n*-BA **44** : PCB-*t*-BA **45** : MPCB-*t*.BA

Figure 2.22 PCBM analogues with improved organization properties.

2.4 Applications

Because of the unique electronic and structural properties of fullerenes, one of the most realistic applications is related to photovoltaics. Widely employed as acceptor in the active layer of bulk heterojunction (BHJ) devices, it faces a few drawbacks. One of them is its lack of solubility; fullerene and its derivatives (e.g., PCBM) tend to segregate into crystalline domains, minimizing the contact surface with the electron donor. However, the main challenge of photovoltaics is to build up highly ordered structures because the presence of defects in the active layer significantly reduces their performance by trapping the charge carriers. A feasible way to induce order is the employment of supramolecular forces. Despite the wide use of H bonding in biological systems by nature, only a few examples of its application with fullerene in devices can be found in the literature [46].

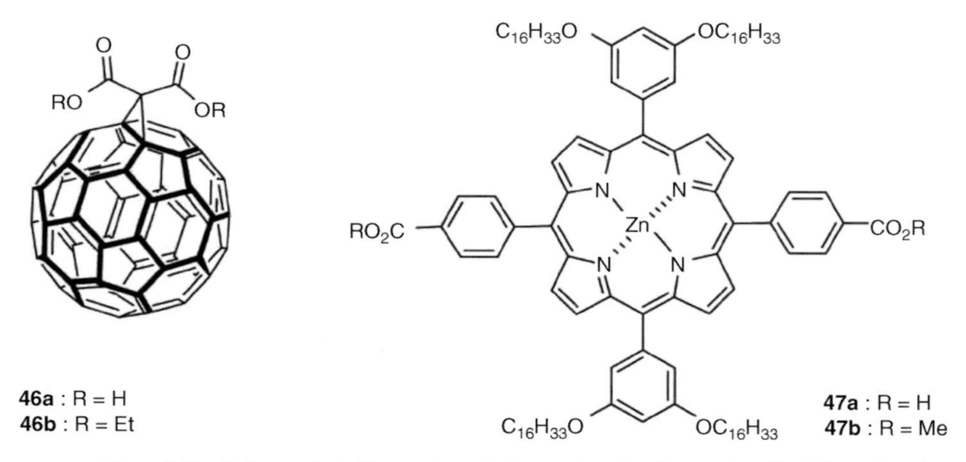

Figure 2.23 Fullerene derivatives and porphyrins employed in dye-sensitized bulk heterojunction solar cells over tin oxide nanostructured electrodes.

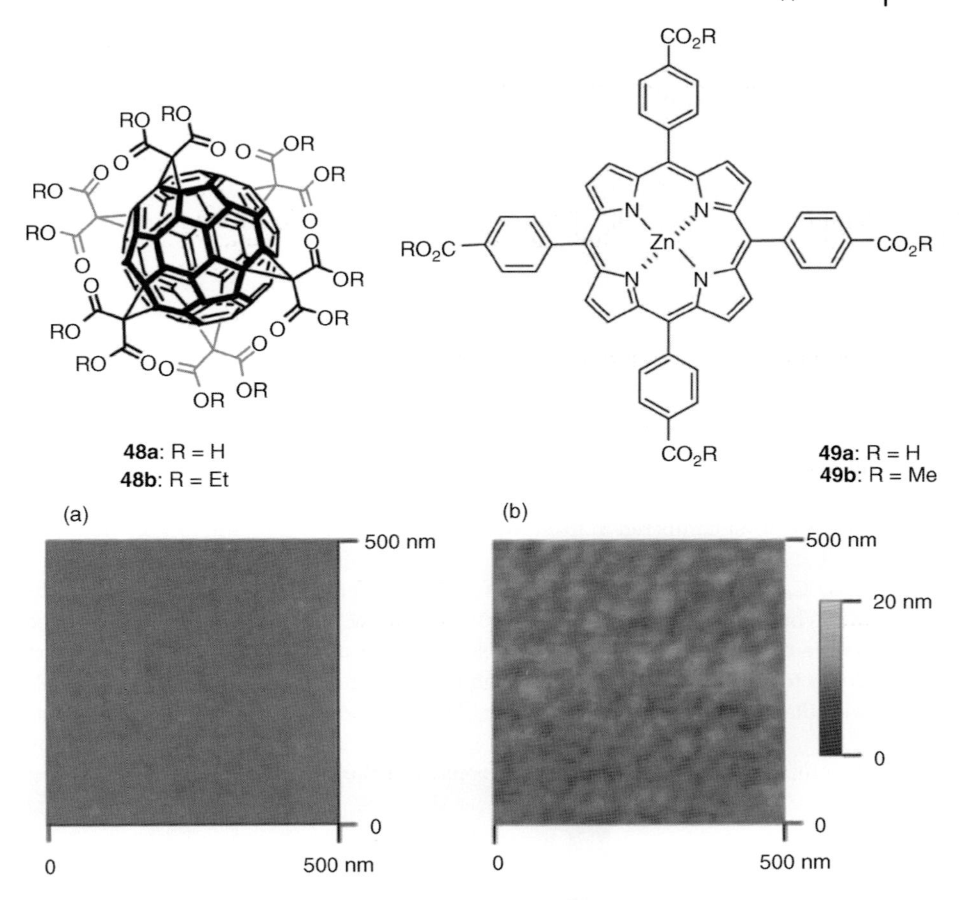

Figure 2.24 (*Top*) Fullerene and porphyrin derivatives used with TiO_2 nanostructured electrode. (*Bottom*) AFM images of (a) ZnP-acid + C_{60}-acid and (b) ZnP-ester + C_{60}-ester on mica substrate. The color scale represents the height topography with bright and dark representing the highest and lowest features, respectively.

Bassani and coworkers bring us an inspired approach toward the use of H bonding to create ordered supramolecular architectures (Figure 2.21) [47]. By using the barbiturate recognition moiety linked to a fullerene acceptor (**40**), linear acceptor–donor–acceptor (ADA) triads are obtained when **40** binds through H bonds to melamines in oligothiophenes (**41**). At the same time, these self-assembled ADA triads form layers over a gold surface coated with different recognition motifs (melamine, methylated barbiturate, and barbiturate), barbiturate **42** being the one with a better photocurrent generation.

The widely used in BHJ solar cells, the fullerene derivative PCBM, as stated above, faces the problem of phase segregation that hinders the performance of the built devices. To avoid this effect, three new [60] methanofullerene analogues of PCBM have recently been synthesized [48]. By replacing the ester group with an amide, as shown in Figure 2.22, these analogues improve their solubility with the polymeric donor (poly-hexil-3-thiophene, P3HT) allowing a homogeneous blend. This fact is reflected in an increased efficiency of the built devices. For example, for the P3HT/PCB-*n*-BA blend without annealing a $\eta_e = 0.78\%$ was found, compared to the 0.59% obtained with P3HT/PCBM blends. These findings show how hydrogen bonding interactions can significantly improve power conversion efficiency by inducing order into the molecular aggregates.

Another example of the potential use of hydrogen bonding to enhance the efficiency of solar cells is given by Imahori *et al.* They have developed several systems to prove how this interaction can increase the photocurrent of the fabricated devices by an effective mixture of their active components. Using different porphyrines and fullerene derivatives bearing carboxylic acids, films were deposited over SnO_2 and TiO_2 nanostructured electrodes, leading to the so-called dye-sensitized bulk heterojunction devices [49]. When mixed films of fullerene **46a** and porphyrin **47a** were prepared with nanostructured SnO_2, electrodes exhibited efficient photocurrent (IPCE = 36%) compared to those without hydrogen bonding interaction (**46b**/**47b**, IPCE < 28%) (Figure 2.23) [49b].

As $C_{60}^{\bullet-}$ (−0.2 V versus NHE) has a higher energy level than SnO_2 (0 V versus NHE) but 0.3 V lower than TiO_2 (−0.5 V versus NHE) [50], the same group developed a system capable of directly injecting electrons in a stepwise way from the porphyrin donor to the electrode passing through the fullerene, by simply replacing tin oxide with titanium oxide [49c]. In this case, the film mixture used a fullerene hexaadduct

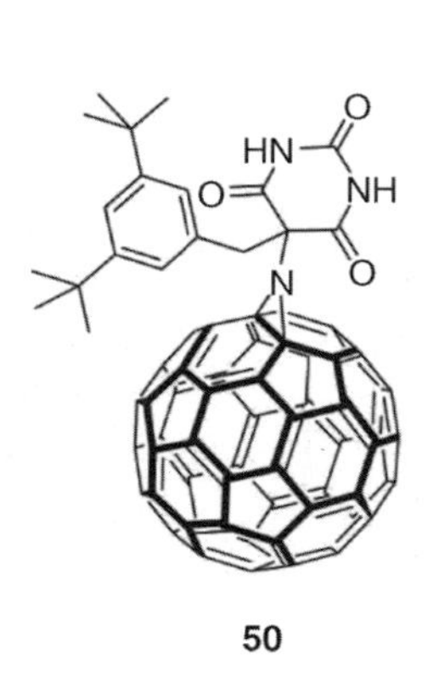

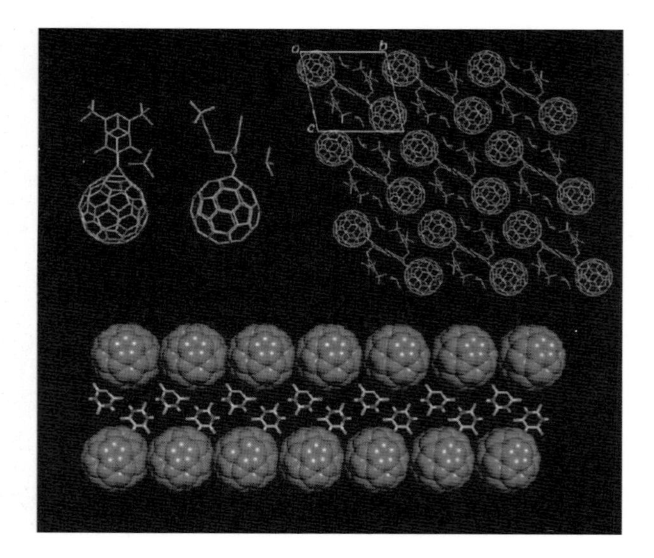

Figure 2.25 (*Left*) Chemical structure of hydrogen bonding barbiturate fullerene **50**. (*Top right*) solid-state crystal structure of **50** and crystal packing viewed along the *a*-axis. (*Bottom right*). Crystal packing viewed along the *c*-axis highlighting the close van der Waals contact between fullerenes in the H bonding ribbon.

(**48**) and a new porphyrin substituted in *meso* positions by four carboxylic groups (**49**). With an increased number of hydrogen bonding motifs, they achieved a homogeneous blend of both fullerene and porphyrin, as shown in the AFM images of Figure 2.24. Such nanostructured TiO_2 electrodes modified with the mixed films of porphyrin and fullerene with hydrogen bonds exhibit efficient photocurrent generation (IPCE value up to 47% for TiO_2/ZnP-acid + C_{60}-acid electrode) compared to the reference systems without hydrogen bonds.

Another research field for hydrogen bonding fullerenes has recently been opened with the preparation of the first OFET device employing a hydrogen bonding fullerene. By combining a solubilizing 3,4-di-*tert*-butylbenzene group and a barbituric acid hydrogen bonding moiety in fullerene derivative **50**, an n-type supramolecular semiconducting fullerene structure was achieved [51]. The experimental evidences show how derivative **50** self-assemblies form well-defined ordered ribbons (see X-ray-resolved crystal structure in Figure 2.25). Contrary to what is observed in pristine fullerene, the distribution of fullerene **50** cages is anisotropic, which is consistent with the determined wire-like semiconductor properties. The charge carrier mobility of the compound was determined from OFET devices and was found to be lower than that of pristine C_{60} (1.2×10^{-4} versus $1.2 \times 10^{-2}\ cm^2/s\ V$); this drop of the bulk charge carrier mobility is attributed to the lower dimensionality of **50** as a conductor. Despite the lower charge transport capabilities, the fabrication of extended linear fullerene polymers through hydrogen bonding self-assembly is the most appealing concept for molecular-level electronics.

Acknowledgments

Financial support by the MEC of Spain (projects CTQ2008-02609/BQU and Consolider-Ingenio 2010C-07-25200) and the CAM (MADRISOLAR project P-PPQ-000225-0505) is acknowledged.

References

1 (a) Lehn, J-.M. (ed.) (1995) *Supramolecular Chemistry: Concepts and Perspectives*, Wiley-VCH Verlag GmbH, Weinheim; (b) Lehn, J.-M. (1993) *Science*, **260**, 1762–1763.

2 (a) Jeffery, G.A. (ed.) (1997) *An Introduction to Hydrogen Bonding*, Oxford University Press, Oxford; (b) Prins, L.J., Reinhoudt, D.N., and Timmerman, P. (2001) *Angew. Chem. Int. Ed.*, **40**, 2382–2426. (c) Cooke, G. and Rotello, V.M. (2002) *Chem. Soc. Rev.*, **31**, 274–283.

3 Sánchez, L., Martín, N., and Guldi, D.M. (2005) *Angew. Chem. Int. Ed.*, **44**, 5374–5382.

4 Goshe, A.J., Steele, I.M., Ceccarelli, C., Rheingold, A.L., and Bosnich, B. (2002) *Proc. Natl. Acad. Sci.*, **99**, 4823–4829.

5 (a) Bibby, T., Nield, J., Partensky, F., and Barber, J. (2001) *Nature*, **413**, 590–590; (b) Abrahams, J.P., Leslie, A.G.W., Lutter, R., and Walker, J.E. (1994) *Nature*, **370**, 621–628.

6 (a) Meyer, T.J. (1989) *Acc. Chem. Res.*, **22**, 163–170; (b) Wasielewski, M.R. (1992) *Chem. Rev.*, **92**, 435–461; (c) Kurreck, H.

and Huber, M. (1995) *Angew. Chem. Int. Ed. Engl.*, **34**, 849–866; (d) Gust, D., Moore, T.A., and Moore, A.L. (2001) *Acc. Chem. Res.*, **34**, 40–48; (e) Holten, D., Bocian, D.F., and Lindsey, J.S. (2002) *Acc. Chem. Res.*, **35**, 57–69; (f) Adams, D.M., Brus, L., Chidsey, E.D., Creager, S., Creutz, C., Kagan, C.R., Kamat, P.V., Lieberman, M., Lindsey, S., Marcus, R.A., Metzger, R.M., Michel-Beyerle, M.E., Miller, J.R., Newton, M.D., Rolison, D.R., Sankey, O., Schanze, K.S., Yardley, J., and Zhu, X. (2003) *J. Phys. Chem. B.*, **107**, 6668–6697; (g) Armaroli, N. (2003) *Photochem. Photobiol. Sci.*, **2**, 73–78; (h) Imahori, H., Mori, Y., and Matano, Y. (2003) *J. Photochem. Photobiol.*, **C4**, 51; (i) Andrews, D.I. (ed.) (2005) *Energy Harvesting Materials*, World Scientific, Singapore.

7 (a) Yazdani, A. and Lieber, C.M. (1999) *Nature*, **401**, 1683–1692; (b) Special issue (2001) *Sci. Am.*, **285**, 32. (c) Anderson, H.L. (2000) *Angew. Chem. Int. Ed.*, **39**, 2451–2453; (d) Bunk, S. (2001) *Nature*, **410**, 127–129.

8 (a) Brabec, C.J., Sariciftci, N.S., and Hummelen, J.C. (2001) *Adv. Funct. Mater.*, **11**, 15–26; (b) Thompson, B.C. and Frèchet, J.M.J. (2008) *Angew. Chem. Int. Ed.*, **47**, 58–77; (c) Delgado, J.L., Bouit, P.-A., Filippone, S., Herranz, M.A., and Martín, N. (2010) *Chem. Commun.*, **46**, 4853–4865; (d) Brédas, J.-L., Beljonne, D., Coropceanu, V., and Cornil, J. (2004) *Chem. Rev.*, **104**, 4971–5004; (e) Würthner, F., and Schmidt, R. (2006) *ChemPhysChem*, **7**, 793–797; (f) Anthony, J.E. (2006) *Chem. Rev.*, **106**, 5028–5048; (g) Mas-Torrent, M. and Rovira, C. (2008) *Chem. Soc. Rev.*, **37**, 827–838; (h) Müllen, K. and Scherf, U. (eds.) (2006) *Organic Light Emitting Devices: Synthesis Properties and Applications*, Wiley-VCH Verlag GmbH, Weinheim; (i) Li, Z. and Meng, H. (eds.) (2007) *Organic Light-Emitting Materials and Devices*, CRC, Boca Raton.

9 (a) Hirsch, A. (ed.) (2005) *The Chemistry of Fullerenes*, Wiley-VCH Verlag GmbH, Weinheim, Germany; (b) Guldi, D.M. and Martín, N. (eds.) (2002) *Fullerenes: From Synthesis to Optoelectronic Properties*, Kluwer Academic Publishers, Dordrecht, The Netherlands; (c) Taylor, R. (ed.) (1999) *Lecture Notes on Fullerene Chemistry: A Handbook for Chemists*, Imperial College Press, London; (d) Langa, F. and Nierengarten, J.-F. (eds.) (2007) *Fullerenes: Principles and Applications*, RSC, Cambridge, UK; (e) Martín, N. (2006) *Chem. Commun.*, 2093–2104.

10 (a) Diederich, F., Echegoyen, L., Gómez-López, M., Kessinger, R., and Fraser Stoddart, J. (1999) *J. Chem. Soc. Perkin Trans.*, **2**, 1577–1586; (b) Brouwer, A.M., Frochot, C., Gatti, F.G., Leigh, D.A., Mottier, L., Paolucci, F., Roffia, S., and Wurpel, G.W.H. (2001) *Science*, **291**, 2124–2128; (c) Keaveney, C.M. and Leigh, D.A. (2004) *Angew. Chem. Int. Ed.*, **43**, 1222–1224.

11 (a) Da Ros, T., Guldi, D.M., Morales, A.F., Leigh, D.A., Prato, M., and Turco, R. (2003) *Org. Lett.*, **5**, 689–693; (b) Mateo-Alonso, A., Fioravanti, G., Marcaccio, M., Paolucci, F., Jagesar, D.C., Brouwer, A.M. and Prato, M. (2006) *Org. Lett.*, **8**, 5173–5176; (c) Mateo-Alonso, A., Guldi, D.M., Paolucci, F., and Prato, M. (2007) *Angew. Chem. Int. Ed.*, **46**, 8120–8126.

12 (a) Beijer, F.H., Kooijman, H., Spek, A.L., Sijbesma, R.P., and Meijer, E.W. (1998) *Angew. Chem. Int. Ed.*, **37**, 75–78; (b) Beijer, F.H., Sijbesma, R.P., Kooijman, H., Spek, A.L., and Meijer, E.W. (1998) *J. Am. Chem. Soc.*, **120**, 6761–6769.

13 (a) Rispens, M.T., Sánchez, L., Knol, J., and Hummelen, J.C. (2001) *Chem. Commun.*, 161–162; (b) González, J.J., González, S., Priego, E., Luo, C., Guldi, D.M., de Mendoza, J., and Martín, N. (2001) *Chem. Commun.*, 163–164.

14 Sánchez, L., Rispens, M.T., and Hummelen, J.C. (2002) *Angew. Chem. Int. Ed.*, **41**, 838–840.

15 Reiriz, C., Brea, R.J., Arranz, R., Carrascosa, J.L., Garibotti, A., Manning, B., Valpuesta, J.M., Eritja, R., Castedo, L., and Granja, J.R. (2009) *J. Am. Chem. Soc.*, **131**, 11335–11337.

16 (a) Yokoyama, T., Yokoyama, S., Kamikado, T., Okuno, Y., and Mashiko, S. (2001) *Nature*, **413**, 619–621; (b) Stepanow, S., Lin, N., Vidal, F., Landa, A., Ruben, M., Barth, J.V., and Kern, K. (2005) *Nano Lett.*, **5**, 901–904; (c) Meier, C., Ziener, U.,

Landfester, K., and Weihrich, P. (2005) *J. Phys. Chem. B*, **109**, 21015–21027; (d) Barth, J.V., Weckesser, J., Cai, C.Z., Günter, P., Bürgi, L., Jeandupeux, O., and Kern, K. (2000) *Angew. Chem. Int. Ed.*, **39**, 1230–1234; (e) Ziener, U., Lehn, J.M., Mourran, A., and Müller, M. (2002) *Chem. Eur. J.*, **8**, 951–957; (f) Lingenfelder, M.A., Spillmann, H., Dmitriev, A., Stepanow, S., Lin, N., Barth, J.V., and Kern, K. (2004) *Chem. Eur. J.*, **10**, 1913–1919; (g) Swarbrick, J.C., Ma, J., Theobald, J.A., Oxtoby, N.S., O'Shea, J.N., Champness, N.R., and Beton, P.H. (2005) *J. Phys. Chem. B*, **109**, 12167–12174; (h) Vidal, F., Delvigne, E., Stepanow, S., Lin, N., Barth, J.V., and Kern, K. (2005) *J. Am. Chem. Soc.*, **127**, 10101–10106; (i) Sánchez, L., Otero, R., Gallego, J.M., Miranda, R., and Martín, N. (2009) *Chem. Rev.*, **109**, 2081–2091.

17 Écija, D., Otero, R., Sánchez, L., Gallego, J.M., Wang, Y., Alcamí, M., Martín, F., Martín, N., and Miranda, R. (2007) *Angew. Chem. Int. Ed.*, **46**, 7874–7877.

18 (a) Harten, U., Lahee, A.M., Toennies, J.P., and Wöll, C. (1985) *Phys. Rev. Lett.*, **54**, 2619–2622; (b) Wöll, C., Chiang, S., Wilson, R.J., and Lippel, P.H., *Phys. Rev. B* (1989) **39**, 7988–7991; (c) Chen, W., Madhavan, V., Jamneala, T., and Crommie, M.F. (1998) *Phys. Rev. Lett.*, **80**, 1469–1472.

19 Wang, Y., Alcamí, M., and Martín, F. (2008) *Chem. Phys. Chem.*, **9**, 1030–1035.

20 Sánchez, L., Sierra, M., Martín, N., Myles, A.J., Dale, T.J., Rebek, J. Jr., Seitz, W., and Guldi, D.M. (2006) *Angew. Chem. Int. Ed.*, **45**, 4637–4641.

21 (a) Guldi, D.M., Ramey, J., Martínez-Díaz, M.V., de la Escosura, A., Torres, T., da Ros, T., and Prato, M. (2002) *Chem. Commun.*, 2774–2775; (b) Martínez-Díaz, M.V., Fender, N.S., Rodríguez-Morgade, M.S., Gómez-López, M., Diederich, F., Echegoyen, L., Stoddart, J.F., and Torres, T. (2002). *J. Mater. Chem.*, **12**, 2095–2099.

22 (a) McClenaghan, N.D., Absalon, C., and Bassani, D.M. (2003) *J. Am. Chem. Soc.*, **125**, 13004–13005; (b) Wessendorf, F., Gnichwitz, J.-F., Sarova, G.H., Hager, K., Hartnagel, U., Guldi, D.M., and Hirsch, A. (2007) *J. Am. Chem. Soc.*, **129**, 16057–16071.

23 Wessendorf, F., Grimm, B., Guldi, D.M., and Hirsch, A. (2010) *J. Am. Chem. Soc.*, **132**, 10786–10795.

24 Wielopolski, M., Atienza, C., Clark, T., Guldi, D.M., and Martín, N. (2008) *Chem. Eur. J.*, **14**, 6379–6390.

25 Atienza-Castellanos, C., Wielopolski, M., Guldi, D.M., van der Pol, C., Bryce, M.R., Filippone, S., and Martín, N. (2007) *Chem. Commun.*, 5146–5166.

26 Gadde, S., Islam, D.-M.S., Wijesinghe, C.A., Subbaiyan, N.K., Zandler, M.E., Araki, Y., Ito, O., and D'Souza, F. (2007) *J. Phys. Chem. C*, **111**, 12500–12503.

27 D'Souza, F., Chitta, R., Gadde, S., Islam, D.-M.S., Schumacher, A.L., Zandler, M.E., Araki, Y., and Ito, O. (2006) *J. Phys. Chem. B*, **110**, 25240–25250.

28 (a) D'Souza, F., Chitta, R., Gadde, S., Zandler, M.E., Sandayanaka, A.S.D., Araki, Y., and Ito, O. (2005) *Chem. Commun.*, 1279–1281; (b) D'Souza, F., Chitta, R., Gadde, S., Zandler, M.E., McCarty, A.L., Sandanayaka, A.S.D., Araki, Y., and Ito, O. (2005) *Chem. Eur. J.*, **11**, 4416–4428.

29 D'Souza, F., Chitta, R., Gadde, S., McCarty, A.L., Karr, P.A., Zandler, M.E., Sandanayaka, A.S.D., Araki, Y., and Ito, O. (2006) *J. Phys. Chem. B*, **110**, 5905–5913.

30 D'Souza, F., El-Khouly, M.E., Gadde, S., Zandler, M.E., McCarty, A.L., Araki, Y., and Ito, O. (2006) *Tetrahedron*, **62**, 1967–1978.

31 D'Souza, F., Chitta, R., Gadde, S., Rogers, L.M., Karr, P.A., Zandler, M.E., Sandanayaka, A.S.D., Araki, Y., and Ito, O. (2007) *Chem. Eur. J.*, **13**, 916–922.

32 Sessler, J.L., Jayawickramarajah, J., Gouloumis, A., Torres, T., Guldi, D.M., Maldonado, S., and Stevenson, K.J. (2005) *Chem. Commun.*, 1892–1894.

33 Torres, T., Gouloumis, A., Sánchez-García, D., Jayawickramarajah, J., Seitz, W., Guldi, D.M., and Sessler, J.L. (2007) *Chem. Commun.*, 292–294.

34 (a) Elhabiri, M., Trabolsi, A., Herschbach, H., Leize, E., Van Dorsselaer, A., Albrecht-Gary, A.-M., and Nierengarten, J.-F. (2005) *Chem. Eur. J.*, **11**, 4793–4798; (b) Hahn, U., Elhabiri, M., Trabolsi, A., Herschbach, H.,

Leize, E., Van Dorsselaer, A., Albrecht-Gary, A.-M., and Nierengarten, J.-F. (2005) *Angew. Chem. Int. Ed.*, **44**, 5338–5341.

35 Nierengarten, J.-F., Hahn, U., Trabolsi, A., Herschbach, H., Cardinali, F., Elhabiri, M., Leize, E., Van Dorsselaer, A., and Albrecht-Gary, A.-M. (2006) *Chem. Eur. J.*, **12**, 3365–3373.

36 (a) Segura, J.L., Martín, N., and Guldi, D.M. (2005) *Chem. Soc. Rev.*, **34**, 31–47; (b) Armaroli, N., Barigelletti, F., Ceroni, P., Eckert, J.-F., Nicoud, J.-F., and Nierengarten, J.-F. (2000) *Chem. Commun.*, 599–600; (c) Segura, J.L., Gómez, R., Martín, N., and Guldi, D.M. (2000) *Chem. Commun.*, 701–702; (d) Guldi, D.M., Swartz, A., Luo, C., Gómez, R., Segura, J.L., and Martín, N. (2002) *J. Am. Chem. Soc.*, **124**, 10875–10886.

37 Rajkumar, G.A., Sandanayaka, A.S.D., Ikeshita, K., Araki, Y., Furusho, Y., Takata, T., and Ito, O. (2006) *J. Phys. Chem. B*, **110**, 6516–6525.

38 (a) Mateo-Alonso, A., Ehli, C., Rahman, G.M.A., Guldi, D.M., Fioravanti, G., Marcaccio, M., Paolucci, F., and Prato, M. (2007) *Angew. Chem. Int. Ed.*, **46**, 3521–3525; (b) Mateo-Alonso, A., Guldi, D.M., Paolucci, F., and Prato, M. (2007) *Angew. Chem. Int. Ed.*, **46**, 8120–8126.

39 Mateo-Alonso, A., Ehli, Ch., Guldi, D.M., and Prato, M. (2008) *J. Am. Chem. Soc.*, **130**, 14938–14939.

40 (a) Segura, J.L. and Martín, N. (2001) *Angew. Chem. Int. Ed.*, **40**, 1372–1409; (b) Segura, M., Sánchez, L., de Mendoza, J., Martín, N., and Guldi, D.M. (2003) *J. Am. Chem. Soc.*, **125**, 15093–15100; (c) Díaz, M.C., Illescas, B.M., Martín, N., Stoddart, J.F., Canales, M.A., Jiménez-Barbero, J., Sarovad, G., and Guldi, D.M. (2006) *Tetrahedron*, **62**, 1998–2002; (d) Martín, N., Sánchez, L., Herranz, M.A., Illescas, B., and Guldi, D.M. (2007) *Acc. Chem. Res.*, **40**, 1015–1024.

41 Illescas, B.M., Santos, J., Díaz, M.C., Martín, N., Atienza, C.M., and Guldi, D.M. (2007) *Eur. J. Org. Chem.*, 5027–5037.

42 (a) Solladié, N., Walther, M.E., Gross, M., Figueira Duarte, T.M., Bourgogne, C., and Nierengarten, J.-F. (2003) *Chem. Commun.*, 2412–2413; (b) Solladié, N., Walther, M.E., Herschbach, H., Leize, E., Van Dorsselaer, A., Figueira Duarte, T.M., and Nierengarten, J.-F. (2006) *Tetrahedron*, **62**, 1979–1987.

43 (a) Pérez, E.M., Sánchez, L., Fernández, G., and Martín, N. (2006) *J. Am. Chem. Soc.*, **128**, 7172–7173; (b) Fernández, G., Pérez, E.M., Sánchez, L., and Martín, N. (2008) *Angew. Chem. Int. Ed.*, **47**, 1094–1097. (c) Fernández, G., Pérez, E.M., Sánchez, L., and Martín, N. (2008) *J. Am. Chem. Soc.*, **130**, 2410–2411; (d) Gayathri, S.S., Wielopolski, M., Pérez, E.M., Fernández, G., Sánchez, L., Viruela, R., Ortí, E., Guldi, D.M., and Martín, N. (2009) *Angew. Chem. Int. Ed.*, **48**, 815–819; (e) Huerta, E., Isla, H., Pérez, E.M., Bo, C., Martín, N., and de Mendoza, J. (2010) *J. Am. Chem. Soc.*, **132**, 5351–5353; (f) Isla, H., Gallego, M., Pérez, E.M., Viruela, R., Ortí, E., and Martín, N. (2010) *J. Am. Chem. Soc.*, **132**, 1772–1773.

44 Santos, J., Grimm, B., Illescas, B.M., Guldi, D.M., and Martín, N. (2008) *Chem. Commun.*, 5993–5995.

45 Brea, R.J., Castedo, L., Granja, J.R., Herranz, M.A., Sánchez, L., Martín, N., Seitz, W., and Guldi, D.M. (2007) *Proc. Natl. Acad. Sci. USA*, **104**, 5291–5294.

46 (a) Thompson, B.C. and Fréchet, J.M.J. (2008) *Angew. Chem. Int. Ed.*, **47**, 58–77; (b) Brédas, J.-L. and Durrant, J.R. (eds.) (2009) Special Issue on Organic Photovoltaics. *Acc. Chem. Res.*, **42**, 1689–1857; (c) Delgado, J.L., Bouit, P.-A., Filippone, S., Herranz, M.A., and Martín, N. (2010) *Chem. Commun.*, 4853–4865.

47 (a) Huang, C.-H., McClenaghan, N.D., Kuhn, A., Hofstraat, J.W., and Bassani, D.M. (2005) *Org. Lett.*, **7**, 3409–3412; (b) Huang, C.-H., McClenaghan, N.D., Kuhn, A., Bravic, G., and Bassani, D.M. (2006) *Tetrahedron*, **62**, 2050–2059.

48 Liu, Ch., Li, Y., Li, C., Li, W., Zhou, Ch., Liu, H., Bo, Z., and Li, Y. (2009) *J. Phys. Chem. C*, **113**, 21970–21975.

49 (a) Imahori, H., Liu, J.-Ch., Hosomizu, K., Sato, T., Mori, Y., Hotta, H., Matano, Y., Araki, Y., Ito, O., Maruyama, N., and Fujitae, S. (2004) *Chem. Commun.*, 2066–2067; (b) Imahori, H., Liu, J.-Ch., Hotta, H., Kira, A., Umeyama, T.,

Matano, Y., Li, G., Ye, S., Isosomppi, M., Tkachenko, N.V., and Lemmetyinen, H. (2005) *J. Phys. Chem. B*, **109**, 18465–18474; (c) Kira, A., Tanaka, M., Umeyama, T., Matano, Y., Yoshimoto, N., Zhang, Y., Ye, S., Lehtivuori, H., Tkachenko, N.V., Lemmetyinen, H., and Imahori, H. (2007) *J. Phys. Chem. C*, **111**, 13618–13626.

50 (a) Eu, S., Hayashi, S., Umeyama, T., Oguro, A., Kawasaki, M., Kadota, N., Matano, Y., and Imahori, H. (2007) *J. Phys. Chem. C*, **111**, 3528–3537; (b) Imahori, H., Hayashi, S., Umeyama, T., Eu, S., Oguro, A., Kang, S., Matano, Y., Shishido, T., Ngamsinlapasathian, S., and Yoshikawa, S. (2006) *Langmuir*, **22**, 11405–11411.

51 Chu, C.-C., Raffy, G., Ray, D., Del Guerzo, A., Kauffmann, B., Wantz, G., Hirsch, L., and Bassani, D.M. (2010) *J. Am. Chem. Soc*, **132**, 12717–12723.

3
Receptors for Pristine Fullerenes Based on Concave–Convex π–π Interactions

Takeshi Kawase

3.1
Introduction

In 1990, Krätchmer and coworkers reported the extraction of fullerenes (C_{60} and C_{70}) from carbon soot using aromatic solvents (Figure 3.1) [1]. After this discovery, receptor molecules for fullerenes have been extensively studied for separation and purification of fullerenes. Supramlecular chemists have found that traditional host molecules composed of electron-rich aromatic rings such as calix[*n*]arenes **1–4**, oxacalix[3]arenes **5**, and cyclotriveratrylenes (CTV) **6** (Figure 3.2) can be employed for the purpose. The crystallographic analyses of these complexes with C_{60} clearly indicated the importance of the π–π interactions between the convex π surface of C_{60} and the concave π surface of the traditional host molecules. On the other hand, except a few examples, the traditional hosts bound fullerenes only in solid state. In order to bind with fullerenes more tightly, two strategies have been presented to construct the well preorganized π cavity so far. First, electron-rich aromatic units are introduced to the host molecule as appendants (Figure 3.2a). Second, two host molecules are linked by appropriate tether(s) to form a dimeric structure with a creft-, ring-, or ball-shaped cavity (Figure 3.2b). The binding abilities of the modified receptors fairly increased in comparison with those of simple traditional hosts. On the other hand, new receptors based on curved conjugated systems such as corannulene **7**, extended TTF (exTTF) **8**, and "carbon nanorings" (Figure 3.3) have been developed recently. These compounds possess a concave π surface matching to the convex surface of C_{60}. Cyclic [6]paraphenyleneacetylene ([6]CPPA) **9** and the related compounds have smooth belt-shaped structure similar to a cut piece of carbon nanotube, and thus may be termed carbon nanorings. These receptors form stable inclusion complexes with fullerenes both in solution and in the solid state, and are good model compounds to explore the nature of the concave–convex π–π interactions. This chapter discusses the fullerene receptors bearing a cavity surrounded by p-orbitals, the so-called "π cavity," the concept of molecular designs, and the nature of π–π interactions. The survey of the fullerene receptors should provide an insight into the supramolecular properties of curved conjugated systems. Although the receptor

Supramolecular Chemistry of Fullerenes and Carbon Nanotubes, First Edition. Edited by Nazario Martin and Jean-Francois Nierengarten.

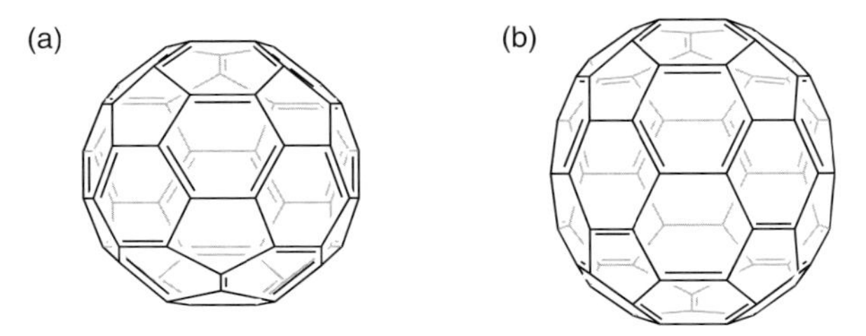

Figure 3.1 Fullerenes: (a) C_{60} and (b) C_{70}.

molecules based on porphyrin π system(s) have occupied an important part of this area, we omitted the topics because the next chapter reviews them in detail.

3.2
Fullerene Receptors Based on Traditional Hosts

3.2.1
Simple Traditional Hosts

The first example of macrocyclic fullerene receptors in solution was reported by Ringsdorf and Diederich in 1992. They found that fullerenes are incorporated into the lipophilic cavity of azacrown ethers such as **10** [2]. Shortly thereafter,

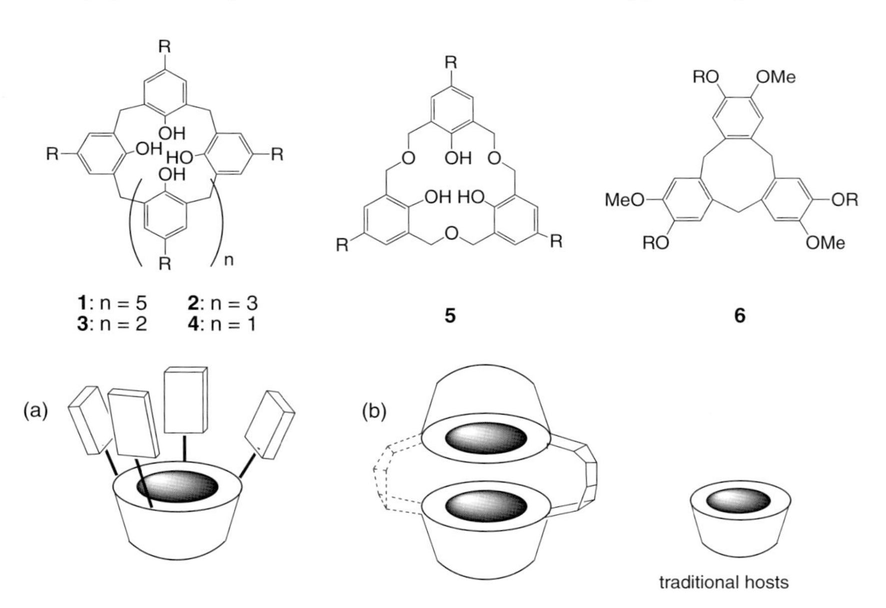

Figure 3.2 Traditional hosts, calix[*n*]arene **1–4**, oxacarix[3]arene **5**, and cyclotriveratrylene **6**, and schematic representations of fullerene receptors: (a) with appendants and (b) with a dimeric structure.

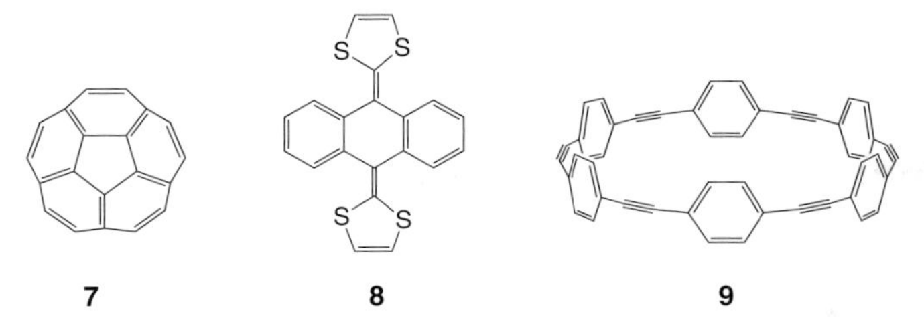

Figure 3.3 Corannulene **7**, extended TTF **8**, and [6]CPPA **9** with curved conjugated systems.

Wennerström and coworkers reported a water-soluble complex of γ-cyclodextrin (γ-CD) **11** with C_{60} [3]. The discovery led to extensive studies on various CD complexes related to tools for nanocomposite [4] and a novel reducing reagent [5]. Geckeler and Constabel reported that cucurbit[7]uril **12** also forms a complex with C_{60} by solid–liquid reaction (Figure 3.4) [6]. In these complexes, the attractive interactions between the heteroatoms as an n-donor and the fullerene surface should be operative [7], but the main driving force of the complexes should be hydrophobic effect. The water-soluble fullerene complexes have attracted increasing attention because of their biological applications [8, 9].

p-tert-Butylcalix[8]arene **1a** selectively precipitated with C_{60} from fullerite (a mixture of C_{60} and C_{70}) and thus led to an efficient purification of C_{60}. The independent discovery by Atwood [10] and Shinkai's [11] groups in 1994 resulted in many studies on the fullerene receptors based on the traditional host molecules. Calix[8]-, -[6]-, and -[5]arenes [12–14], oxacalix[3]arene [15], and CTV [16] derivatives formed a variety of inclusion complexes with fullerenes. The crystallographic analyses revealed that the complex between the traditional hosts with C_{60} construct a "ball and socket" nanostructure in the crystals [17]. Komatsu reported that *p*-halooxacalix[3]arenes, **5a** and **5b** (R = I or Br), can serve as an efficient receptor for C_{70}. The receptors accomplished preferential precipitation of C_{70} from a toluene solution of C_{60} and C_{70} [18]. The complexation between the calixarene receptors with fullerenes depends primarily both on the size of the cavity and on the function groups at the upper rim of the hosts. Thus, calix[4]arene (R = I or Br) **4a,b** [19] and

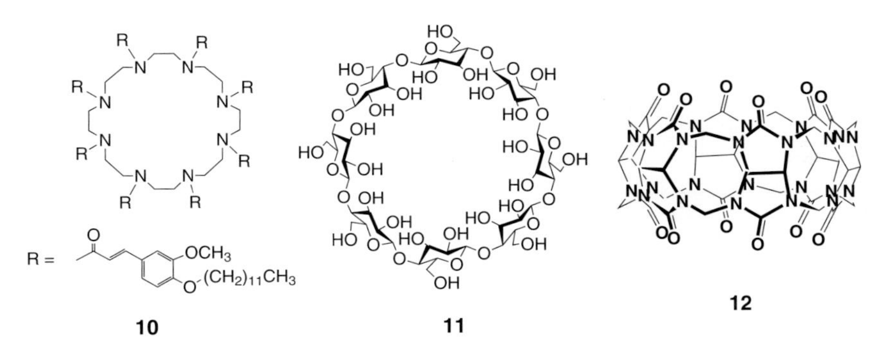

Figure 3.4 Azacrown **10**, γ-cyclodextrin **11**, and cucurbit[7]uril **12**.

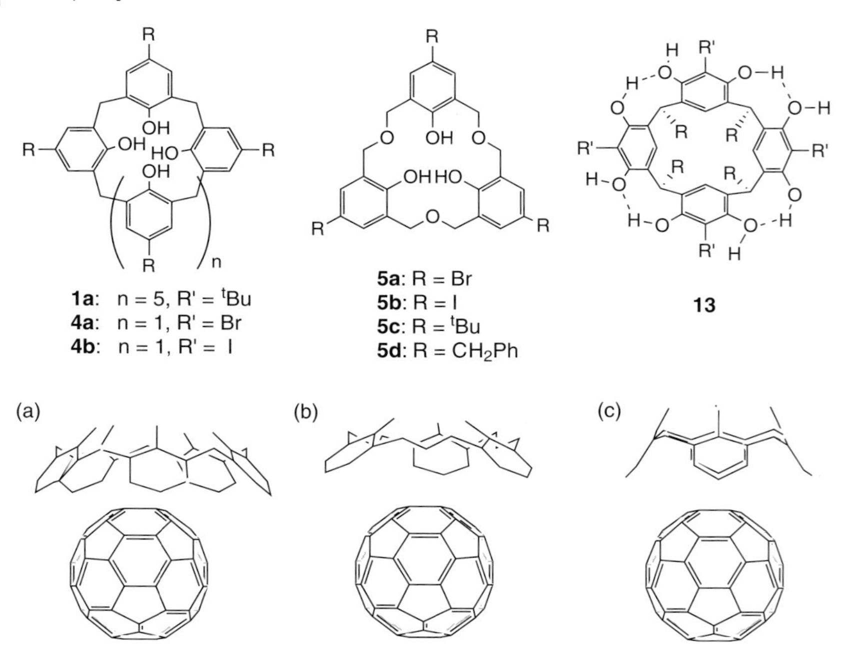

Figure 3.5 Calixarene based fullerene receptors; schematic molecular models of (a) calix[5]arene, (b) homooxacalix[3]arene, and (c) calix[4]arene complexes with C_{60}.

calix[4]resorcinarene **13** [20], with relatively small cavities and fourfold symmetry do not cover the fullerene molecule in the crystals. The theoretical studies clearly indicate the disadvantage of calix[4]arene derivatives, compared to other calixarenes (Figure 3.5) [21].

The traditional host molecules form inclusion complexes with fullerenes in solid state; however, the binding ability in nonpolar organic solvents is generally poor. Contrary to the results, Fukazawa and coworkers have first found that OH-unsubstituted calix[5]arenes **3c** form rather stable complexes in solution [22]. The association constants (K_a) of calix[5]arene derivatives **3a-f** summarized in Table 3.1 are in the range 2–3 × 10^3 M^{-1} (Figure 3.6) [23–25]. Owing to the C_{5v} cone conformation, the cavity

Table 3.1 Association constants (K_a, M^{-1}) of selected calixarenes and C_{60} at 25 °C in toluene.

Receptors	K_a	Reference	Receptors	K_a	Reference
3a	30 ± 2	[30]	**3f**	3000 ± 200	[32]
3b	1673 ± 70	[29]	**14**	676 ± 28	[33]
3c	2120 ± 110	[29]	**15**	296 ± 9	[34]
3d	292 ± 15	[30]	**5c**	35.6 ± 0.3	[22]
3e	2800 ± 200	[31]	**5d**a)	~100	[31]

a) **5c** formed a 2 : 1 complex with C_{60}.

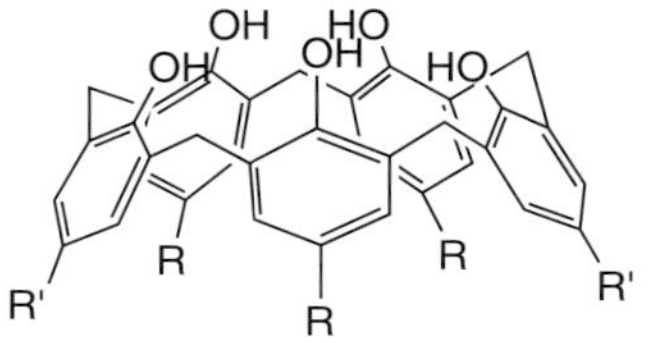

Figure 3.6 Calix[5]arene **3a** (R = R′ = H), **3b** (R = R′ = Me), **3c** (R = Me, R′ = I), **3d** (R = R′ = allyl), **3e** (R = R′ = CH2Ph), and **3f** (R = R′ = Ph).

size and curvature are complementary to fullerenes (Figure 3.5a). Georghiou *et al.* have prepared calix[4]naphthalene **14** [26] and trioxacalix[3]naphthalene **15** [27] derivatives. These hosts with deep cavities afford more stable complexes than the corresponding calixarenes (Table 3.1). A cavitand **16** with a deep cavity was also prepared by Rebek's group (Figure 3.7). Although the K_a values of **16** for fullerenes are not so large ($900 \pm 250\,M^{-1}$ for C_{60} in toluene), the selectivity between C_{60} and C_{70} is relatively high probably due to the rigidity of the cavity [28].

Boyd and Reed have demonstrated that graphitic and typical arene/arene distances are in the range 3.3–3.5 Å, fullerene/arene approaches lie in the range 3.0–3.5 Å, and fullerene/fullerene separations are typically approximately 3.2 Å [29]. The close approach is proposed to reflect an attractive π–π interaction. The counterbalance between fullerene/fullerene and fullerene/host interactions would play an important role in the packing structure of inclusion complexes [13]. On the basis of recent experimental studies, Raston has proposed to ascribe the frequently observed short distance to the polarization of fullerene in the crystal packing [30].

3.2.2
Modified Traditional Host Molecules

Shinkai and coworkers designed calix[6]arene derivatives bearing electron-rich aniline or 1,3-diaminobenzene units, expecting that the CT interaction would act

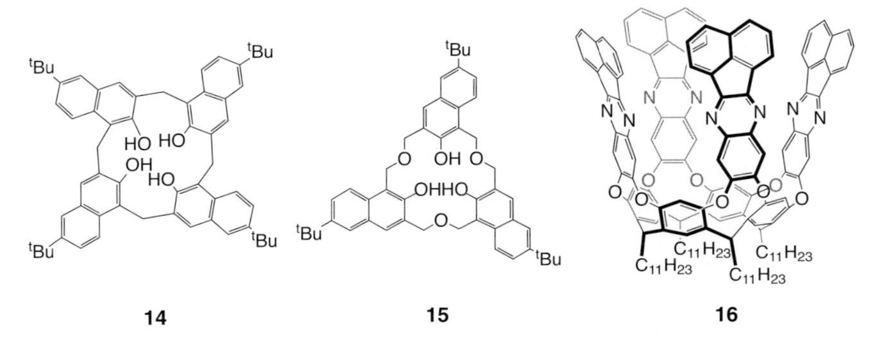

Figure 3.7 Calix[4]naphthalene **14**, oxacalix[3]naphthalene **15**, and a cavitand **16**.

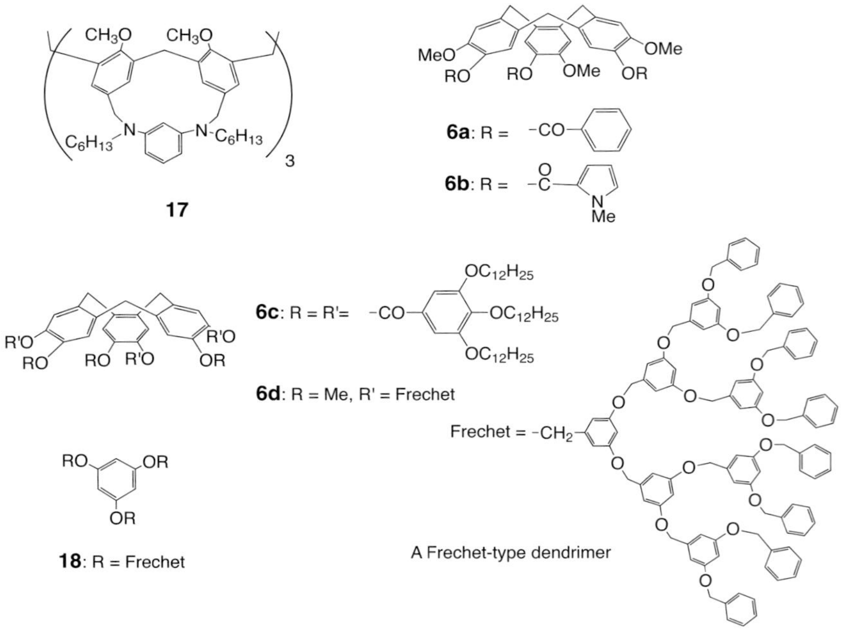

Figure 3.8 Modified calix[6]arene **17**, modified CTV **6a–6d**, and dendrimer **18**.

as a driving force [31]. The K_a value of **17** for C_{60} increased to $1.1 \times 10^2\,M^{-1}$. Matsubara and coworkers applied the concept and prepared CTV-based receptors **6a,b** having benzoyl or *N*-methylpyrrole pendants [32]. The K_a value of **6b** for C_{60} is $4.8 \times 10^4\,M^{-1}$ (Figure 3.8). Nierengarten and coworkers also synthesized a CTV derivative **6c** bearing long alkyl chains. The C_{60} complex of **6c** has a mesomorphic property [33]. The Frechet-type dendrimers provide the space size comparable with C_{60}. Thus, **6d** bearing the Frechet-type of dendrons acts as a receptor for fullerenes. The K_a values are significantly increased as the generation number of the dendric substituents is increased [34]. Shinkai's group independently found the receptor property of the dendrimers with a phloroglucinol core **18** for C_{60}. But, whether the CTV unit exists or not, the K_a value of the dendrimer is not so high ($<100\,M^{-1}$) [35].

Introduction of strong electron donating groups as appendants led to an increase in the affinity of the receptors for fullerenes. In this context, host molecules composed of triarylamines should exhibit high affinities for fullerenes in comparison with the corresponding phenol-based receptors. Actually, the K_{SV} values of a dendrimer **19** [36], azacalix[3]arene[3]pyridine **20** [37], and azacalix[n]pyridine **21** ($n = 4$–9) [38] are considerably high (K_{SV} for $\mathbf{19}\cdot C_{60} = 180000 \pm 80000 \times 10^5\,M^{-1}$, K_{SV} for $\mathbf{20}\cdot C_{60} = 70680 \pm 2060\,M^{-1}$, and K_{SV} for $\mathbf{20}\cdot C_{70} = 136620 \pm 3770\,M^{-1}$) (Figure 3.9). The participation of additional attractive CT interaction should play an important role in the complexation. The Stern–Volmer constants (K_{SV}) would be

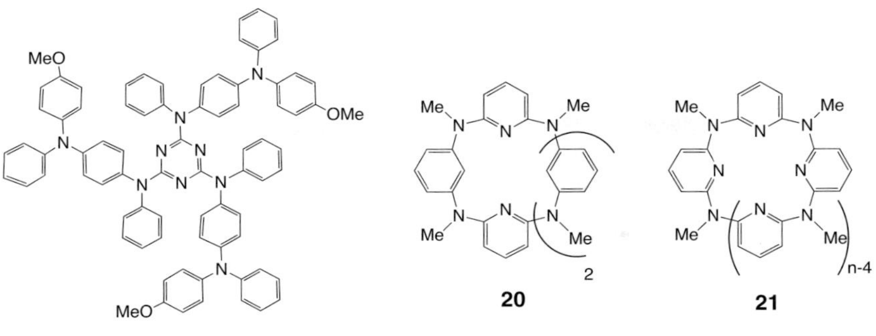

Figure 3.9 A triarylamine-based dendron **19** and azacalix[*n*]arenes **20** and **21**.

informative to evaluate the relative stability of the receptors [39]. However, the values are not direct measures of the complexation because of the quenching mechanisms. Care should be taken to avoid several trivial artifacts [40].

Simple calix[4]pyrrole derivatives did not act as a fullerene receptor probably due to their small cavity. But, some π-extended derivatives can be employed for unique fullerene receptors. Sessler and Jeppesen's group constructed a tetrathiafulvalene-functionalized calix[4]pyrrole **22**. When the compound is treated with chloride anion in solution, it produces a bowl-like receptor that is able to encapsulate C_{60} in a 2 : 1 barrel-like manner. The green color of the complex **22**·Cl^{-1}·C_{60} clearly indicates favorable CT interactions between the TTF donor units and the C_{60} moiety [41]. Lee and coworkers also reported that calix[1]pyreno[3]pyrrole **23** exhibits considerably high binding affinity for C_{70} in the presence of F^- ($K_{SV} = (5.4 \pm 0.53) \times 10^6\ M^{-1}$) (Figure 3.10) [42].

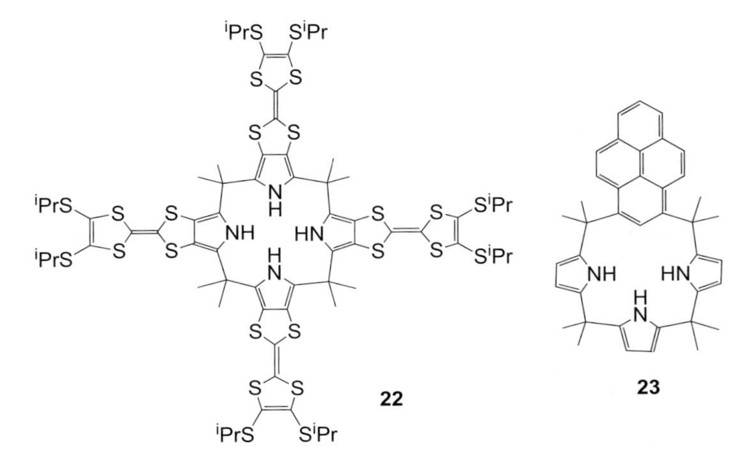

Figure 3.10 Calix[4]pyrrole derivatives **22** and **23**.

3.2.3
Receptors Bearing a Dimeric Structure of Traditional Host Molecules

Fukazawa and Haino found that calix[5]arene **3b** forms a 1 : 1 complex in solution, but it does form a 1 : 2 complex in the solid state [43]. The crystallographic analyses reveal that one C_{60} molecule is sandwitched by two calix[5]arene cavities. In this context, they designed and prepared new fullerene receptors **24** with a calix[5]arene-dimer structure. The K_a values of **24a** bearing a 1,3-diethynylphenylene tether for fullerenes are about 100 times larger than those of monomers [44]. The dimeric host **24b** bearing two diacetylenic tethers show high binding abilities for higher fullerenes (C_{70}, C_{76}, C_{78}, etc.) corresponding to their expanded cavities [45]. Several dimer receptors based on calix[5]- and -[6]arenes [23], a calix[4]arene [46], and CTV [47] were prepared and their association constants for fullerenes were up to 10^4–10^5 (Figure 3.11). In comparison with these calixarene dimers, **24a** shows the highest affinity for C_{60}. The results clearly indicate the efficiency of the calix[5]arene framework.

Cage-shaped host molecules for the encapsulation of fullerene can also be constructed either by coordination bonds of transition metal ions or by hydrogen bonds in place of covalent bonds. Shinkai's group first reported a C_{60} complex with a self-assembled cage, the oxacalix[3]arene dimer **25** cross-linked by three Pd(II) complexes (Figure 3.12) [48]. Claessens, and Diederich and Dalcanale's groups also described self-assembled cage-shaped molecules based on two subphthalocyanine units **26** [49] and two cavitand units **27** [50], respectively. Calix[5]arene derivatives prepared by Fukazawa's group construct the dimeric structure **28** in the presence of silver (I) or copper (II) ions to catch a C_{60} molecule in the resulting cavity [51, 52]. These cage molecules involving metal cation(s) show relatively small K_a values probably due to the high molecular mobility. Two calix[5]arenes with urea functionality **29** (Figure 3.13) form a ternary complex with C_{60} [53]. Polar functional groups such as urea and amide play an important role in assembling subunits with hydrogen

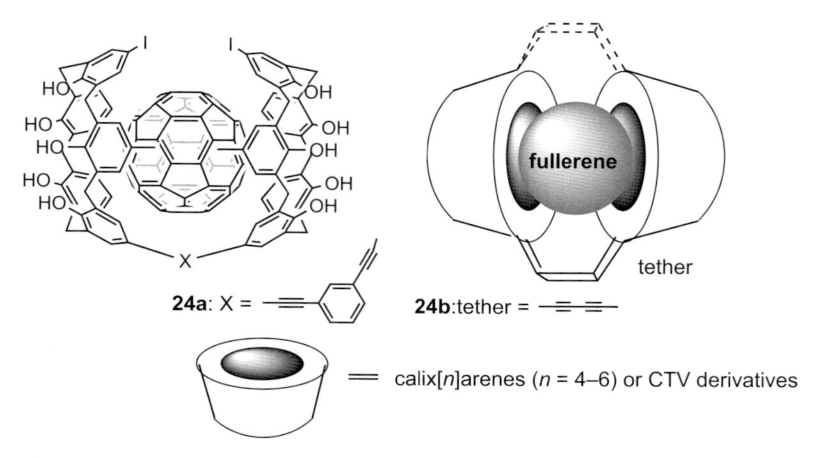

Figure 3.11 Structures of calix[5]arene dimer **24**, and schematic representations of dimeric hosts.

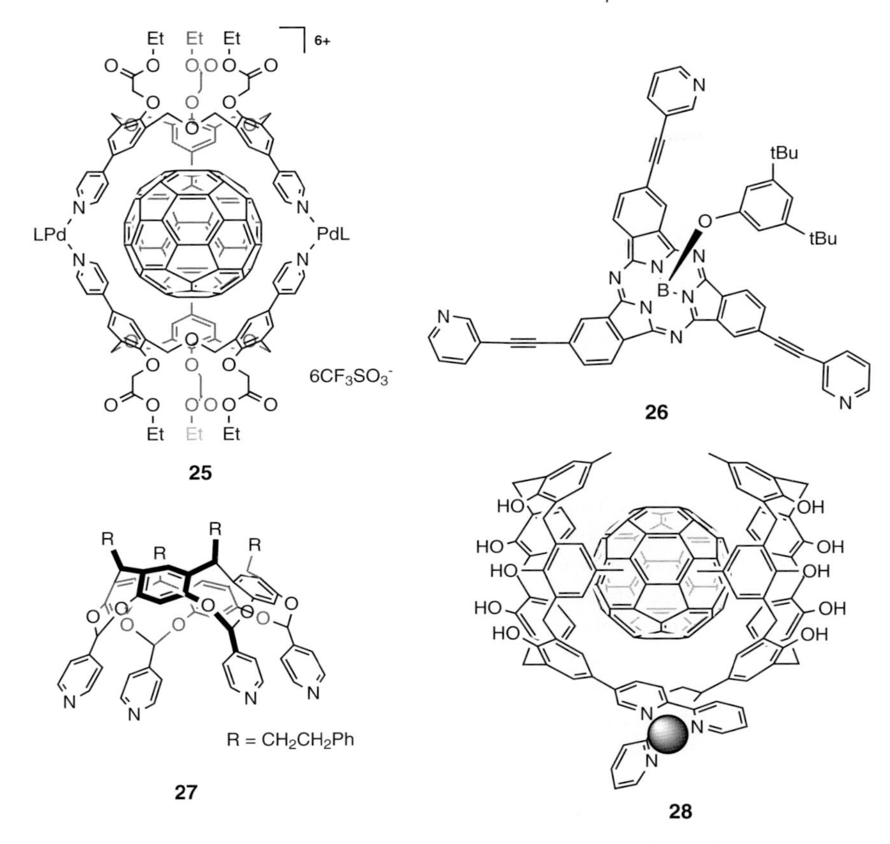

Figure 3.12 Dimeric host molecules **25** and **28**, and the components **26** and **27**.

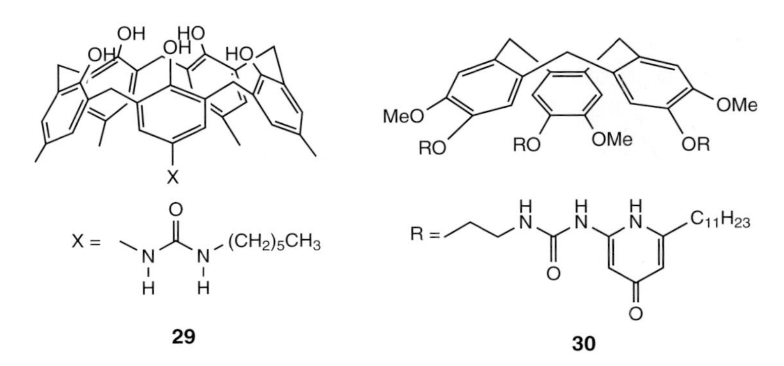

Figure 3.13 Calix[5]arene and CVT-based hydrogen-bonded dimeric receptors **29** and **30**.

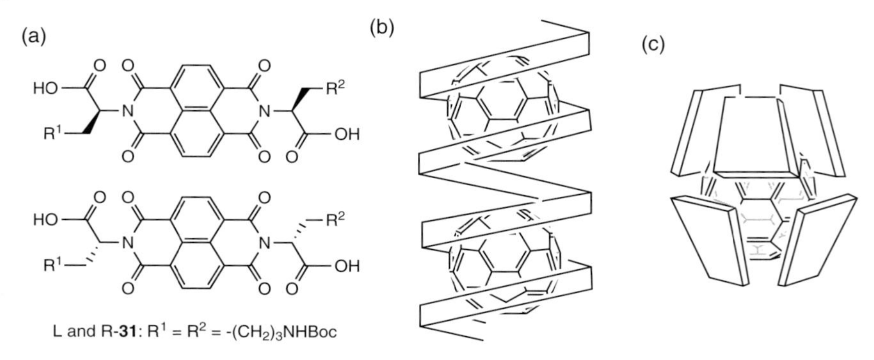

Figure 3.14 (a) Naphthalene imide **31**, (b) helical suprastructure of **31** and C_{60}, and (c) C_{70} receptor composed of **31** and C_{70}.

bonds. A CTV derivative **30** with ureidopyrimidinone units prepared by de Mendoza forms a dimeric hydrogen-bonded assembly that encapsulates a fullerene molecule within its large cavity (Figure 3.13). The system displays a remarkable selectivity for the encapsulation of C_{70} over C_{60} [54]. The receptor also formed a 2 : 1 complex with C_{84}, and the association constant is about 10 times higher than those of the C_{60} and C_{70} complexes [55]. A calixresorcinol-based metal-assembled host with rigid and sizeable cavity forms extremely stable inclusion complexes with fullerenes ($K_a > 10^5\ M^{-1}$) [56]. Naphthalene diimides (NDI) **31** functionalized with amino acids construct self-assembled helical nanotubes using hydrogen bondings. The helical superstructures possess tubular cavities with a mean diameter of 12.4 [Å] and are capable of accommodating a string of C_{60} (Figure 3.14b) [57]. On the other hand, the NDIs **31** spontaneously form a new C_{70} receptor by changing the hydrogen bonding network (Figure 3.14c) [58]. The counterbalance between van der Waals interaction and hydrogen bonding arrangements play an important role in the formation of the different superstructures.

3.3
Hydrocarbon Receptors

Hydrocarbon molecules with a rigid framework tend to form inclusion complexes with fullerenes. The crystallographic analyses of triptycene **32** and azatriptycene **33** revealed that one included C_{60} molecule is sandwiched by two molecules of **32** and is surrounded by three molecules of **33** in the crystals (Figure 3.15) [59]. The results intuitively suggest that the face-to-face-type interaction (π–π interaction) would be operative as a dominant force in the complexation. The concave shape of these molecules allows their efficient packing with C_{60} surface to gain wide van der Waals contact [60].

Analogous to the traditional host molecules, hydrocarbon receptors are weak receptors and did not form complexes in the solution phase. Appropriate modification to expand the π cavity should be needed. Receptor molecules based on a

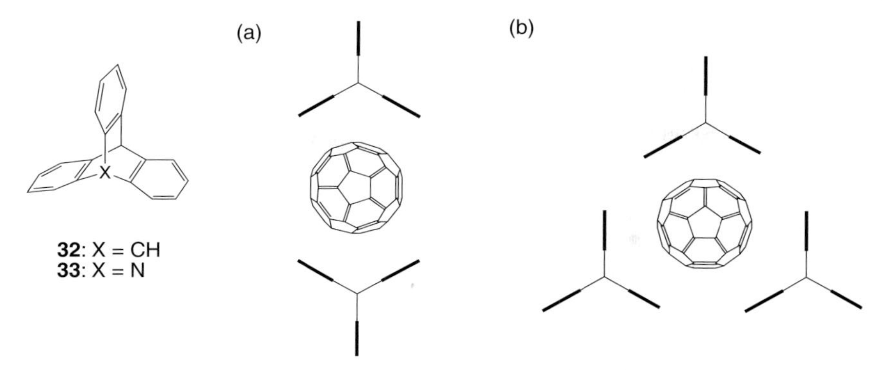

Figure 3.15 Molecular structures of **32** and **33**; packing arrangement: (a) the complexes of C_{60} and **32** and (b) the complexes of C_{60} and **33**.

tribenzotriquinacene skeleton bearing dithia- or diaza-heteroaromatic rings **34** and **35** were designed and prepared by Georghiou and Kuck's and Volkmer's groups (Figure 3.16) [61, 62]. The concave shape of these molecules allows their efficient packing with C_{60} surface to gain wide van der Waals contact. These hosts exhibit moderately high K_a values ($\sim 10^3\,M^{-1}$) in solutions. Recently, Kobayashi and Kawashima's group synthesized the triarylmethane and triarylphosphate-based host molecules with C_3 symmetry. In **36** and **37**, three anthryl groups construct a π cavity suitable for a C_{60} molecule. They act as fluorescent host molecules. Their K_{SV} values were estimated to be $(4.6 \pm 0.2) \times 10^4\,M^{-1}$ for **36** and $(3.5 \pm 0.2) \times 10^4\,M^{-1}$ for **37**, respectively (Figure 3.16) [63, 64].

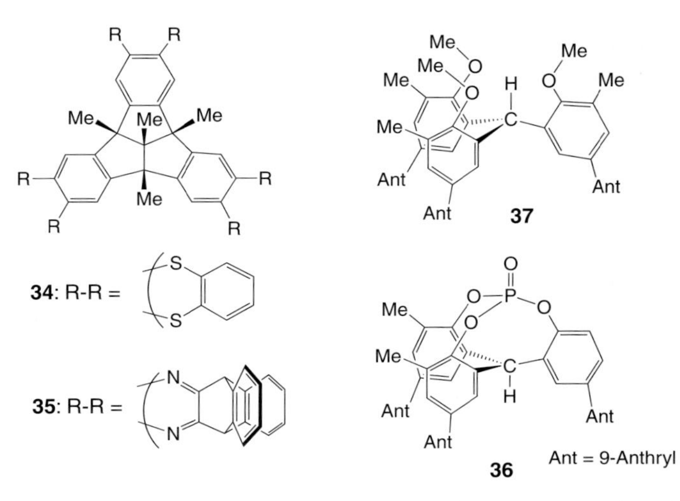

Figure 3.16 Triquinacene-based receptors **34** and **35**, and triarylmethane-based receptors **36** and **37**.

3.4
Receptors Bearing a Curved Conjugated System

3.4.1
Receptors Based on Bowl-Shaped Conjugated Systems

Corannulene **7** synthesized by Barth and Lowton in 1966 has been known as the first bowl-shaped conjugated system. The discovery of fullerenes has prompted a general interest in bowl-shaped aromatic hydrocarbons. Then, simple and practical syntheses of **7** have been developed by Scott, Rabideau, and Siegel's groups [65]. The curved π surfaces of C_{60} and **7** are geometrically well matched; it forms a stable complex with $(C_{60})^+$ in the gas phase [66]. However, unsubstituted corannulene shows no evidence for complexation with fullerenes both in solution and in solid state. Georghiou and Scott's group have designed and synthesized corannulene derivatives **38** and **39** bearing expanded π systems. They act as a fullerene receptor. The K_a values of **38** and **39** for C_{60} is ~300 M^{-1} and ~1400 M^{-1} in toluene, respectively [67, 68]. A double concave hydrocarbon Buckycatcher **40** composed of two corannulene moieties prepared by Sygula includes a C_{60} molecule in the concave cavity in the solid and solution state (K_a = 8600 M^{-1}) (Figure 3.17) [69].

Extended TTF (exTTF) **8** adopts a nonplanar structure, and the concave form of the anthracene moiety is a perfect match to the convex surface of C_{60}; however, no experimental evidence of association was obtained in this case too. Martin's group designed tweezers-shaped receptors **41** and **42** composed of two exTTF and an isophthalate or a tetrphthalate diester. The receptors exhibit high affinity for C_{60}. The stoichiometry of **42** to C_{60} varies a 1 : 1 complex to a 2 : 2 complex, depending upon the solvents [70]. Upon photoexcitation at the corresponding CT band of the complexes, photoinduced electron transfer (PET) from the receptors to fullerene readily took place in benzonitrile to form a fully charge-separated $C_{60}\cdot^-$ exTTF^{+-} state [71]. They investigated the relative contributions of the concave–convex π–π interaction to bind C_{60} using the host **41** and the other tweezers-shaped hosts **43–45** with various curvatures (Figure 3.18). The order of K_a values of the complexes for C_{60} is **41** > **43** > **44**, and no sign of association of receptor **45** was observed. The results support the perceptible contribution of concave–convex complementarity to the stabilization of supramolecular associates [72]. A concave tetrathiafulvalene-type

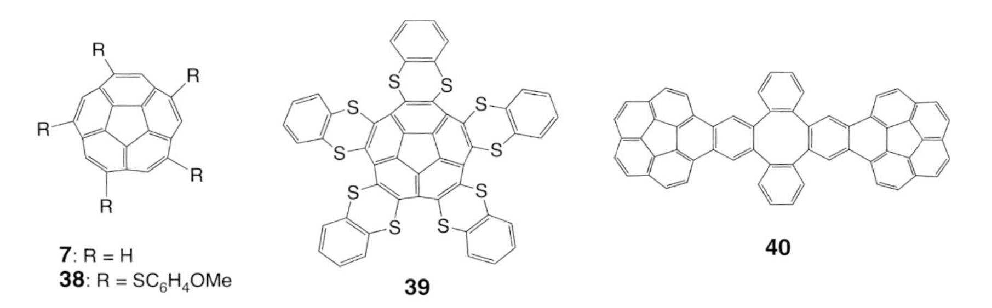

Figure 3.17 Corannulene-based receptors, **38–40**.

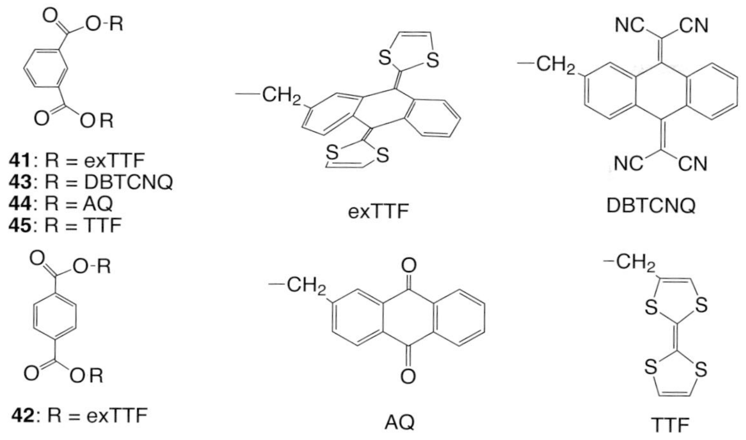

41: R = exTTF
43: R = DBTCNQ
44: R = AQ
45: R = TTF

42: R = exTTF

exTTF

DBTCNQ

AQ

TTF

Figure 3.18 ExTTF-based receptors **41** and **42**, and the related compounds **43–45**.

donor **46** has two concave faces composed of three aromatic and dithiole rings. It forms a 1 : 1 inclusion complex with fullerenes in solution ($K_a = (1.2 \pm 0.3) \times 10^3$ M^{-1} in $CDCl_3/CS_2$). A theoretical calculation suggests that the association preferentially occurs on the aromatic face of **46** [73]. Martin and Mendoza's group constructed the most efficient exTTF-based receptor **47** for fullerenes, where three exTTF subunits are linked to a CTV scaffold through short ether linkages. The concave surfaces of both the CTV and the exTTF subunits should nicely wrap around the entrapped fullerene guest. The log K_a values for C_{60} and C_{70} are $5.3 \pm 0.2\ M^{-1}$ and $6.3 \pm 0.6\ M^{-1}$, respectively (Figure 3.19) [74].

3.4.2
Receptors Bearing a Cylindrical Cavity

Analogous to the relationship between crown ethers and podands, in general, cyclic receptors have higher binding ability than the corresponding acyclic receptors. Macrocyclic compounds **48** and **49** bearing ether oxygen atoms synthesized by

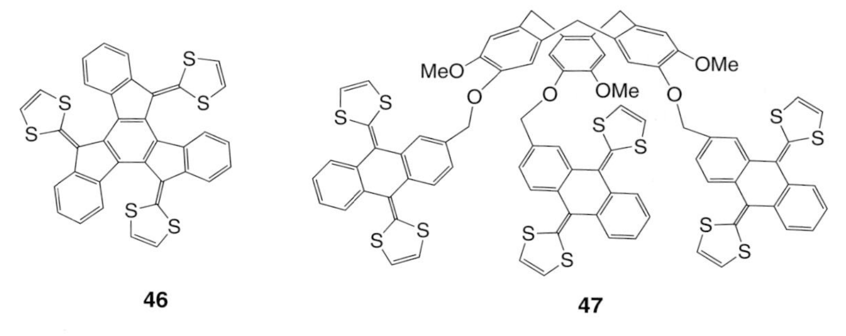

Figure 3.19 ExTTF-based receptors **46** and **47**.

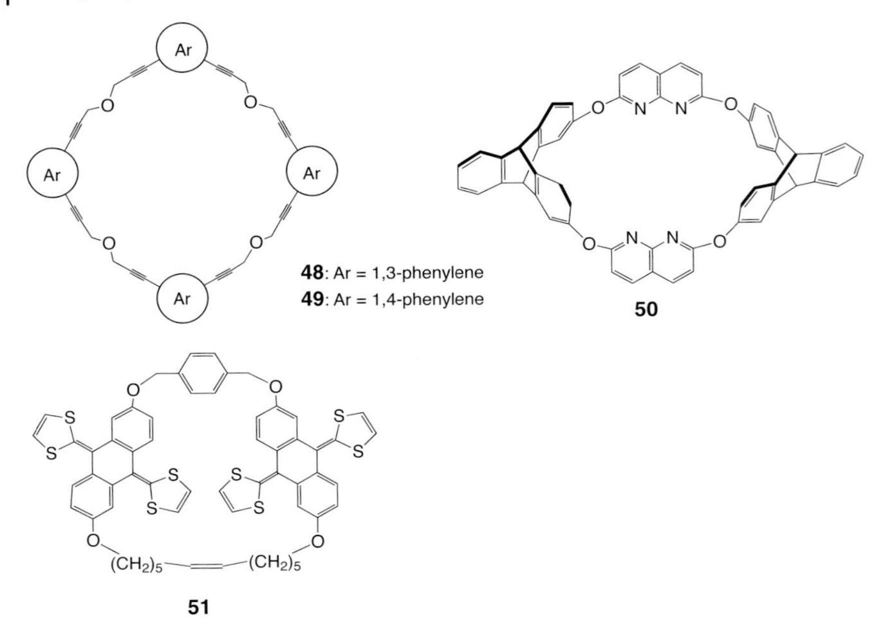

Figure 3.20 Cyclic receptors **48–51**.

Yoshida and coworkers form rather stable complexes with C_{60} (**48**: $K_a = 3410 \pm 1510\,M^{-1}$ and **49**: $K_a = 5430 \pm 370\,M^{-1}$ in toluene) [75]. Triptycene-derived oxacalixarene **50** prepared by Hu and Chen also forms fullerene complexes ($K_{SV} = 75400 \pm 2900\,M^{-1}$ for C_{60} and $K_{SV} = 89600 \pm 3100\,M^{-1}$ for C_{70} in toluene) [76]. Martin's group synthesized a bis-exTTF macrocyclic receptor **51**. Although the host molecule would possess a rather flexible cavity, the evaluated association constants for fullerenes are considerably high ($\log K_a = 6.5 \pm 0.5\,M^{-1}$ for C_{60} in PhCl, and the value for C_{70} was unmeasurably large). The results would be due to high affinity of exTTF moieties for fullerene π surface (Figure 3.20) [77].

3.4.3
Carbon Nanorings

The belt-shaped conjugated systems, in which the p-orbitals are aligned horizontally on a rigid surface, have been regarded as attractive synthetic targets because of the geometrical beauty. These compounds may be termed "carbon nanorings" that serve as good model compounds for carbon nanotube. Taking into account the ready formation of "fullerene peapod," receptor properties of carbon nanorings for fullerenes are highly promising. Cyclacenes **52** and cyclophenacenes **53** are representative of the family of carbon nanorings [78–81]; however, receptors based on these compounds have not been synthesized yet. Very recently, cyclic [n]paraphenylenes (CPP) **54** ($n = 7$ to 18) were prepared by three groups indepen-

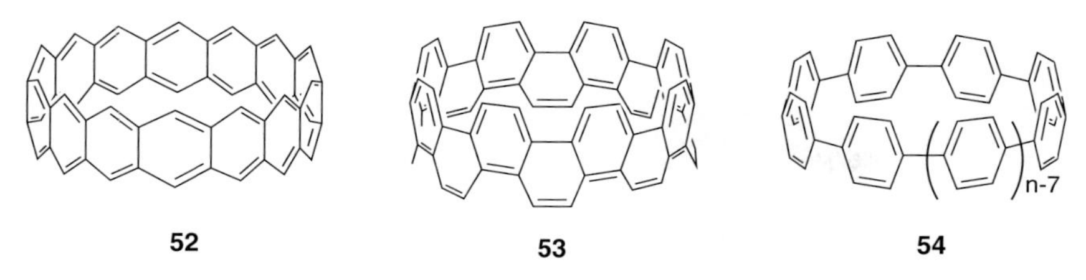

Figure 3.21 Carbon nanorings **52–54**.

dently (Figure 3.21) [81–83]. Particularly, [10] CPP having a 1.38 nm of cavity size forms a stable inclusion complex with C_{60} ($K_{SV} = (4.34 \pm 0.04) \times 10^6\ M^{-1}$) [84].

Kawase *et al.* have synthesized [*n*]CPPAs ($n = 5{\sim}9$) **55–59** and the related compounds having naphthylene rings **60–62** (Figure 3.22). According to the theoretical calculations, cyclic [*n*]paraphenyleneacetylenes ([*n*]CPPAs) possessing 1,4-phenylene and ethynylene units alternately adopt rigid and belt-shaped structures with well-defined cavities [85]. Actually, crystallographic analysis of [6]CPPA **56** revealed that the cavity size is almost comparable to that of (10,10)carbon nanotube [86]. The carbon nanorings **56**, **57**, and **60–62** form stable inclusion complexes with fullerenes both in solution and in solid state [87–89]. The molecular structure of the complex of [6]CPPA **56** and methanofullerene derivative **63** (Figure 3.23) revealed that the concave–convex π–π interaction is operative between the host and the guest as a major driving force [87]. However, relatively small spectral changes in 450–650 nm range of absorption were observed in the titration experiments. The results clearly indicate little contribution of the CT interaction in the complexation. On the other hand, the stability of complexes correlates well with the van der Waals contact between the host and the guest (Table 3.2) [88, 89]. The K_{SV} value of the

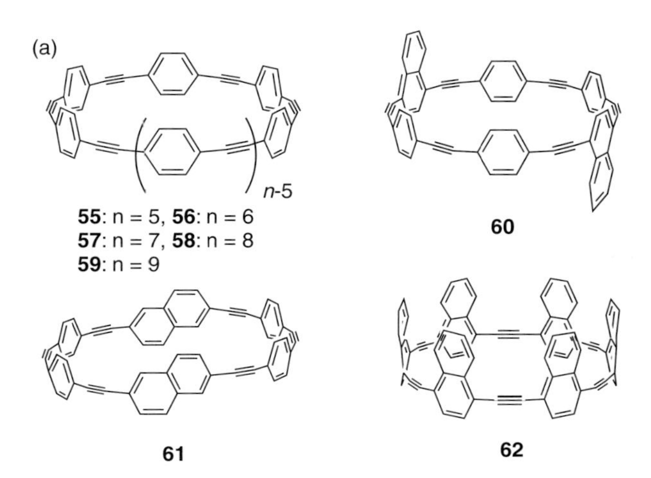

Figure 3.22 Cyclic [*n*]paraphenyleneacetylene ([*n*]CPPA) **55–59** and the related compounds **60–62**.

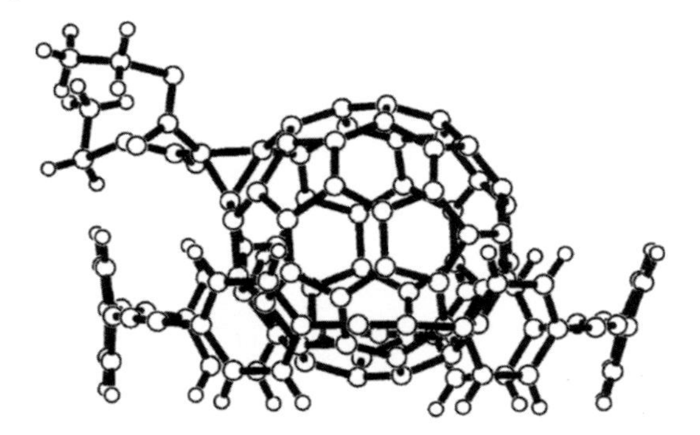

Figure 3.23 Molecular structure of [6]CPPA **56** and methanofullerene **63**.

complex **62**·C_{60} is the largest in the known receptors for fullerenes despite a hydrocarbon receptor [89]. Moreover, a solid-to-liquid extraction experiment proved the considerably high selectivity of **62** for C_{70} against C_{60} (>10 : 1) [88]. In contrast to pristine C_{60}, its derivatives possess perturbed electronic properties. The properties correlate well with the electronegativity of the attached atoms. For example, the incorporation of an electron-positive silicon atom increases the electron density of the π systems of fullerenes. To evaluate the stability of these complexes, the values of Gibbs activation energies ($\Delta G^{\ddagger}_{dis}$) for dissociation of the complexes of silylated fullerenes, **64** and **65**, methanofullerene **63**, and C_{60} with **56** determined by the VT-NMR experiments were recorded (Figure 3.24). The $\Delta G^{\ddagger}_{dis}$ values are on the order of **65** (8.5 ± 0.2 kcal/mol) < **64** (8.8 ± 0.2 kcal/mol) < **63** (9.4 ± 0.2 kcal/mol) < C_{60} (9.9 ± 0.3 kcal/mol). Therefore, an increase in electron density weakens the binding between the host and the guest [90].

Table 3.2 Diameters (Φ)[a] of the cavity of hosts, K_a, K_{SV}, and $\Delta G^{\ddagger}$ values[b] of the complexes.

Complex	Φ[a]	K_a ($\times 10^{-4}$)[b]	K_{SV} ($\times 10^{-4}$)[b]	$\Delta G^{\ddagger}$ (CD_2Cl_2)
56·C_{60}	1.31	1.6 ± 0.2	7.0	9.9 ± 0.2
56·C_{70}		1.8 ± 0.2	14	9.6 ± 0.2
60·C_{60}	1.31	2.5∼6.0	27	10.8 ± 0.3
60·C_{70}		—[c]	26	10.1 ± 0.2
61·C_{60}	1.41	∼10	26	< 9
61·C_{70}		∼100	430	11.9 ± 0.8
62·C_{60}	1.31	—[c]	770	14.1 ± 0.3
62·C_{70}		—[c]	1000	13.3 ± 0.3
57·C_{60}	1.53	—	5.6	< 9
57·C_{70}		—	21	< 9

a) nm, evaluated by AM1 calculations.
b) M^{-1}, in C_6H_6.
c) Undeterminable.

63: X = C(COOEt)$_2$

64: X = SiR_2, R = 2,6-diethylphenyl

65: X = $(SiR_2)_2$, R = 2,4,6-trimethylphenyl

Figure 3.24 Fullerene derivatives **63–65**.

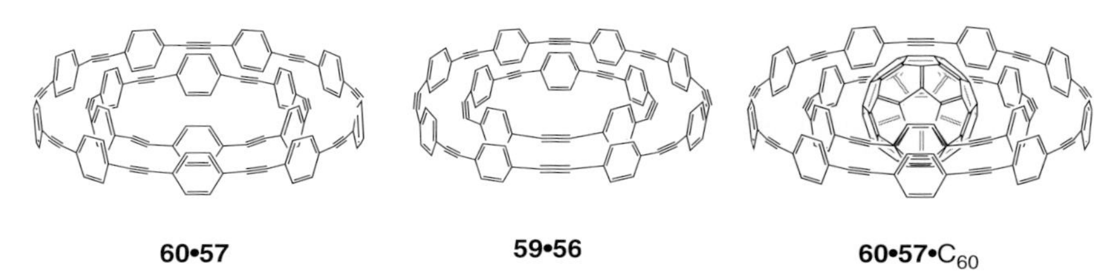

Figure 3.25 Ring-in-ring complexes of carbon nanorings and fullerene.

When the van der Waals distance (0.34 nm) is taken into account, [8]- and [9] CPPAs, **58** and **59**, have almost perfect complementarity to [5]- and [6]CPPA, **55** and **58**, respectively. Actually these carbon nanorings construct ring-in-ring complexes [91, 92]. Moreover, **59** and **56** form an onion-type supramolecular structure with a C_{60} (Figure 3.25) [91]. The K_a values and thermodynamic parameters of these complexes are summarized in Table 3.3. The similar K_a values of **59·56** in the presence and the absence of C_{60} suggest that a C_{60} molecule affords little electronic and structural perturbation to the host **57**. Together with the small ΔS value for the formation of **59·56**, the complex is characteristic of a van der Waals complex. On the other hand, the K_a value at 30 °C for the formation of **58·55** is about 200 times larger than for the formation of **59·56**, despite the fact that **58·55** has a relatively small contact area. Moreover, the ΔS value for the formation of **58·55** is significantly large. These results prove the substantial participation of the electrostatic interaction prior to the dispersion force. The formation of fullerene complexes is known to be accompanied by large solvent and entropy effects [93]. Planar phenylacetylene

Table 3.3 The K_a values[a)] and thermodynamic parameters of ring-in-ring complexes in $CDCl_3$.

	K_a	ΔH[b)]	ΔS[c)]
59·56	~40	~−3.0	~−2.4
58·55	9200 ± 1400	−0.75	16

a) at 303 K, M^{-1}.
b) kcal/mol.
c) kcal/(mol K).

macrocycles that do not contain electron-withdrawing aromatic substituents do not aggregate in nonpolar solvent because π–π stacking between planar aromatic hydrocarbons causes an electrostatic repulsive force [94]. On basis of the results, we proposed that concave–convex π–π interaction can vary from repulsive to attractive depending on the strain present in the component π electron systems. The drastic increase in the K_a values going from **59·56** to **56**·C_{60} or **58·55** can be rationalized in terms of the contribution of additional electrostatic interaction caused by this type of increased polarity of the π systems.

3.5 Conclusions

In general, nonbonded interaction between π-conjugated systems can be considered on the basis of the following three factors: the van der Waals (VDW) interaction, the electrostatic (ES) interaction, and the charge transfer (CT) interaction [95]. Fullerenes can be regarded as an electron acceptor, and C_{60} exhibits a CT absorption band in the 400–650 nm range in aromatic solvents [96, 97]. The complexation of receptors based on the traditional hosts and carbon nanorings with fullerenes causes slight changes in the CT absorptions. Thus, the CT interaction would be negligible in the complexation. The VDW forces alone do not provide a sufficiently strong driving force. Combined with these results, ES interactions likely serve as the most probable driving force for the complexation. On the other hand, strong electron donors such as *N,N*-dimethylaniline provide exciplexes causing the hyperchromic shift in a 450–650 nm range of absorption [98]. Thus, the participation of additional attractive CT interaction should play an important role in the high affinity of triarylamine-based receptors for fullerenes. The supramolecular properties of fullerenes can be understood in terms of the possession of positive electronic potentials in their convex surfaces because the concave face of the fullerene receptors discussed here apparently possesses a considerably larger negative potential [99]. Recent theoretical studies suggest that the convex and concave faces of curved conjugated systems should be oppositely polarized because of the unsymmetrical nature of their p-orbitals. However, these studies have still led to controversial predictions. Further experimental and theoretical studies on these complexes and related substances will deepen our understanding on the novel nature of fullerene and other curved π electron systems.

References

1 Krätchmer, W., Lamb, L.D., Fostiropoulos, K., and Huffman, D.R. (1990) Solid C_{60}: a new form of carbon. *Nature*, **347**, 354–358.

2 Diederich, F., Effing, J., Jonas, L., Jullien, L., Plesnivy, T., Ringsdorf, H., Thilgen, C., and Weinstein, D. (1992) C_{60} and C_{70} in a Basket?: investigations of mono- and multilayers from azacrown compounds and fullerenes. *Angew. Chem. Int. Ed. Engl.*, **31**, 1599–1602.

3 Andersson, T., Nilsson, K., Sundahl, M., Westman, G., and Wennerström, O. (1992) C_{60} embedded in γ-cyclodextrin: a water-soluble fullerene. *J. Chem. Soc. Chem. Commun.*, 604–606.

4 Liu, Y., Wang, H., Liang, P., and Zhang, H.-Y. (2004) Water-soluble supramolecular fullerene assembly mediated by metallobridged β-cyclodextrins. *Angew. Chem. Int. Ed.*, **43**, 2690–2694.

5 Nishibayashi, Y., Saito, M., Uemura, S., Takekuma, S., Takekuma, H., and Yoshida, Z. (2004) Buckminsterfullerenes: a non-metal system for nitrogen fixation. *Nature*, **428**, 279; Takekuma, S., Takekuma, H., and Yoshida, Z. (2005). Reducing ability of supramolecular C_{60} dianion toward C=O, C=C and N−N bonds. *Chem. Commun.*, 1628–1630.

6 Constabel, F. and Geckeler, K.E. (2004) Solvent-free self-assembly of C_{60} and cucurbit[7]uril using high-speed vibration milling. *Tetrahedron Lett.*, **45**, 2071.

7 Yoshida, Z., Takekuma, H., Takekuma, S., and Matsubara, Y. (1994) Molecular recognition of C_{60} with γ-cyclodextrin. *Angew. Chem. Int. Ed. Engl.*, **33**, 1597–1599; Constable, E.C. (1994) Taking fullerenes from large molecules to supramolecules. *Angew. Chem. Int. Ed Engl.*, **33**, 2269–2271.

8 Da Ros, T. and Proto, M. (1999) Medicinal chemistry with fullerenes and fullerene derivatives. *Chem. Commun.*, 663–669.

9 Nakamura, E. and Isobe, H. (2003) Functionalized fullerenes in water. The first 10 years of their chemistry, biology, and nanoscience. *Acc. Chem. Res.*, **36**, 807–815.

10 Atwood, J.L., Koutsantonis, G.A., and Raston, C.L. (1994) Purification of C_{60} and C_{70} by selective complexation with calixarenes. *Nature*, **368**, 229–231; Raston, C.L., Atwood, J.L., Nichols, P.J., and Sudria, I.B.N. (1996) Supramolecular encapsulation of aggregates of C_{60}. *Chem. Commun.*, 2615–2616.

11 Suzuki, T., Nakashima, K., and Shinkai, S. (1994) Very convenient and efficient purification method for fullerene (C_{60}) with 5,11,17,23,29,35,41,47-octa-*tert*-butylcalix[8]arene-49,50,51,52,53,54,55,56-octol. *Chem. Lett.*, 699–702.

12 Raston, C.L., Atwood, J.L., Nichos, P.J., and Sudria, I.B.N. (1996) Supramolecular encapsulation of aggregates of C_{60}. *Chem. Commun.*, 2615–2616.

13 Atwood, J.L., Barbour, L.J., Raston, C.L., and Sudria, I.B.N. (1998) C_{60} and C_{70} compounds in the pincerlike jaws of calix[6]arene. *Angew. Chem. Int. Ed.*, **37**, 981.

14 Atwood, J.L., Barbour, L.J., Heaven, M.W., and Raston, C.L. (2003) Controlling van der Waals contacts in complexes of fullerene C_{60}. *Angew. Chem. Int. Ed.*, **36**, 3254–3257; Atwood, J.L., Barbour, L.J., Heaven, M.W., and Raston, C.L. (2003) Association and orientation of C_{70} on complexation with calyx[5]arene. *Chem. Commun.*, 2270–2271.

15 Tsubaki, K., Tanaka, K., Kinoshita, T., and Fuji, K.;. (1998) Complexation of C_{60} with hexahomooxacalix[3]arenes and supramolecular structures of complexes in the solid state. *Chem. Commun.*, 895–896.

16 Steed, J.W., Junk, P.C., Atwood, J.L., Barnes, M.J., Raston, C.L., and Burkhalter, R.S. (1994) Ball and socket nanostructures: new supramolecular chemistry based on cyclotriveratrylene. *J. Am. Chem. Soc.*, **116**, 10346–10347.

17 Hardie, M.J. and Raston, C.L. (1999) Confinement and recognition of icosahedral main group cage molecules: fullerene C_{60} and *o*-, *m*-, *p*-dicarbadodecaborane(12). *Chem. Commun.*, 1153–1163; Diederich, F. and Gomez-Lopez, M. (1999) Supramolecular fullerene chemistry. *Chem. Soc. Rev.*, **28**, 263–277.

18 Komatsu, N. (2003) Preferential precipitation of C_{70} over C_{60} with *p*-halohomooxacalix[3]arenes. *Org. Biomol. Chem.*, **1**, 204–209.

19 Barbour, L.J., Orr, G.W., and Atwood, J.L. (1997) Supramolecular interaction of C_{60} into a calixarene bilayer: a well-ordered solid-state structure dominated by

van der Waals contacts. *Chem. Commun.*, 1439–1440.

20 Rose, K.N., Barbour, L.J., Orr, G.W., and Atwood, J.L. (1998) Self-assembly of carcerand-like dimers of calyx[4]resorcinarene facilitated by hydrogen bonded solvent bridges. *Chem. Commun.*, 407–408; Garcia, M.M., Uribe, M.I.C., Palacios, E.B., Ochoa, F.L., Toscano, A., Cogordan, J.A., Rios, S., and Cruz-Almanza, R. (1999) Host–guest complexation of 1,8,15,22-tetraphenyl[14] metacyclophan-4,11,18,25- tetramethyl-3,5,10,12,17,19,24,26-octol with C_{60}. *Tetrahedron*, **55**, 6019.

21 Araki, K., Akao, K., Ikeda, A., Suzuki, T., and Shinkai, S. (1996) Molecular design of calixarene-based host molecules for inclusion of C_{60} in solution. *Tetrahedron Lett.*, **37**, 73–77; Shinkai, S. and Ikeda, A. (1999) Novel interactions of calixarene p-systems with metal ions and fullerenes. *Pure Appl. Chem.*, **71**, 275–280.

22 Haino, T., Yanase, M., and Fukazawa, Y. (1997) New supramolecular complex of C_{60} based on calix[5]arene: its structure in the crystal and in solution. *Angew. Chem. Int. Ed. Engl.*, **36**, 259–260.

23 Wang, J. and Gutsche, C.D. (1998) Complexation of fullerenes with bis-calix [n]arenes synthesized by tandem Claisen rearrangement. *J. Am. Chem. Soc.*, **120**, 12226–12231.

24 Atwood, J.L., Barbour, L.J., Nichols, P.J., Raston, C.L., and Sandoval, C.A. (1999) Symmetry-aligned supramolecular encapsulation of C_{60}: $[C_{60} \subset (L)_2]$, L = *p*-benzylcalix[5]arene or *p*-benzylhexahomooxacalix[3]arene. *Chem. Eur. J.*, **5**, 990–996.

25 Makha, M., Hardie, M.J., and Raston, C.L. (2002) Inter-digitation approach to encapsulation of C_{60}: [C_{60} ⊂ (*p*-phenylcalix[5]arene)$_2$]. *Chem. Commun.*, 1446–1447.

26 Georghiou, P.E., Tran, A.H., Stroud, S.S., and Thompson, D.W. (2006) Supramolecular complexation studies of [60]fullerene with calix[4]naphthalenes: a reinvestigation. *Tetrahedron*, **62**, 2036–2044.

27 Mizyed, S., Ashram, M., Miller, D.O., and Georghiou, P.E. (2001) Supramolecular complexation of [60]fullerene with hexahomotrioxacalix[3]naphthalenes: a new class of naphthalene-based calixarenes. *J. Chem. Soc. Perkin Trans.*, **2**, 1916.

28 Tucci, F.C., Rudkevich, D.M., and Rebek, J. Jr. (1999) Deeper Cavitands. *J. Org. Chem.*, **64**, 4555–4559.

29 Sun, Y., Drovetskaya, T., Bolskar, R.D., Bau, R., Boyd, P.D.W., and Reed, C.A. (1997) Fulleride of pyrrolidine-functionalized C_{60}. *J. Org. Chem.*, **62**, 3642–3649.

30 Grey, I.E., Hardie, M.J., Ness, T.J., and Raston, C.L. (1999) Octaphenylcyclotetrasiloxane confinement of C_{60} into double columnar arrays. *Chem. Commun.*, 1139–1140; Hardie, M.J., Torrens, R., and Raston, C.L. (2003) Characterization of a new 1 : 1 $(C_{60})(CHBr_3)$ intercalation complex. *Chem. Commun.*, 1854–1855.

31 Araki, K., Akao, K., Ikeda, A., Suzuki, T., and Shinkai, S. (1996) Molecular design of calixarene-based host molecules for inclusion of C_{60} in solution. *Tetrahedron Lett.*, **37**, 73–77; Shinkai, S. and Ikeda, A. (1999) Novel interactions of calixarene p-systems with metal ions and fullerenes. *Pure Appl. Chem.*, **71**, 275–280.

32 Matsubara, H., Hasegawa, A., Shiwaku, K., Asano, K., Uno, M., Takahashi, S., and Yamamoto, K. (1998) Supramolecular inclusion complexes of fullerenes using cyclotriveratrylene derivatives with aromatic pendants. *Chem. Lett.*, 923–924.

33 Felder, D., Heinrich, B., Guillon, D., Nicoud, J.-F., and Nierengarten, J.-F. (2000) A liquid crystalline supramolecular complex of C_{60} with a cyclotriveratrylene derivative. *Chem. Eur. J.*, **6**, 3501–3507.

34 Nierengarten, J.-F., Oswald, L., Eckert, J.-F., Nicoud, J.-F., and Armaroli, N. (1999) Complexation of fullerenes with dendritic cyclotriveratrylene derivatives. *Tetrahedron Lett.*, **40**, 5681–5684.

35 Numata, M., Ikeda, A., Fukuhara, C., and Shinkai, S. (1999) Dendrimers can act as a host for [60]fullerene. *Tetrahedron Lett.*, **40**, 6945–6948.

36 Schuster, D.I., Rosenthal, J., MacMahon, S., Jarowski, P.D., Alabi, C.A., and Guldi, D.M. (2002) Formation and photophysics of a stable

concave–convex supramolecular complex of C_{60} and a substituted *s*-triazine derivative. *Chem. Commun.*, 2538–2539.

37 Wang, M.-X., Zhang, X.-H., and Zheng, Q.-Y. (2004) Synthesis, structure, and [60]fullerene complexation properties of azacalix[*m*]arene[*n*]pyridines. *Angew. Chem. Int. Ed.*, **43**, 838–843.

38 Zhang, E.-X., Wang, D.-X., Zheng, Q.-Y., and Wang, M.-X. (2008) Synthesis of large macrocyclic azacalix[*n*]arenes ($n = 6 - 9$) and their complexation with fullerenes C_{60} and C_{70}. *Org. Lett.*, **10**, 2565–2568.

39 Mustafizur Rahman, A.F.M., Bhattacharya, S. Peng, X., Kimura, T., and Komatsu, N. (2008). Unexpectedly large binding constants of azulenes with fullerenes. *Chem. Commun.*, 1196–1198.

40 Stella, L., Capodilupo, A.L., and Bietti, M. (2008) A reassessment of the association between azulene and [60]fullerene. Possible pitfalls in the determination of binding constants through fluorescence spectroscopy. *Chem. Commun.*, 4744–4746.

41 Nielsen, K.A., Cho, W.-S., Sarova, G.H., Petersen, B.M., Bond, A.D., Becher, J., Jensen, F., Guldi, D.M., Sessler, J.L., and Jeppesen, J.O. (2006) Supramolecular receptor design: anion-triggered binding of C_{60}. *Angew. Chem. Int. Ed.*, **45**, 6848–6853.

42 Yoo, J., Kim, Y., Kim, S.-J., and Lee, C.-H. (2010) Anion-modulated, highly sensitive supramolecular fluorescence chemosensor for C_{70}. *Chem. Commun.*, **46**, 5449–5451.

43 Haino, T., Yanase, M., and Fukazawa, Y. (1997) Crystalline supramolecular complexes of C_{60} with calix[5]arenes. *Tetrahedron Lett.*, **38**, 3739–3742.

44 Haino, T., Yanase, M., and Fukazawa, Y. (1998) Fullerenes enclosed in bridged calix [5]arenes. *Angew. Chem. Int. Ed.*, **37**, 997; Haino, T., Yanase, M., Fukunaga, C., and Fukazawa, Y. (2006) Fullerenes encapsulation with calix[5]arenes. *Tetrahedron*, **62**, 2025–2035.

45 Haino, T., Fukunaga, C., and Fukazawa, Y. (2006) A new calix[5]arene-based container: selective extraction of higher fullerenes. *Org. Lett.*, **8**, 3545–3548.

46 Iglesias-Sanchez, J.C., Fragoso, A., de Mendoza, J., and Prados, P. (2006) Aryl–aryl linked bi-5,5′-*p*-*tert*-butylcalix[4]arene tweezer for fullerene complexation. *Org. Lett.*, **8**, 2571.

47 Matsubara, H., Shimura, T., Hasegawa, A., Semba, M., Asano, K., and Yamamoto, K. (1999) Syntheses of novel fullerene tweezers and their supramolecular inclusion complex of C_{60}. *Chem. Lett.*, 1099–1110.

48 Ikeda, A., Yoshimura, M., Uzdu, H., Fukuhara, C., and Shinkai, S. (1999) Inclusion of [60]fullerene in a homooxacalix[3]arene-based dimeric capsules cross-linked by a Pd^{II}-pyridine interaction. *J. Am. Chem. Soc.*, **121**, 4296–4297.

49 Claessens, C.G. and Torres, T. (2004) Inclusion of C_{60} fullerene in a M_3L_2 subphthalocyanine cage. *Chem. Commun.*, 1298–1299.

50 Pirondini, L., Bonifazi, D., Cantadori, B., Braiuca, P., Campagnolo, M., Zorzi, R.D., Geremia, S., Diederich, F., and Dalcanale, E. (2006) Inclusion of methano[60] fullerene derivatives in cavitand-based coordination cages. *Tetrahedron*, **62**, 2008–2015.

51 Haino, T., Araki, H., Yamanaka, Y., and Fukazawa, Y. (2001) Fullerene receptor based on calyx[5]arene through metal-assisted self-assembly. *Tetrahedron Lett.*, **42**, 3203–3206.

52 Haino, T., Yamanaka, Y., Araki, H., and Fukazawa, Y. (2002) Metal-induced regulation of fullerene complexation with double-calix[5]arene. *Chem. Commun.*, 402–403.

53 Yanase, M., Haino, T., and Fukazawa, Y. (1999) A Self-assembling molecular container for fullerenes. *Tetrahedron Lett.*, **40**, 2781–2784.

54 Huerta, E., Metselaar, G.A., Fragoso, A., Santos, E., Bo, C., and de Mendoza, J. (2007) Selective binding and easy separation of C_{70} by nanoscale self-assembled capsules. *Angew. Chem. Int. Ed.*, **46**, 202–205.

55 Huerta, E., Cequier, E., and de Mendoza, J. (2007) Preferential separation of fullerene[84] from fullerene mixtures by encapsulation. *Chem. Commun.*, 5016–5018.

56 Fox, O.D., Cookson, J., Wilkinson, E.J.S., Drew, M.G.B., MacLean, E.J., Teat, S.J., and Beer, P.D. (2006) Nanosized polymetallic resorcinarene-based host assemblies that strongly bind fullerenes. *J. Am. Chem. Soc.*, **128**, 6990–7002.

57 Dan Pantos, G., Wietor, J.-L., and Sanders, J.K.M. (2007) Filling helical nanotubes with C_{60}. *Angew. Chem. Int. Ed.*, **46**, 2238–2240.

58 Weitor, J.-L., Dan Pantos, G., and Sanders, J.K.M. (2008) Templated amplification of an unexpected receptor for C_{70}. *Angew. Chem. Int. Ed.*, **47**, 2689–2692.

59 Veen, E.M., Postma, P.M., Jonkman, H.T., Spek, A.L., and Feringa, B.L. (1999) Solid state organisation of C_{60} by inclusion crystallisation with triptycenes. *Chem. Commun.*, 1709–1710.

60 Tanaka, K. and Caira, M.R. (2002) Supramolecular complex of C_{60} with tetrakis(*p*-iodophenyl)ethylene. *J. Chem. Res. (S)*, **2002**, 642.

61 Bredenköuer, H., Henne, S., and Volkmer, D. (2007) Nanosized ball joints constructed from C_{60} and tribenzotriquinacene sockets: synthesis, component self-assembly and structural investigations. *Chem. Eur. J.*, **13**, 9931–9938.

62 Georghiou, P.E., Dawe, L.N., Tran, H.-A., Strübe, J., Neumann, B., Stammler, H.-G., and Kuck, D. (2008) C_{3v}-symmetrical tribenzotriquinacenes as hosts for C_{60} and C_{70} in solution and in the solid state. *J. Org. Chem.*, **73**, 9040–9047.

63 Kobayashi, J., Domoto, Y., and Kawashima, T. (2010) Synthesis of a C_3 symmetric host molecule for C_{60}bearing a bicyclic triarylphosphate framework. *Chem. Lett.*, **39**, 134–135.

64 Kobayashi, J., Domoto, Y., and Kawashima, T. (2009) Inclusion of C_{60} into the hexagonal columnar space formed by intra- and intermolecular CH···π interactions. *Chem. Commun.*, 6186–6188.

65 Wu, Y.-T. and Siegel, J.S. (2006) Aromatic molecular-bowl hydrocarbons: synthetic derivatives, their structures, and physical properties. *Chem. Rev.*, **106**, 4843–4867; Tsefrikas, V.M. and Scott, L.T. (2006) Geodesic polyarenes by flash vacuum pyrolysis. *Chem. Rev.*, **106**, 4868–4884; Rabideau, P.W. and Sygula, A. (1996) Buckybowls: polynuclear aromatic hydrocarbons related to the buckminsterfullerene surface. *Acc. Chem. Res.*, **29**, 235–242.

66 Becker, H., Javahery, G., Petrie, S., Cheng, P.-C., Schwarz, H., Scott, L.T., and Bohme, D.K. (1993) Gas-phase ion/molecular reactions of corannulene, a fullerene subunit. *J. A m. Chem. Soc.*, **115**, 11636–11637.

67 Mizyed, S., Georghiou, P.E., Bancu, M., Cuadra, B., Rai, A.K., Chen, P., and Scott, L.T. (2001) Embracing C_{60} with multiarmed geodesic partners. *J. Am. Chem. Soc.*, **123**, 12770.

68 Georghiou, P.E., Tran, A.H., Mizyed, S., Bancu, M., and Scott, L.T. (2005) Concave polyarenes with sulfide-linked flaps and tentacles: new electron-rich hosts for fullerenes. *J. Org. Chem.*, **70**, 6158–6163.

69 Sygula, A., Fronczek, F.R., Sygula, R., Rabideau, P.W., and Olmstead, M.M. (2007) A double concave hydrocarbon buckycatcher. *J. Am. Chem. Soc.*, **129**, 3842–3843.

70 Perez, E.M., Sanchez, L., Fernadez, G., and Martin, N. (2006) exTTF as a building block for fullerene receptors unexpected solvent-dependent positive homotropic cooperativity. *J. Am. Chem. Soc.*, **128**, 7172–7173.

71 Gayathri, S.S., Wielopolski, M., Perez, E.M., Fernandez, G., Sanchez, L., Viruela, R., Orti, E., Guldi, D.M., and Martin, N. (2009) Discrete supramolecular donor–acceptor complexes. *Angew. Chem. Int. Ed.*, **48**, 815–819.

72 Perez, E.M., Capodilupo, A.L., Fernandez, G., Sanchez, L., Viruela, P.M., Viruela, R., Orti, E., Bietti, M., and Martin, N. (2008) Weighting non-covalent forces in the molecular recognition of C_{60}. Relevance of concave–convex complementarity. *Chem. Commun.*, 4567–4569.

73 Perez, E.M., Sierra, M., Sanchez, L., Torres, M.R., Viruela, R., Viruela, P.M., Orti, E., and Martin, N. (2007)

Concave tetrathiafulvalene-type donors as supramolecular partners for fullerenes. *Angew. Chem. Int. Ed.*, **46**, 1847–1851.

74 Huerta, E., Isia, H., Perez, E.M., Bo, C., Martin, N., and de Mendoza, J. (2010) Tripodal exTTF-CTV hosts for fullerenes. *J. Am. Chem. Soc.*, **132**, 5351–5353.

75 Yamaguchi, Y., Kobayashi, S., Amita, N., Wakamiya, T., Matsubara, Y., Sugimoto, K., and Yoshida, Z. (2002) Creation of nanoscale oxaarenecyclynes and their C_{60} complexes. *Tetrahedron Lett.*, **43**, 3277–3280.

76 Hu, S.-Z. and Chen, C.-F. (2010) Triptycene-derived oxacalixarene with expanded cavity: synthesis, structure and its complexation with fullerenes C_{60} and C_{70}. *Chem. Commun.*, 4199–4201.

77 Isia, H., Gallego, M., Perez, E.M., Viruela, R., Orti, E., and Martin, N. (2010) A bis-exTTF macrocyclic receptor that associates C_{60} with micromolar affinity. *J. Am. Chem. Soc.*, **132**, 1772–1773.

78 Nakamura, E., Tahara, K., Matsuo, Y., and Sawamura, M. (2003) Synthesis, structure, and aromaticity of a hoop-shaped cyclic benzenoid [10] cyclophenacene. *J. Am. Chem. Soc.*, **125**, 2834–2835.

79 Matsuo, Y. and Nakamura, E. (2008) Selective multiaddition of organocopper reagents to fullerenes. *Chem. Rev.*, **108**, 3016–3028.

80 Kawase, T. and Kurata, H. (2006) Carbon nanorings and their complexing abilities: exploration of the concave–convex π–π interaction. *Chem. Rev.*, **106**, 5250–5273.

81 Jasti, R., Bhattacharjee, J., B. Neaton, J.B., and Bertozzi, C.R. (2008) Synthesis, characterization, and theory of [9]-, [12]-, and [18]cycloparaphenylene: Carbon nanohoop structures. *J. Am. Chem. Soc.*, **130**, 17646–17647; Sisto, T.J., Golder, M.R., Hirst, E.S., and Jasti, R. (2011) Selective synthesis of strained [7] cycloparaphenylene: an orange-emitting fluorophore. *J. Am. Chem. Soc.* **133**, 15800–15802.

82 Tanaka, H. Omachi, H., Yamamoto, Y., Bouffard, J., and Itami, K. (2009) Selective Synthesis of [12]Cycloparaphenylene, *Angew. Chem. Int. Ed.*, **48**, 6112–6116; Omachi, H., Matsuura, S., Segawa, Y., and Itami, K. (2010) A Modular and size-selective synthesis of [n] cycloparaphenylenes: A step toward the selective synthesis of [n,n] single-walled carbon nanotubes. *Angew. Chem. Int. Ed.*, **49**, 10202–10205.

83 Yamago, S., Watanabe, Y., and Iwamoto, T. (2010) Synthesis of [8] cycloparaphenylene from a square-shaped tetranuclear platinum complex, *Angew. Chem. Int. Ed.*, **49**, 757–759. Iwamoto, T., Watanabe, Y., Sakamoto, Y., Suzuki, T., and Yamago, S. (2011) Selective and random syntheses of [n]cycloparaphenylenes ($n = 8$–13) and size dependence of their electronic properties. *J. Am. Chem. Soc.* **133**, 8354–8361.

84 Iwamoto, T. Watanabe, Y., Sadahiro, T., Haino, T., and Yamago, S. (2011) Size-selective encapsulation of C_{60} by [10] cycloparaphenylene: formation of the shortest fullerene-peapod. *Angew. Chem. Int. Ed.* **50**, 8342–8344.

85 Kawase, T., Darabi, H.R., and Oda, M. (1996) Cyclic [6]- and [8]paraphenylacetylenes. *Angew. Chem. Int. Ed. Engl.*, **35**, 2662–2664.

86 Kawase, T., Seirai, Y., Darabi, H.R., Oda, M., Sarakai, Y., and Tashiro, K. (2003) All-hydrocarbon inclusion complexes of carbon nanorings: cyclic [6]- and [8]paraphenylacetylenes. *Angew. Chem. Int. Ed.*, **42**, 1621–1624.

87 Kawase, T., Tanaka, K., Darabi, H.R., and Oda, M. (2003) Complexation of a carbon nanoring with fullerenes. *Angew. Chem. Int. Ed.*, **42**, 1624–1627.

88 Kawase, T. and Oda, M. (2006) Complexation of carbon nanorings with fullerenes. *Pure Appl. Chem.*, **78**, 831–839.

89 Kawase, T., Tanaka, K., Seirai, Y., Shiono, N., and Oda, M. (2003) Complexation of carbon nanorings with fullerenes. Novel supramolecular dynamics and structural tuning for a fullerene sensor. *Angew. Chem. Int. Ed.*, **42**, 5597–5560.

90 Kawase, T., Fujiwara, N., Tsutsumi, M., Oda, M., Maeda, Y., Wakahara, T., and Akasaka, T. (2004) Supramolecular dynamics of cyclic [6]paraphenyleneacetylene complexes with [60]- and [70]fullerene derivatives: electronic and structural effect to the complexation. *Angew. Chem. Int. Ed.*, **43**, 5060–5062.

91 Kawase, T., Tanaka, K., Shiono, N., Seirai, Y., and Oda, M. (2004) Onion-type complexation based on carbon nanorings and a buckminsterfullerene. *Angew. Chem. Int. Ed.*, **43**, 1722–1724.

92 Kawase, T., Nishiyama, Y., Nakamura, T., Ebi, T., Matsumoto, K., Kurata, H., and Oda, M. (2007) Cyclic [5]paraphenyleneacetylene: synthesis, properties and formation of a ring-in-ring complex showing considerably large association constant and entropy effect. *Angew. Chem. Int. Ed.*, **46**, 1086–1088.

93 Yanase, M., Matsuoka, M., Tatsumi, Y., Suzuki, M., Iwamoto, H., Haino, T., and Fukazawa, Y. (2000). Thermodynamic Study on supramolecular complex formation of fullerene with calyx[5]arenes in organic solvents. *Tetrahedron Lett.*, **41**, 493–497.

94 Zhao, D. and Moore, J.S. (2003) Shape-persistent aryl ethynylene macrocycles: syntheses and supramolecular chemistry. *Chem. Commun.*, 807–818.

95 Hunter, C.A. (1994) Meldola lecture: the role of aromatic interactions in molecular recognition. *Chem. Soc. Rev.*, **23**, 101; Hunter, C.A., Lawson, K.R., Perkins, J., and Urch, C.J. (2001) Aromatic interactions. *J. Chem. Soc. Perkin Trans.*, **2**, 651.

96 Bensasson, R.V., Bienvenue, E., Dellinger, M., Leach, S., and Seta, P. (1994) C_{60} in model biological systems: a visible–UV absorption study of solvent-dependent parameters and solute aggregation. *J. Phys. Chem.*, **98**, 3492–3500.

97 Catalan, J., Saiz, J.L., Laynez, J.L., Jagerovic, N., and Elguero, J. (1995) The colors of C_{60} solutions. *Angew. Chem. Int. Ed. Engl.*, **34**, 105–107.

98 Sun, Y.-P., Bunker, C.E., and Ma, B. (1994) Quantitative studies of ground and excited state charge transfer complexes of fullerenes with *N,N*-dimethylaniline and *N,N*-diethylaniline. *J. Am. Chem. Soc.*, **116**, 9692–9699.

99 Kamieth, M., Klärner, F.-G., and Diederich, F. (1998) Modeling the supramolecular properties of aliphatic-aromatic hydrocarbons with convex–concave topology. *Angew. Chem. Int. Ed.*, **37**, 3303–3306.

4
Cooperative Effects in the Self-Assembly of Fullerene Donor Ensembles

Jean-François Nierengarten

4.1
Introduction

Research focused on the use of C_{60} as the acceptor in covalently bound donor–acceptor pairs has received considerable attention for the preparation of photoactive molecular devices displaying photoinduced energy and/or electron transfer processes [1]. Related examples of fullerene-containing noncovalent systems have also been described [2]. The assembly of the two molecular components by using supramolecular interactions rather than covalent chemistry is indeed particularly attractive since the range of systems that can be investigated is not severely limited by the synthetic route. As part of this research, examples include the self-assembly of C_{60} derivatives bearing an ammonium unit with crown ethers [3] and the apical coordination of C_{60}-pyridine derivatives to metalloporphyrins [4]. Such interactions are, however, weak [3, 4]. Consequently, the binding constants are rather low and only a small fraction of the two components are effectively associated in solution. The stability of the complexes can, however, be dramatically increased when additional recognition elements are present [5]. Such systems are actually perfectly suited to illustrate cooperative effects in the self-assembly of supramolecular systems. Indeed, cooperativity is a major concept for understanding molecular recognition events [6–8]. It is also a key mechanism that regulates the behavior of complex molecular systems in biology [6]. Cooperativity arises from the interplay of two or more interactions, so that the system as a whole behaves differently from expectations based on the properties of the individual interactions acting in isolation [7]. Coupling of interactions can lead to positive or negative cooperativity, depending on whether one interaction favors or disfavors another. The basic issues concerning the definition and assessment of different types of cooperativity have been recently addressed in two articles [7, 8]. Two types of cooperativity, namely, allosteric and chelate cooperativity, have been distinguished. In this chapter, both types of cooperativity will be discussed and illustrated with supramolecular fullerene-containing systems based on the self-assembly of C_{60} derivatives bearing an ammonium unit with crown ethers and of C_{60}-pyridine with metalloporphyrins.

Supramolecular Chemistry of Fullerenes and Carbon Nanotubes, First Edition. Edited by Nazario Martin and Jean-Francois Nierengarten.

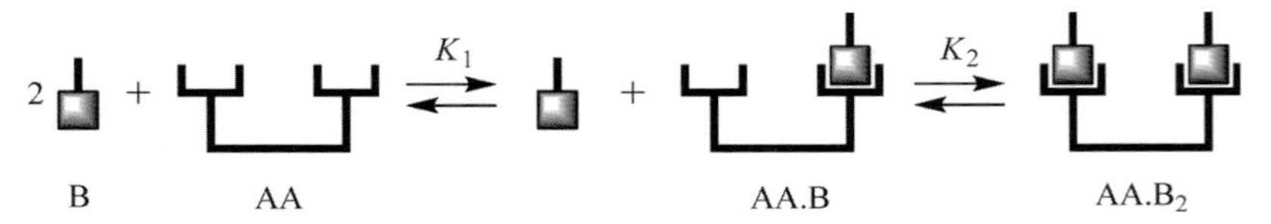

Figure 4.1 Interaction of a monovalent ligand B with a ditopic receptor AA.

4.2 Allosteric Cooperativity

4.2.1 General Principle

Allosteric cooperativity is the type of cooperativity that is best understood, and we will only briefly consider the archetypal case that involves the binding of a monovalent ligand B to a divalent receptor AA (Figure 4.1).

The receptor has three possible states: free AA, partially bound AA·B, and fully bound $AA{\cdot}B_2$. The equilibria are characterized by two association constants K_1 and K_2, which are defined by Eqs. (4.1) and (4.2). Some authors prefer to consider the microscopic association constants K_1' and K_2' that take into account the statistical factor reflecting the degeneracy of the partially bound intermediate [7, 8].

$$AA + B \overset{K_1}{\leftrightarrows} AA \cdot B \qquad K_1 = 2K_1' = \frac{[AA \cdot B]}{[AA] \times [B]} \tag{4.1}$$

$$AA \cdot B + B \overset{K_2}{\leftrightarrows} AA \cdot (B)_2 \qquad K_2 = 1/2\, K_2' = \frac{[AA \cdot (B)_2]}{[B] \times [AA \cdot B]} \tag{4.2}$$

The cooperativity of the system can be conveniently deduced from the interaction parameter α, which is defined by Eq. (4.3).

$$\alpha = \frac{K_2}{K_1} \tag{4.3}$$

In the *absence of cooperativity*, $K_2' = K_1' = K$ and $\alpha = 0.25$. The reference constant K can be evaluated by studying the binding of the monovalent ligand B to a monovalent model of the receptor A, or alternatively, by directly taking the value of the constant K_1' as the reference constant [8]. In the case of *negative cooperativity* ($K_1' > K_2'$ and $\alpha < 0.25$), the intermolecular interaction in the intermediate AA·B is stronger than in the fully bound state $AA{\cdot}B_2$. Formation of the fully bound complex takes place over a wider concentration range than for a corresponding one-site reference receptor [7]. The intermediate AA·B is the dominant species at intermediate concentrations, and in the limit of $\alpha \ll 0.25$, the fully bound state is never populated. Finally, in the case of *positive cooperativity* ($K_1' < K_2'$ and $\alpha > 0.25$), the interactions in the fully bound state are more favorable than in the intermediate. In other words, AA·B is a better supramolecular receptor for B compared to AA.

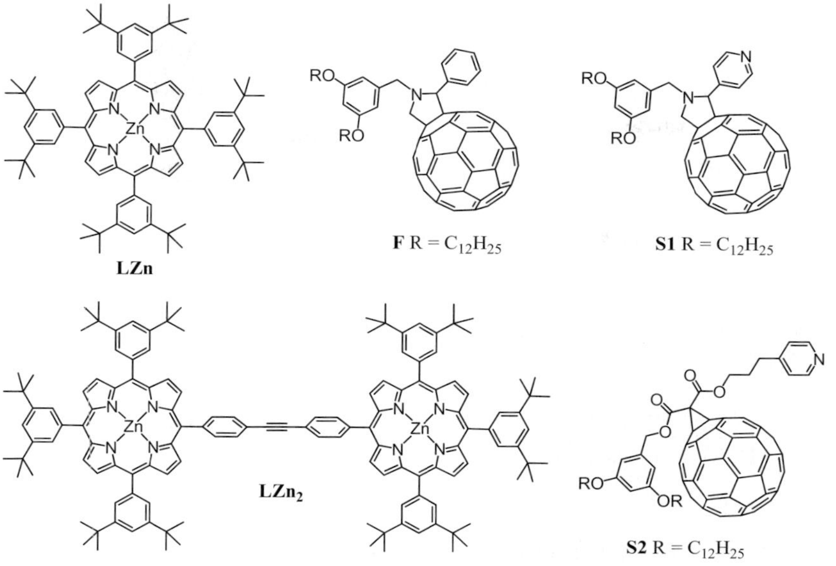

Figure 4.2 Chemical structure of **LZn**, **LZn_2**, **F**, **S1**, and **S2**.

For high α values ($\alpha \gg 0.25$), the intermediate is never populated, and all-or-nothing, two-state behavior is observed [7].

4.2.2
Allosteric Cooperativity in Supramolecular Fullerene Donor Ensembles

Allostreric cooperativity has been evidenced in multicomponent supramolecular ensembles resulting from the self-assembly of fullerene–pyridine substrates with multi-Zn(II)-porphyrin receptors [9]. The different building blocks used in this study are shown in Figure 4.2.

The association of **LZn** with fullerene substrates **S1** and **S2** has been studied in CH_2Cl_2 at 25 °C by UV–vis binding studies. The addition of **S1** or **S2** to **LZn** resulted in bathochromic shift of the Zn(II)-porphyrin absorption bands in agreement with the axial ligation of the pyridyl moieties (Figure 4.3) [4]. Luminescence titrations were also carried out. Indeed, a strong quenching of the porphyrin emission was observed upon addition of the C_{60}-pyridine derivatives to CH_2Cl_2 solutions of **LZn** (Figure 4.3). At this point, it must be emphasized that both intramolecular and intermolecular (collisions and reabsorption events) quenching processes can occur. In order to have a suitable reference, all the investigations on mixtures of **LZn** and **S1** or **S2** have been carried out in parallel with mixtures of the porphyrin receptor and model fullerene derivative **F** unable to form a supramolecular complex with Zn(II)-porphyrins (i.e., a C_{60} derivative with no pyridine unit) [9]. Since a comparison with the reference solution is always done, the intermolecular quenching processes can be ignored, and the difference in emission intensity between the two solutions only accounts for

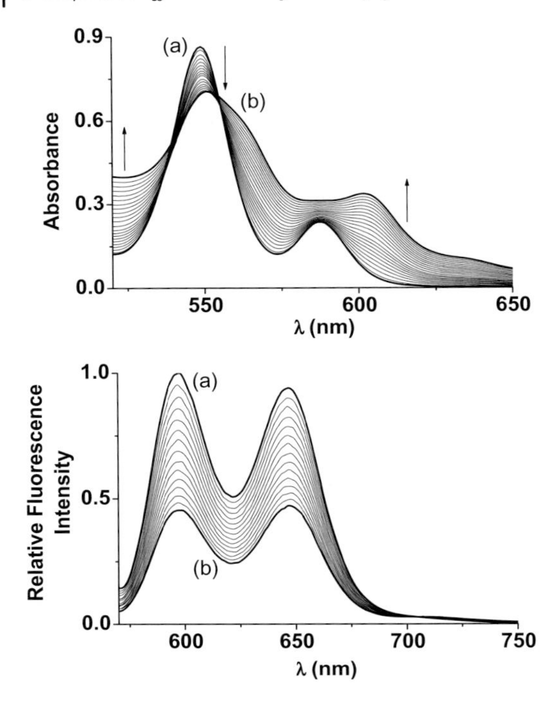

Figure 4.3 (*Top*) UV–visible absorption spectrophotometric titration of **LZn** with **S1**; $l = 0.2$ cm; (a) $[\mathbf{LZn}]_{tot} = 1.85 \times 10^{-4}$ M; (b) $[\mathbf{S1}]_{tot}/[\mathbf{LZn}]_{tot} = 6.8$. (*Bottom*) Luminescence spectrophotometric titration of **LZn** with **S1**; $\lambda_{exc} = 559$ nm; emission and excitation slit widths 15 nm and 20 nm, respectively; (a) $[\mathbf{LZn}]_{tot} = 1.79 \times 10^{-6}$ M; (b) $[\mathbf{S1}]_{tot}/[\mathbf{LZn}]_{tot} = 63$. Solvent: CH_2Cl_2; $T = 25.0(2)$ °C.

intramolecular quenching. The titrations were performed at constant concentrations of porphyrin **LZn**. The spectral changes observed in the emission spectra upon addition of **S1** or **S2** were recorded. The excitation occurred at an isosbestic point at a wavelength where both complexed and uncomplexed species exhibit the same molar absorption coefficient. In any case, the porphyrin emission bands were quenched and redshifted upon successive addition of **S1** or **S2** to **LZn** (Figure 4.3).

Finally, for the sake of comparison and to emphasize the role of the fullerene units in the final conjugates, the binding properties of pyridine (**Py**) were also examined. In all these cases, the titrations allowed the characterization of a single supramolecular complex: [(**LZn**)(**py**)] ($\log K_1 = 3.55(4)$), [(**LZn**)(**S1**)] ($\log K_1 = 3.50(8)$), and [(**LZn**)(**S2**)] ($\log K_1 = 3.66(8)$). The binding constants are not strongly dependent on the nature of the substrate, thus pointing out the absence of additional interactions between the porphyrinic π system and the C_{60} unit within the supermolecules obtained from **S1** and **S2**.

The binding behavior of **Py**, **S1**, and **S2** to bis-metalloporphyrin $\mathbf{LZn_2}$ was also investigated by UV–vis absorption and luminescence in CH_2Cl_2. The results of the

Table 4.1 Stability constants determined for LZn_2 and substrates **S1**, **S2**, and **Py** by UV–vis and luminescence binding studies.[a)]

Substrate	$\log K_1$	$\log K_2$	$\log \beta_2$	K_2/K_1
Py	4.1(1)[b)]	3.4(3)[b)]	7.4(3)[b)]	0.3(1)
	4.1(3)[c)]	3.6(9)[c)]	7.7 (9)[c)]	
S1	—	—	8.4(5)[b)]	3.2(7)
	4.02(2)[c)]	4.5(2)[c)]	8.5(2)[c)]	
S2	—	—	8.6(1)[b)]	10.0(3.3)
	3.7(3)[c)]	4.7(3)[c)]	8.4(3)[c)]	

a) Solvent: CH_2Cl_2; $T = 25.0(2)\,°C$; error $= 3\sigma$.
b) Determined from the UV–visible absorption titration.
c) Determined from the luminescence titration.

thermodynamic studies of $\mathbf{LZn_2}$ with the different substrates are summarized in Table 4.1.

Both UV–vis and luminescence studies of $\mathbf{LZn_2}$ with substrate **Py** evidence the presence of two complexes: $[(\mathbf{LZn_2})(\mathbf{Py})]$ and $[(\mathbf{LZn_2})(\mathbf{Py})_2]$. In contrast, only the global stability constants of the complexes $[(\mathbf{LZn_2})(\mathbf{S1})_2]$ and $[(\mathbf{LZn_2})(\mathbf{S2})_2]$ were deduced from the UV–vis titrations suggesting that the 1 : 1 complexes are minor species under our experimental conditions. This observation is a first indication of positively cooperative interactions in the 1 : 2 associates. Processing of the luminescence data allowed to precisely determine the successive stability constants defined by Eqs. (4.4) and (4.5).

$$\mathbf{LZn_2} + \mathbf{S}n \overset{K_1}{\rightleftarrows} [(\mathbf{LZn_2})(\mathbf{S}n)] \qquad K_1 = \frac{[(\mathbf{LZn_2})(\mathbf{S}n)]}{[(\mathbf{LZn_2}) \times (\mathbf{S}n)]} \tag{4.4}$$

$$[(\mathbf{LZn_2})(\mathbf{S}n)] + \mathbf{S}n \overset{K_2}{\rightleftarrows} [(\mathbf{LZn_2})(\mathbf{S}n)_2] \qquad K_2 = \frac{[(\mathbf{LZn_2})(\mathbf{S}n)_2]}{[(\mathbf{LZn_2})] \times [(\mathbf{LZn_2})(\mathbf{S}n)]} \tag{4.5}$$

A thorough examination of these thermodynamic parameters thus confirmed the strong positive cooperativity for the self-assembly of the 1 : 2 ensembles. For the binding of **S1** and **S2** to $\mathbf{LZn_2}$, the K_2/K_1 values summarized in Table 4.1 are significantly larger than 0.25, which is the value expected for a statistical model and clearly indicates positive intramolecular interactions in the 1 : 2 associates. The observed cooperativity may be ascribed to strong intramolecular fullerene–fullerene interactions between the two guests within $[(\mathbf{LZn_2})(\mathbf{S1})_2]$ and $[(\mathbf{LZn_2})(\mathbf{S2})_2]$ (Figure 4.4). This hypothesis is further supported by the absence of any positive interactions for the 2 : 1 complex obtained from $\mathbf{LZn_2}$ and **Py** for which the K_2/K_1 ratio $\approx 0.3(1)$. It is also important to highlight that the K_2/K_1 ratio is significantly increased for substrate **S2** with respect to **S1**. Indeed, the chain connecting the C_{60} moiety to the pyridine binding unit in **S2** gives a larger degree of flexibility, thus allowing optimized contacts between the two C_{60} spheres in the 1 : 2 complex.

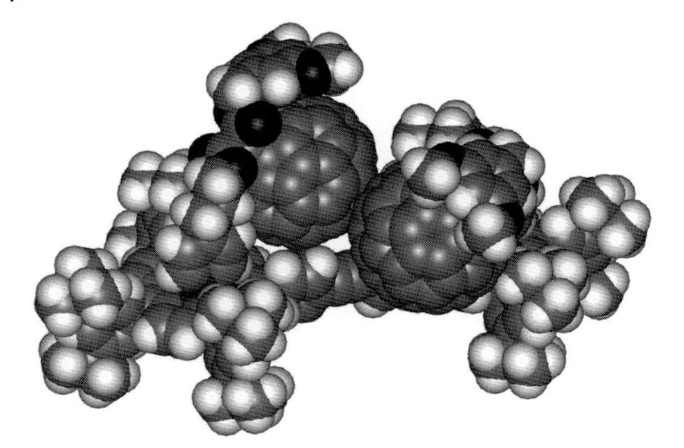

Figure 4.4 Calculated structure of the supramolecular complex [(**LZn$_2$**)(**S2**)$_2$] showing the intramolecular fullerene–fullerene interactions (the dodecyl chains have been replaced by methyl groups in the calculations).

Similar binding studies have also been carried out from hexaporphyrin **LZn$_6$** (Figure 4.5). Upon addition of **Py**, the UV–vis spectra of **LZn$_6$** change substantially and the observed bathochromic shifts of the B and Q bands are in full agreement with the apical coordination of the **Py** substrate onto the Zn(II)-porphyrin moieties of **LZn$_6$**. As shown in Figure 4.6, Job's plot revealed a 1 : 6 stoichiometry for the complex of **Py** with **LZn$_6$**. The UV–vis titration results are also presented in Figure 4.6. Interestingly, clear isosbestic points are observed. This result may reflect that the Zn (II)-porphyrin units in **LZn$_6$** behave as independent binding sites. Indeed, the

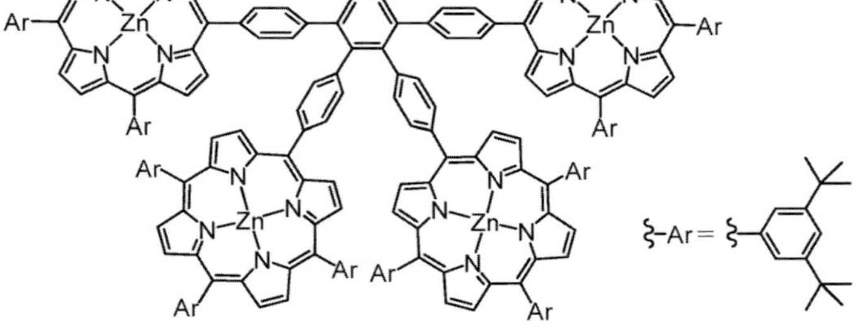

LZn$_6$

Figure 4.5 Chemical structure of **LZn$_6$**.

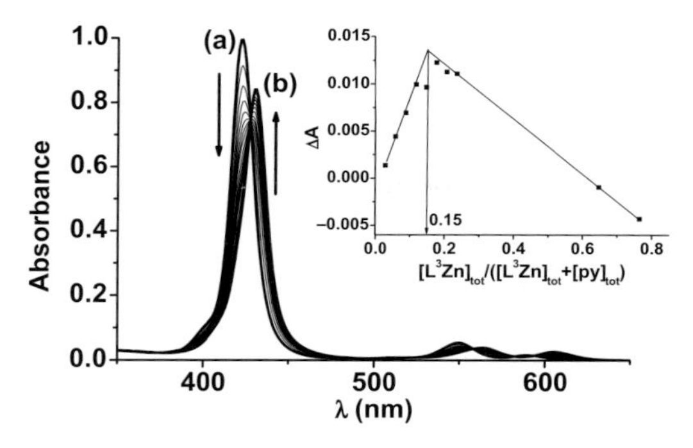

Figure 4.6 UV–vis spectrophotometric titration of $\mathbf{LZn_6}$ with substrate **Py**. $l = 0.5$ cm; (a) $[\mathbf{LZn_6}]_{tot} = 1.17 \times 10^{-6}$ M; (b) $[\mathbf{Py}]_{tot}/[\mathbf{LZn_6}]_{tot} = 735$. Solvent: CH_2Cl_2; $T = 25.0(2)$ °C. *Inset*: Job's plots ($\Delta A/\Delta A_{max}$ at 564 nm) upon mixing $\mathbf{LZn_6}$ and **Py**; $([\mathbf{LZn_6}]_{tot} + [\mathbf{Py}]_{tot}) = 3.0 \times 10^{-5}$ M.

absorption and emission spectra of $\mathbf{LZn_6}$ are similar to those of **LZn** and $\mathbf{LZn_2}$, thus providing further evidence that the porphyrin subunits behave independently in $\mathbf{LZn_6}$.

The same studies were carried out with compounds **S1** and **S2**. In both cases, Job's plot provided evidence for 1 : 6 complex formation and the UV–vis titrations are closely similar to that discussed for compound **Py**. The apparent association constants [10] deduced from the spectrophotometric titrations of $\mathbf{LZn_6}$ with **Py**, **S1**, and **S2** are summarized in Table 4.2.

The complexation of $\mathbf{LZn_6}$ with ligands **Py**, **S1**, and **S2** was further investigated by luminescence experiments. The apparent association constants derived from the emission data (Table 4.2) are higher about one order of magnitude for the fullerene-substituted ligands **S1** and **S2** with respect to those obtained from the UV–vis titrations. These differences have been ascribed to the partial quenching of the emission of free Zn(II)-porphyrin units in $\mathbf{LZn_6}$ upon the binding of the first fullerene ligand. However, the binding studies clearly revealed a significant increase in the binding constants for the fullerene-functionalized substrates relative to **Py**. As discussed for $\mathbf{LZn_2}$ (*vide supra*), the latter observation can be rationalized by the existence of intramolecular π–π interactions between the different fullerene subunits within the complexes resulting from the association of $\mathbf{LZn_6}$ with **S1** or **S2**. This positive cooperative effect is more pronounced for substrate **S2**. As seen for the

Table 4.2 Apparent stability constants ($\log K^*$) determined for $\mathbf{LZn_6}$ and substrates **Py**, **S1**, and **S2** by UV–vis and luminescence binding studies.[a]

Py	S1	S2
3.7(3)[b]	4.1(2)[b]	4.5(5)[b]
4.11(8)[c]	5.0(1)[c]	5.49(9)[c]

a) Solvent: CH_2Cl_2; $T = 25.0(2)$ °C; error = 3σ.
b) Determined from the UV–visible absorption titration.
c) Determined from the luminescence titration.

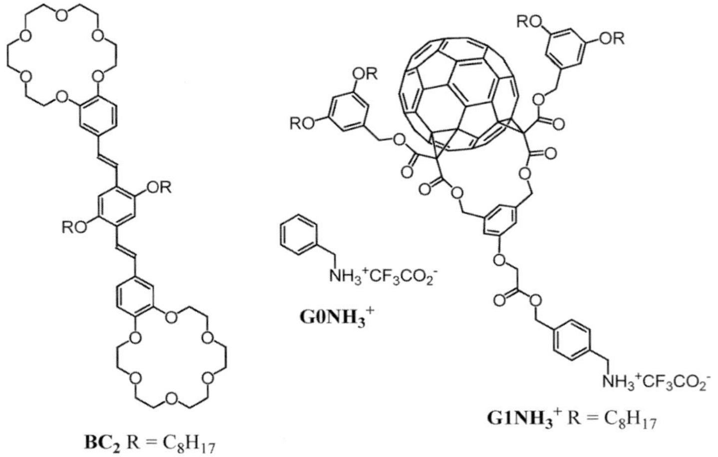

Figure 4.7 Chemical structure of $\mathbf{BC_2}$, $\mathbf{G0NH_3}^+$, and $\mathbf{G1NH_3}^+$.

binding on the ditopic receptor $\mathbf{LZn_2}$, the spacer connecting the C_{60} unit to the pyridine moiety in **S2** brings sufficient flexibility to optimize the fullerene–fullerene interactions.

Similar positive cooperative effects have been observed for supramolecular systems based on the self-assembly of C_{60} derivative $\mathbf{G1NH_3}^+$ bearing an ammonium unit with a ditopic benzocrown ether receptor $\mathbf{BC_2}$ (Figure 4.7) [11]. To quantify the interactions between the bis-benzocrown ether host and the ammonium guest, the complexation between $\mathbf{G1NH_3}^+$ and $\mathbf{BC_2}$ was investigated in CH_2Cl_2 by UV/vis absorption binding studies. For comparison purposes, binding studies were also performed with a reference unsubstituted benzylammonium guest ($\mathbf{G0NH_3}^+$).

The spectral changes occurring upon successive addition of $\mathbf{G(0–1)NH_3}^+$ to a CH_2Cl_2 solution of $\mathbf{BC_2}$ were monitored and the binding constants deduced from the resulting data are summarized in Table 4.3. The processing of the spectrophotometric data led to the characterization of both 1 : 1 and 2 : 1 supramolecular edifices and the two successive binding constants defined by Eqs. (4.6) and (4.7) were determined.

$$K_1 = \frac{[(\mathbf{BC_2})(\mathbf{G}n\mathbf{NH_3}^+)]}{[(\mathbf{BC_2})] \times [(\mathbf{G}n\mathbf{NH_3}^+)]} \tag{4.6}$$

$$K_2 = \frac{[(\mathbf{BC_2})(\mathbf{G}n\mathbf{NH_3}^+)_2]}{[(\mathbf{BC_2})] \times [(\mathbf{BC_2})(\mathbf{G}n\mathbf{NH_3}^+)]} \tag{4.7}$$

Due to the rather weak complexation-induced absorption changes upon addition of the ammonium guests to the solutions of $\mathbf{BC_2}$, the binding constants were obtained with high errors. Thus, the authors further investigated the binding of $\mathbf{G1NH_3}^+$ to

Table 4.3 Stability constants determined by the UV–vis and luminescence binding studies.[a]

	BC_2		
	$\log K_1$	$\log K_2$	K_2/K_1
$G0NH_3^+$	4.5(9)[b]	3.4(1.8)[b]	0.08(0.12)
	nd	nd	
$G1NH_3^+$	5.6(8)[b]	6.5(2)[b]	4.0(1.2)
	5.0(1)[c]	5.6(1)[c]	
$G2NH_3^+$	5.8(6)[b]	6.7(8)[b]	9(3)
	5.33(1)[c]	6.3(1)[c]	
$G3NH_3^+$	nd	$\log\beta_2 = 12.6(9)$[b]	16(4)
	5.28(7)[c]	6.48(7)[c]	

a) All the measurements have been carried out in CH_2Cl_2 at $25 \pm 0.2\,°C$. The errors correspond to standard deviations given as 3σ. nd: not determined.
b) Determined from the UV–visible absorption titration.
c) Determined from the indirect luminescence titration.

BC_2 by luminescence studies. Indeed, the strong emission of the π-conjugated system of **BC_2** ($\lambda_{em} = 450$ nm, $\phi = 0.65 \pm 0.06$) is dramatically quenched upon binding of the fullerene-containing ammonium derivative **$G1NH_3^+$**. The association constants determined from the emission binding studies are summarized in Table 4.3. The log *K* values determined from the luminescence titrations are lower with respect to those obtained from the spectrophotometric titrations (Table 4.3). These differences could originate from the high errors of the log *K* values deduced from the UV–vis binding studies.

For the 2 : 1 complex obtained from **BC_2** and the model ammonium derivative **$G0NH_3^+$**, the K_2/K_1 ratio of 0.08(0.12) shows a negative cooperative effect. The latter may be ascribed to the electrostatic repulsion for the binding of two positively charged guests. In contrast, the K_2/K_1 ratio is significantly larger than 0.25 in the case of **$G1NH_3^+$**. The latter observation clearly indicates that the stability of the supramolecular complex [(**$G1NH_3^+$**)$_2$(**BC_2**)] is significantly higher than that of its analogue [(**$G1NH_3^+$**)(**BC_2**)] despite the electrostatic repulsion. The thermodynamic data suggest a structure in which the two C_{60} units of [(**$G1NH_3^+$**)$_2$(**BC_2**)] strongly interact through π–π stacking interactions as observed in the calculated structure of the 2 : 1 supramolecular complex depicted in Figure 4.8.

The positive cooperative effect evidenced for the self-assembly of the 2 : 1 complex [(**$G1NH_3^+$**)(**BC_2**)] prompted the same authors to increase the number of C_{60} units attached on the ammonium building block in order to generate additional possible intramolecular interactions between the two guests in the 2 : 1 assembly (Figure 4.9) [12].

For the 2 : 1 noncovalent arrays obtained from **$G2NH_3^+$** and **$G3NH_3^+$**, the K_2/K_1 ratio (Table 4.3) are significantly larger than 0.25 clearly indicating positive intramolecular interactions in the 2 : 1 assemblies. The latest observation may be

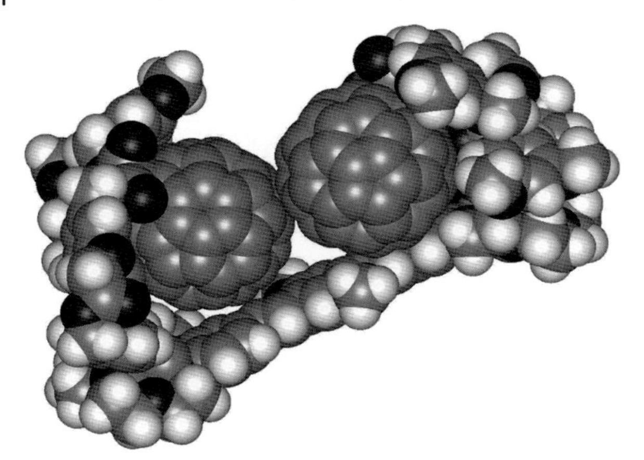

Figure 4.8 Calculated structure of the supramolecular complex [(**BC**$_2$)(**G1NH**$_3$$^+$)$_2$] showing the intramolecular fullerene–fullerene interactions (the dodecyl and octyl chains have been replaced by methyl groups in the calculations).

again ascribed to strong intramolecular fullerene–fullerene interactions between the two guests within the 2 : 1 supramolecular assembly. This hypothesis is also supported by the absence of any positive interactions for the 2 : 1 complex obtained from **BC**$_2$ and an ammonium derivative lacking the fullerene subunits (**G0NH**$_3$$^+$). Finally, it is also important to highlight that the K_2/K_1 ratio is significantly increased when the size of the dendritic branches is increased. In other words, the positive cooperativity is more and more effective when the number of C_{60} units is increased. This positive dendritic effect further confirms that intramolecular fullerene–fullerene interactions must be at the origin of the observed positive cooperative effect.

4.3 Chelate Cooperativity

4.3.1 General Principle

While allosteric cooperativity is well recognized, the assessment of chelate cooperativity is a more complicated issue [7, 8]. A detailed theoretical analysis that consistently assesses chelate cooperativity is far beyond the scope of this chapter in which cooperativity is described as a tool to obtain stable supramolecular assemblies. Thus, only a simplified description will be given. A detailed discussion on chelate cooperativity can be found in a recent article by Ercolani and Schiaffino [8].

To describe chelate cooperativity, we will consider only the simple case involving two molecules having two binding sites as illustrated in Figure 4.10. This system is

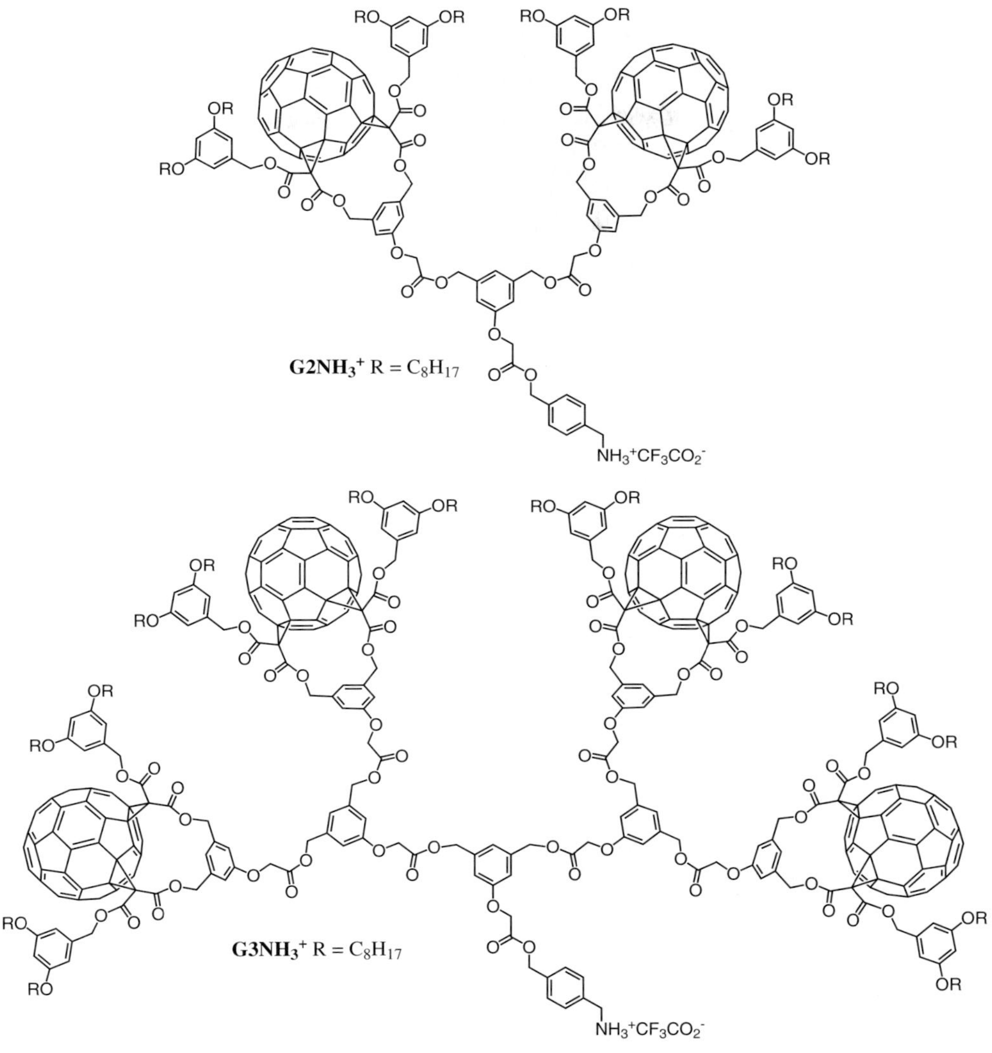

Figure 4.9 Chemical structure of $\mathbf{G2NH_3^+}$ and $\mathbf{G3NH_3^+}$.

more complicated than the allosteric system in Figure 4.1 because there are more possible bound states. However, if the ligand is present in large excess relative to the receptor, then complexes involving more than one receptor can be ignored because they will not be significantly populated. Under these conditions, there are only four states for the receptor (Figure 4.10c): free AA, two 1 : 1 complexes (the partially bound open intermediate o-AA·BB and the fully bound cyclic complex c-AA·BB), and the 2 : 1 complex $(AA)_2.BB_2$.

(a)

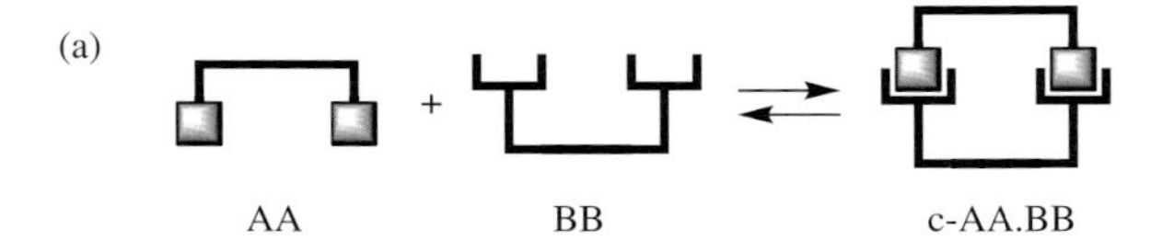

(b)

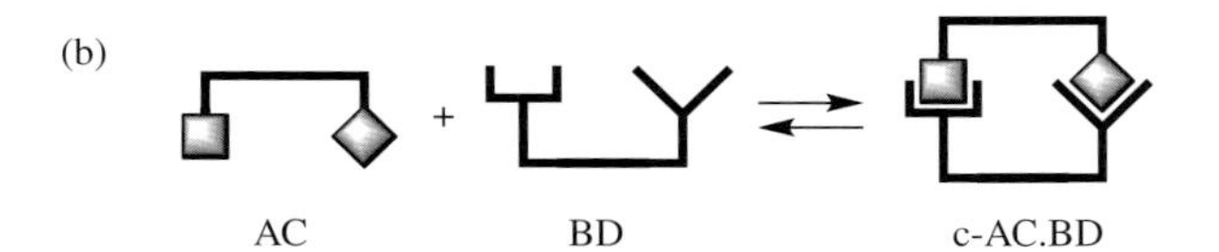

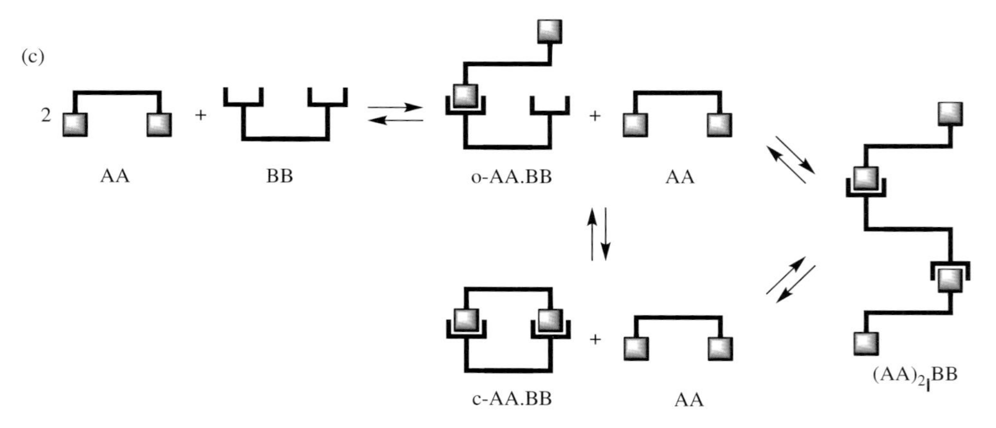

Figure 4.10 (a) Binding of a divalent ligand AA to a divalent receptor BB. (b) Binding of a divalent asymmetric ligand AC to a complementary receptor BD. (c) When the ligand AA is in large excess compared to the receptor BB ($[AA]_0 \gg [BB]_0$), complexes involving more than one receptor molecules can be neglected and only four states are possible for the ditopic receptor.

At the molecular level, the key feature that defines the properties of this system is the intramolecular binding interaction that leads to the cyclic 1 : 1 complex c-AA·BB. In the case of *negative cooperativity*, the partially bound intermediate is more stable than the cyclic complex. The system is unaffected by the presence of the cyclic complex, and the behavior is identical to that found for monovalent ligands [7, 8]. In contrast, in the case of *positive cooperativity*, the cyclic complex is more stable than the partially bound intermediate, and c-AA·BB is the major species over a wide concentration range. The open partially bound intermediate o-AA·BB is barely populated, and formation of the 2 : 1 complex AA·$(BB)_2$ is suppressed compared to the situation with the corresponding monovalent ligands. In other words, the cooperative assembly of the complex is driven by the difference in strength between the intermolecular and the intramolecular interactions, and is a consequence of the molecular architecture [7, 8].

4.3.2
Binding of a Divalent Ligand AA to a Divalent Receptor BB

Chelate cooperativity has been used to produce stable supramolecular complex from the bis-porphyrin receptor $\mathbf{LZn_2}$ and the bis-pyridine fullerene ligand $\mathbf{S_3}$ [13]. Owing to the perfect complementarity of the two components, the bis-pyridine substrate $\mathbf{S_3}$ can be clicked on the ditopic porphyrin derivative $\mathbf{LZn_2}$, thus leading to a noncovalent macrocyclic 1 : 1 complex (Figure 4.11).

The ability of bis-porphyrin $\mathbf{LZn_2}$ to form a supramolecular complex with bis-pyridine $\mathbf{S_3}$ was first evidenced by ^{1}H-NMR binding studies performed at 298 K in $CDCl_3$. Complexation-induced changes in chemical shifts were observed upon addition of 1 equiv. of **LZn2** to a solution of $\mathbf{S_3}$. In particular, dramatic upfield shifts were seen for the chemical shift of the protons belonging to the pyridine moieties ($\Delta\delta > 4.5$ ppm). When coordination on the Zn(II) ions took place, the pyridine moieties of $\mathbf{S_3}$ are located atop the porphyrin cores and the important shift is a result of the ring current effect of the porphyrin macrocycles. Importantly, changes in the chemical shift were also evidenced for all the protons of the spacer between the two pyridine units of **2** ($\Delta\delta = 0.2$–0.6 ppm). The latter observation suggests that all these protons must be close to the π-conjugated system of $\mathbf{LZn_2}$ in the associate that is in good agreement with the formation of a macrocyclic 1 : 1 supramolecular complex

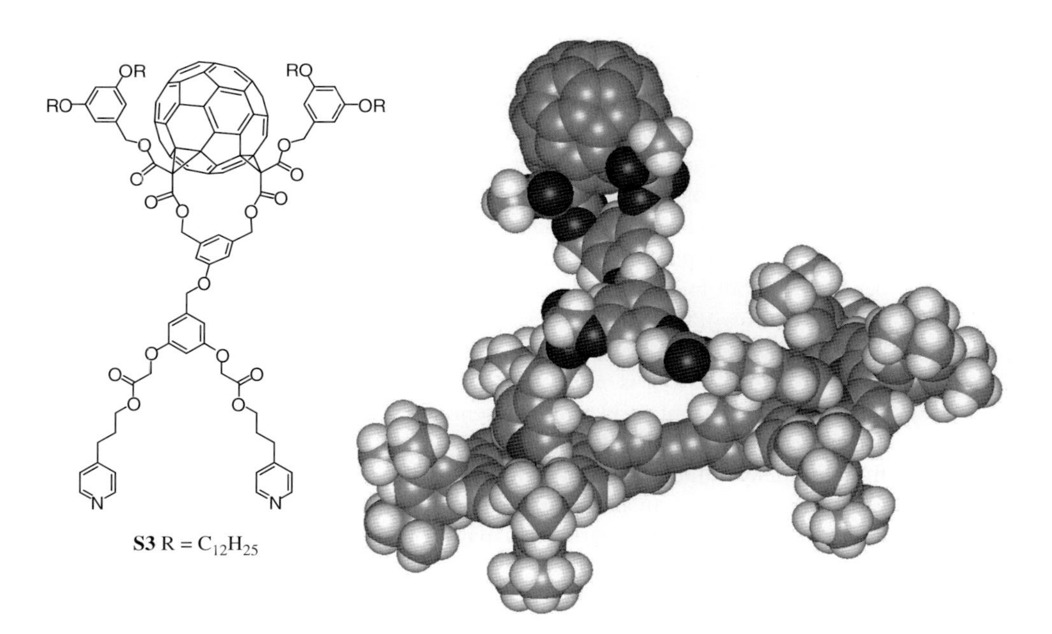

Figure 4.11 Chemical structure of **S3** and calculated structure of the supramolecular macrocyclic complex [($\mathbf{LZn_2}$) (**S3**)] (the dodecyl chains have been replaced by methyl groups in the calculations).

(Figure 4.11). To quantify the interactions between **LZn_2** and **S_3**, the complexation was studied in CH_2Cl_2 by UV–vis and fluorescence binding studies. The processing of the absorption and luminescence data revealed the formation of a single species, the macrocyclic complex, with a binding constant $\log K_1 = 5.05 \pm 0.09$. When compared to the supramolecular complex obtained from **LZn_1** and **S_2** (Figure 4.2, $\log K_1 = 3.78 \pm 0.02$), the increased stability of the macrocyclic complex [(**LZn_2**)(**S_3**)] may result from the simultaneous coordination of the two Zn centers of **LZn_2** by the two pyridine moieties of **S_3**. In other words, chelate cooperativity is at the origin of the high stability of the noncovalent macrocyclic complex [(**LZn_2**)(**S_3**)].

Aida and coworkers have reported the preparation of fullerene-rich dendritic structures resulting from the apical coordination of C_{60} derivatives bearing two pyridyl moieties to dendritic molecules appended with multiple Zn(II) porphyrin units [14]. For example, compound **LZn_{24}** bound **S_4** strongly to form stable [(**LZn_{24}**)(**S_4**)$_{12}$] (Figure 4.12). Upon titration with **S_4** in $CHCl_3$ at 25 °C, **LZn_{24}** displayed a large spectral change in the Soret and Q bands, characteristic of the axial coordination of zinc porphyrins, with a clear saturation profile at a molar ratio **S_4/LZn_{24}** exceeding 12. The average binding affinity (K), as estimated by simply assuming a one-to-one coordination between the individual zinc porphyrin and the pyridine units, is $1.2 \times 10^6\ M^{-1}$. This value is more than two orders of magnitude greater than association constants reported for monodentate coordination between zinc porphyrins and pyridine derivatives. The sizeable increase in stability can be ascribed to the simultaneous coordination of two Zn centers of **LZn_{24}** by the two pyridine moieties of **S_4** as in the case of the supramolecular system resulting from the axial coordination of the bis-Zn(II)-porphyrinic receptor **LZn_2** to substrate **S_3** bearing two pyridine subunits.

Supramolecular assembly [(**LZn_{24}**)(**S_4**)$_{12}$] combining C_{60} units and porphyrin moieties is also a photochemical molecular device. Indeed, the photophysical properties of this system have been studied in detail and an almost quantitative intramolecular photoinduced electron transfer from the photoexcited porphyrins to the C_{60} units evidenced by means of steady-state emission spectroscopy and nanosecond flash photolysis measurements. Excited state dynamic studies have been carried out to investigate both charge separation and charge recombination events in [(**LZn_{24}**)(**S_4**)$_{12}$]. The charge separation rate constants (k_{CS}) and the charge recombination rate constants (k_{CR}) have been thus deduced. Importantly, the k_{CS}/k_{CR} ratio for [(**LZn_{24}**)(**S_4**)$_{12}$] is more than an order of magnitude greater than those reported for precedent porphyrin–fullerene supramolecular dyads and triads [15]. It is obvious that a larger number of the fullerene units in [(**LZn_{24}**)(**S_4**)$_{12}$] can enhance the probability of the electron transfer from the zinc porphyrin units. However, in addition to this, one can also presume that an efficient energy migration along the densely packed Zn(II) porphyrin array may enhance the opportunity for this electron transfer.

The preparation of a stable macrocyclic noncovalent array has also been described from a bis-crown ether receptor **BC_2** and a complementary bis-ammonium fullerene ligand (**$L(NH_3^+)_2$**, Figure 4.13) [16].

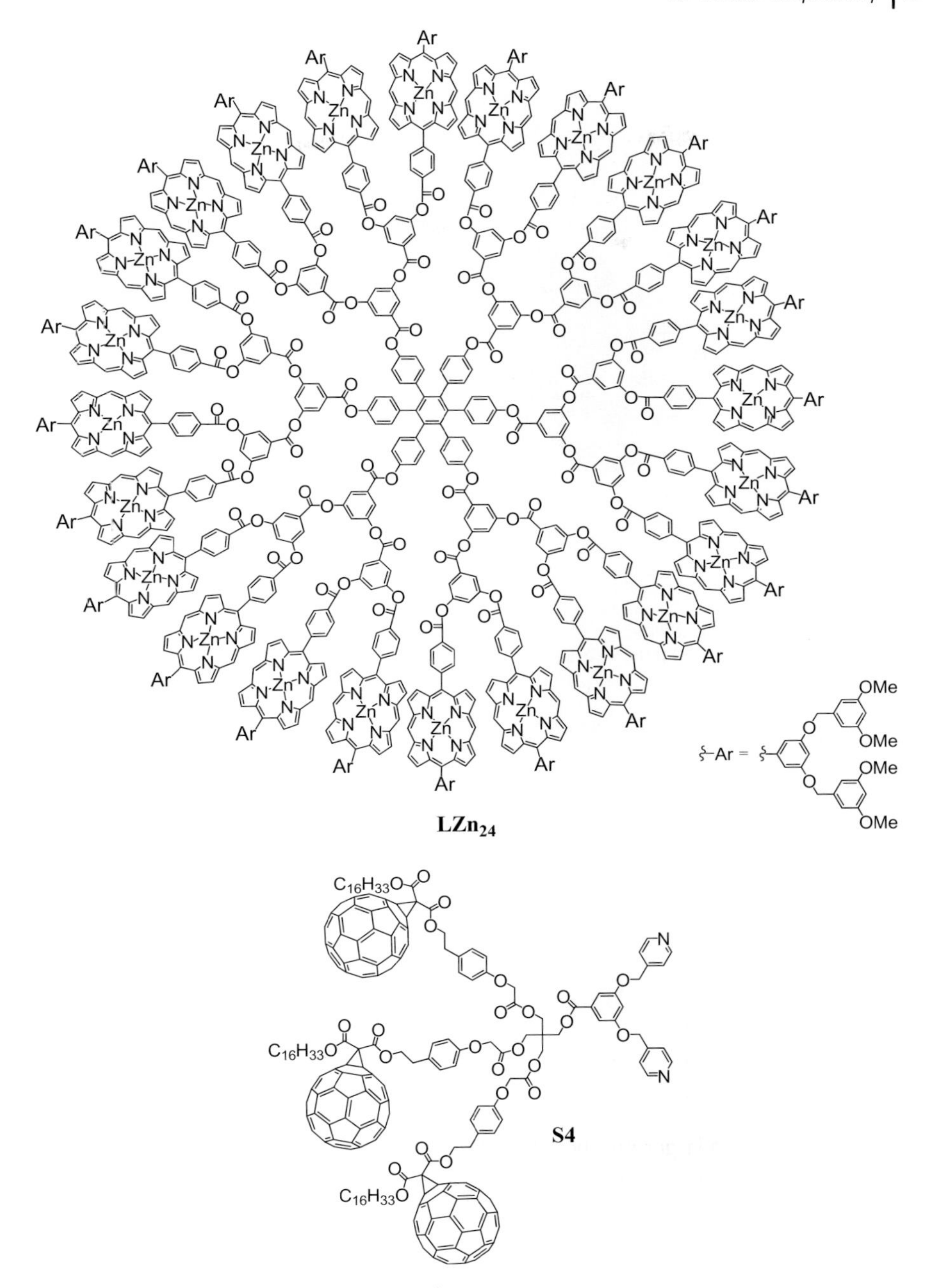

Figure 4.12 Chemical structure of **LZn$_{24}$** and **S4**.

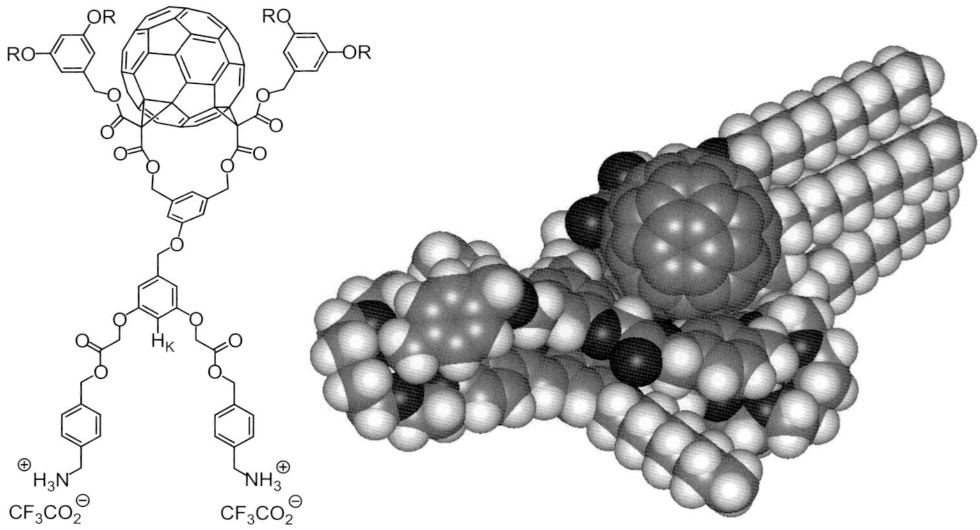

$L(NH_3^+)_2$ R = $C_{12}H_{25}$

Figure 4.13 Chemical structure of **L(NH$_3^+$)$_2$** and calculated structure of the supramolecular macrocyclic complex [(**L(NH$_3^+$)$_2$**)(**BC$_2$**)].

The binding behavior of bis-ammonium **L(NH$_3^+$)$_2$** to the bis-crown ether receptor **BC$_2$** was first investigated by ^{1}H NMR in $CDCl_3$ at 298 K. As depicted in Figure 4.14, the comparison between the ^{1}H NMR spectra of **L(NH$_3^+$)$_2$**, **BC$_2$**, and an equimolar mixture of both components revealed complexation-induced changes in chemical shifts. In particular, a dramatic downfield shift is seen for the signal of H_K in

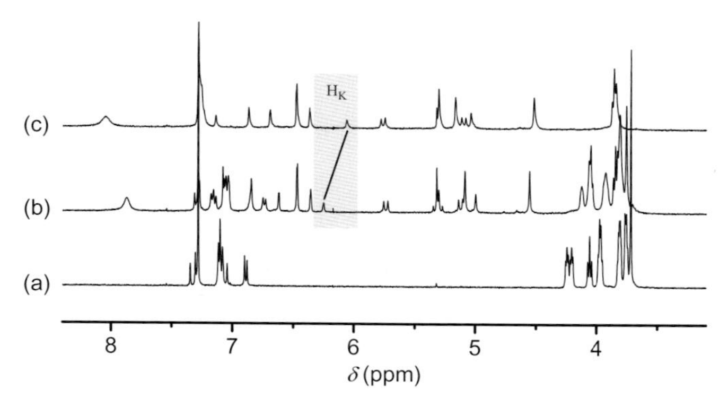

Figure 4.14 ^{1}H NMR (400 MHz, 25 °C) of equimolar $CDCl_3$ solutions of **BC$_2$** (a), **L(NH$_3^+$)$_2$** (c) and a 1 : 1 mixture of **BC$_2$** and **L(NH$_3^+$)$_2$** (b); the downfield shift for the signal of proton H_K (see Figure 4.13) upon addition of **BC$_2$** shows that this proton must be close to the π-conjugated system of **BC$_2$** in the supramolecular complex, which is in good agreement with the proposed structure.

$\mathbf{L(NH_3^+)_2}$ upon addition of $\mathbf{BC_2}$. The latter observation is in good agreement with the formation of a macrocyclic 1 : 1 supramolecular complex in which proton H_K is located on top of the π-conjugated system of $\mathbf{BC_2}$.

The formation of a 1 : 1 macrocyclic complex was further evidenced by electrospray mass spectrometry (ESMS). The positive ES mass spectrum recorded under mild conditions (extracting cone voltage $V_c = 200$ V) from a 1 : 1 mixture of $\mathbf{L(NH_3^+)_2}$ and $\mathbf{BC_2}$ is dominated by a doubly charged ion peak at $m/z = 1723.1$ that can be assigned to the 1 : 1 complex after loss of the trifluoroacetate counteranions (calculated $m/z = 1723.11$). The complexation was further investigated in CH_2Cl_2 by UV/Vis binding studies. The processing of these data allowed the characterization of a single supramolecular complex $[(\mathbf{L(NH_3^+)_2})(\mathbf{BC_2})]$ in solution ($\log K_1 = 6.3 \pm 0.4$), which is in excellent agreement with the ESMS and NMR data. The comparison with thermodynamic data available in the literature for closely related systems (crown ether and ammonium derivatives) [3, 17] shows a coordination stronger by more than three orders of magnitude. This is mainly associated with the two-center host–guest topography, in other words, to chelate cooperativity.

4.3.3 Binding of a Divalent Asymmetric Ligand AC to a Complementary Receptor BD

Supramolecular C_{60}–porphyrin conjugates have been mainly obtained from C_{60} derivatives bearing a pyridyl moiety and metalloporphyrins through coordination to the metal ion [18]. Owing to the apical binding on the porphyrin subunit, the attractive van der Waals interaction of the fullerene sphere with the planar π surface of the porphyrin seen for several covalent C_{60}–porphyrin derivatives [19] or in the solid-state structures of porphyrin/fullerene cocrystals [20] is not possible in such arrays. Interestingly, during the studies of the noncovalent C_{60}–porphyrin ensemble obtained from methanofullerene derivative $\mathbf{LNH_3^+}$ bearing an ammonium function and porphyrin–crown ether conjugate **PBC** (Figure 4.15), it was found that π stacking interactions between the two chromophores can have a dramatic effect on the recognition interactions [21].

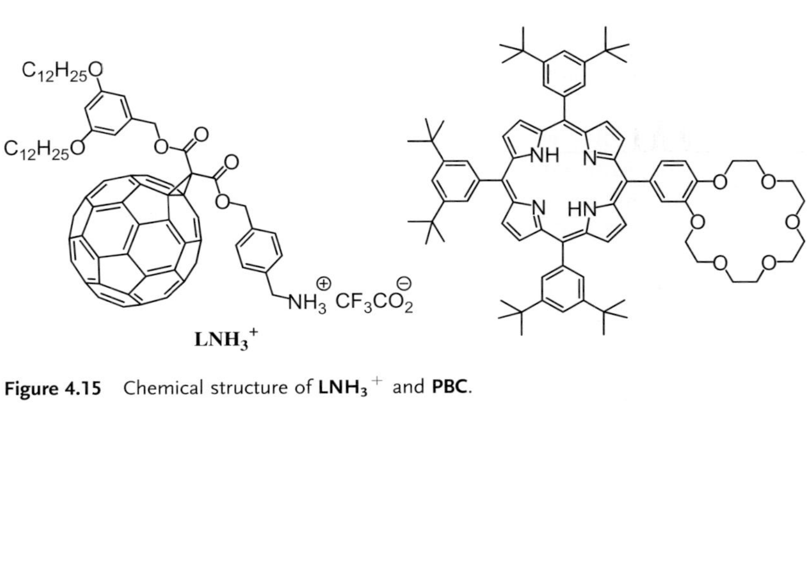

Figure 4.15 Chemical structure of $\mathbf{LNH_3^+}$ and **PBC**.

The ability of porphyrin **PBC** to form a supramolecular complex with ammonium $\mathbf{LNH_3}^+$ was first evidenced by ^{1}H-NMR binding studies performed at 298 K in $CDCl_3$. As depicted in Figure 4.16, the comparison between the ^{1}H-NMR spectra of $\mathbf{LNH_3}^+$, **PBC**, and mixtures of both components revealed new sets of signals and complexation-induced changes in chemical shifts. This is particularly visible for the pyrrolic protons of the porphyrin moiety and the signals arising from the crown ether subunit. The latter observations suggest the existence of two conformers for the supramolecular complex obtained upon association of the two components, one being in fast exchange (**A**) with uncomplexed $\mathbf{LNH_3}^+$ and **PBC** on the NMR timescale, the other one (**B**) in slow exchange as shown in Figure 4.16.

On the one hand, the initial ammonium–crown ether association leading to conformer **A** is responsible for the observed complexation-induced changes in chemical shifts upon addition of ammonium $\mathbf{LNH_3}^+$ to solutions of **PBC**. On the other hand, the new sets of signals seen in the ^{1}H-NMR spectra correspond to conformer **B**. The dramatic upfield shift observed for the pyrrolic protons in **B** must be the result of the close proximity of the fullerene sphere suggesting that supramolecule [($\mathbf{LNH_3}^+$)(**PBC**)] adopts a conformation in which the C_{60} subunit is located atop the porphyrin macrocycle. Further evidence for C_{60}–porphyrin interactions came from UV–vis measurements. Addition of increasing amount of $\mathbf{LNH_3}^+$ to a CH_2Cl_2 solution of **PBC** causes a redshift of the Soret band, no further spectral changes being observed beyond the addition of ~5 equiv. of **1** ($\lambda_{max} = 421$ and 427 nm before and after addition of $\mathbf{LNH_3}^+$, respectively). The latter observation confirmed the existence of the C_{60}–porphyrin interactions in [($\mathbf{LNH_3}^+$)(**PBC**)]. Effectively, redshifts in the Soret band have been observed for covalent C_{60}–porphyrin conjugates due to intramolecular π stacking of the two chromophores [22]. Finally, the formation of supramolecular complex [($\mathbf{LNH_3}^+$)(**PBC**)] was also evidenced in the gas phase by electrospray mass spectrometry (ES-MS) [23]. The spectrum obtained from an equimolar mixture of $\mathbf{LNH_3}^+$ and **PBC** displayed only one peak at $m/z = 2587.2$ corresponding to the 1 : 1 complex after loss of the trifluoroacetate counteranion (calculated $m/z = 2587.24$). The K_a value for the binding of $\mathbf{LNH_3}^+$ to **PBC** was determined by a fluorescence titration. A surprisingly high value for the association constant $K_a = 3.75 \times 10^5\ M^{-1}$ was obtained. Effectively, the association constant between porphyrin **PBC** and fullerene $\mathbf{LNH_3}^+$ is unexpectedly increased by two orders of magnitude compared to the K_a value found for the complexation of $\mathbf{LNH_3}^+$ with commercially available benzo-18-crown-6 ($K_a = 2.1 \times 10^3\ M^{-1}$, determined in $CDCl_3$ by NMR titration). Such a stabilization of the supramolecular complex formed between $\mathbf{LNH_3}^+$ and **PBC**, which can be attributed to an additional intramolecular interaction, provides further evidence for the π stacking of the fullerene moiety and the porphyrin subunit in [($\mathbf{LNH_3}^+$)(**PBC**)] suggested by the NMR studies. As depicted in Figure 4.17, the calculated structure of supramolecular complex [($\mathbf{LNH_3}^+$)(**PBC**)] reveals that the linker between the ammonium group and the fullerene sphere has the perfect length to allow intramolecular stacking of the fullerene moiety and the porphyrin subunit in full agreement with the experimental observations. Similar findings have also been reported by Martin and coworkers for the binding of $\mathbf{LNH_3}^+$ to an extended tetrathiafulvalene derivative bearing a crown ether subunit [24].

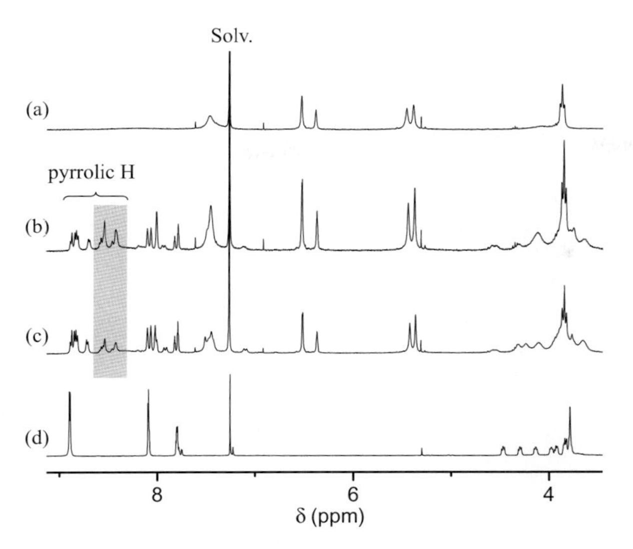

LNH_3^+ + PBC

Fast

A

$CF_3CO_2^\ominus$

Slow

B

$CF_3CO_2^\ominus$

Figure 4.16 (*Top*) ^{1}H-NMR spectra ($CDCl_3$, 300 MHz, 298 K) of **LNH_3^+** (a), **PBC** (d), a 2 : 1 mixture of **LNH_3^+** and **PBC** (b), and a 1 : 1 mixture of **LNH_3^+** and **PBC** (c); the signals of the pyrrolic protons of **B** are highlighted. (*Bottom*) Schematic representation of the two conformers of the supramolecular complex obtained upon association of **LNH_3^+** and **PBC**. **A** is in fast exchange with uncomplexed **LNH_3^+** and **PBC** on the NMR timescale and **B** in slow exchange.

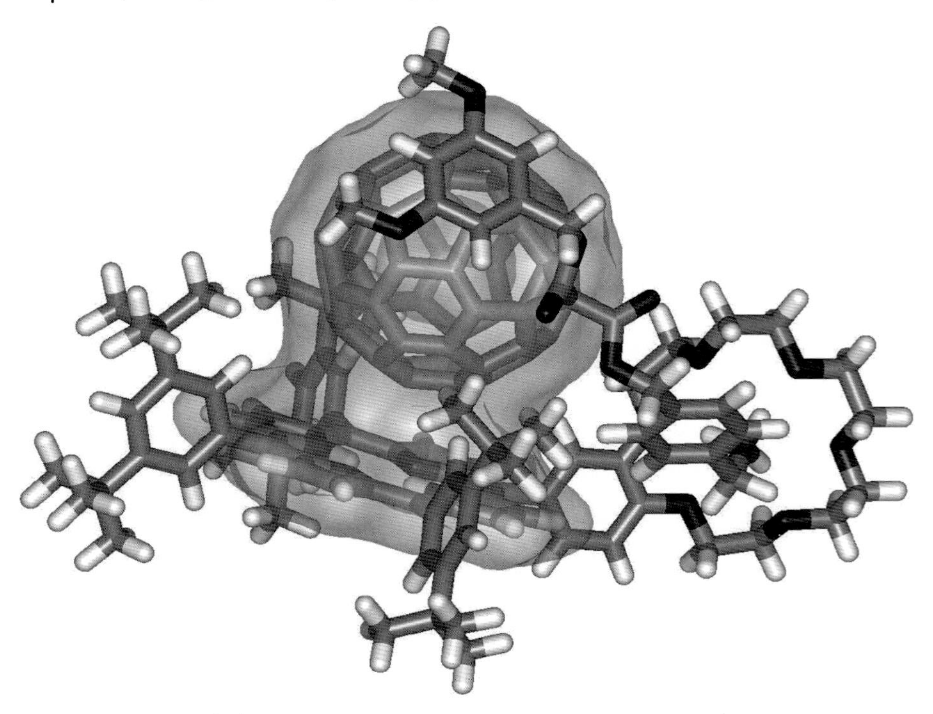

Figure 4.17 Calculated structure of the supramolecular complex [(**LNH_3^+**)(**PBC**)] (molecular modeling performed with *Spartan*, the dodecyl chains have been replaced by methyl groups in the calculations).

Two-point bound porphyrin–fullerene conjugates via simultaneous axial coordination and cation–crown ether complexation have been reported by D'Souza and coworkers [25]. As a typical example, the supramolecular complex obtained from **$SPyNH_3^+$** and **ZnPCB** is depicted in Figure 4.18. The magnitude of the binding constant ($4.48 \times 10^4\,M^{-1}$) determined by luminescence binding studies revealed a stable self-assembly in polar benzonitrile. It is worth to point out that supramolecular metalloporphyrin–fullerene dyads are usually not obtained in this solvent as benzonitrile is also able to form apical complexes with Zn(II)-porphyrins. The latter observation nicely highlights the contribution of the second interaction to the overall stability of the self-assembled dyad.

4.4
Conclusions

Fullerene-containing supramolecular acceptor–donor systems have been selected to illustrate the two distinct types of cooperativity that determine the speciation in supramolecular and biological systems: allosteric and chelate cooperativity.

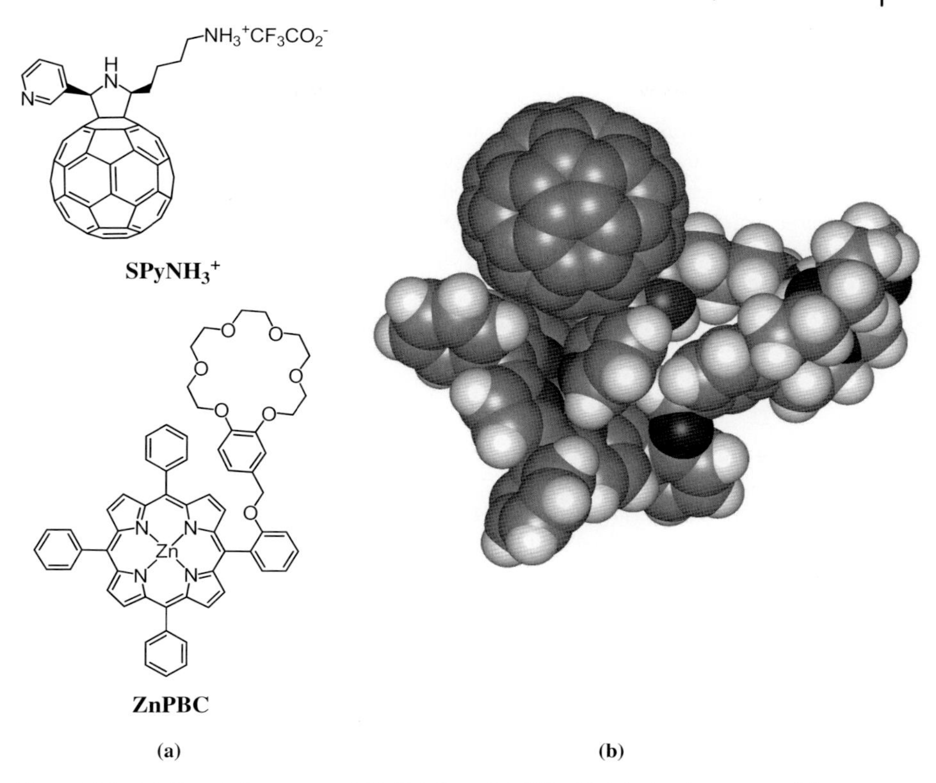

Figure 4.18 (a) Chemical structure of **$SPyNH_3^+$** and **ZnPCB**. (b) Calculated structure of the supramolecular macrocyclic complex [(**$SPyNH_3^+$**)(**ZnPCB**)].

Such effects are important for the design of stable supramolecular ensembles. In the case of allosteric cooperativity, the examples discussed in this chapter demonstrate clearly that increasing the number of building blocks in supramolecular architectures can increase their stability due to an increasing number of possible intramolecular secondary interactions (π–π stacking and hydrophobic interactions). On the other hand, stable macrocyclic supramolecular ensembles can be obtained by taking profit of chelate cooperativity. Both approaches are modular and appear easily applicable to a wide range of functional groups for the preparation of new stable supramolecular architectures with tunable structural and electronic properties.

4.5 Experimental Details

The UV/Vis and luminescence binding studies have been carried out under standard conditions [9]. As typical examples, the experimental procedures used to determine the *K* values reported in Tables 4.1 and 4.2 are described in this section.

4.5.1
General

All the binding studies were carried out with spectroscopic grade dichloromethane (E. Merck, 99.9% for spectroscopy). To prevent any photochemical degradation, all the solutions were protected from daylight exposure. All stock solutions were prepared using an AG 245 Mettler Toledo analytical balance (precision 0.01 mg), and the complete dissolution in dichloromethane was achieved using an ultrasonic bath. The concentrations of the stock solutions of the receptors and the substrates ($\approx 10^{-4}$ M) were calculated by quantitative dissolution of solid samples in dichloromethane.

4.5.2
UV–Visible Titrations

The spectrophotometric titrations of **LZn**, **LZn$_2$**, and **LZn$_6$** with **S1** ($[\mathbf{LZn}]_{tot} = 1.84 \times 10^{-4}$ M; $[\mathbf{LZn_2}]_{tot} = 5.65 \times 10^{-5}$ M and $[\mathbf{LZn_6}]_{tot} = 2.93 \times 10^{-6}$ M) and **S2** ($[\mathbf{LZn}]_{tot} = 5.82 \times 10^{-5}$ M; $[\mathbf{LZn_2}]_{tot} = 5.65 \times 10^{-5}$ M and $[\mathbf{LZn_6}]_{tot} = 3.04 \times 10^{-6}$ M) were carried out in a Hellma quartz optical cell ($l = 0.2$ cm). To evaluate the influence of the fullerene units on the binding constants, spectrophotometric titrations of the same receptors with **Py** were conducted under similar experimental conditions ($[\mathbf{LZn}]_{tot} = 1.79 \times 10^{-6}$ M, $l = 1$ cm; $[\mathbf{LZn_2}]_{tot} = 1.13 \times 10^{-6}$ M, $l = 1$cm; $[\mathbf{LZn_6}]_{tot} = 1.17 \times 10^{-6}$ M, $l = 0.5$ cm). Microvolumes of a concentrated solution of **S1**, **S2**, or **Py** were added to 2, 1, or 0.4 mL of **LZn**, **LZn$_2$**, or **LZn$_6$** with microliter Hamilton syringes (#710 and #750). The $[(\text{substrate})]_{tot}/[(\text{receptor})]_{tot}$ ratios were varied within ranges from 0 to 863 for **Py**, from 0 to 108 for **S1**, and from 0 to 24.5 for **S2**. Special care was taken to ensure that complete equilibration was attained. The corresponding UV–visible spectra were recorded from 290 to 700 nm on a Cary 300 (Varian) spectrophotometer maintained at 25.0(2) °C by the flow of a Haake NB 22 thermostat. The spectrophotometric data were processed with SPECFIT program [26], which adjust the stability constants and the corresponding extinction coefficients of the species formed at equilibrium (for typical examples, see Figures 4.3 and 4.6). SPECFIT uses factor analyses to reduce the absorbance matrix and extract the eigenvalues prior to the multiwavelength fit of the reduced data set according to the Marquardt algorithm [27].

4.5.3
Luminescence Titrations

Luminescence titrations were carried out on solutions of **LZn**, **LZn$_2$**, and **LZn$_6$** with absorbances smaller than 0.1 at wavelengths $\geq \lambda_{exc}$ in order to avoid any errors due to the inner filter effect. The titrations of 2 mL of **LZn**, **LZn$_2$**, and **LZn$_6$** with **Py** ($[\mathbf{LZn}]_{tot} = 1.79 \times 10^{-6}$ M; $[\mathbf{LZn_2}]_{tot} = 1.13 \times 10^{-6}$ M and $[\mathbf{LZn_6}]_{tot} = 1.17 \times 10^{-6}$ M), **S1** ($[\mathbf{LZn}]_{tot} = 1.79 \times 10^{-6}$ M; $[\mathbf{LZn_2}]_{tot} = 1.13 \times 10^{-6}$ M and $[\mathbf{LZn_6}]_{tot} = 4.91 \times 10^{-8}$ M), and

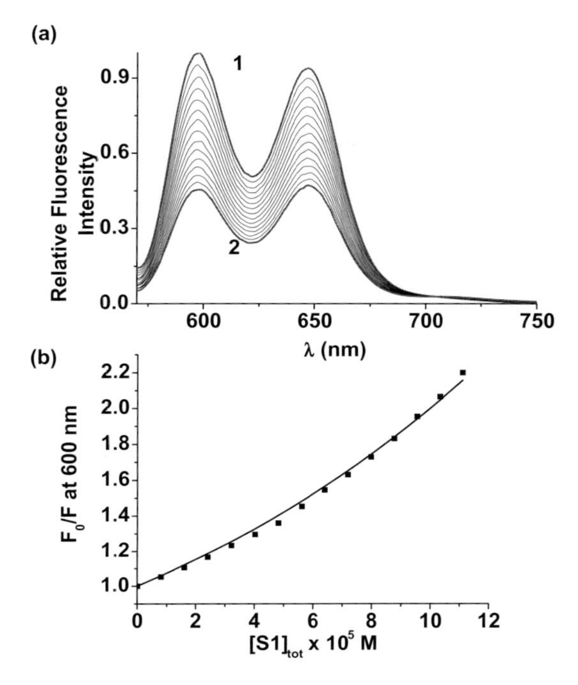

Figure 4.19 (a) Luminescence spectrophotometric titration of **LZn** with **S1**. $\lambda_{exc} = 559$ nm; emission and excitation slit widths 15 nm and 20 nm, respectively; (1) $[\mathbf{LZn}]_{tot} = 1.79 \times 10^{-6}$ M; (2) $[\mathbf{S1}]_{tot}/[\mathbf{LZn}]_{tot} = 63$. Solvent: CH_2Cl_2; $T = 25.0(2)$ °C. (b) Variation of F^0/F at 600 nm versus the concentration of **S1**. The trend line is the result of the nonlinear least-square fit of the experimental data according to Eq. (4.8).

S2 ($[\mathbf{LZn}]_{tot} = 1.90 \times 10^{-6}$ M; $[\mathbf{LZn_2}]_{tot} = 1.16 \times 10^{-6}$ M and $[\mathbf{LZn_6}]_{tot} = 4.91 \times 10^{-8}$ M) were carried out in a 1 cm Hellma quartz optical cell by addition of known microvolumes of solutions of **S1**, **S2**, and **Py** with microliter Hamilton syringes (#710 and #750). The $[(\text{substrate})]_{tot}/[(\text{receptor})]_{tot}$ ratios were varied within ranges from 0 to 531 for **Py**, from 0 to 220 for **S1**, and from 0 to 277 for **S2**. Special care was taken to ensure that complete equilibration was attained. The excitation wavelengths were set at 557(1) or 559(1) nm for **LZn**, at 557(1) or 559(1) nm for $\mathbf{LZn_2}$, and 558(1) or 428(1) nm for $\mathbf{LZn_6}$, respectively, and correspond in most of the cases to the isosbestic points between the electronic spectra of the free Zn(II) porphyrinic receptors and of the corresponding pentacoordinated complexes. The Zn(II) porphyrin-centered luminescence spectra were recorded from 500 to 800 nm on a Perkin-Elmer LS-50B maintained at 25.0(2) °C by the flow of a Haake FJ thermostat. The slit widths were set at 15 and 20 nm for the emission and excitation, respectively. Luminescence titrations of **LZn**, $\mathbf{LZn_2}$, and $\mathbf{LZn_6}$ receptors were conducted, under precise identical experimental conditions, by a model fullerene derivative unable to form complexes (**F**) in order to separate the variation in the luminescence intensity that results from dynamic and reabsorption phenomena. For **Py** substrate, the spectrofluorimetric data were processed with SPECFIT program [26]. For guests **S1** and **S2**, the

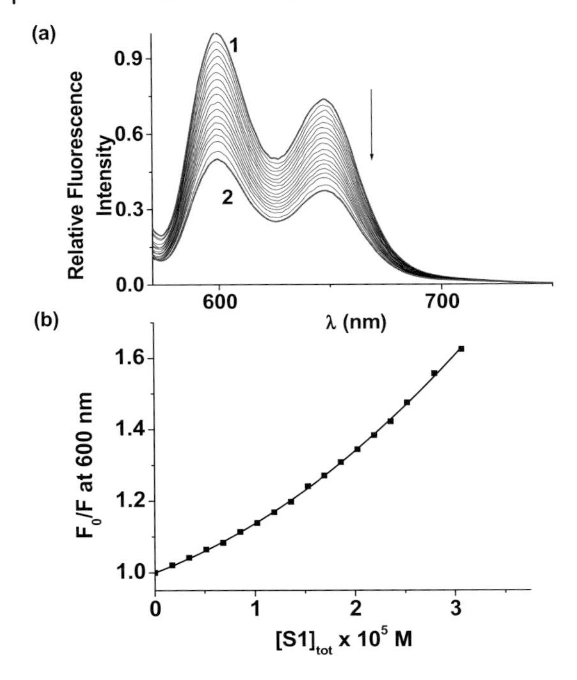

Figure 4.20 (a) Luminescence spectrophotometric titration of $\mathbf{LZn_2}$ with **S1**. $\lambda_{exc} = 559$ nm; emission and excitation slit widths 15 and 20 nm, respectively; (1) $[\mathbf{LZn_2}]_{tot} = 1.13 \times 10^{-6}$ M; (2) $[\mathbf{S1}]_{tot}/[\mathbf{LZn}^2]_{tot} = 27.2$. Solvent: CH_2Cl_2; $T = 25.0$ (2) °C. (b) Variation of F^0/F at 600 nm versus the concentration of **S1**. The trend line is the result of the nonlinear least-square fit of the experimental data according to Eq. (4.9).

luminescence data sets were analyzed [28] with Microcal Origin program [29] according to the modified Stern–Volmer equations (4.8) and (4.9).

$$F^0/F = (1 + K_{SV}[(\mathbf{S}n)])(1 + K_1[(\mathbf{S}n)]) \tag{4.8}$$

$$F^0/F = (1 + K_{SV}[(\mathbf{S}n)])(1 + K_1[(\mathbf{S}n)] + K_1K_2[(\mathbf{S}n)]^2) \tag{4.9}$$

where F^0 is the normalized fluorescence intensity of the porphyrin derivative (**LZn** or $\mathbf{LZn_2}$) in the absence of the **S*n*** (**S*n*** = **S1** or **S2**), F is the fluorescence intensity of the porphyrin derivative (**LZn** or $\mathbf{LZn_2}$) in the presence of **S*n***, [(**S*n***)] the molar concentration of fullerene derivative **S*n***, and K_{SV} is the pseudo-Stern–Volmer constant. The K_{SV} values for both **LZn** and $\mathbf{LZn_2}$ were obtained from the luminescence titrations carried out under the same experimental conditions with fullerene derivative **F** unable to form a supramolecular complex with Zn(II)-porphyrins and determined according to classical Stern–Volmer treatment (Eq. (4.10)).

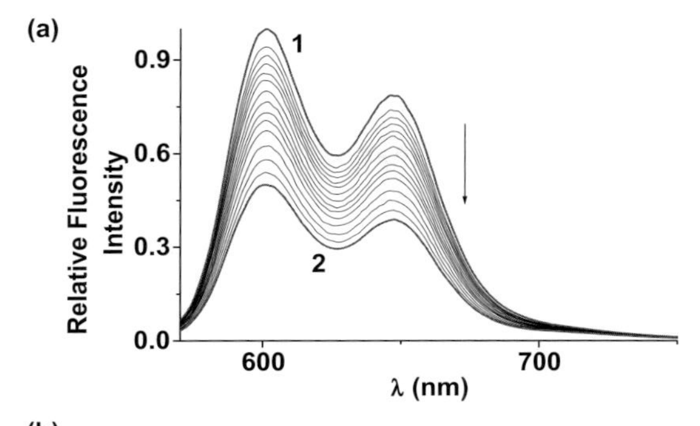

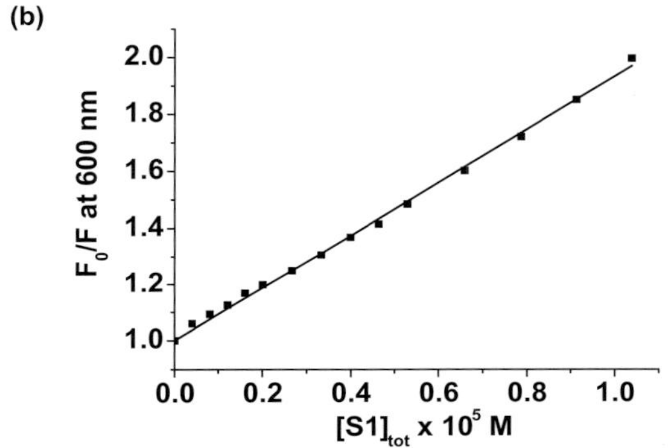

Figure 4.21 (a) Luminescence titration of $\mathbf{LZn_6}$ with **S1**. $\lambda_{ex} = 428$ nm; emission and excitation slit widths 15 and 20 nm, respectively; (1) $[\mathbf{LZn_6}]_{tot} = 4.91 \times 10^{-8}$ M; (2) [s1]/ $[\mathbf{LZn_6}]_{tot} = 220$. Solvent: CH_2Cl_2; $T = 25.0(2)$ °C. (b) Variation of F^0/F at 600 nm versus the concentration of **S1**. The trend line is the result of the linear least-square fit of the experimental data according to Eq. (4.11).

$$F^0/F = (1 + K_{SV}[(\mathbf{F})]) \tag{4.10}$$

where F^0 is the normalized fluorescence intensity of the porphyrin derivative (**LZn** or $\mathbf{LZn_2}$) in the absence of the **F**, F is the fluorescence intensity of the porphyrin derivative (**LZn** or $\mathbf{LZn_2}$) in the presence of **F**, [(**F**)] the molar concentration of fullerene derivative **F**, and K_{SV} is the pseudo-Stern–Volmer constant.

Typical examples of experimental data are depicted in Figures 4.19 and 4.20.

For host $\mathbf{LZn_6}$, the apparent stability constants were determined from the luminescence data sets according to the modified Stern–Volmer equation (4.11).

$$F^0/F = (1 + K_{SV}[(\mathbf{S}n)])(1 + K^*[(\mathbf{S}n)]) \tag{4.11}$$

where F^0 is the normalized fluorescence intensity of $\mathbf{LZn_6}$ in the absence of the **S*n*** (**S*n*** = **S1** or **S2**), F is the fluorescence intensity of $\mathbf{LZn_6}$ in the presence of **S*n***, [(**S*n***)] the molar concentration of fullerene derivative **S*n***, and K_{SV} is the pseudo-Stern–Volmer constant. A typical example is depicted in Figure 4.21.

Acknowledgments

This research was supported by the CNRS and the University of Strasbourg. I warmly thank all my coworkers and collaborators for their outstanding contributions, and their names are cited in the references.

References

1 Martín, N., Sanchez, L., Illescas, B., and Perez, I. (1998) *Chem. Rev.*, **98**, 2527; (b) Nierengarten, J.-F. (2004) *New J. Chem.*, **28**, 1177; Martín, N. (2006) *Chem. Commun.*, 2093; Figueira-Duarte, T.M., Gégout, A., and Nierengarten, J.-F. (2007) *Chem. Commun.*, 109; Accorsi, G. and Armaroli, N. (2010) *J. Phys. Chem. C*, **114**, 1385.
2 Diederich, F. and Gómez-López, M. (1999) *Chem. Soc. Rev.*, **28**, 263; Guldi, D.M. and Martin, N. (2002) *J. Mater. Chem.*, **12**, 1978; Sanchez, L., Martin, N., and Guldi, D.M. (2005) *Angew. Chem. Int. Ed.*, **44**, 5374; Schuster, D.I., Li, K., and Guldi, D.M. (2006) *C. R. Chimie*, **9**, 892; Hahn, U., Cardinali, F., and Nierengarten, J.-F. (2007) *New J. Chem.*, **31**, 1128; Perez, E.M., Illescas, B.M., Angeles Herranz, M., and Martin, N., *New J. Chem.* (2009) **33**, 228; Martin, N. and Nierengarten, J.-F. (eds) (2006) Supramolecular Chemistry of Fullerenes, Tetrahedron Symposium-in-Print, vol. 62 (9), pp. 1905–2132.
3 Gutiérrez-Nava, M., Nierengarten, H., Masson, P., Van Dorsselaer, A., and Nierengarten, J.-F. (2003) *Tetrahedron Lett.*, **44**, 3043.
4 Armaroli, N., Diederich, F., Echegoyen, L., Habicher, T., Flamigni, L., Marconi, G., and Nierengarten, J.-F. (1999) *New J. Chem.*, **23**, 77.
5 Nierengarten, J.-F., Hahn, U., Figueira Duarte, T.M., Cardinali, F., Solladie, N., Walther, M.E., Van Dorsselaer, A., Herschbach, H., Leize, E., Albrecht-Gary, A.-M., Trabolsi, A., and Elhabiri, M. (2006) *C. R. Chimie*, **9**, 1022.
6 Ben-Naim, A. (2001) *Cooperativity and Regulation in Biochemical Processes*, Kluwer, New York.
7 Hunter, C.A. and Anderson, H.L. (2009) *Angew. Chem. Int. Ed.*, **48**, 7488.
8 Ercolani, G. and Schiaffino, L. (2011) *Angew. Chem. Int. Ed.*, **50**, 1762.
9 Trabolsi, A., Urbani, M., Delgado, J.L., Ajamaa, F., Elhabiri, M., Solladie, N., Nierengarten, J.-F., and Albrecht-Gary, A.-M. (2008) *New J. Chem.*, **32**, 159.
10 Hou, J.L., Yi, H.P., Shao, X.B., Li, C., Wu, Z.Q., Jiang, X.K., Wu, L.Z., Tung, C.H., and Li, Z.T. (2006) *Angew. Chem. Int. Ed.*, **45**, 796.
11 Elhabiri, M., Trabolsi, A., Cardinali, F., Hahn, U., Albrecht-Gary, A.-M., and Nierengarten, J.-F. (2005) *Chem. Eur. J.*, **11**, 4793.
12 Nierengarten, J.-F., Hahn, U., Trabolsi, A., Herschbach, H., Cardinali, F., Elhabiri, M., Leize, E., Van Dorsselaer, A., and Albrecht-Gary, A.-M. (2006) *Chem. Eur. J.*, **12**, 3365.
13 Trabolsi, A., Elhabiri, M., Urbani, M., Delgado de la Cruz, J.L., Ajamaa, F., Solladie, N., Albrecht-Gary, A.-M., and Nierengarten, J.-F. (2005) *Chem. Commun.*, 5736.
14 Li, W.-S., Kim, K.S., Jiang, D.-L., Tanaka, H., Kawai, T., Kwon, J.H., Kim, D., and Aida, T. (2006) *J. Am. Chem. Soc.*, **128**, 10527.
15 Gust, D., Moore, T.A., and Moore, A.L. (2001) *Acc. Chem. Res.*, **34**, 40; Guldi, D.M. (2002) *Chem. Soc. Rev.*, **31**, 22.
16 Hahn, U., Elhabiri, M., Trabolsi, A., Herschbach, H., Leize, E., Van Dorsselaer, A., Albrecht-Gary, A.-M., and Nierengarten, J.-F. (2005) *Angew. Chem. Int. Ed.*, **44**, 5338.
17 Lehn, J.-M. (1995) *Supramolecular Chemistry, Concepts and Perspectives*, Wiley-VCH Verlag GmbH, Weinheim, Germany.

18 Mateo-Alonso, A., Sooambar, C., and Prato, M. (2006) *C. R. Chimie*, **9**, 944; Chitta, R. and D'Souza, F. (2008) *J. Mater. Chem.*, **18**, 1440.

19 Schuster, D.I., Jarowski, P.D., Kirschner, A.N., and Wilson, S.R. (2002) *J. Mater. Chem.*, **12**, 2041; Iehl, J., Vartanian, M., Holler, M., Nierengarten, J.-F., Delavaux-Nicot, B., Strub, J.-M., Van Dorsselaer, A., Wu, Y., Mohanraj, J., Yoosaf, K., and Armaroli, N. (2011) *J. Mater. Chem.*, **21**, 1562.

20 Boyd, P.D.W. and Reed, C.A. (2005) *Acc. Chem. Res.*, **38**, 235.

21 Solladie, N., Walther, M.E., Gross, M., Figueira Duarte, T.M., Bourgogne, C., and Nierengarten, J.-F. (2003) *Chem. Commun.*, 2412.

22 Armaroli, N., Marconi, G., Echegoyen, L., Bourgeois, J.-P., and Diederich, F. (2000) *Chem. Eur. J.*, **6**, 1629.

23 Solladie, N., Walther, M.E., Herschbach, H., Leize, E., Van Dorsselaer, A., Figueira Duarte, T.M., and Nierengarten, J.-F. (2006) *Tetrahedron*, **62**, 1979.

24 Santos, J., Grimm, B., Illescas, B.M., Guldi, D.M., and Martin, N. (2008) *Chem. Commun.*, 5993.

25 D'Souza, F., Chitta, R., Gadde, S., Zandler, M.E., Sandanayaka, A.S.D., Araki, Y., and Ito, O. (2005) *Chem. Commun.*, 1279; D'Souza, F., Chitta, R., Gadde, S., Zandler, M.E., McCarty, A.L., Sandanayaka, A.S.D., Araki, Y., and Ito, O. (2005) *Chem. Eur. J.*, **11**, 4416.

26 Gampp, H., Maeder, M., Meyer, C.J., and Zuberbühler, A.D. (1985) *Talanta*, **32**, 95; Rossoti, F.J.C., Rossoti, H.S., and Whewell, R.J. (1971) *J. Inorg. Nucl. Chem.*, **33**, 2051; Gampp, H., Maeder, M., Meyer, C.J., and Zuberbühler, A.D. (1985) *Talanta*, **32**, 257; Gampp, H., Maeder, M., Meyer, C.J., and Zuberbühler, A.D. (1986) *Talanta*, **33** 943.

27 Marquardt, D.W. (1963) *J. Soc. Indust. Appl. Math.*, **11**, 431; Maeder, M. and Zuberbühler, A.D. (1990) *Anal. Chem.*, **62**, 2220.

28 Marshall, A.G. (1978) *Biophysical Chemistry*, John Wiley & Sons, Inc., New York, p. 70;Freifelder, D.M. (1982) *Physical Biochemistry*, W. H. Freeman and Co., New York, p. 659;Scatchard, G. (1949) *Ann. N. Y. Acad. Sci.*, **51**, 660; Benesi, H.A. and Hildebrand, J.H. (1949) *J. Am. Chem. Soc.*, **71**, 2703; Barra, M., Bohne, C., and Scaiano, J.C. (1990) *J. Am. Chem. Soc.*, **112**, 8075; Htun, H. (2004) *J. Fluoresc.*, **14**, 217.

29 Microcal Software, Microcal™ Origin™ 7.0, Microcal Software, Inc., Northampton, USA, 2002.

5
Fullerene-Containing Rotaxanes and Catenanes

Aurelio Mateo-Alonso

5.1
Introduction

Rotaxanes and catenanes are mechanically interlocked molecules [1–3], which consist of two or more submolecular components mechanically bonded to each other that cannot be separated without breaking a covalent bond. In particular, rotaxanes consist of a dumbbell-shaped component (also called thread or axl), which is threaded through a macrocycle (Figure 5.1). In a rotaxane, thread and macrocycle are interlocked to each other since two stoppers are present at both ends of the thread, which prevent dissociation (unthreading) as the stoppers are larger than the internal diameter of the macrocycle. While in a pseudorotaxane at least one of the stoppers is not present (Figure 5.1), thus macrocycle and ring can dissociate depending on the conditions. Although pseudorotaxanes are not truly interlocked architectures, they are equally important as in some cases they are precursors to the synthesis of rotaxanes. Catenanes consist of two or more interlocked macrocycles (Figure 5.1).

5.1.1
Synthetic Strategies

5.1.1.1 Rotaxanes

Besides the low yielding statistical synthetic strategies [4], based on ring-closing reactions in the presence of excess of thread hoping that some of the rings will form around it, efficient yields can be obtained only by preorganizing the components utilizing supramolecular interactions such as metal coordination, π stacking, and hydrogen bonding. The three most common strategies to synthesize rotaxane are capping, clipping, and slipping (Figure 5.2).

Synthesis via the capping strategy relies on template-mediated synthesis. The thread is preassembled within the macrocycle by noncovalent interactions between both components. The formed pseudorotaxane is then converted to the rotaxane by reacting the ends of the threaded guest with large groups preventing disassociation.

Supramolecular Chemistry of Fullerenes and Carbon Nanotubes, First Edition. Edited by Nazario Martin and Jean-Francois Nierengarten.

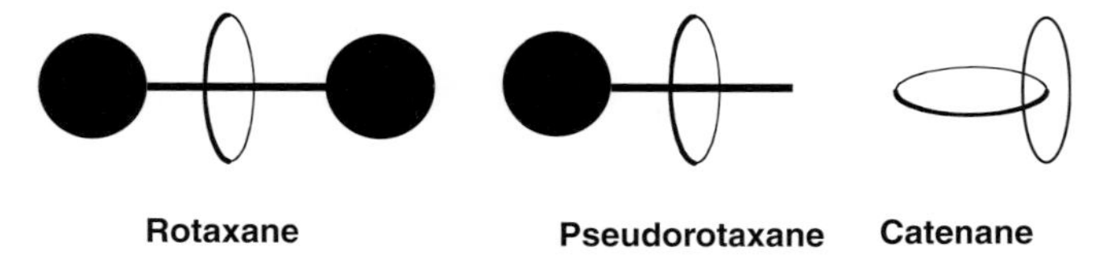

Figure 5.1 Schematic representation of rotaxanes, pseudorotaxanes, and catenanes.

The clipping strategy is similar to the capping reaction except that in this case the thread with the two stoppers is bound to a partial macrocycle. The partial macrocycle then undergoes a ring-closing reaction around a template present in the thread forming the rotaxane.

The method of slipping exploits the kinetic stability of the rotaxane. If the end groups of the dumbbell are of an appropriate size, it will be able to reversibly thread the macrocycle through by increasing the temperature. By cooling the dynamic complex, the macrocycle is kinetically trapped within the thread.

5.1.1.2 **Catenanes**

There are two primary approaches to the organic synthesis of catenanes. As in the case of rotaxanes, the statistical approach is based on performing a ring-closing reaction in the presence of one of the rings with the hope that some of the new rings

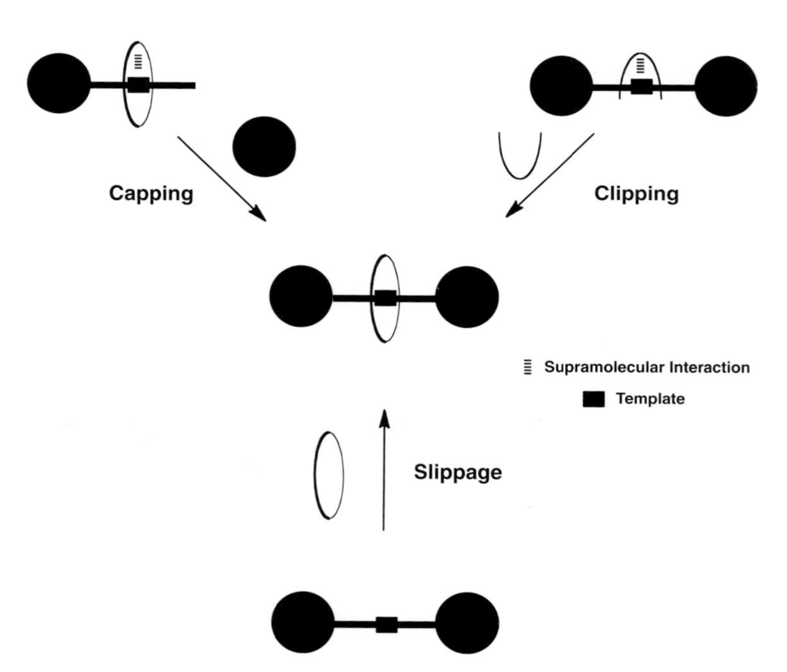

Figure 5.2 Synthetic strategies for the synthesis of rotaxanes.

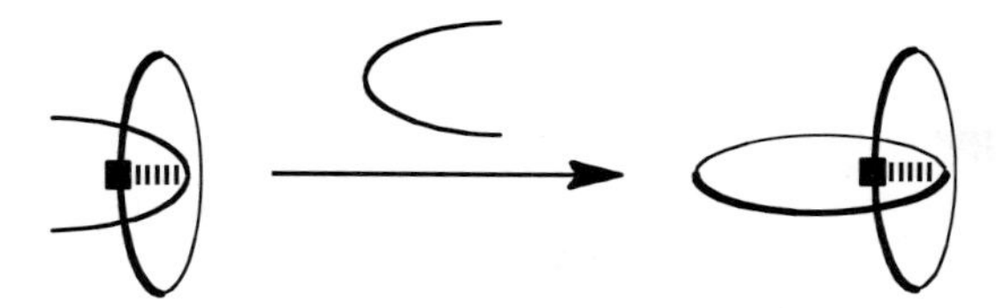

Figure 5.3 Templated synthesis of catenanes.

will form around those present in solution giving the desired catenane product [5]. The second approach relies on supramolecular preorganization of the macrocyclic precursors by template-directed synthesis utilizing supramolecular interactions that favor the prearrangement of a partial ring around a complete ring to form the desired catenane upon the final ring-closing reaction (Figure 5.3).

5.1.2
Bistable Rotaxanes and Catenanes

Molecules built through mechanical bonds are intrinsically multistable as their interlocked components can adopt multiple co-conformations. The interlocked components of rotaxanes and catenanes have a large mobility and they can be subjected to different types of submolecular motion (Figure 5.4). These features have several important implications both in fundamental science and molecular nanotechnology. For example, in bistable rotaxanes (or molecular shuttles) the macrocycle can be positioned at two different positions over the thread (stations) by applying an external stimulus that ranges from illumination [6, 7] and variation of the electrochemical potential [8–10] to a change in solvent [11] or pH [12, 13]. There is a great deal of interest in manipulating the relative position of the components as this change can be used to modify the physicochemical properties of molecules [11, 14, 15]. In the same way in bistable catenanes, one of the rings can circumrotate in different directions to different positions of the other ring [16, 17].

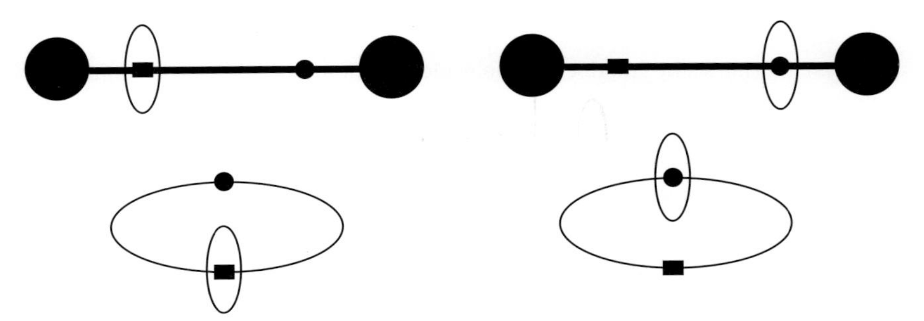

Figure 5.4 Bistable rotaxanes and catenanes.

5.2 Fullerene Rotaxanes and Catenanes

The synthetic methods used to prepare mechanically interlocked molecules functionalized with fullerenes have been adapted from well-established methodologies, such as those developed by Sauvage [18], Stoddart [19, 20], Vogtle [21], Leigh [22], and Sanders [23].

Fullerenes and especially C_{60} have been widely studied and today are valuable building blocks for the preparation of materials with potential applicability in different technological fields including photovoltaics, nonlinear optics, optoelectronics, and medicine [24–27].

The properties of fullerenes derive from the three-dimensional structure of their extended π system that makes them excellent electron acceptors and chromophores. Fullerenes are known to exhibit a sizeable nonlinear optical response [28, 29] due to their large π-conjugated surface and the extensive charge delocalization [30].

Fullerenes have been widely used as electron acceptors in the preparation of donor–acceptor systems [31] (that mimic a key step in photosynthesis) and in the preparation of organic solar cells [32]. The introduction of fullerenes in mechanically interlocked molecular architectures permits the study of donor–acceptor geometries that cannot be assessed by covalent or supramolecular chemistry [33]. In general, these donor–acceptor systems present the fullerene electron acceptor in one submolecular fragment and the electron donor is present in the opposite component in order to study through-space electron transfer.

Also, fullerenes have been used as stoppers in rotaxanes not only because they are large and bulky but also because they have very well-defined properties that can be used to probe the motion of the ring [33].

Lately, it has been shown that fullerenes can be used to induce submolecular motion, expanding the applications of fullerenes as active component molecular machines [33].

All the above-mentioned topics will be discussed in detail in this chapter in several sections organized by the supramolecular forces used to assemble the different components.

5.2.1 Metal Coordination

The first mechanically interlocked molecule functionalized with a fullerene was reported by Diederich and Sauvage in 1995 (Scheme 5.1) [34, 35]. A rotaxane displaying two fullerene stoppers was prepared by metal coordination (Scheme 5.1). Pseudorotaxane **1** was assembled by coupling a macrocycle and a thread with both phenanthroline moieties through the formation of a tetrahedral bisphenanthroline Cu(I) complex. Pseudorotaxane **1** was then capped with two C_{60} stoppers functionalized with acetylenes (**2**) leading to the formation of the desired rotaxane (**3**).

Following this strategy, rotaxanates and catenates functionalized with porphyrin or ferrocene electron donors and C_{60} electron acceptors have been synthesized. Irradiation of such dyads lead to the formation of long-lived charge-separated state that can be followed spectroscopically (Scheme 5.2) [36]. The lifetimes of the charge-

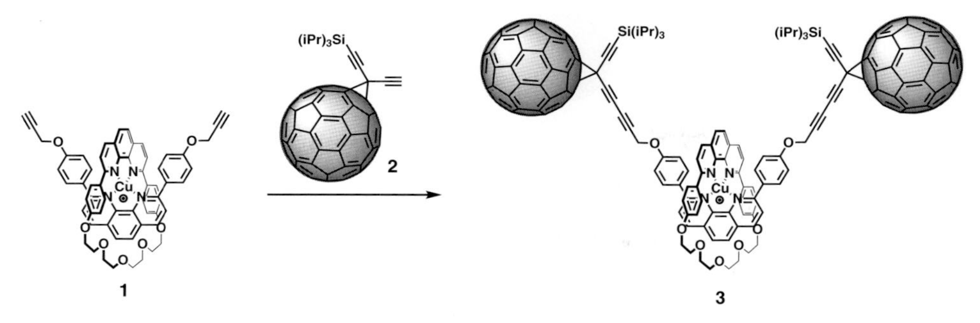

Scheme 5.1 Fullerene-stoppered rotaxanes assembled by metal coordination.

separated state vary drastically with the relative position of the donor and acceptor. Lifetimes of 32 μs were observed when rotaxanate **4** was studied (Scheme 5.2), which presents two fullerene stoppers and a porphyrin electron donor on the macrocycle [37, 38]. Conversely, in the case of rotaxanate **5**, which presents porphyrin stoppers and C_{60} on the macrocycle, lifetimes of 1.17 μs were measured [39]. Rotaxane **5** can be transformed into catenate **6** by complexing both porphyrin stoppers with bypyridine [40]. Very recently, a route has been reported to cap this type of rotaxanates [41] and catenates [42] by click chemistry.

5.2.2
π Stacking Interactions

The first catenane containing a fullerene (**10**) was reported by the groups of Diederich and Stoddart in 1997 (Scheme 5.3) [43]. Catenane **10** was synthesized by π–π stacking interactions using the recognition between electron-poor and electron-rich aromatics. Macrocycle **6** displays an electron-rich hydroquinone template. The reaction of **7** with *p*-xylylene dibromide (**8**) clipped an electron-deficient cyclobis-(paraquat-*p*-phenylene) ($CBPQT^{4+}$) macrocycle around the hydroquinone template.

Following this methodology, a bistable rotaxane containing a fullerene has been synthesized and studied (Scheme 5.4) [44]. The thread presents two electron-rich units, 1,5-dihydroxynaphthalene (DNP) and tetrathiafulvalene (TTF). In a similar fashion, the electron-deficient $CBPQT^{4+}$ macrocycle was clipped around an electron-rich template. In this case the TTF unit that acts as template as it is more electron-rich than DNP. As a matter of fact the macrocycle stays over the TTF station (coconformer **11A**) after the clipping reaction. Oxidation of the TTF to the radical cation ($TTF^{\bullet +}$) leads to the translocation of the macrocycle from the TTF to the DNP unit because of the electrostatic repulsion between the positively charged $CBPQT^{4+}$ and $TTF^{\bullet +}$ (coconformer **11B**).

Other types or electron-rich and electron-deficient aromatics have been used to prepare fullerene-functionalized rotaxanes and catenanes (Scheme 5.5) [45]. Pseudorotaxane **12** was assembled *in situ* by complexation of a thread, bearing one fullerene stopper and a 1,4,5,8-naphthalenetetracarboxylic diimide template, with a ring, bearing two electron-rich DNPs. Depending on the reaction conditions pseudorotaxane **12** evolves either into catenane **13** by allowing the terminal malonate of

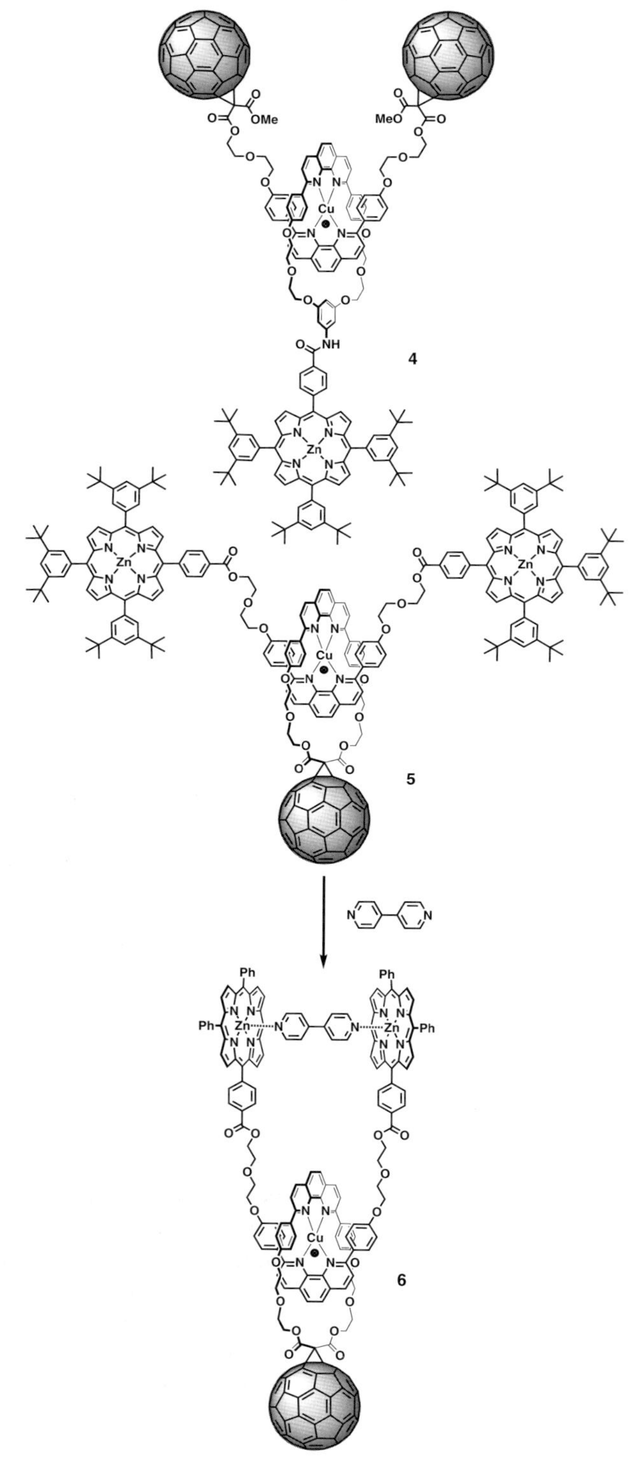

Scheme 5.2 Rotaxane and catenane dyads assembled by metal coordination.

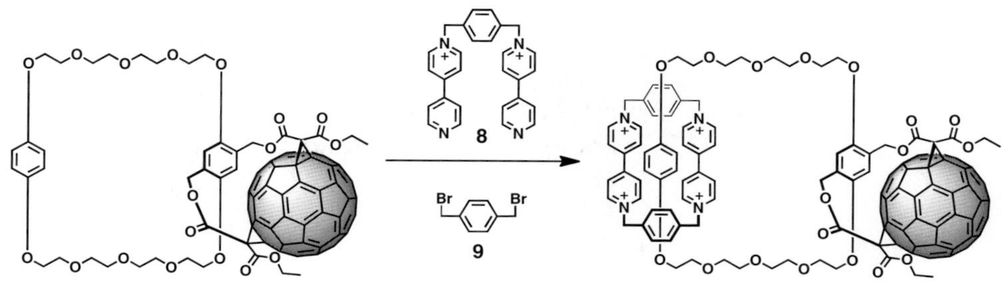

Scheme 5.3 Synthesis of a fullerene–catenane by π–π interactions.

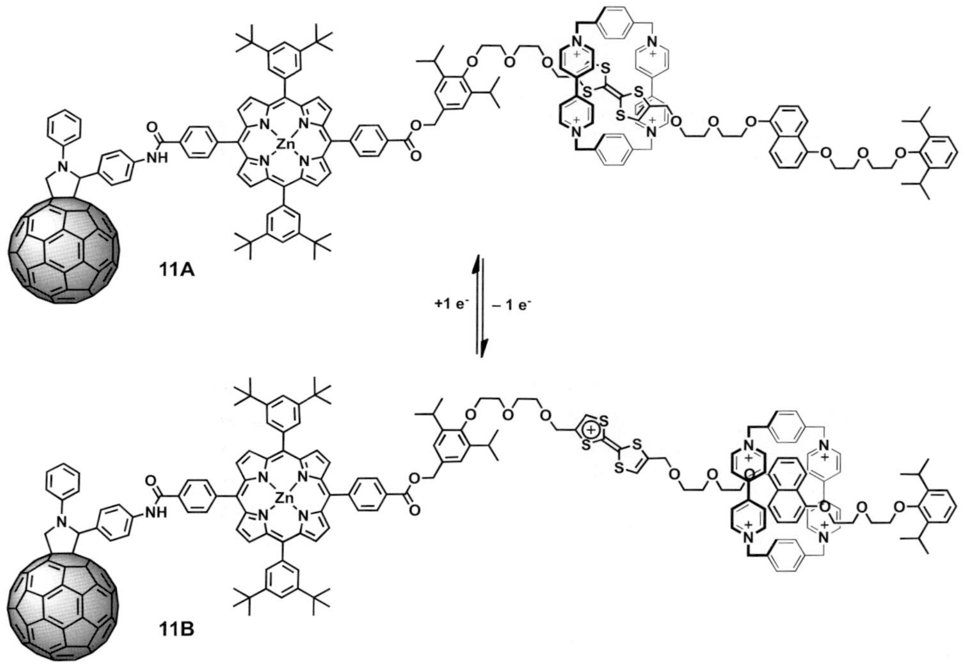

Scheme 5.4 Bistable rotaxane assembled by π–π interactions.

the thread to react with the fullerene stopper at the other end or into rotaxane **14** by capping the terminal malonate with an additional C_{60} stopper [45].

5.2.3
Hydrogen Bonds

The hydrogen bond recognition between ammonium salts and crown ethers has been used to prepare pseudorotaxanes [46] and rotaxanes [47, 48]. The complexation

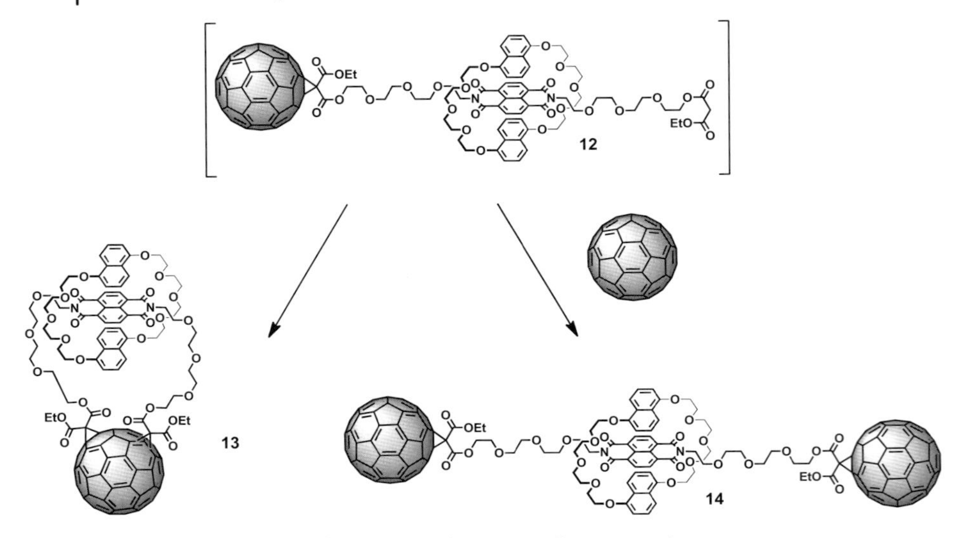

Scheme 5.5 Synthesis of rotaxanes and catenanes from a pseudorotaxane.

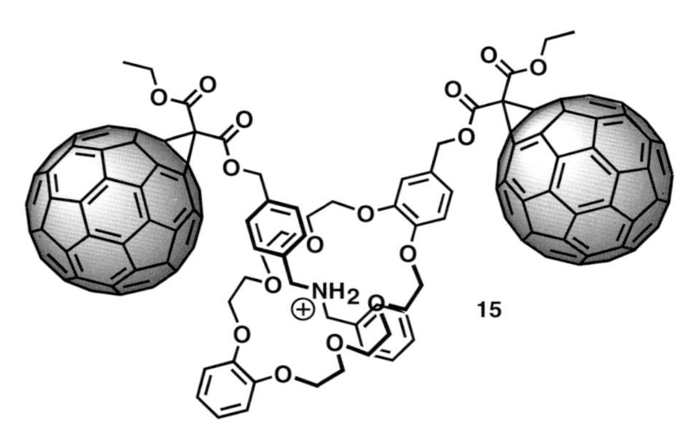

Figure 5.5 Pseudorotaxane assembled by hydrogen bonding.

of two independent fullerene derivatives, one functionalized with a crown-ether and the other one functionalized with an ammonium salt, into pseudorotaxane **15** has been achieved by mixing two solutions of both components (Figure 5.5) [46].

Rotaxanes can also be synthesized through this strategy by cappping a preassembled pseudorotaxane rotaxane (Scheme 5.6) [47, 48]. Pseudorotaxanes showing either a fullerene on the macrocycle (**16**) or a terminal sulfine group (**19**) have been prepared by mixing the corresponding thread and crown ether. Pseudorotaxane **16** can be capped with isocyanate **17** to give rotaxane **18** with a fullerene on the macrocycle. Meanwhile pseudorotaxane **19** can be capped with C_{60} *in situ* via Diels–Alder reaction to give fullerene-stoppered rotaxane **20** [49].

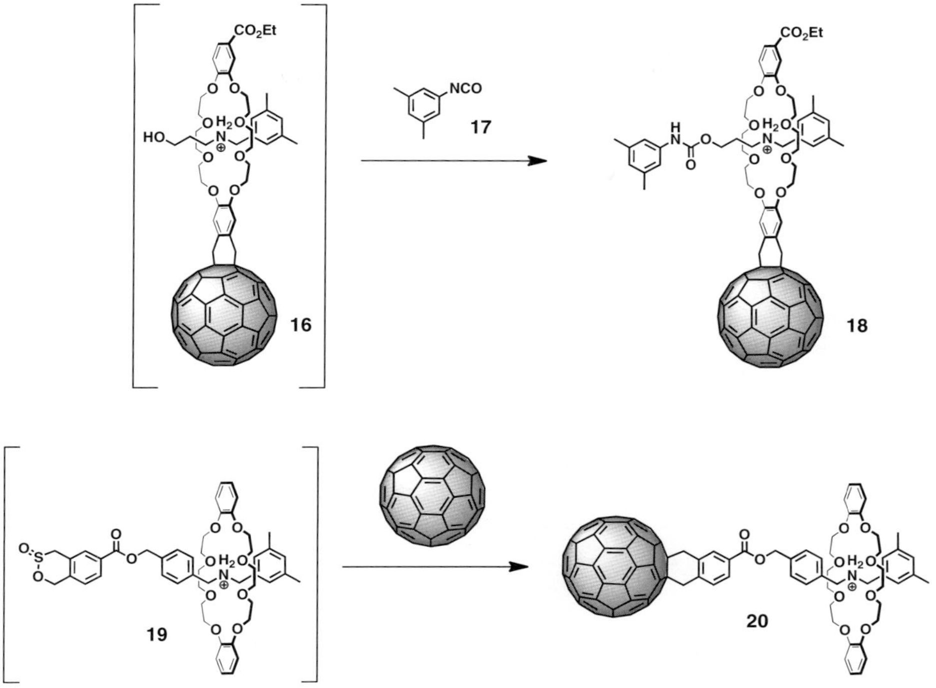

Scheme 5.6 Rotaxanes assembled by hydrogen bonding.

Pseudorotaxanes with one or two phthalocyanines fused to a crown ether were complexed to an axl with a single fullerene stopper and an ammonium salt (Figure 5.6) [50, 51]. The assembly of both components in solution leads to the formation of dyad **21**, in which radical-pairs with lifetimes of 1.5 μs were found upon irradiation. [2]Pseudorotaxanes [52] and [3]pseudorotaxanes [52, 53] have also been

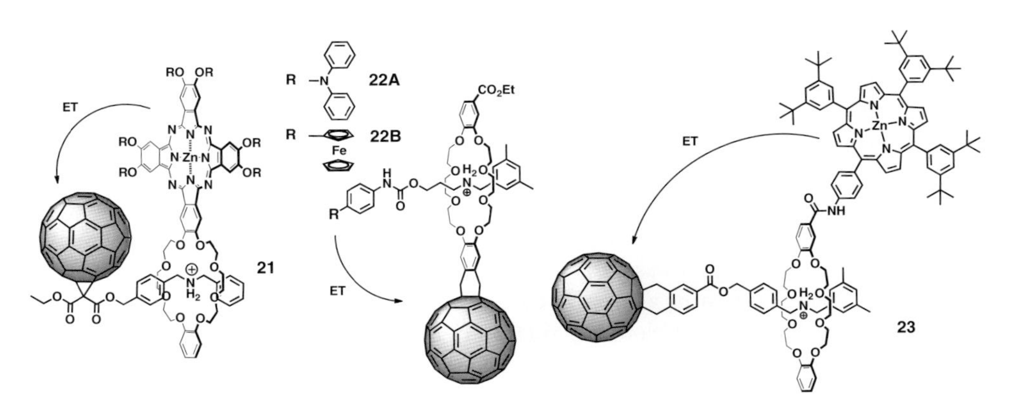

Figure 5.6 Donor–acceptor systems assembled by hydrogen bonds.

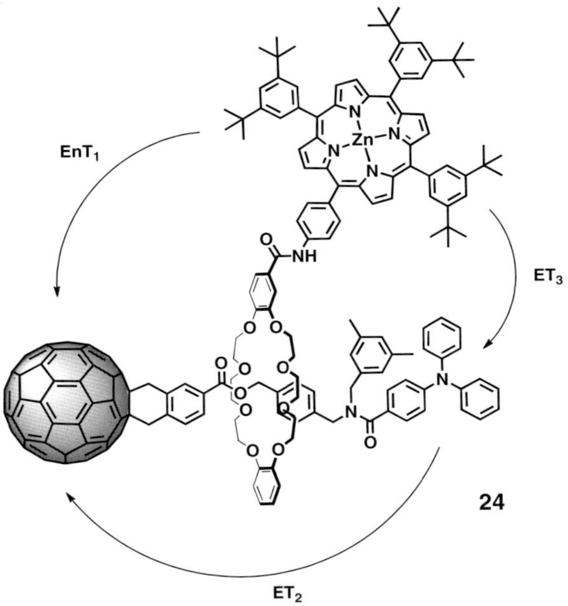

Figure 5.7 Rotaxane triad.

prepared with extended TTF derivatives with one or two ammonium groups to bind one or two fullerene derivatives with a crown ether functionality. Several rotaxanes [47, 48] have been synthesized displaying a fullerene on the macrocycle, and either triphenylamine (**22A**) [54] or ferrocene (**22B**) [55, 56] electron donors as stoppers. Photophysical studies revealed the formation of a charge-separated state with lifetimes of 360 and 13 ns, respectively, for **22A** and **22B**. Similarly, dyad **23** with a fullerene stopper and a porphyrin electron donor on the macrocycle [49] gave radical pairs with a lifetime of 180 ns [57] after photoexcitation.

Triads can be prepared from rotaxane **23** (Figure 5.6) by acylating the ammonium salt encapsulated by the crown ether with an additional chromophore (Figure 5.7). When the acylating agent was a ferrocene derivative, electron transfer from the porphyrin on the macrocycle to the fullerene was observed [58]. However, a charge shift from the porphyrin to the ferrocene was not detected. When a triarylamine derivative was used to acylate the ammonium salt (**24**) [59, 60], photoexcitation of the porphyrin led to energy transfer (EnT_1) from porphyrin to the fullerene, which was then followed by electron transfer (ET_2) from the triphenylamine moiety to the fullerene. Subsequently, a charge shift (ET_3) was observed from triphenylamine to porphyrin.

Rotaxanes bearing a fullerene on the macrocycle can also be synthesized by using the methodology developed by Vogtle [21]. In fact, the groups of Takata and Ito have reported the synthesis of rotaxane **25** with a sulfolene group on the macrocycle (Scheme 5.7). Rotaxane **25** decomposes under thermal conditions into a 1,4-diene that can react with C_{60} to give rotaxane **26** with a fullerene on the macrocycle.

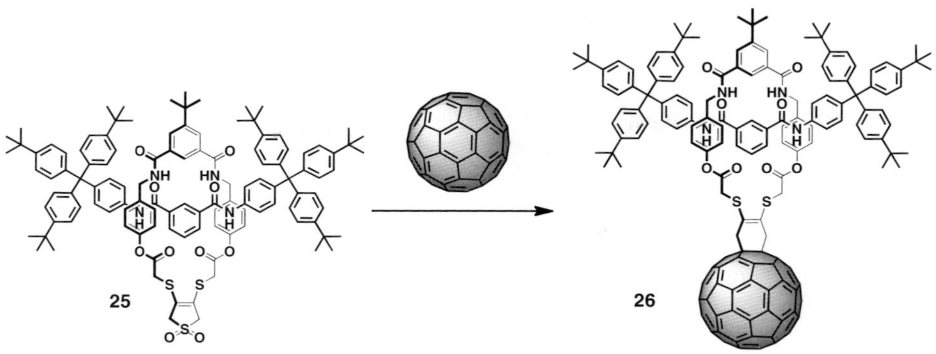

Scheme 5.7 Synthesis of a rotaxane with a fullerene on the macrocycle.

The use of porphyrins as stoppers leads to the formation of dyads (Figure 5.8). Photophysical studies demonstrated that ET occurred between the porphyrin and the C_{60} after photoexcitation. The lifetime of the radical pair can be tuned by changing the length of the axl [47, 48, 61–63]. The lifetime of the charge-separated state increases with the axl length as rotaxanes **27A**, **27B**, and **27C** present a lifetime of 180, 230, and 625 ns, respectively [62].

Rotaxanes can also be prepared by hydrogen bonds by means of Leigh-type clipping reactions [22]. This approach has allowed the preparation of rotaxanes bearing a

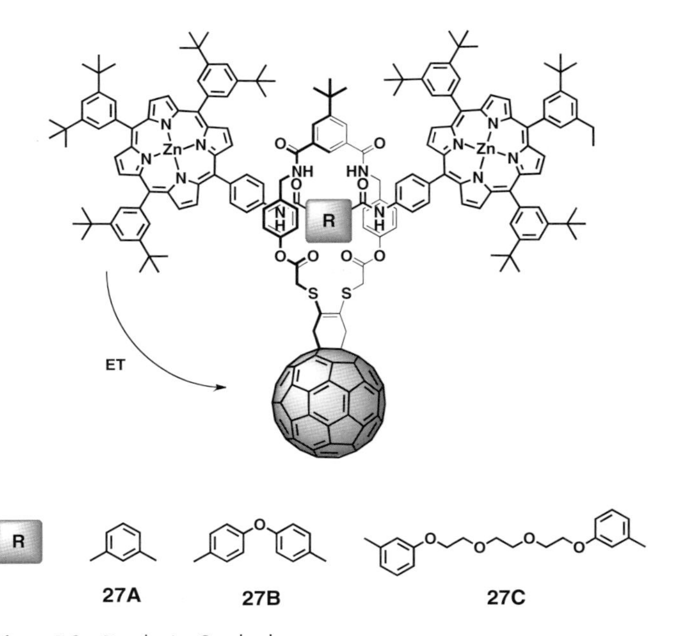

Figure 5.8 Porphyrin–C_{60} dyads.

fullerene stopper in a single step (Scheme 5.8) [33, 64, 65]. The benzylic amide macrocycle is clipped around the thread assisted by a 1,4 diamide template that can form four hydrogen bonds with the macrocycle. The addition of the precursors of the macrocycle, isophthaloyl chloride (**29**) and *p*-xylylene diamine (**30**), to thread **28** under high dilution conditions leads to rotaxane **31**.

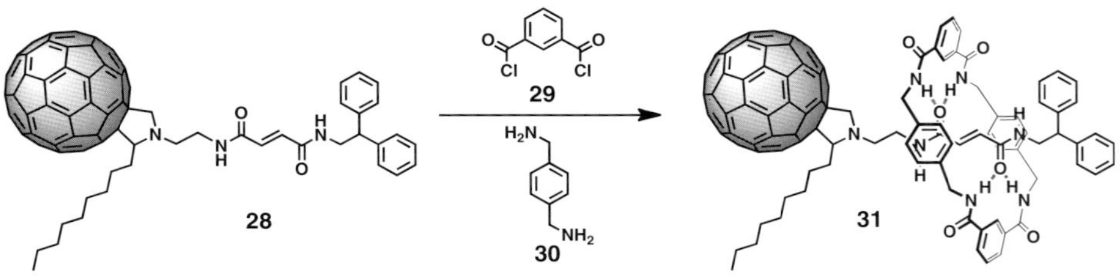

Scheme 5.8 Synthesis of a fullerene-stoppered rotaxane with a benzylic amide macrocycle.

Ferrocene electron donors can be introduced in the macrocycle by a modification of this protocol leading also to rotaxanated charge transfer dyes (Figure 5.9) [66–69]. Photophysical characterization of dyad **32** revealed ET from ferrocene to C_{60} yielding a radical pair with a lifetime of 9 ns. The lifetime of the photogenerated radical pair increases to 26 ns by increasing the relative distance between the 1,4-diamide template on the thread and the fullerene by using a triethyleneglycol spacer.

Also, triads have been prepared by functionalizing the thread with a pyridyl moiety that allows the introduction of an additional metalloporphyrin electron donor/acceptor on the rotaxane (Figure 5.9) [68]. Triad **33** is comprised of a C_{60} electron

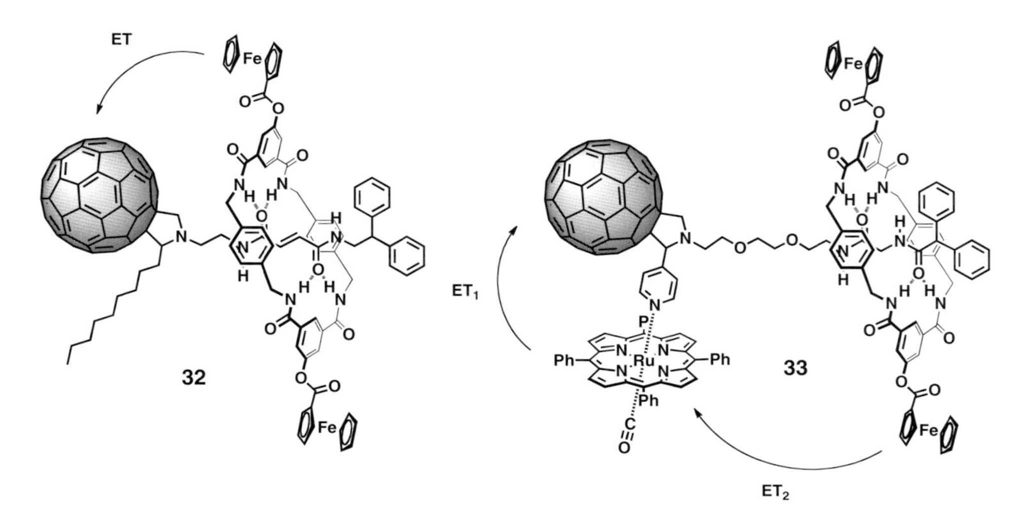

Figure 5.9 Rotaxane dyad and triad.

acceptor and two ferrocene electron donors coupled by hydrogen bonds in a rotaxane fashion. A ruthenium carbonyl tetraphenylporphyrin (RuTPP) can be coordinated to the pending pyridyl group present between the C_{60} and the ferrocene units. The excitation of the central RuTPP unit results in electron transfer from RuTPP to C_{60}, followed by charge shift from RuTPP to ferrocene.

The first bistable rotaxane containing a fullerene was reported in 2003 (Scheme 5.9) [70]. Rotaxane **34** is composed of a fullerene stopper and a benzylic amide macrocycle. In solvents that do not disturb hydrogen bonds, such as CH_2Cl_2 and $CHCl_3$, the macrocycle resides preferentially over the glycylglycine template present on the thread by complementary hydrogen bond recognition (coconformation **34A**). Instead, in solvents that strongly disturb hydrogen bonds, such as DMSO, the macrocycle resides over the alkyl chain (coconformation **34B**) due to the solvation of the hydrogen bond binding sites. The photophysical properties of the fullerene were used to monitor the position of the macrocycle along the thread. The triplet–triplet features of the fullerene are nearly independent of the solvent in the case of the thread, which are similar to those of rotaxane **34** in DMSO. When rotaxane **34** was studied in CH_2Cl_2 the proximity of the macrocycle to the fullerene sphere induced visible differences in the triplet–triplet spectra and the triplet lifetime, shorter by a factor of 1.7.

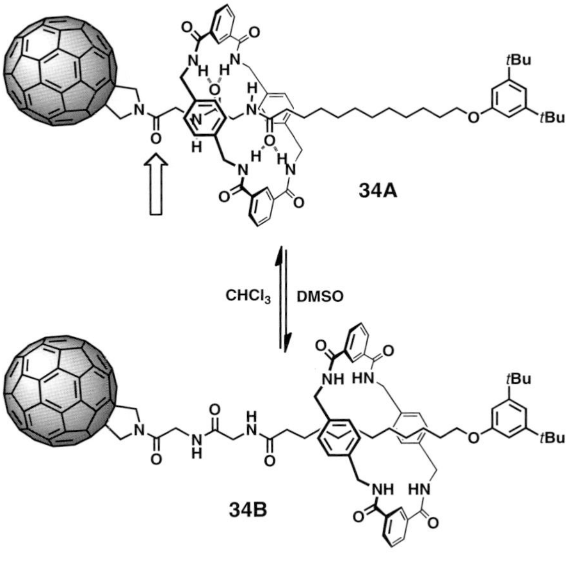

Scheme 5.9 Solvent switchable molecular shuttle.

Rotaxane **35** [71] (Scheme 5.10) was designed with an analogous structure to that of rotaxane **34**. However, when rotaxane **35** was studied in DMSO, the macrocycle switched in the opposite direction. Reverse shuttling was explained as a result of

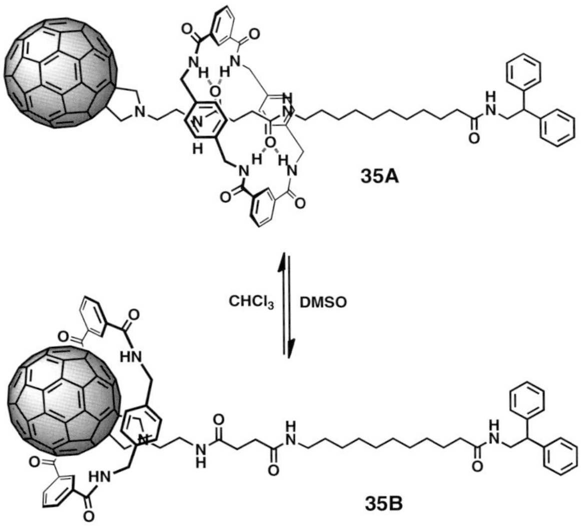

Scheme 5.10 Reverse shuttling.

solvation and π–π interactions between the macrocycle and the fullerene. Rotaxane **34** presents an extra amide between the template and the fullerene (indicated by an arrow on Scheme 5.9). The solvation of the three amides of rotaxane **34**, positioned next to the fullerene, does not allow the macrocycle to get close to the fullerene after decomplexation. Hence, the macrocycle switches toward the alkyl chain. However, in rotaxane **35**, the succinamide template is connected to the fulleropyrrolidine stopper by two methylene groups, hence no additional amides are present in-between the template and the fullerene. Solvation of the succinamide amides of rotaxane **35** by DMSO molecules allows the macrocycle to switch in both directions, either toward the alkyl chain or toward the fullerene. In the case of rotaxane **35**, the macrocycle switches preferentially toward the fullerene to adopt coconformation **35B** because now solvation allows the macrocycle to establish π–π interactions with the fullerene. The existence of π–π interactions was demonstrated by a series of electrochemical experiments that revealed the stabilization of the electrochemically generated anions of the fullerene with the proximity of the macrocycle to the fullerene stopper.

Rotaxane **36** [72, 73] was designed to exploit the above-mentioned π–π interactions between macrocycle and fullerene to induce shuttling over long distances (Scheme 5.11). The hydrogen bond binding station was placed far away from the fullerene through a triethylene glycol spacer. In solvents that do not disturb the hydrogen bonds (CH_2Cl_2 and $CHCl_3$), the macrocycle stays preferentially on the peptidic station (coconformer **36A**), far away from the fullerene. Instead, in solvents such as DMSO and DMF that weaken the hydrogen bonds, π–π interactions between the macrocycle and the fullerene are allowed, which induce a large positional change of

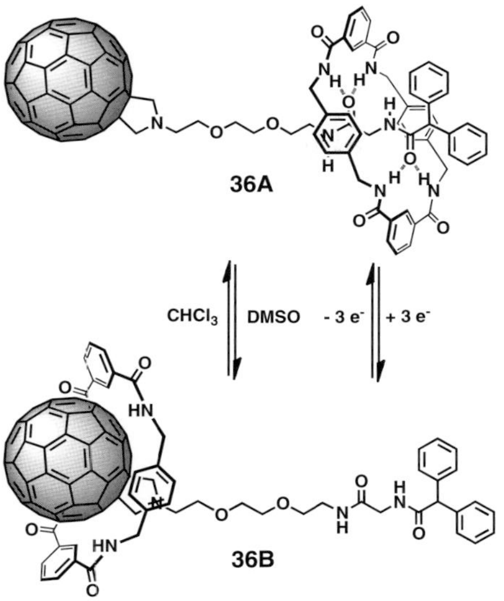

Scheme 5.11 Fullerene-driven molecular shuttle.

the macrocycle that assumes coconformation **36B**. Absorption, emission, and cyclic voltammetry experiments confirmed that shuttling takes place through π–π interactions between the fullerene and the macrocycle. The emission of the fullerene in the thread is nearly independent of solvent, matching the emission of rotaxane **36** in nonpolar solvents (coconformation **36A**). Measurements carried out on rotaxane **36** in DMSO showed that the residual fluorescence of the fullerene is 51% of that in CH_2Cl_2. An important feature of rotaxane **36** is that the translocation of the macrocycle can also be achieved electrochemically by the reduction of the fullerene to its trianion. A coconformational change takes place, transforming $\mathbf{36A}^{3-}$ into $\mathbf{36B}^{3-}$, benefiting the stabilization of the negative charge present on the fullerene. A switching efficiency of 89% was calculated by combination of cyclic voltammetry and digital simulations.

The displacement of the macrocycle can be used to change the chemical reactivity of functional groups [74]. Fulleropyrrolidine *N*-oxides are thermally unstable, which complicates their study and characterization. In fact, such *N*-oxides are stable only for days even if stored at −20 °C. Their instability is patent from the facile reversibility of the oxidation reaction. Fulleropyrrolidine *N*-oxide **37** (Scheme 5.12) can be cleanly deoxygenated to the parent fulleropyrrolidine by heating in the presence of protic solvents. When rotaxane **35** (Scheme 5.10) was *N*-oxidized to rotaxane **38** (Scheme 5.12), the macrocycle switched from the succinamide template to bind the *N*-oxide and one of the amides of the succinamide. Remarkably, the encapsulation of the *N*-oxide by intrarotaxane hydrogen bonding increases its chemical stability and inhibits its deoxygenation. These results indicate that such enhanced stability is a function of both intramolecular hydrogen bonding and encapsulation of the *N*-oxide.

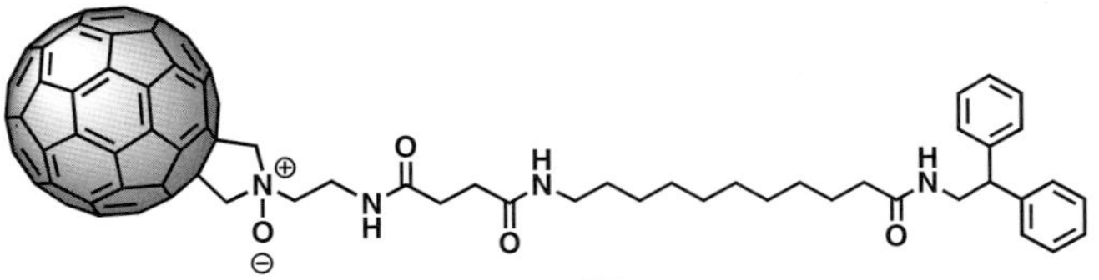

37

Unstable Fulleropyrrolidine N-Oxide

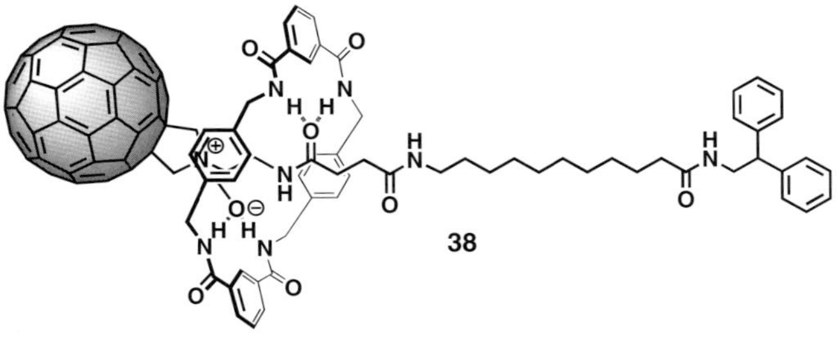

38

Stable Fulleropyrrolidine N-Oxide

Scheme 5.12 Stabilization of fulleropyrrolidine *N*-oxides.

The introduction of ferrocene electron donors on the macrocycle of a bistable fullerene-stoppered rotaxane (**39**) can be used to study the effects of submolecular displacement on electron transfer (Scheme 5.13) [66]. Molecular shuttle **39** behaves exactly as the analogous **36**, the translocation of the macrocycle is triggered through a solvent change or through changes on the redox state of the fullerene by cyclic voltammetry. Steady-state and transient absorption photophysical measurements revealed through-space photoinduced electron transfer between the fullerene stopper and the ferrocenes on the macrocycle. In CH_2Cl_2, the lifetime of the charge-separated state was 26.2 ns. Addition of hexafluoroisopropanol to **39** shortens the lifetime to 13.0 ns, a consequence imposed by weakening the hydrogen bonds that decreases the relative spatial separation between the donor and the acceptor, while increasing the shuttling rate. A series of photophysical experiments on a set of reference compounds was carried out to conclude that the displacement of the macrocycle contributes roughly in a 28% to the effects observed on the lifetime of the photoinduced radical pair, demonstrating that the displacement of the macrocycle can be used to modulate ET.

The nonlinear optical (NLO) response of fullerenes can also be modulated by shuttling [67]. The microscopic NLO responses of rotaxanes **36** and **39** were measured by means of Z-scan technique. The measured values clearly show that shuttling rules the NLO response. Indeed, the translocation of the macrocycle on both rotaxanes **36** and **39** gives rise to a twofold difference of the NLO response when comparing coconformers **A** and **B**.

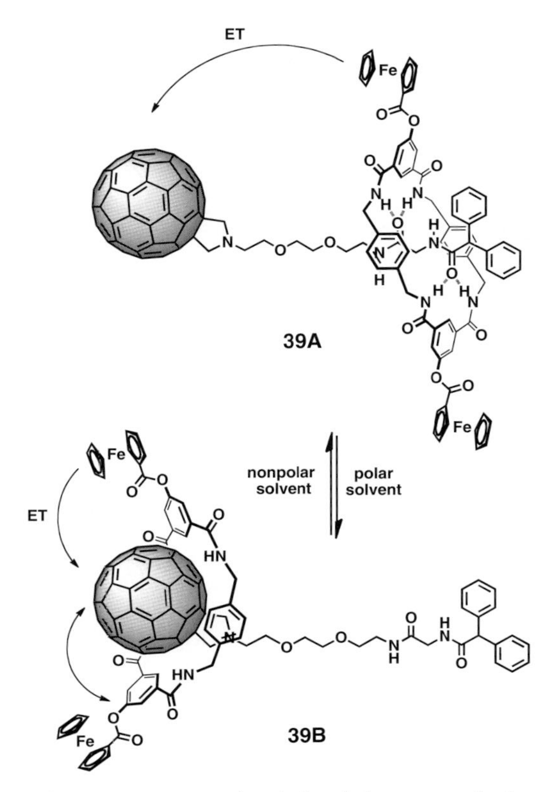

Scheme 5.13 Tuning photoinduced electron transfer through shuttling.

5.3 Conclusions

Several methodologies have shown to be effective in introducing fullerenes into rotaxanes, pseudorotaxanes, and catenanes. A series of donor–acceptor systems have been synthesized and studied, in which the fullerenes have been used as an electron acceptor. The use of mechanical bonds to couple the fullerene electron donor with an electron acceptor facilitates the study of donor–acceptor geometries that cannot be achieved by covalent or supramolecular chemistry, promotes through-space electron transfer, and also allows the modulation of the electron transfer imposed by the multistability of rotaxanes and catenanes. In addition, other properties of the fullerenes can be modulated such as their emission, nonlinear optical response, and chemical stability of its derivatives, demonstrating that submolecular motion can be used as an effective means to modulate the properties of materials at the molecular level. The biggest impact of fullerenes on mechanically interlocked molecular architectures relies on the use of fullerenes as multitask components in molecular machines that do not only exploit their size and properties but, most importantly, also use them as units to induce submolecular motion. Since, in principle, there are no limitations to the number of co-conformations that rotaxanes and catenanes can adopt, there are plenty of opportunities for design and synthesis of new fullerene-based molecular devices that present multiple properties depending on the relative organization of their components, paving the way for the preparation of multiproperty molecular materials with unlimited applications.

References

1 Sauvage, J.P. and Dietrich-Buchecker, C. (eds) (1999) *Catenanes, Rotaxanes, and Knots*, Wiley-VCH Verlag GmbH.

2 Sauvage, J.P. and Gaspard, P. (eds) (2010) *From Non-Covalent Assemblies to Molecular Machines*, Wiley-VCH Verlag GmbH.

3 Stoddart, J.F. (2009) *Chem. Soc. Rev.*, **38**, 1802–1802.

4 Harrison, I.T. and Harrison, S. (1967) *J. Am. Chem. Soc.*, **89**, 5723–5724.

5 Wasserman, E. (1960) *J. Am. Chem. Soc.*, **82**, 4433–4434.

6 Brouwer, A.M., Frochot, C., Gatti, F.G., Leigh, D.A., Mottier, L., Paolucci, F., Roffia, S., and Wurpel, G.W.H. (2001) *Science*, **291**, 2124–2128.

7 Balzani, V., Clemente-Leon, M., Credi, A., Ferrer, B., Venturi, M., Flood, A.H., and Stoddart, J.F. (2006) *Proc. Natl. Acad. Sci. USA*, **103**, 1178–1183.

8 Tseng, H.R., Vignon, S.A., and Stoddart, J.F. (2003) *Angew. Chem. Int. Ed.*, **42**, 1491–1495.

9 Alteri, A., Gatti, F.G., Kay, E.R., Leigh, D.A., Martel, D., Paolucci, F., Slawin, A.M.Z., and Wong, J.K.Y. (2003) *J. Am. Chem. Soc.*, **125**, 8644–8654.

10 Durola, F. and Sauvage, J.P. (2007) *Angew. Chem. Int. Ed.*, **46**, 3537–3540.

11 Leigh, D.A., Morales, M.A.F., Perez, E.M., Wong, J.K.Y., Saiz, C.G., Slawin, A.M.Z., Carmichael, A.J., Haddleton, D.M., Brouwer, A.M., Buma, W.J., Wurpel, G.W.H., Leon, S., and Zerbetto, F. (2005) *Angew. Chem. Int. Ed.*, **44**, 3062–3067.

12 Elizarov, A.M., Chiu, S.H., and Stoddart, J.F. (2002) *J. Org. Chem.*, **67**, 9175–9181.

13 Keaveney, C.M. and Leigh, D.A. (2004) *Angew. Chem. Int. Ed.*, **43**, 1222–1224.

14 Bottari, G., Leigh, D.A., and Perez, E.M. (2003) *J. Am. Chem. Soc.*, **125**, 13360–13361.

15 Perez, E.M., Dryden, D.T.F., Leigh, D.A., Teobaldi, G., and Zerbetto, F. (2004) *J. Am. Chem. Soc.*, **126**, 12210–12211.

16 Collier, C.P., Mattersteig, G., Wong, E.W., Luo, Y., Beverly, K., Sampaio, J., Raymo, F.M., Stoddart, J.F., and Heath, J.R. (2000) *Science*, **289**, 1172–1175.

17 Hernandez, J.V., Kay, E.R., and Leigh, D.A. (2004) *Science*, **306**, 1532–1537.

18 Dietrich-Buchecker, C.O. and Sauvage, J.P. (1987) *Chem. Rev.*, **87**, 795–810.

19 Griffiths, K.E. and Stoddart, J.F. (2008) *Pure Appl. Chem.*, **80**, 485–506.

20 Cantrill, S.J., Pease, A.R., and Stoddart, J.F. (2000) *J. Chem. Soc. Perkin Trans.*, **2**, 3715–3734.

21 Vögtle, F., Dünnwald, T., and Schmidt, T. (1996) *Acc. Chem. Res.*, **29**, 451–460.

22 Kelly, T., Kay, E., and Leigh, D. (2005) *Molecular Machines*, vol. 262, Springer, Berlin, pp. 133–177.

23 Raehm, L., Hamilton, D.G., and Sanders, J.K. (2002) *Synlett*, **2002**, 1743–1761.

24 Mateo-Alonso, A., Tagmatarchis, N., and Prato, M. (2006) *Nanomaterials Handbook* (ed. Y. Gogotsi), CRC Press, Boca Raton, FL.

25 Mateo-Alonso, A., Bonifazi, D., and Prato, M. (2006) *Carbon Nanotechnology* (ed. L. Dai), Elsevier, Amsterdam.

26 Mateo-Alonso, A., Sooambar, C., and Prato, M. (2006) *Org. Biomol. Chem.*, **4**, 1629–1637.

27 Campidelli, S., Mateo-Alonso, A., and Prato, M. (2007) *Fullerenes: Principles and Applications* (eds F. Langa and J.F. Nierengarten), Royal Society of Chemistry, London.

28 Couris, S., Koudoumas, E., Ruth, A.A., and Leach, S. (1995) *J. Phys. B At. Mol. Opt.*, **2**, 4537–4554.

29 Koudoumas, E., Konstantaki, M., Mavromanolakis, A., Couris, S., Fanti, M., Zerbetto, F., Kordatos, K., and Prato, M. (2003) *Chem. Eur. J.*, **9**, 1529–1534.

30 Guldi, D.M. (2000) *Chem. Commun.*, 321–327.

31 Guldi, D.M., Rahman, G.M.A., Sgobba, V., and Ehli, C. (2006) *Chem. Soc. Rev.*, **35**, 471–471.

32 Brabec, C.J., Dyakonov, V., Parisi, J., and Sariciftci, N.S. (2003) *Organic Photovoltaics: Concepts and Realization*, Springer.

33 Mateo-Alonso, A., Guldi, D.M., Paolucci, F., and Prato, M. (2007) *Angew. Chem. Int. Ed.*, **46**, 8120–8126.

34 Diederich, F., Dietrichbuchecker, C., Nierengarten, J.F., and Sauvage, J.P. (1995) *J. Chem. Soc. Chem. Commun.*, 781–782.

35 Armaroli, N., Diederich, F., Dietrich-Buchecker, C.O., Flamigni, L., Marconi, G., Nierengarten, J.F., and Sauvage, J.P. (1998) *Chem. Eur. J.*, **4**, 406–416.

36 Schuster, D.I., Li, K., and Guldi, D.M. (2006) *C. R. Chim.*, **9**, 892–908.

37 Li, K., Schuster, D.I., Guldi, D.M., Herranz, M.A., and Echegoyen, L. (2004) *J. Am. Chem. Soc.*, **126**, 3388–3389.

38 Jakob, M., Berg, A., Rubin, R., Levanon, H., Li, K., and Schuster, D.I. (2009) *J. Phys. Chem. A*, **113**, 5846–5854.

39 Li, K., Bracher, P.J., Guldi, D.M., Herranz, M.A., Echegoyen, L., and Schuster, D.I. (2004) *J. Am. Chem. Soc.*, **126**, 9156–9157.

40 Schuster, D.I., Li, K., Guldi, D.M., and Ramey, J. (2004) *Org. Lett.*, **6**, 1919–1922.

41 Megiatto, J.D., Spencer, R., and Schuster, D.I. (2009) *Org. Lett.*, **11**, 4152–4155.

42 Megiatto, J.D., Schuster, D.I., Abwandner, S., de Miguel, G., and Guldi, D.M. (2010) *J. Am. Chem. Soc.*, **132**, 3847–3861.

43 Ashton, P.R., Diederich, F., GomezLopez, M., Nierengarten, J.F., Preece, J.A., Raymo, F.M., and Stoddart, J.F. (1997) *Angew. Chem. Int. Ed.*, **36**, 1448–1451.

44 Saha, S., Flood, A.H., Stoddart, J.F., Impellizzeri, S., Silvi, S., Venturi, M., and Credi, A. (2007) *J. Am. Chem. Soc.*, **129**, 12159–12171.

45 Nakamura, Y., Minami, S., Iizuka, K., and Nishimura, J. (2003) *Angew. Chem. Int. Ed.*, **42**, 3158–3162.

46 Diederich, F., Echegoyen, L., Gomez-Lopez, M., Kessinger, R., and Stoddart, J.F. (1999) *J. Chem. Soc. Perkin Trans.*, **2**, 1577–1586.

47 Sasabe, H., Ikeshita, K., Rajkumar, G.A., Watanabe, N., Kihara, N., Furusho, Y., Mizuno, K., Ogawa, A., and Takata, T. (2006) *Tetrahedron*, **62**, 1988–1997.

48 Sasabe, H. and Takata, T. (2007) *J. Porphyr. Phthalocya.*, **11**, 334–341.

49 Sasabe, H., Kihara, N., Furusho, Y., Mizuno, K., Ogawa, A., and Takata, T. (2004) *Org. Lett.*, **6**, 3957–3960.

50 Martinez-Diaz, M.V., Fender, N.S., Rodriguez-Morgade, M.S., Gomez-Lopez, M., Diederich, F., Echegoyen, L., Stoddart, J.F., and Torres, T. (2002) *J. Mater. Chem.*, **12**, 2095–2099.

51 Guldi, D.M., Ramey, J., Martinez-Diaz, M.V., de la Escosura, A., Torres, T., Da Ros, T., and Prato, M. (2002) *Chem. Commun.*, 2774–2775.

52 Diaz, M.C., Illescas, B.M., Martin, N., Stoddart, J.F., Canales, M.A., Jimenez-Barbero, J., Sarova, G., and Guldi, D.M. (2006) *Tetrahedron*, **62**, 1998–2002.

53 Illescas, B.M., Santos, J., Diaz, M.C., Martin, N., Atienza, C.M., and Guldi, D.M. (2007) *Eur. J. Org. Chem.*, 5027–5037.

54 Sandanayaka, A.S.D., Sasabe, H., Araki, Y., Furusho, Y., Ito, O., and Takata, T. (2004) *J. Phys. Chem. A*, **108**, 5145–5155.

55 Sandanayaka, A.S.D., Sasabe, H., Araki, Y., Kihara, N., Furusho, Y., Takata, T., and Ito, O. (2006) *Aust. J. Chem.*, **59**, 186–192.

56 Rajkumar, G.A., Sandanayaka, A.S.D., Ikeshita, K., Araki, Y., Furusho, Y., Takata, T., and Ito, O. (2006) *J. Phys. Chem. B.*, **110**, 6516–6525.

57 Sasabe, H., Sandanayaka, A.S.D., Kihara, N., Furusho, Y., Takata, T., Araki, Y., and Ito, O. (2009) *Phys. Chem. Chem. Phys.*, **11**, 10908–10915.

58 Maes, M., Sasabe, H., Kihara, N., Araki, Y., Furusho, Y., Mizuno, K., Takata, T., and Ito, O. (2005) *J. Porphyr. Phthalocya.*, **9**, 724–734.

59 Sandanayaka, A.S.D., Sasabe, H., Araki, Y., Kihara, N., Furusho, Y., Takata, T., and Ito, O. (2010) *J. Phys. Chem. A*, **114**, 5242–5250.

60 Sasabe, H., Furusho, Y., Sandanayaka, A.S.D., Araki, Y., Kihara, N., Mizuno, K., Ogawa, A., Takata, T., and Ito, O. (2006) *J. Porphyr. Phthalocya.*, **10**, 1346–1359.

61 Watanabe, N., Kihara, N., Furusho, Y., Takata, T., Araki, Y., and Ito, O. (2003) *Angew. Chem. Int. Ed.*, **42**, 681–683.

62 Sandanayaka, A.S.D., Watanabe, N., Ikeshita, K.I., Araki, Y., Kihara, N., Furusho, Y., Ito, O., and Takata, T. (2005) *J. Phys. Chem. B*, **109**, 2516–2525.

63 Sandanayaka, A.S.D., Ikeshita, K., Watanabe, N., Araki, Y., Furusho, Y., Kihara, N., Takata, T., and Ito, O. (2005) *Bull. Chem. Soc. Jpn.*, **78**, 1008–1017.

64 Mateo-Alonso, A. and Prato, A.M. (2006) *Tetrahedron*, **62**, 2003–2007.

65 Mateo-Alonso, A., Rahman, G.M.A., Ehli, C., Guldi, D.M., Fioravanti, G., Marcaccio, M., Paolucci, F., and Prato, M. (2006) *Photochem. Photobiol. Sci.*, **5**, 1173–1176.

66 Mateo-Alonso, A., Ehli, C., Rahman, G.M.A., Guldi, D.M., Fioravanti, G., Marcaccio, M., Paolucci, F., and Prato, M. (2007) *Angew. Chem. Int. Ed.*, **46**, 3521–3525.

67 Mateo-Alonso, A., Iliopoulos, K., Couris, S., and Prato, M. (2008) *J. Am. Chem. Soc.*, **130**, 1534–1535.

68 Mateo-Alonso, A., Ehli, C., Guldi, D.M., and Prato, M. (2008) *J. Am. Chem. Soc.*, **130**, 14938–14939.

69 Mateo-Alonso, A. and Prato, M. (2010) *Eur. J. Org. Chem.*, 1324–1332.

70 Da Ros, T., Guldi, D.M., Morales, A.F., Leigh, D.A., Prato, M., and Turco, R. (2003) *Org. Lett.*, **5**, 689–691.

71 Mateo-Alonso, A., Fioravanti, G., Marcaccio, M., Paolucci, F., Jagesar, D.C., Brouwer, A.M., and Prato, M. (2006) *Org. Lett.*, **8**, 5173–5176.

72 Mateo-Alonso, A., Fioravanti, G., Marcaccio, M., Paolucci, F., Rahman, G.M.A., Ehli, C., Guldi, D.M., and Prato, M. (2007) *Chem. Comm.*, 1945–1947.

73 Mendoza, S.M., Berna, J., Perez, E.M., Kay, E.R., Mateo-Alonso, A., De Nadai, C., Zhang, S., Baggerman, J., Wiering, P.G., Leigh, D.A., Prato, M., Brouwer, A.M., and Rudolf, P. (2008) *J. Electron Spectrosc.*, **165**, 42–45.

74 Mateo-Alonso, A., Brough, P., and Prato, M. (2007) *Chem. Commun.*, 1412–1414.

6
Biomimetic Motifs Toward the Construction of Artificial Reaction Centers

Bruno Grimm and Dirk M. Guldi

6.1
Introduction

One of the key aspects of switching to carbon-free/innovative sources of energy is the research on solar energy conversion. The Sun provides planet Earth with energy that corresponds to almost 6000 times the current global consumption of primary energy (14 TW) [1–3]. Considering advantages like inexhaustibility and worldwide accessibility, solar energy conversion has the potential to become the major component of innovative carbon-free energy sources. Since the discovery of the photovoltaic effect by Becquerel in 1839 and the introduction of the first photovoltaic cell in the 1950s, much effort has been directed to pioneering the use of solar radiation as a primary energy source [4]. To this end, an evolution of solar technologies has started with a large variety of solar technologies being available [5–7]. However, the market for primary energy is still dominated by fossil fuels, and high costs for present solar technologies hamper a broader accessibility to commercial energy markets [8]. Therefore, research efforts are mainly focused on low-cost and/or high efficiency with some solar technologies that have already reached a mature stage. In the following discussion, three general approaches in solar energy conversion are considered: (i) photon-to-electric energy conversion, (ii) photon-to-thermal-to-electric energy conversion, and (iii) photon-to-chemical energy conversion.

Photovoltaic devices directly convert energy following the absorption of light. For example, a semiconductor as photoactive material absorbs light of energy that is equal to or higher than its bandgap. The photon absorption promotes an electron from the valence band to the conduction band to generate an electron–hole pair. To succeed in energy capturing and current generation, the electron–hole pair recombination needs to be suppressed. For photovoltaic devices based on inorganic semiconductors as photoactive material, the charges (i.e., electrons and holes) are separated by an electric field at the p- and n-doped junction. However, such a type of charge separation also carries the disadvantages of inorganic semiconductors. On the one hand, the efficiency is decisively controlled by bulk properties such as crystallinity and chemical purity. On the other hand, the efficiency is directly

Supramolecular Chemistry of Fullerenes and Carbon Nanotubes, First Edition. Edited by Nazario Martin and Jean-Francois Nierengarten.

affected by the number of charge carriers that migrate to the n–p junction before recombining [5–7, 9].

Regarding organic photovoltaic devices (OPVs), the processes of light excitation and charge separation differ fundamentally. For organic semiconductor materials, the electronic and optical properties are illustrated by the summation of the individual atomic orbitals. Generally, the energy levels of the highest occupied molecular orbital (HOMO) and the lowest unoccupied molecular orbital (LUMO) define the bandgap energy. The latter influences the molecule's properties. When small molecules or conjugated polymers are irradiated, excitons, that is, electron–hole pairs bound together by electrostatic and/or lattice interactions, are generated. A way to separate these excitons and to yield free electrons and holes is to engineer suitable electron donor–acceptor interfaces. When, for example, the LUMO of the electron acceptor-level resides below the electron donor's LUMO level, an excited state is generated at the junction between both materials that easily dissociates. This dissociation creates electrons on one side of the junction, separated from the holes on the other side of the junction. Implicit in it is a photoinduced energy gradient at the interface, which then drives the photovoltaic effect. The efficiencies of organic photovoltaic devices are restricted by several requirements. First, excitons have to reach the electron donor–acceptor interface before recombining. Second, free electrons and holes need to be efficiently transferred to the electrodes before a back transfer from the LUMO of the electron acceptor to the HOMO of the electron donor occurs [10, 11].

Improvements and optimizations in terms of higher light absorption, faster charge separation, better transport, and more efficient collection go hand in hand with higher overall efficiencies. Therefore, novel paradigms in both inorganic and organic photovoltaic technologies are progressing, namely, setups based on inorganic semiconductor materials such as silicon (c-Si, pc-Si, or α-Si) [6, 12, 13], III–V compounds (GaAs, InP) [14, 15], and chalcogenides (CdTe, CIGS) [16, 17], and various organic-based thin films [18, 19].

Another *modus operandi* for photoelectrochemical cells is dye-sensitized solar cells (DSSC) [20, 21]. In this type of solar system, an organic dye – placed on the surface of a nanostructured metal oxide semiconductor substrate (TiO_2) – absorbs light, which, in turn, gives rise to an excited state that deactivates via electron injection into the conduction band of TiO_2. A redox couple in solution, such as ascorbate or I^-/I_3^-, scavenges the simultaneously generated holes. Further improvements were made by depositing molecules to form stacked devices (i.e., multilayer organic devices) and in the field of bulk heterojunctions. To this end, organic electron donor and electron acceptor materials are blended on the nanoscale. Synthetic efforts, on the other hand, were made to create artificial photosynthetic macromolecular structures, where absorption of light and charge separation are realized by separate complexes within the same structure.

All the above-mentioned photovoltaic devices successfully convert solar energy to electric energy. Still they differ substantially in their overall efficiencies and technological limitations. For instance, inorganic photovoltaic devices are generally limited in their efficiency by the overlap between the solar spectrum and the semiconductor bandgap, some optical losses due to reflection off the cell surface, recombination of electron–hole pairs, and the resistance of the metal–semiconductor contact [22]. However, silicon photovoltaic devices are promising candidates for high-performance solar energy conversion. Efficiencies exceeding 20% for c-Si-based photolithographically passivated emitter solar cells (PESC) have been reported [23]. Other examples, such as the back point contact cell with 22.3% [24] and the passivated emitter, rear locally diffused (PERL) with 24.7% efficiencies [25], underline the successful development. Inorganic chalcogenide-based $CdTe/CdS/SnO_2$ devices can be deposited by means of spray pyrolysis, electrodeposition, vapor deposition, and close space sublimation, reaching conversion efficiencies of 16.5% [26]. Notably, the sensitivity of these materials against moisture constitutes a major drawback to cell stability. The fabrication and synthetic methods of organic photovoltaic devices vary substantially and show for that reason quite variable characteristics in efficient solar energy conversion. All these characteristics have to be tuned and optimized, starting from the synthesis and design of molecular building blocks to general fabrication routes. Key features for an optimized organic photovoltaic device efficiency are the open-circuit voltage (V_{OC}) and the short-circuit current (I_{SC}). V_{OC} relates mainly to the difference between the LUMO of the electron acceptor and the HOMO of the electron donor. I_{SC} should be at its maximum, achievable by photon absorption enhancement (e.g., using NIR sensitive, low bandgap organic semiconductors or active electron acceptor materials) or by an increase in charge mobility and charge collection [27].

In the photon-to-thermal-to-electric energy conversion approach, the concentrated solar radiation energy is transformed to thermal energy, which is then converted to electric energy. Here, heat storage tanks (molten salt tanks or concrete blocks) are used to store heat during the day to power steam turbines during the night or during peak demand. The so-called concentrating thermal power (CSP) plants are online – see, for example, Desertec Foundation – and supply power continuously for 24 h a day. To this date, photovoltaic devices are lacking this "on-demand" feature. In fact, they fall behind owing to their comparable high investment costs [28, 29].

In photon-to-chemical energy conversion, chemical fuels with a focus on energetic advantages compared to conventional technical processes are produced. In general, the fuel production requires a source of energy, a material that can be oxidized to produce electrons, and a material that can be reduced by those electrons to yield a fuel. Frequently discussed chemical fuels are methane, ammonia, and hydrogen. But on a scale that meets our demand, the most promising fuel is hydrogen [30]. Here, water may act as a source of electrons, which produces oxygen and hydrogen ions. Protons are convenient materials for reduction to produce hydrogen gas. Green plants, some bacteria, and some protistans master this approach. In particular, the energy of the sun is used to oxidize water and to generate electrons with a potential negative enough to produce adenosine triphosphate (ATP and/or β-nicotinamide adenine dinucleotide phosphate (NADPH). Antenna molecules in pigments absorb light in the visible light region and transfer the excitation energy to the photosynthetic reaction center (PRC). Cascades of energy and electron transfer reactions between electron donors and electron acceptors that feature exceptional low reorganization energies (λ) guarantee that long-lived charge-separated states are generated. The latter allows migration of the reducing and oxidizing equivalents to the catalytic sites for water oxidation and fuel production. Importantly, the catalytic sites for water oxidation are identical in all of the known photosynthetic active organisms. Redox couples, on the other hand, transport the reducing equivalents, and a transmembrane redox gradient powers the synthesis of ATP and NADPH. All the steps in photosynthesis are exergonic, which are regulated by various catalysts. These catalysts control kinetically all the processes and prevent the production of reactive and dangerous intermediates that could harm the cell environment.

The ultimate goal of contemporary research is the successful implementation of photobiological photon-to-chemical energy processes into artificial photosynthetic fuel production systems. To date, practical applications are still far fetched. In recent years, considerable progress has been made in the development of separate components. Nevertheless, for functionally linking these components, more fundamental research on the molecular mechanisms, the structure and functionality of enzyme catalysts, and the kinetics of biological hydrogen metabolism is necessary [31–36].

All the aforementioned approaches to convert solar energy are helpful in harvesting the abundance of energy around us on a daily basis. All the described technologies face, however, comparable challenges, in particular, the aspects related to the steps involved in the conversion of photon energy into electricity, that is, photon absorption, charge separation, and charge transport. Therefore, understanding the nature of electron transfer reactions is key to developing a comprehensive picture of how photosynthetic charge separation, charge transport, and charge storage occur. Basic findings will allow for the design of bioinspired blueprints for fuel production through artificial photosynthesis and, moreover, will result in substantial improvements in efficient charge separation and migration for the fabrication of photovoltaic devices. Such breakthroughs and technological advances are essential to overcome the barriers of conversion efficiency and manufacturing costs and pave the way for solar energy to be a major contribution to a future carbon-free energy portfolio.

6.2 Supramolecular Architectures for Solar Energy Conversion

6.2.1 General Considerations

But what are the key requirements for the successful design of the so-called artificial photosynthetic reaction center? The overall starting point for artificial photosynthesis is a light-induced electron transfer. For mimicking electron transfer reactions, the reaction center should consist of a light absorbing chromophore that acts as electron donor and an additional electron-accepting moiety or chromophore. The number of photons that are absorbed by the chromophore governs the electron transfer efficiency. Therefore, much effort has been directed to the design of integrative building blocks, which harvest light throughout the entire wavelength region of the solar spectrum. To this end, suitable molecular building blocks with absorption cross sections in the visible light region are fullerenes. Since the initial discovery by Curl, Kroto, and Smalley in 1985 [37–40] and the large-scale synthetic accessibility by the Krätschmer and Huffman carbon arc method [41], chemists and physicists have studied their appealing properties. The latter range from superconductivity in the solid state [42] to light-activated antimicrobial agents in living cells [43]. In the context of photoinduced electron transfer reactions, fullerenes emerge as excellent electron acceptors. Fullerenes readily accept up to six electrons in electrochemical experiments, and the extraordinary small reorganization energy in electron transfer reactions has led to significant breakthroughs in the field of mimicking photosynthesis [44–49]. Light absorbing electron donors are virtually all around us including natural chlorophylls and bacteriochlorophylls. In the context of optimizing photosynthetic model systems, diverse electron donors, namely, porphyrins, phthalocyanines, and tetrathiafulvalenes, have been linked to fullerenes. All the aforesaid electron donor acceptor systems show rich and extensive absorptions in the solar spectrum, and together with their excellent redox properties meet the requirements as essential components in solar energy conversion.

The ultimate goal in the design of photosynthetic mimicking is an ultrafast electron transfer combined with a slow back electron transfer. Factors such as the nature of the photo- and redox-active units, solvent, electron donor–acceptor distance, and molecular topology and the nature of the connecting link affect strongly the photophysical outcome. An elegant approach to address all the aforementioned factors is the use of supramolecular motifs to assemble electron donor and electron acceptor moieties. For the construction of new ensembles, biomimetic motifs such as electrostatics, hydrogen bonding, π–π stacking, and metal coordination provide fundamental assets. First, supramolecular binding is reversible and the binding strength is tunable by means of the binding motif and temperature. This is in sharp contrast to covalent motifs linked. Second, nature relies on supramolecular motifs to assemble individual building blocks. The surrounding protein matrix secures the photosynthetic active components in a well-defined spatial arrangement and mediates the energy/electron transfer processes [50, 51].

Although the major part of this chapter focuses on supramolecular model systems for electron transfer reactions, the first examples of fullerene-based electron transfer assemblies were made in the field of covalently linked electron donor–acceptor conjugates. In this regard, Imahori and coworkers reported a porphyrin–fullerene conjugate, where a porphyrin and a fullerene are connected via an amido bond at the aryl ring *para* position (**1**) [52, 53]. Owing to the double-bond character of the amido bond, the

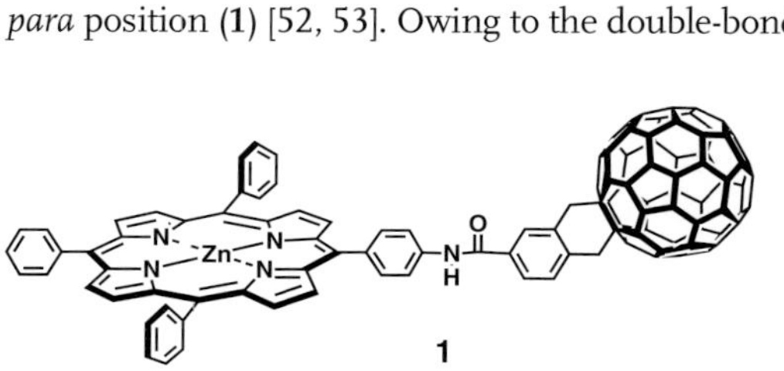

N Zn N N N N H O

1

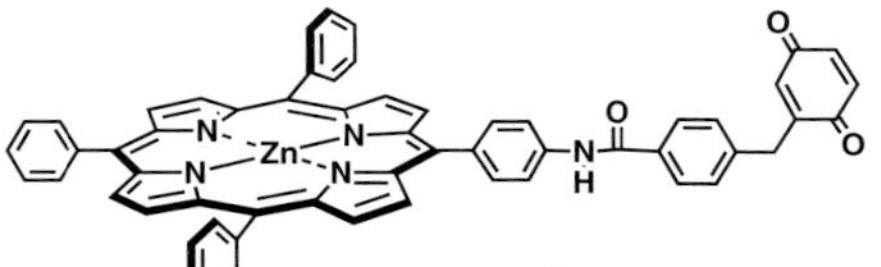

2

entire conjugate is considered to be rigid with an edge-to-edge distance (or center-to-center distance) of about 12.6 Å (18.6 Å). As shown by absorption, fluorescence, and transient absorption measurements, an electron transfer occurs from the porphyrin singlet excited state to the fullerene. As a matter of fact, the formation of the radical cation species of the porphyrin and radical anion species of the fullerene evolves. Many electron donor fullerene systems have been reported to this day. Importantly, the size and shape effect of fullerenes on electron transfer processes was not established at the time of Imahori's study. Therefore, a structurally related porphyrin-quinone **2** with similar features regarding electron acceptor strength and electron donor–acceptor linkage (edge-to-edge distance is 12.7 Å) was designed. Subsequently, an accurate evaluation of fullerene and quinone-based electron donor–acceptor systems was performed. Porphyrin-quinone **2** resembles the fullerene-based **1** in terms of electron transfer from the porphyrin singlet excited state to the electron-accepting unit. A closer look revealed, however, noticeable differences. For instance, the rate constant for charge separation in **1** is 6 times faster than that in **2**, whereas the rate of charge recombination in **1** is 25 times slower relative to the analogous rate constants noted for **2**. The driving force for electron transfer is in the porphyrin–quinone conjugate 0.28 eV higher than the driving force in the porphyrin–fullerene conjugate.

Nevertheless, the resulting differences are rationalized within the framework of the Marcus theory of nonadiabatic electron transfer. The semiclassical theory deals with the rate constants of charge separation and charge recombination as dependence on the free energy changes of the reaction, namely, ΔG^0_{CS} and ΔG^0_{CR} [54–56]. The electron transfer rate constant is equal to the product of a preexponential term including the matrix element V, which describes the electronic coupling between electron donor and electron acceptor, and an exponential term. The exponential term is given by the reorganization energy (λ) and the free energy change ($-\Delta G^0$). The reorganization energy refers to changes in the internal structure of the building blocks and the organization of the surrounding solvent system when going from the initial to the final state.

$$k_{ET} = \sqrt{\left(\frac{4\pi^3}{h^2 \lambda k_B T}\right)} \times V^2 \times \exp\left(-\frac{(\lambda + \Delta G^0)^2}{4\lambda k_B T}\right) \qquad (6.1)$$

With these parameters in hand, the correlation of electron transfer rate and free energy change falls into three different regimes. First, the electron transfer rate increases with increasing thermodynamic driving force ($-\Delta G^0 < \lambda$), known as the normal region of the bell-shape relationship. Second, the driving force is equal to the reorganization energy ($-\Delta G^0 \sim \lambda$), the reaction rate is at its maximum and is governed mainly by the magnitude of the electronic coupling (V). Third, going

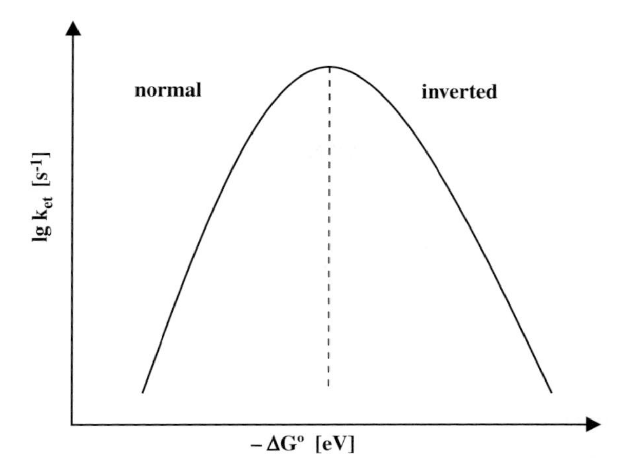

Figure 6.1 Parabolic dependence of electron transfer reactions versus thermodynamic driving force.

beyond the thermodynamic maximum, the exothermic region of the parabola ($-\Delta G^0 > \lambda$) is reached. Here, an additional increase in the driving force yields an actual slowdown of the reaction rate. The latter is the so-called Marcus inverted region (Figure 6.1).

Upon recasting the porphyrin-fullerene **1** and porphyrin-quinone **2**, the accelerated charge separation and decelerated charge recombination in **1** – compared to **2** – are explained by the smaller reorganization energy of fullerenes compared to that known for quinones. The smaller reorganization energy forces the charge separation into the normal region of the Marcus parabola ($-\Delta G^0 \sim \lambda$), while the charge recombination occurs in the Marcus inverted region ($-\Delta G^0 > \lambda$). A plausible explanation for this observation relates to the "three-dimensional" π system of fullerenes. The spherical structure delocalizes charges over the whole framework, whereas quinone with its confined "two-dimensional" π system localizes the charge mainly on the oxygen atoms. As a consequence, the charge density at each fullerene carbon atom – in the photoreduced fullerene – is much smaller than in the photoreduced quinone. In fact, this defines the smaller reorganization energy of fullerenes. This pioneering work is one of the inceptions for an unlimited number of fullerene-based electron donor–acceptor conjugates and hybrids. A variety of organic electron donor molecules were integrated via covalent bonds to modulate the spatial arrangement of electron donor and fullerene. By this means, the free energy changes for charge separation were decreased, on the one hand, and the energy gap for the charge recombination process was increased, on the other hand.

In terms of artificial photosynthesis, efficient and long-lived charge separation in photosynthetic bacteria is achieved by implementing electron transfer cascades. Tuning the reorganization energy is key to minimizing the loss of free energies and to force the charge recombination into the Marcus inverted region. In both theoretical calculations and experiments, reorganization energies between 0.2 and 0.3 eV (i.e., charge separation mechanism from the chlorophyll dimer to the accessory chlorophyll) were obtained. Such outstandingly small reorganization energies originate from the spatially large redox centers delocalizing their charges, similar to what is proposed for fullerenes [53].

In this light, artificial photosynthetic model systems with suitable redox centers, free energies, and reorganization energies could emerge as potential candidates toward a photosynthetic system that features efficient and long-lived charge-separated state. But the spatial arrangement of the redox building blocks plays a critical role.

The supramolecular approach serves as an elegant tool to assemble the essential building blocks and is represented in the following in order of increasing selectivity. Selectivity – along with affinity– has always been an important goal in supramolecular chemistry [57, 58]. Binding motifs such as Coulomb interactions and π–π stacking with bonding energies of 100–350 kJ/mol and 0–50 kJ/mol, respectively, reveal reasonable binding affinities to assemble electron donor–acceptor conjugates. They lack, however, selectivity and sensitivity. More directional and selective with bonding energies of 4–120 kJ/mol are hydrogen bonds. A way to overcome poor stability single-point hydrogen bonding lies in the successful implementation of multiple point hydrogen bonding. Similar structural control over supramolecular conjugates is attained in the metal–ligand coordination approach with bonding energies of 50–200 kJ/mol. All the aforesaid binding motifs differ substantially in terms of binding affinity, sensitivity, and selectivity, but in combination they facilitate the assembly of the individual building blocks into architectures favorable to solar energy conversion.

6.2.2
Coulomb Interactions

In natural photosynthetic systems, Coulomb interactions are of paramount importance and can be found in discriminative facets of biological systems. They affect the electrostatic environment of proteins, which, in turn, determines their binding affinities and functions in biological systems. It comes as no surprise that Coulomb interactions modify the function and conformation of proteins and molecules involved in the photosynthetic reactions and, as a consequence, control the electron transfer events. Therefore, the interest in the context of supramolecular electron transfer reactions is focused on the generation of Coulomb complexes comprising dendritic fullerene carboxylates (i.e., **4** and **5**) and octapyridinium ZnP salt **3** [59–61]. Job's plot experiments confirmed a 1 : 1 stoichiometry for **3·4**, for which an appreciably high association constant of $10^8\,M^{-1}$ was calculated. Such a high complex stability originates from a combination of electrostatic and charge transfer interactions that are operative between the electron accepting C_{60} and C_{70} oligocarboxylate and the electron donating ZnP. In particular, a fluorescence intensity decrease in the octapyridinium ZnP salt as the dendritic C_{60} oligocarboxylate concentrations are increased and evidence for a new, short-lived emissive component suggests a static quenching event inside **3·4**. From steady-state fluorescence titrations, a value of $(1.1 \pm 0.1) \times 10^8\,M^{-1}$ was derived for the association constant of the ZnP/C_{60} (**3·4**) hybrid at pH 7.2. Stability assays revealed that the complex disassembles at higher ionic strength with little dependence on the nature of the anions. Coulomb interactions obviously play a major role. Unambiguous evidence for an intracomplex electron transfer quenching, namely, formation of $ZnP^{\bullet +}/C_{60}^{\bullet -}$ (3·4) and $ZnP^{\bullet +}/C_{70}^{\bullet -}$ (3·5) radical ion pairs, was gathered in time-resolved transient absorption measurements. Importantly, the lifetimes of the corresponding radical ion pairs range from nanoseconds to a few microseconds.

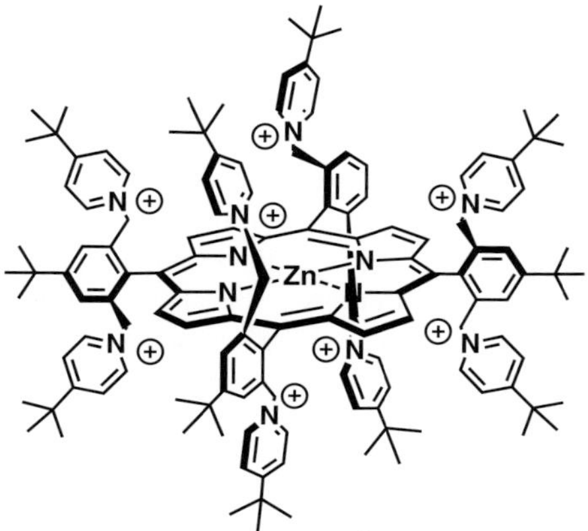

3

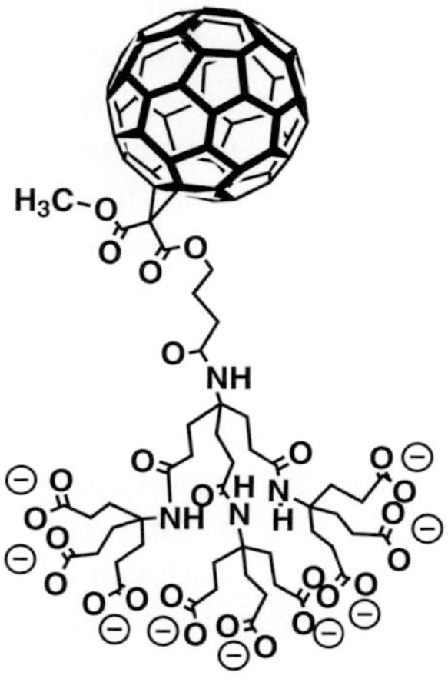

4

As a complement, the same dendritic C_{60} oligocarboxylate was assembled with cytochrome *c* (Cyt*c*) via electrostatic interactions [62]. The zinc analogue of Cyt*c* was prepared according to a modified procedure and helped to accomplish the photophysical investigations. Fluorescence spectroscopy and transient absorption spectroscopy confirmed the association of ZnCyt*c* and the dendritic C_{60} oligocarboxylate ($\sim 10^5\,M^{-1}$), and a photoinduced electron transfer within the ZnCtyc/C_{60}, which evolves from the photoexcited protein to the fullerene. Extensive molecular dynamics (MD) simulations and circular dichroism supported these findings.

In a more recent example, the dendritic C_{60} oligocarboxylates **4** and **7** were used to construct sapphyrin-based electron transfer model systems [63–65]. Sapphyrin (**6**) is characterized by a first excited singlet state that is about 0.20 eV lower than those of porphyrins, which renders them of particular interest as a potential photodonor. The changes associated with the sapphyrin's Soret band absorption were used to perform

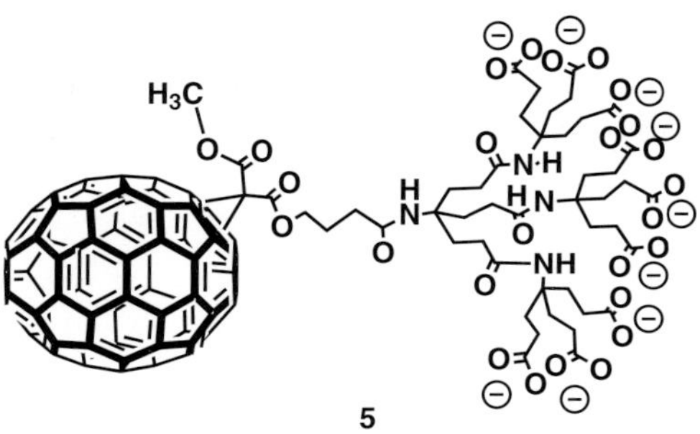

5

Job's plot experiments in phosphate buffer (pH 7.2, ionic strength = 0.012). In particular, they provide clear proof for a 1 : 1 stoichiometry for **6·4** and a 2 : 1 stoichiometry for **(6)$_2$·7**. Association constants of $10^6\,M^{-1}$ for **6·4** and $10^6/10^5\,M^{-1}$ for **(6)$_2$·7** were calculated for the stepwise formation of the 1 : 1 and 2 : 1 complexes of **(6)$_2$·7**, respectively. In additional experiments, the nature of the interactions in **6·4** and **(6)$_2$·5** was probed by increasing the ionic strength, while different amounts of NaCl were added, and a drop in binding strength was detected for **6·4** and **(6)$_2$·7**. Such a trend is consistent with interactions that are mediated by anion binding, rather than, for instance, solvophobic or direct C_{60}/sapphyrin electron donor–acceptor interactions. Time-resolved transient absorption measurements gave conclusive evidence for the proposed electron transfer process. Irradiation with 468 nm laser pulses gave rise to charge-separated states that feature charge separation lifetimes of 470 and 600 ps in the case of **6·4** and **(6)$_2$·7**, respectively. In particular, the charge-separated state of **(6)$_2$·7** is notably stabilized compared to that of **6·4**. Charge delocalization between the two sapphyrin moieties **6** might be responsible for this difference.

Earlier work by Wasielewski and Sessler was able to demonstrate energy transfer within a carboxylate anion-linked porphyrin/sapphyrin hybrid ensemble **8** [64]. In this particular case, the sapphyrin moiety functions as an energy acceptor rather than as a photodonor. Unfortunately, the electrostatic bonding motif that is imperative between the carboxylate anions and the sapphyrin is rather weak – $10^3\,M^{-1}$ in

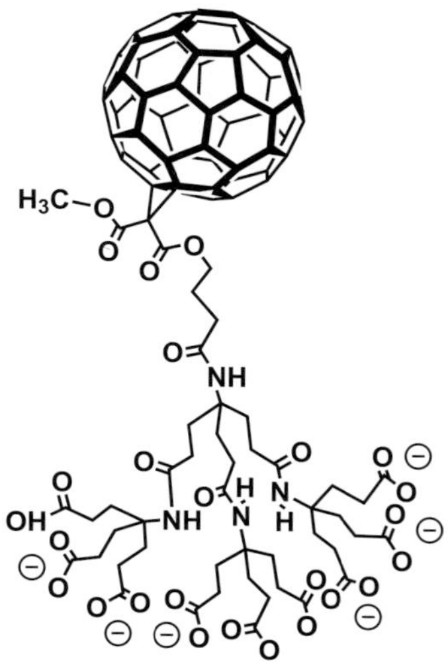

4

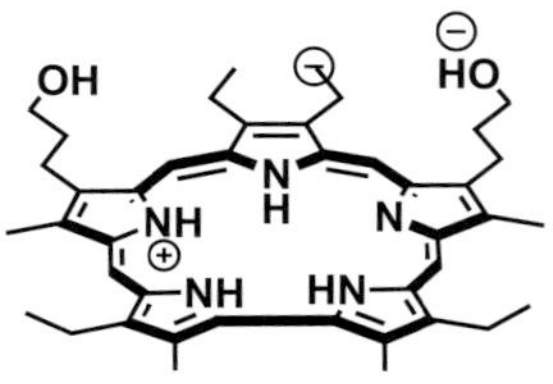

6

CH_2Cl_2. This, in fact, rendered it very difficult to construct well-defined electron donor–acceptor hybrids. However, time-resolved transient absorption measurements on the picosecond timescale with excitation pulses at 417 nm unravel a singlet–singlet energy transfer from the porpyhrin to sapphyrin with kinetics that are consistent with a Förster-type mechanism.

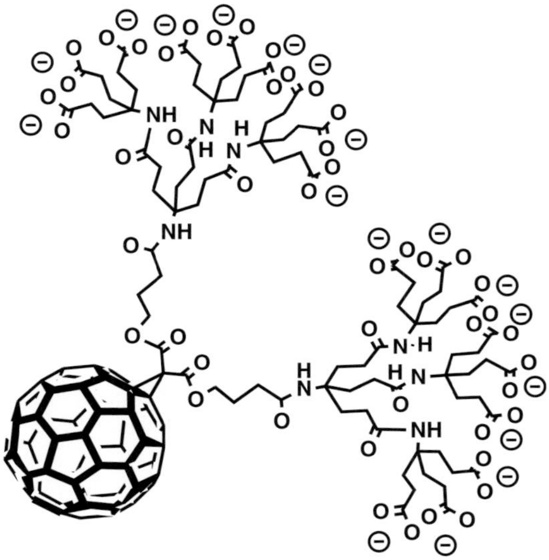

7

6.2.3
π–π Stacking

In the absence of alternative bonding motifs such as Coulomb interactions, complementary size and maximizing the number of points of interactions are key factors in devising stable fullerene architectures. The careful control over the competition between host–host, guest–host, and guest–guest interactions is important in determining the structure of supramolecular ensembles. In fullerene chemistry, for instance, C_{60}–C_{60} interactions play a major role [51]. Thus, a porphyrin "cyclic dimer" (**9**) [66, 67] and a porphyrin "jaw" (**10**) [68] were developed to gain topological control in π–π associates [69]. The electron-rich walls of the porphyrins and their considerable contact with incumbent C_{60} encouraged experiments, and strong interactions were indeed detected [70–73]. In both constructs, discrete van der Waals complexes are realized with a core of two porphyrins (i.e., PdP, palladium 3-pyridiyltriphenylporphyrin, or ZnP, zinc biphenyltetrahexylporphyrin) controlling the selective C_{60} incorporation. The close proximity in **9** and **10** is the inception to π electronic donor–acceptor interactions. Such effects are detectable in the shifts of the absorption bands in the spectra of the resulting composites. In comparison to the model porphyrin systems, redshifts of the Soret- and Q-band transitions, accompanied by lower extinction coefficients, are consistently observed and attest to the mutual perturbation of the π systems. Concomitantly, the chromophore's emission gives rise to a progressive quenching after addition of variable C_{60} concentrations.

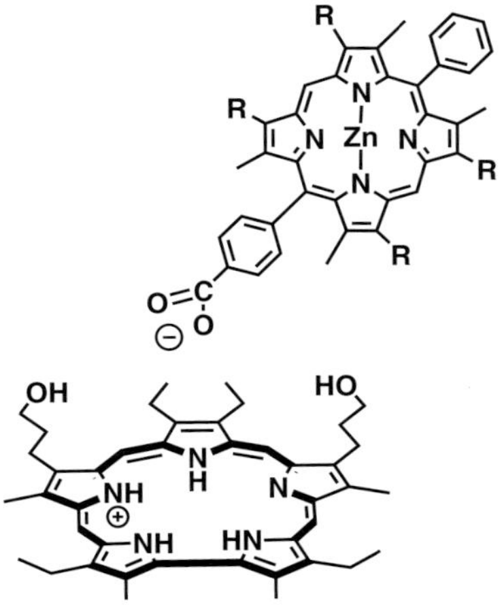

8

More recently, a calixarene scaffold bearing two porphyrins – bisporphyrins **11** or **12** – emerged as a supramolecular host for the efficient inclusion of C_{60}. In this approach, the systematic variation of (i) the linkage, (ii) the type of porphyrin, and (iii) the solution properties were considered to be unraveling factors, which affect the binding of C_{60}. Interestingly, the differences in binding constants point to a strong dependence on different solvation energies and show that the desolvation of C_{60} is one of the major keys in controlling the inclusion [74].

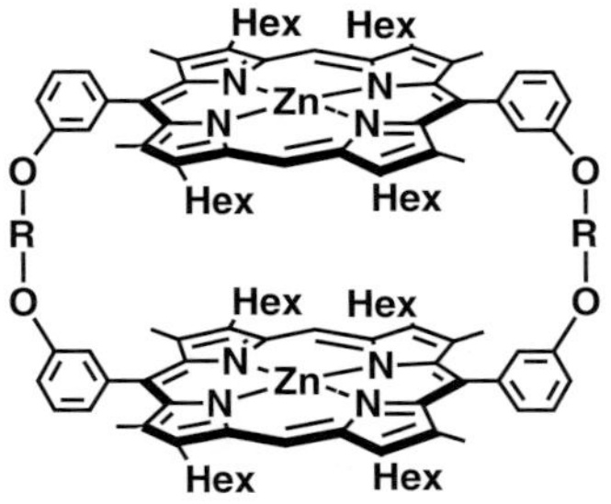

9

On the basis of the findings by Boyd and coworkers [74], the complexation chemistry of **11** and **12** was used to afford host/guest inclusion complexes with a number of different fullerenes (i.e., C_{60} (**13**), $Sc_3N@C_{80}$(**14**), and

$Lu_3N@C_{80}$(**15**)) [75]. In this respect, the pocket-like structure defined by the calixarene bisporphyrin hosts is ideally suited to bind fullerenes and afford interesting host–guest structures. By conducting absorption/fluorescence assays, the supramolecular inclusion of the aforementioned building blocks were probed. The observed exponential concentration/absorption and concentration/fluorescence relationships allowed a precise characterization of the 1 : 1 binding process with resulting K_a values that are on the order of 10^3–10^5 M^{-1}. Femtosecond transient absorption spectroscopy unraveled the nature of excited state interactions between the bisporphyrins (**11** and **12**) and the fullerenes (**13**, **14**, and **15**). In the case of **12·13**, **12·14**, and **12·15**, charge separation dominates the intrinsic deactivation pathways to give the one-electron oxidized form of **12** and the one-electron reduced species of **13**, **14**, and **15**. In particular, charge separation arises in ~100 ps with lifetimes for the charge-separated state of 534 ps and 781 ps for **12·13** and **12·14**, respectively. Turning to the differential absorption changes that evolve upon photoexciting **11·14**, instead of seeing the spectral markers of the one-electron oxidized form of **11** and the one-electron reduced form of **14**, it is the one-electron reduced form of **11** and the one-electron oxidized form of **14** that develop within the first 200 ps. Conclusive evidence for this electron transfer process came from spectroelectrochemistry measurements, wherein under oxidative and reductive conditions the characteristic maxima of oxidized **14** and reduced **11** could be noted. The close agreement with the photolysis experiments corroborates the $\mathbf{11}^{\bullet-}\cdot\mathbf{14}^{\bullet+}$ ion pair formation. The characteristic transient bands were used to determine the lifetime of the charge-separated state (229 ps) via multiwavelength analysis.

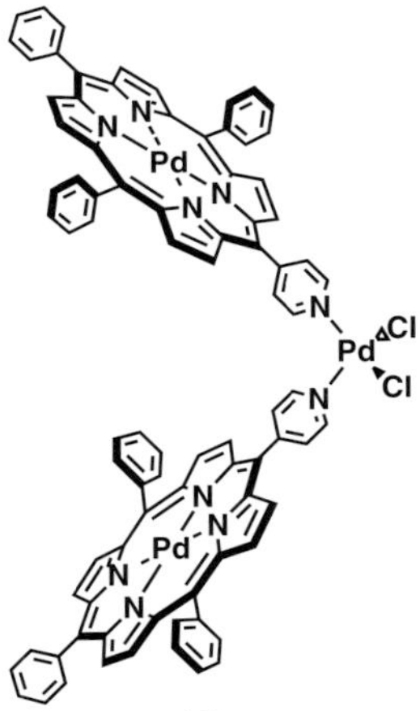

10

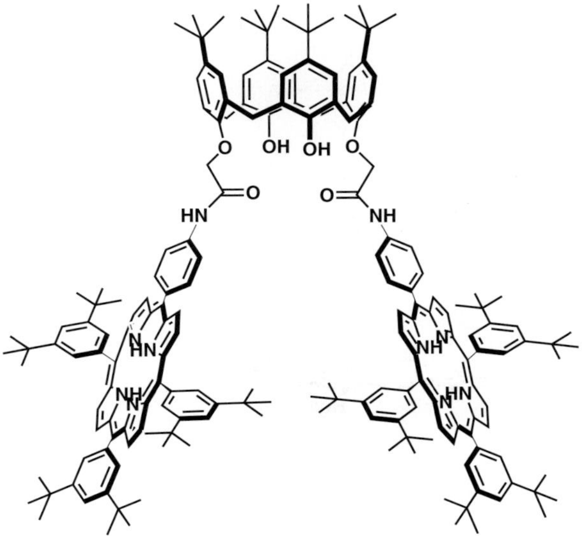

11

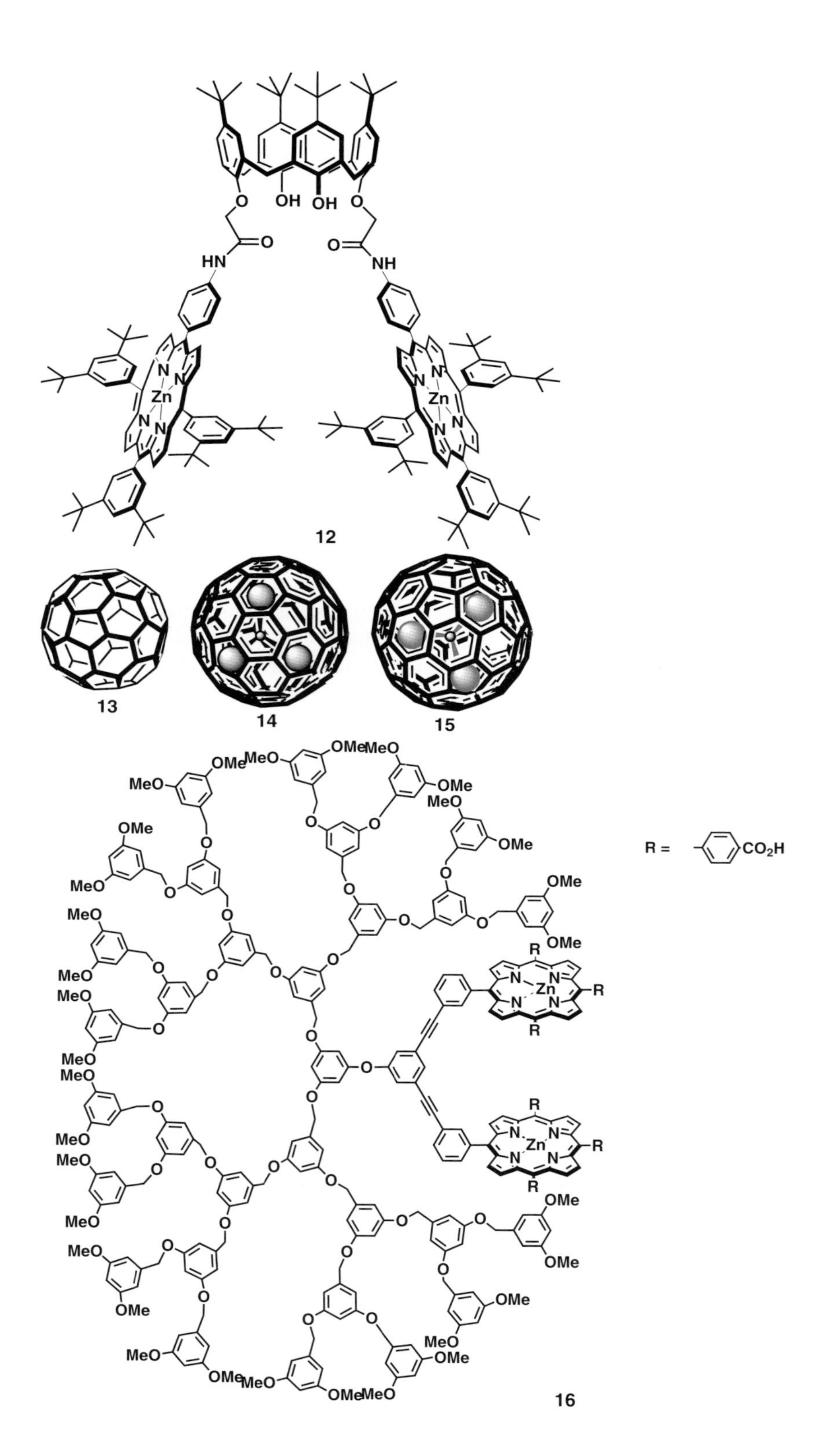

OH
OH
O
O
HN
NH
N
N
Zn
N
N
12
13
14
15
MeO
OMe
R =
CO2H
R
Zn
16

The affinity of cyclic dimers of metalloporphyrins toward fullerenes, such as C_{60} and C_{70}, detailed in the preceding paragraph, was used in related work as a platform to form "supramolecular peapods" composed of a hydrogen-bonded ZnP nanotube and fullerenes [76]. According to the authors' protocol, a 1,1,2,2-tetrachloroethane (TCE) solution of a mixture of **16** (1.2×10^{-5} M) and C_{60} (2.4×10^{-5} M) was heated once at 120 °C and then allowed to stand at 40 °C for 4 days. TEM observation of the resulting mixture showed the presence of very long (i.e., > 1 μm) fibers with a uniform diameter of 15 nm. Interestingly, the ZnP dimer forms in the absence of fullerenes only a heavily entangled, irregular assembly. In contrast with **16**, an ester version of the acyclic dimer without hydrogen bonding capability hardly interacts with fullerenes.

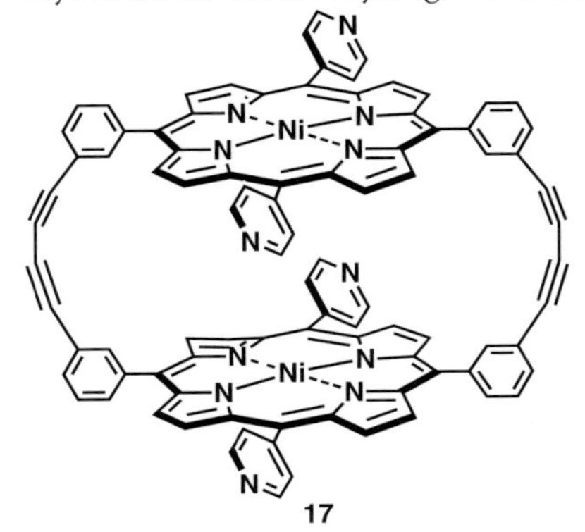

17

Another example is worth noting, wherein the first X-ray crystal structure determination of a supramolecular peapod was reported on a porphyrin nanotube from the cyclic porphyrin dimer **17** with self-assembling pyridyl substituents [77]. The tube is constructed by stacking the cyclic molecules through unique C–H···N hydrogen bonds and π–π interactions of the pyridyl groups in the crystal. The existence of a 1 : 1 inclusion complex of C_{60} with **17** was confirmed both in the solution and in the crystal.

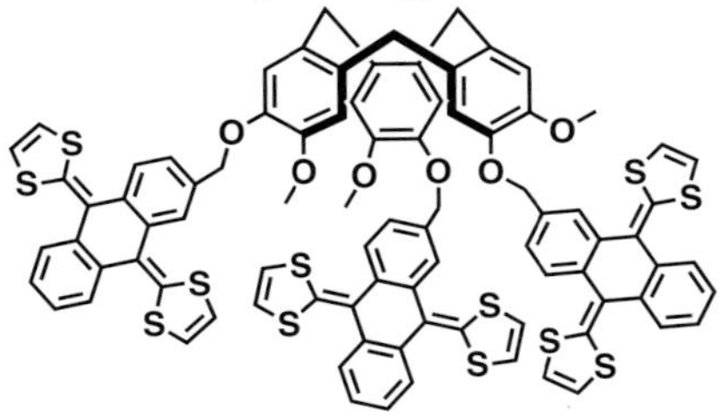

18

In another elegant approach, the design of stable noncovalent fullerene architectures was accomplished on the basis of π–π interactions by using exTTF as fullerene receptors. For instance, the concave aromatic surface of 2-[9-(1,3-dithiol-2-ylidene) anthracen-10(9*H*)-ylidene]-1,3-dithiole (exTTF) was exploited for the molecular recognition of C_{60}. Maximizing the number of exTTF hosts goes hand in hand with increased affinity to C_{60}. In a recent work, Martin and coworkers used cyclotriveratrylene as a scaffold to increase the number of exTTF binding sites (**18**) [78]. The concave surfaces of both the CTV and the exTTF subunits wrap around the entrapped fullerene to yield stable complexes with binding constants of $2.0 \times 10^5\ M^{-1}$ for C_{60} and $2.0 \times 10^6\ M^{-1}$ for C_{70}. Light-induced ESR experiments were used as a proof for the existence of a photoinduced electron transfer from exTTF to C_{60}, illustrated by an albeit weak signal ($g = 2.0014$; $\Delta H = 5.9$ G) that increases in intensity with irradiation time and decreases rapidly when the light is switched off.

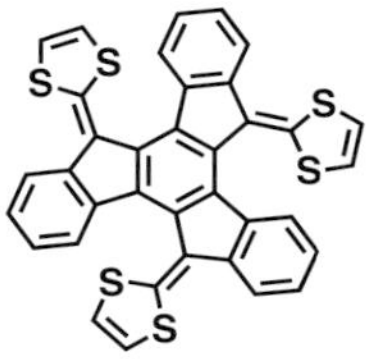

19

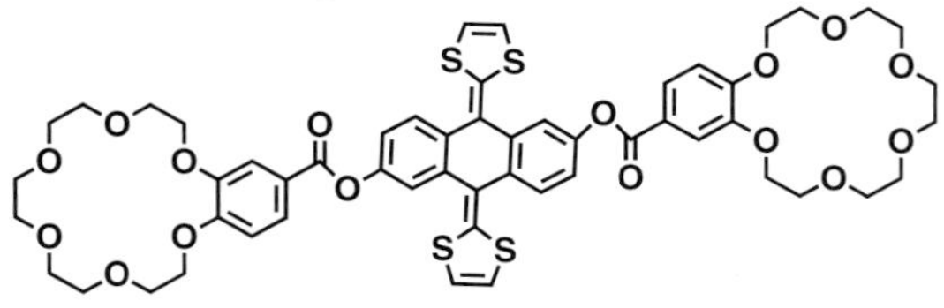

20

Another well-suited scaffold decorated with thiole units that has emerged as an excellent host for fullerene is truxene **19** [79, 80]. The extended π-delocalized aromatic truxene core provides a large aromatic surface with which fullerenes establish favorable noncovalent interactions. In fact, the association of **19** and fullerenes in solution was investigated by ^1H NMR titrations with C_{60} and C_{70} as guests affording binding constants of $1.2 \times 10^3\ M^{-1}$ and $8.0 \times 10^3\ M^{-1}$ for C_{60} and C_{70}, respectively. DFT (MPWB1K/6–31G** level) calculations assigned this difference in binding to the increase in surface when going from C_{60} to C_{70}.

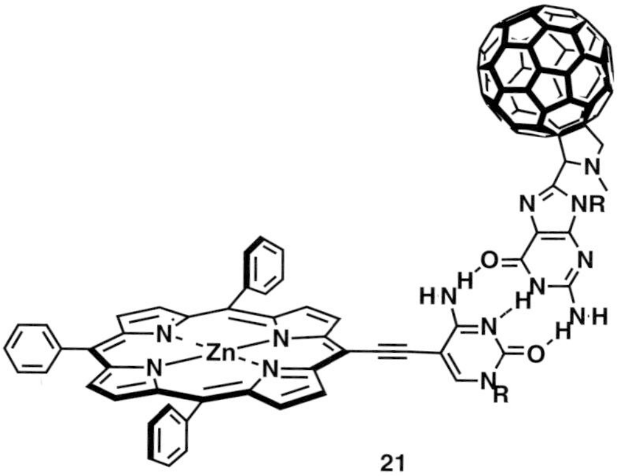

21

In the context of cooperative binding, an exTTF-based receptor was designed to include C_{60} and C_{70} by means of the introduction of two crown ethers as recognition element (**20**) [81]. First examples of supramolecular binding between C_{60} and crown ethers were based on aza-crown ethers. In these studies, several crown ethers have been shown to form inclusion complexes with C_{60} and C_{70}. Rather strong binding arises from interactions between the C_{60}/C_{70} π electron system and the crown ether π electrons [82–85]. In **20**, however, the cooperativity of n–π and π–π interactions stemming from the crown ether and exTTF, respectively, with C_{60} guarantee stable binding in several solvents including benzonitrile, chlorobenzene, and toluene. Noticeable binding strengths, which are found to be on the order of $10^6\ M^{-1}$, relate to distinct 1 : 1 complexes. Support for the existence of 1 : 1 complexes with C_{60} came from Job's plot titration experiments. In absorption assays, a new broad transition at 480 nm in benzonitrile emerged at the expense of the 438 nm band of exTTF. This new band is ascribed to a redistribution of charge density. In particular, **20** donates electron density to the electron accepting C_{60} to form correspondingly $\mathbf{20}^{\delta+}\cdot\mathbf{C_{60}}^{\delta-}$. The differential absorption changes obtained upon photoexciting $\mathbf{20}\cdot\mathbf{C_{60}}$ into, for example, the charge transfer features at 480 nm reveal the instantaneous formation of photoexcited $\mathbf{20}^{\delta+}\cdot\mathbf{C_{60}}^{\delta-}$. The $\mathbf{20}^{\delta+}\cdot\mathbf{C_{60}}^{\delta-}$ state transforms over a time period of 6.5 ps into the charge-separated state $\mathbf{20}^{\bullet+}\cdot\mathbf{C_{60}}^{\bullet-}$. Charge separation and charge recombination processes are considerably fast leading to a charge-separated state that recombines with a lifetime of 50 ps to the energetically lower lying singlet and triplet excited state of C_{60}. For $\mathbf{20}\cdot\mathbf{C_{70}}$, the corresponding $\mathbf{20}^{\bullet+}\cdot\mathbf{C_{70}}^{\bullet-}$ charge-separated state – with a signature of the one-electron reduced radical anion of C_{70} at 1290 nm – has a lifetime of 80 ps in benzonitrile.

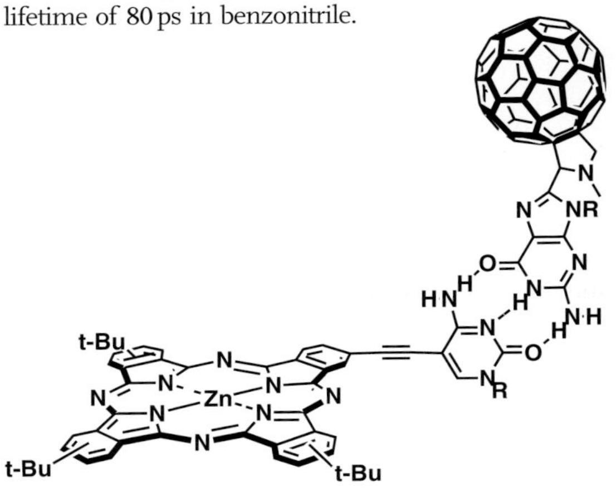

22

6.2.4
Hydrogen Bonding

Coulomb interactions and π–π stacking are appropriate tools to engineer multimolecular three-dimensional arrays of nanometer dimensions. However, such supramolecular interactions lack selectivity and sensitivity. Hydrogen bonding, with a high degree of directionality and sensitivity, emerge as attractive organization principles to govern the confined assembly of electron transfer systems for solar energy conversion. Single-point hydrogen bonding shows binding energies between 4 and 120 kJ/mol. One approach to overcome the poor stability of hydrogen bonding relies on the design of multiple point hydrogen bonds. Inspired by the pioneering work on Watson–Crick hydrogen bonding, a three-point guanosine–cytidine couple has been systematically pursued to design two- and three-dimensional nanostructures with well-defined geometries [86]. In particular, studies on a ZnP/C_{60} hybrid (**21**) pointed out that Watson–Crick base pairing provides a powerful tool in constructing self-assembled systems, with binding constant of $5.1 \times 10^4\ M^{-1}$ and radical ion pair state lifetime of 2.02 μs [87]. Although it is believed that devising supramolecular systems on the basis of Watson–Crick base pairing is promising, the topology of **21** is far from being optimal. In particular, through-space interactions must be assumed to control charge separation and charge recombination dynamics. When replacing ZnP in ZnP–C_{60}(**21**) with ZnPc to give ZnPc/C_{60}(**22**), a significantly shortened lifetime of 3.0 ns is thought to reflect the pronounced coupling between ZnPc and C_{60} [88]. The surprisingly large association constant $1.7 \times 10^7\ M^{-1}$ provides further evidence for this finding.

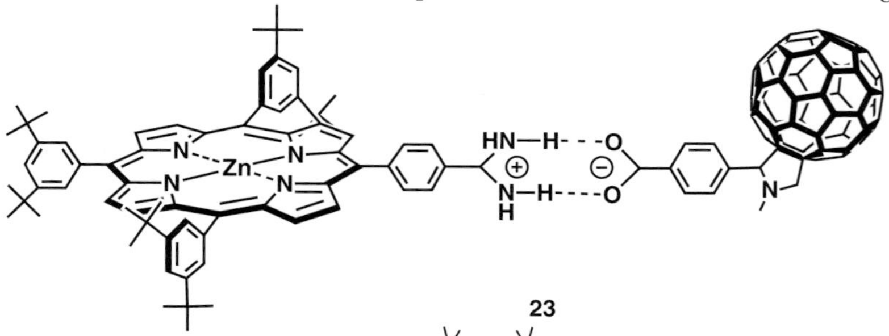

23

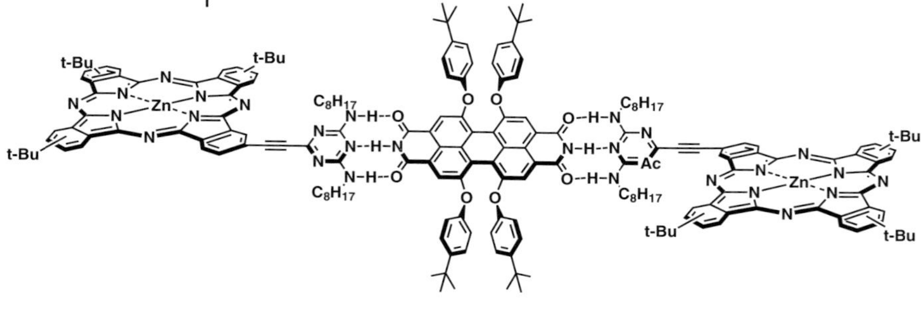

24

A linear arrangement in hydrogen bonding-assisted ZnP/C_{60} systems is obtained when a two-point amidinium–carboxylate bonding motif is employed – see, for example, **23** [89]. This bonding motif guarantees an extraordinarily high stabilization. In fact, association constants that reach up to $10^7\ M^{-1}$ prompt to an outstanding electronic coupling, which, in turn, fosters an efficient formation of a long-lived radical ion pair state (i.e. ~10 μs in THF). Interestingly, such remarkable radical ion pair lifetimes surpass previously reported systems based on nonamidinium-carboxylate binding motifs by several orders of magnitude [90–92].

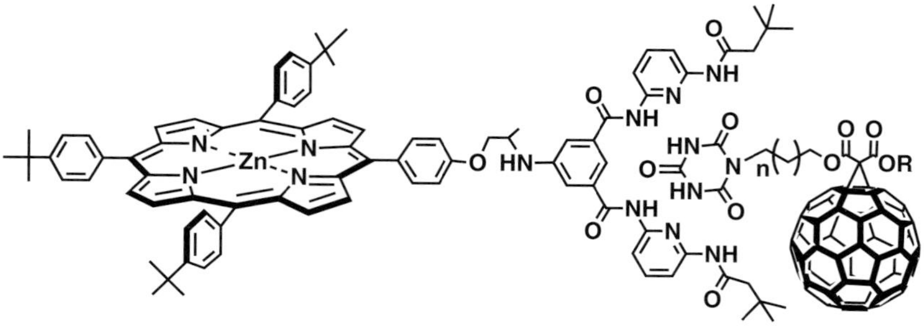

25

The use of a three-point hydrogen bonding between a ditopic ZnPc melamine moiety and a complementary bifunctional PDI is another compelling way to assemble supramolecular electron donor–acceptor hybrids [93]. Binding constants for **24** reach $2.0 \times 10^5\,M^{-1}$ and $7.0 \times 10^4\,M^{-1}$ in THF and benzonitrile, respectively. Evidence for a singlet energy transfer from the PDI singlet excited state to the energetically lower lying ZnPc singlet excited state was gathered by means of fluorescence titrations, time-resolved fluorescence measurements, and femtosecond transient absorption spectroscopy. However, no spectroscopic proof for an electron transfer reaction, that is, the one-electron oxidized form of ZnPc and the one-electron reduced form of PDI with maxima at 520/840 nm and 700 nm, respectively, was seen to develop. In other words, high-energy photons are absorbed by the highly photoluminescent PDI chromophore, which undergoes Förster resonant energy transfer to ZnPc.

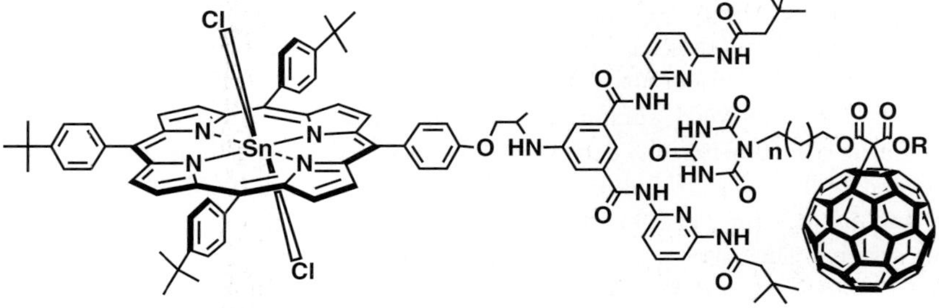

26

Another successful approach to overcome the low stability of one-point hydrogen bonding is the arrangement of a six-point hydrogen bonding that emerges between a Hamilton Receptor and a cyanuric acid moiety [94]. In this context, new libraries of

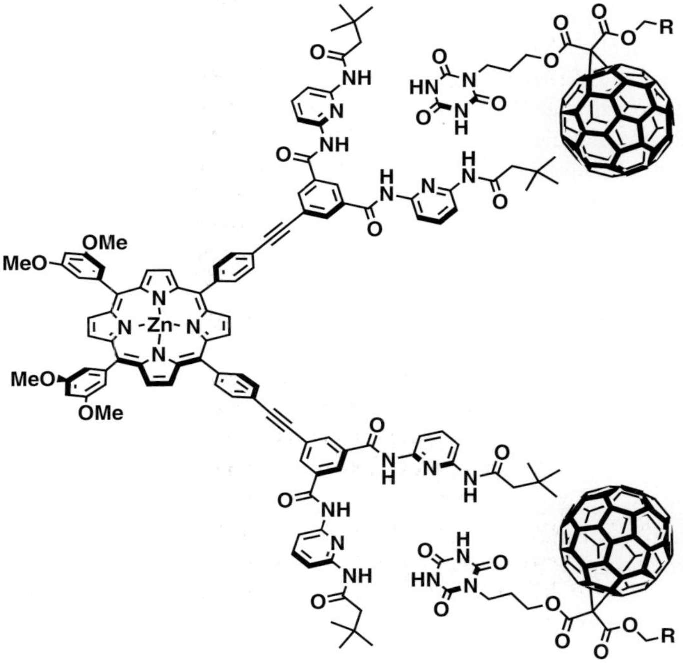

27

C_{60} monoadducts containing cyanuric acid side chains and ZnP (**25**)/SnP (**26**) involving the complementary Hamilton receptor unit have been synthesized [95]. This approach leads to comparatively strong binding with association constants K_a that range in apolar solvents (i.e., toluene, *ortho*-dichlorobenzene, and dichloromethane) between 10^3 and $10^5\,M^{-1}$. Significantly, the association constants increase when hexylene, instead of propylene, spacers are used, indicating the release of steric repulsion between the binding motifs. Transient absorption studies revealed photoinduced electron transfer from ZnP to C_{60} in the corresponding complexes, which generate radical ion pair states that are persistent well beyond the nanosecond timescale. In the case of the complexes involving SnP, energy transfer instead of electron transfer was observed.

An additional increase in binding strength of about one order of magnitude is reached by employing a ditopic Hamilton receptor – realized in **27** and **28** [96]. Here, *cis*- and *trans*-configured Hamilton receptor-functionalized ZnP were assembled with a cyanuric acid bearing C_{60} derivative. For **27** and **28**, association constants are obtained for the two-step binding process. K_a values of 10^3 and $10^7\,M^{-1}$ for **27**, and of 10^4 and $10^6\,M^{-1}$ for **28**, show a prominent positive cooperativity that is present in both systems. The better and stronger self-assembly for **27** relies on the *cis* pattern of the Hamilton receptors. As reported by Vögtle *et al.* [94], Hamilton receptors adopt three different conformations at room temperature, namely, *cis–cis*, *cis–trans*, and *trans–trans*. The complexation of the Hamilton receptor with complementary cyanuric derivatives causes the remaining free binding site to adopt the planar *cis–cis* conformation, which is the most favorable conformation for the subsequent binding steps. A series of photophysical experiments – ranging from steady state and time-resolved fluorescence experiments to transient absorption measurements on the femtosecond and nanosecond timescales – were conducted. These provide decisive confirmation about electron donor–acceptor interactions, as they are operative between ZnP and C_{60}. In **27**, kinetic analyses of the charge-separated species, that is, the one-electron oxidized ZnP and the one-electron reduced C_{60}, gave rise to a charge-separated state lifetime of 1.4 ns.

Notable is that in the case of the aforementioned **27** and **28**, the flexible alkyl chains, which were introduced as spacers, may have well been the inception to inter- and/or intramolecular folding. Conjugated spacers (i.e., *p*-phenylene-ethynylene, *p*-phenylene-vinylene, *p*-ethynylene, and fluorene), on the one hand, provide wire-like behavior in terms of electron transfer/electron transport and, on the other hand, restrict the flexibility in the resulting Hamilton receptor/cyanuric acid binding. Intramolecular electron transfer/electron transport along π-conjugated oligomers, such as *o*-phenylenevinylene (oPV), has been tested in several electron donor–acceptor conjugates with porphyrins [97, 98], anilines [99, 100], or ferrocenes [101] as electron donors and C_{60} as electron acceptor. In these studies, the following factors were considered: (i) small attenuation factors β that guarantees efficient electron conduction over large distances, (ii) good contacts with the termini consisting of electron donors and electron acceptors, and (iii) good orbital mixing between electron donor and electron acceptor states [102–106].

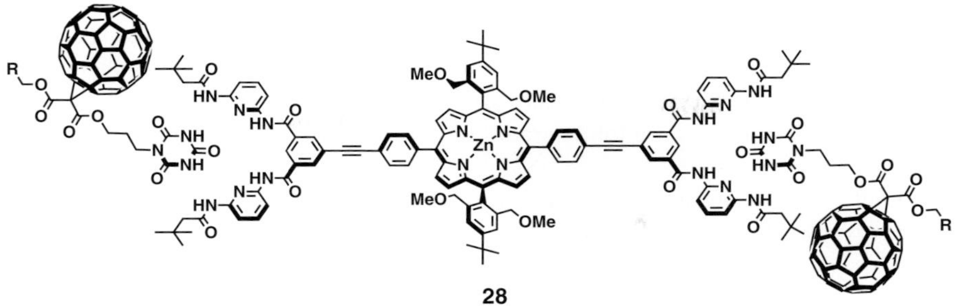

28

In a novel series of supramolecular assembled ZnP/C_{60} hybrids (**29**·**30**, **29**·**31**, **29**·**32**, **29**·**33**, and **29**·**34**), all these aspects were combined [107]. In particular, *p*-phenylene-ethynylene (**30** and **31**), *p*-ethynylene (**32**), *p*-phenylene-vinylene (**33**), or fluorene (**34**) are used as spacers to link C_{60} to the Hamilton receptor, which *p*-phenylene-ethynylene bridged to ZnP and the cyanuric acid (**29**). Detailed studies demonstrate that the electronic communication in the rigid electron donor–acceptor hybrids is governed by the nature of the conjugated spacer moieties. Spacers with good charge transfer properties such as *p*-phenylene-vinylene and fluorene (i.e., **29**·**33** and **29**·**34**) with low β-values facilitate charge transfer along the supramolecular bridge. In addition, they cause an apparent increase in the binding constants with values of 2.15×10^4 and $1.89 \times 10^4\,M^{-1}$ for **29**·**33** and **29**·**34**, respectively. Interestingly, the latter contradicts the results from ^1HNMR experiments, where only the genuine nature of the six-point hydrogen bonding is reflected, without considering the electronic communication between the terminal electron donors and electron acceptors. Spacers with relatively large β-values such as *p*-phenylene-ethynylene in **29**·**30** or **29**·**31** impede charge transfer along the supramolecular bridge. This is also pointed out in the binding constants. They are 5.24×10^3 and $5.00 \times 10^2\,M^{-1}$ for **29**·**30** and **29**·**31**, respectively, and about 1–2 magnitudes lower than those for **29**·**33** and **29**·**34**. Transient absorption measurements on the femtosecond and nanosecond timescale provided further insights into the electron transfer activity. Upon photoexciting **29** with 420 nm laser pulses, the characteristic fingerprint of the charge-separated species, that is, the one-electron oxidized form of ZnP and the one-electron reduced form of C_{60}, developed on a timescale of hundreds of picoseconds. Conjugates with *p*-phenylene-vinylene and fluorene spacers (**29**·**30** or **29**·**31**) showed electron donor–acceptor behavior with charge-separated state lifetimes of 40 and 53 ns, respectively. When exciting the *p*-phenylene-ethynylene-bridged complexes

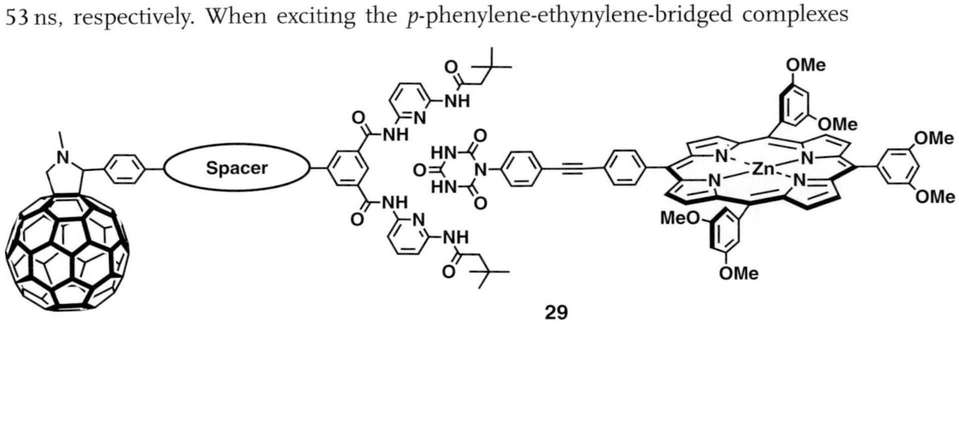

29

with 420 nm laser pulses, the radical ion pair emerges within 322 ps and decays via charge recombination on the nanosecond timescale (71 ns) to the electronic ground state in the case of **29·30**. Interestingly, introduction of a second *p*-phenylene-ethynylene repeat unit (i.e., **29·30** versus **29·31**) eliminates the charge transfer process with no evidence for any radical ion pair state formation.

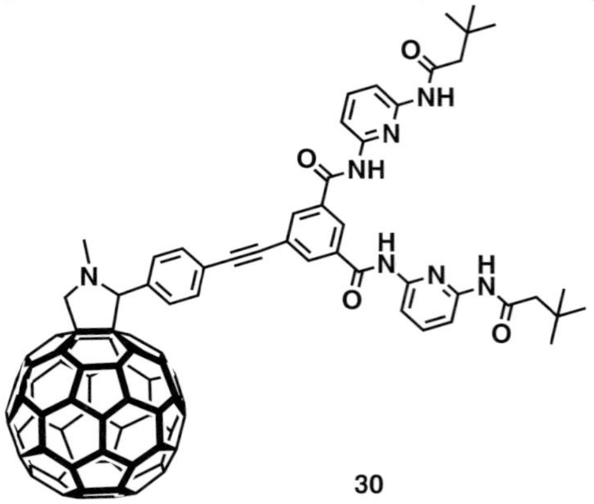

30

Molecular recognition principles, in the form of crown ether complexation, are the inception for realizing molecularly organized thin-film assemblies and nanoarchitectures [108–110]. In this light, the cooperativity between hydrogen bonding and π–π interactions was used to design a highly stable supramolecular complex formed by C_{60} and H_2P (**35**) [111, 112]. In particular, the weak hydrogen bonding based on ammonium–crown ether recognition motif was introduced as a preorganizing factor. In general, threading ammonium salts through crown ethers is governed by binding constants that are typically on the order of $\sim 10^3\,M^{-1}$. By virtue of an additional recognition element, namely, π–π stacking, the stabilization increased by several orders of magnitudes reaching a value of $3.7 \times 10^5\,M^{-1}$. π–π and charge transfer interactions are by far the dominating contribution, while hydrogen bonding interactions with a maximum strength of $10^3\,M^{-1}$ play a secondary role in the overall

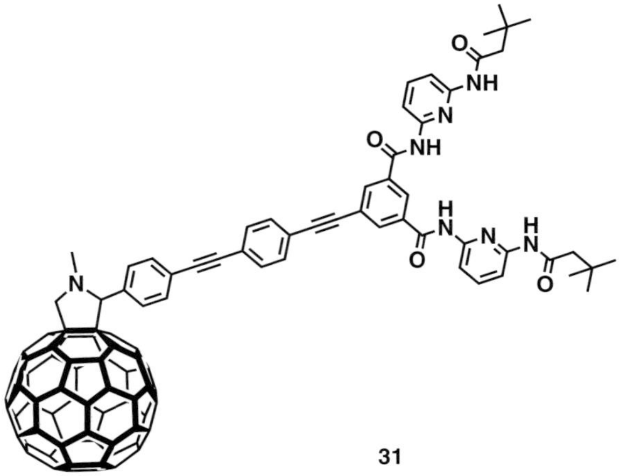

31

interactions. Since the emission of the H_2P-crown is fully recovered upon addition of base, which destroys the supramolecular assembly, H_2P-crown/C_{60} appears to be an excellent candidate as a molecular switch operated via control of recognition by changing the pH of the solution. On similar grounds, a recognition motif was chosen that relies on threading a dibenzylammonium unit through a dibenzo-24-crown-8 macrocycle, affording reasonably stable pseudorotaxane-like ZnPc/C_{60} complexes with binding constants of $1.4 \times 10^4\,M^{-1}$ (**36**) and $1.9 \times 10^4\,M^{-1}$ (**37**) [113, 114]. Microsecond-lived charge-separated states, $ZnPc^{\bullet +}/C_{60}^{\bullet -}$, are the products of efficient intracomplex electron transfer events.

A similar recognition motif was used in **38**. Here, the ammonium crown ether binding motif was chosen to preorganize and to assemble exTTF and C_{60}. By means of cooperativity between hydrogen bonds and π–π interactions a binding constant on the order of $\sim 10^6\,M^{-1}$ was obtained. Such strong interactions also impose a notable impact on the charge separation and charge recombination dynamics. Once the radical ion pair is formed, it decays rather rapidly (9.2 ps). Threading the benzylammonium salt through the crown ether in **38** facilitates the close proximity between exTTF and C_{60} by π–π interactions between the benzene rings of exTTF and C_{60}, similar to what is seen in related tweezers complexes [78, 80, 115–119].

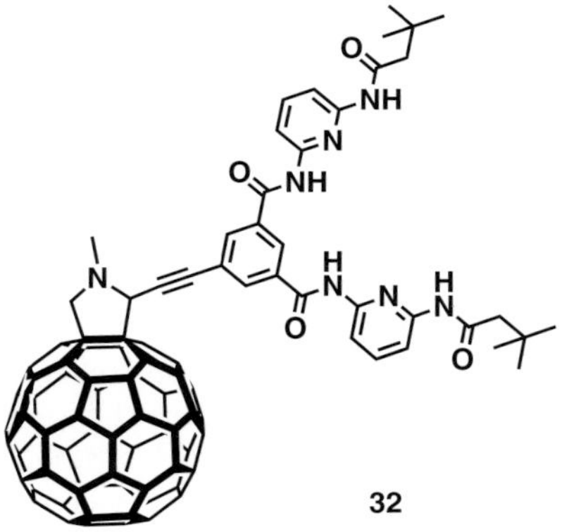

32

D'Souza *et al.* realized a multicomponent supramolecular ZnPc/C_{60} hybrid by a step-wise self-assembly. In the first step, a "special pair" donor – a cofacial ZnPc dimer – was formed via potassium ion-induced dimerization of ZnPc (**39**) [120]. The dimer was subsequently self-assembled with functionalized fullerenes via "two-point" binding involving axial coordination and crown ether–alkyl ammonium cation complexation. The binding constant for the ZnPc/C_{60} complexation calculated by constructing the Benesi – Hildebrand plot was found to be $2.5 \times 10^3\,M^{-1}$, suggesting a moderate stability. On the basis of a detailed kinetic analysis, that is, time-resolved emission and transient absorption spectroscopies on different time-scales, it was found that the charge separation occurs from the ZnPc triplet excited state to C_{60} rather than from the ZnPc singlet excited state to afford a long-lived CS state. The role of the cofacial ZnPc dimer, which is stabilizing the CS state in electron donor–acceptor hybrid, could be attested by reference experiments, in which the lifetimes of charge-separated state of the ZnPc monomer and the ZnPc dimer with values 4.8 and 6.7 μs, respectively, were compared. In earlier work, the same group realized a ZnP tetrad by potassium ion-induced dimerization and subsequent self-assembling with functionalized fullerenes via axial coordination and crown ether–alkyl ammonium cation complexation. It is worth noting that the electron transfer is originating here from the ZnP singlet excited state rather than from the ZnP triplet excited state to yield a 50 ns lived charge-separated state [121].

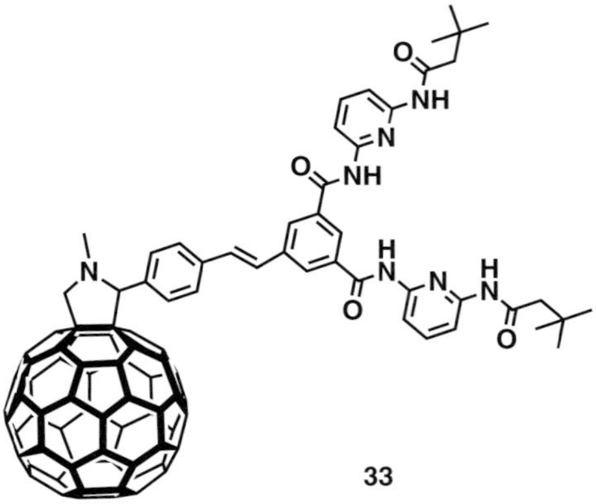

33

6.2.5 Metal–Ligand Coordination

Although supramolecular assemblies are easily accessible, they often provide imprecise structural control. Perfect control over the geometric arrangement is realized by metal to ligand coordination. In an early work, pyridyl-substituted fulleropyrrolidine derivatives were used to form stable linear complexes with ZnP (**40**) via metal–ligand coordination [122]. The axial symmetry of the complex discloses such a strong binding affinity ($7.4 \times 10^4\,M^{-1}$) that different supramolecular electron donor–accep-

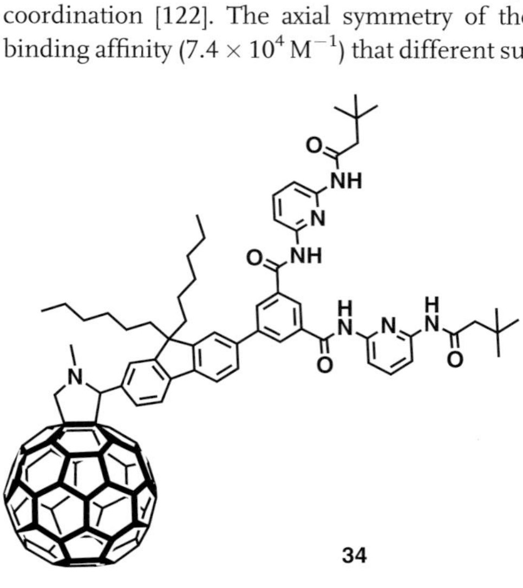

34

tor hybrids were formed with the characteristic metal–ligand coordination as bonding motif.

In light of this precedence, a Bodipy unit is tethered to the peripheral position of a phthalocyanine core and forms a conjugate that interacts through coordination to *N*-pyridylfulleropyrrolidine (**41**) [123]. Interestingly, the panchromatic conjugate **41** absorbs in two disparate but complementary sections of the solar spectrum (Bodipy: 525 nm; ZnPc: 680 nm). Upon being subject to photoexcitation at 480 nm, an intramolecular transduction of singlet excited state energy occurs. The latter funnels the excitation from the light absorbing and energy donating Bodipy (2.3 eV) to the energy-accepting ZnPc (1.8 eV) subunit. The net result is the photosensitization of the ZnPc moiety by means of Bodipy photoexcitation. Due to the presence of the electron accepting *N*-pyridylfulleropyrrolidine, intraconjugate electron transfer produces a long-lived (39.9 ns) charge-separated state, Bodipy–$ZnPc^{\bullet +}/C_{60}^{\bullet -}$, which ultimately decays to produce the ground state.

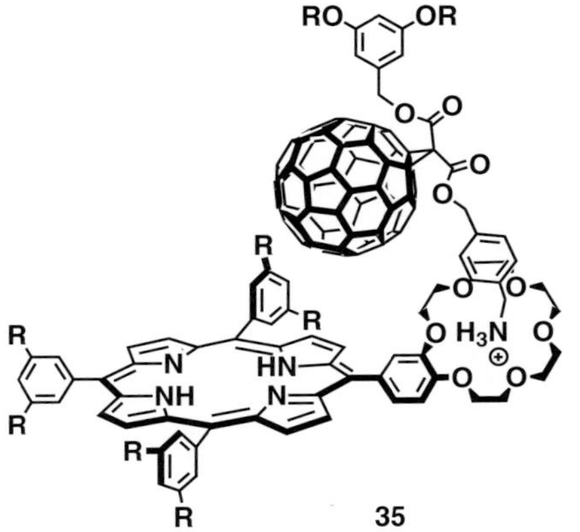

35

In a recent work, metal–ligand coordination between lanthanide and two phthalocyanine macrocycles is used as platform to achieve **42**, a double-decker lanthanide (III) bis(phthalocyaninato)–C_{60} conjugate [124]. Neither absorption nor electrochemistry measurements could provide evidence for any appreciable ground-state inter-

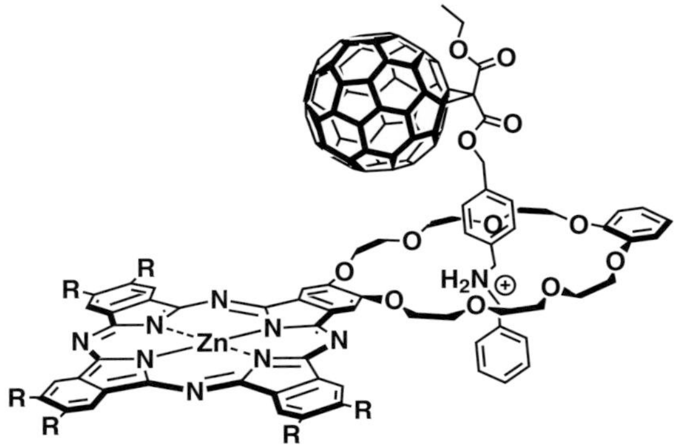

36

actions. Photophysical studies with **42** indicate that only after irradiation at 387 nm, which excites both C_{60} and $[Ln^{III}(Pc)(Pc')]$ components, a rapid photoinduced (30 ps) and long-lived (i.e., beyond the instrument's time resolution of 3 ns) electron transfer from the $[Ln^{III}(Pc)\ (Pc')]$ to C_{60} occurs.

In another approach, the metal coordination was used as a synthetic approach to assemble RuPc and different C_{60} derivatives [125]. Owing to the rich coordination chemistry of ruthenium, several electron donor–acceptor hybrids (**43–45**) were realized. Photophysical studies allowed determining the influence of the donor and acceptor ratio on the electron transfer chemistry. In fact, the charge transfer behavior is dominated by the mono- or hexakis substitution of C_{60} rather than by the ratio of the electron donors or electron acceptors. Compounds **43** and **44**, for example, afford the formation of $RuPc^{\bullet+}/C_{60}^{\bullet-}$, whereas the presence of the hexakis-substituted C_{60} in **45** does not exert a charge transfer from the electron donating RuPc to the electron accepting C_{60}. In this particular case, the charge-separated state (1.94 eV) is located above that of the RuPc singlet excited state (1.88 eV), which renders an intramolecular charge transfer endothermic.

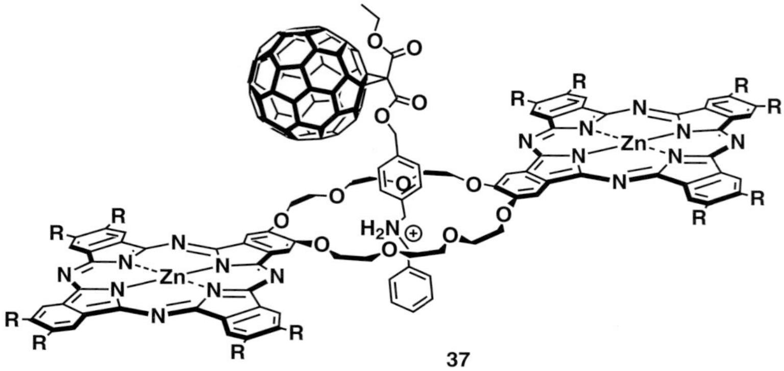

37

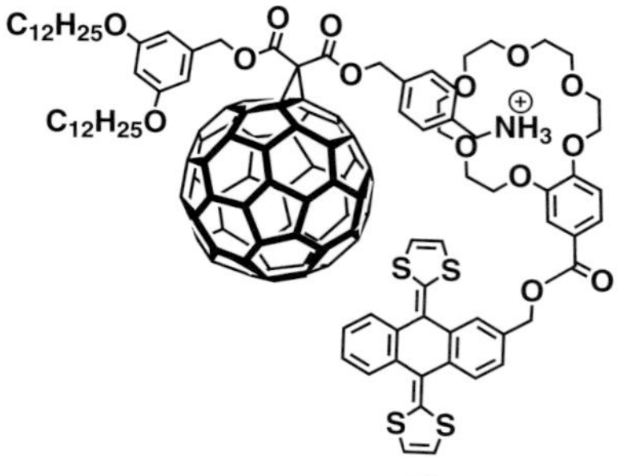

38

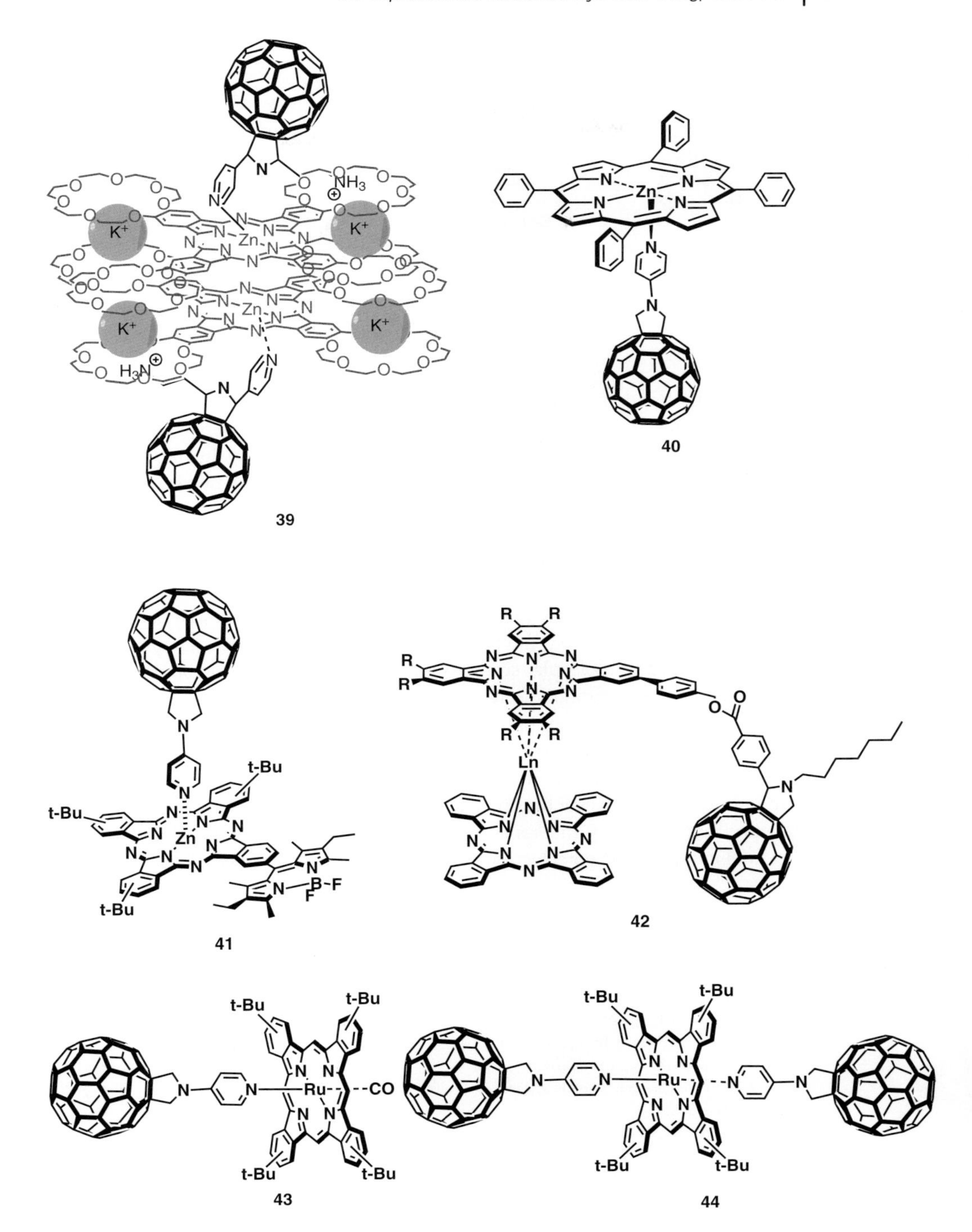
39
40
41
42
43
44

45

6.3 Outlook

This chapter has demonstrated the effectiveness of employing fullerenes in biomimetic strategies for devising thermodynamic stable but kinetic labile 1D, 2D, or 3D networks. These strategies (Coulomb interactions, π–π stacking, hydrogen bonding, and metal–ligand coordination) provide the means for an evident trend toward the facile preparation of precise structures never before accomplished by conventional synthetic chemistry. The most important aspect in this area is the regulation of the inherently weak forces seen in biomimetic organization principles on a molecular basis. In this context, the current concepts illustrate that relating size and shape to the function of the resulting composites led to composites with new and original properties. Unquestionably, the peculiar shape and unique electronic properties of fullerenes – as stiff molecular scaffolds – has a significant bearing on the design of novel, well-ordered supramolecular arrays. Despite some remarkable recent successes, it is clear that the examples discussed in this chapter represent only the tip of the iceberg. More research in this area is needed to fully explore the possibilities offered by these materials, for example, in the production of both active and passive devices. This underlines more than ever the great need for creative synthetic chemistry, which will ensure that eventually fullerenes become an important building block of future technologies, such as optoelectronics, batteries, and photovoltaics.

References

1 Hamakawa, Y. (2002) *Sol. Energy Mater. Sol. Cells*, **74**, 13–23.
2 Kruse, O., Rupprecht, J., Mussgnug, J.H., Dismukes, G.C., and Hankamer, B. (2005) *Photochem. Photobiol. Sci.*, **4**, 957–669.
3 Crabtree, G.W. and Lewis, N.S. (2007) *Phys. Today*, **60**, 37–42.
4 Becquerel, A.E. (1839) *Comt. Rend. Acad. Sci.*, **9**, 561–567.
5 Goetzberger, A., Knobloch, J., and Voss, B. (1998) *Crystalline Silicon Solar Cells*, John Wiley & Sons, Inc., New York.
6 Green, M.A. (1995) *Silicon Solar Cells: Advanced Principles and Practice;*, Bridge Printery, Sidney.
7 Goetzberger, A., Hebling, C., and Schock, H.-W. (2003) *Mater. Sci. Eng. R.*, **40**, 1–46.
8 European Photovoltaic Industry Association/Greenpeace (2001) Solar Generation Report. European Photovoltaic Industry Association/Greenpeace.
9 Bergmann, R.B. (1999) *Appl. Phys. A*, **69**, 187–194.

10 McGehee, M.D. and Heeger, A.J. (2000) *Adv. Mater.*, **12**, 1655–1668.

11 Nunzi, J.-M. (2002) *C. R. Phys.*, **3**, 523–542.

12 Carlson, D.E. and Wronski, C.R. (1976) *Appl. Phys. Lett.*, **28**, 671–673.

13 Schultz, O., Glunz, S.W., and Willeke, G.P. (2004) *Prog. Photovolt. Res. Appl.*, **12**, 553–558.

14 Miles, R.W., Zoppi, G., and Forbes, I. (2007) *Mater. Today*, **10**, 20–27.

15 Bosi, M. and Pelosi, C. (2007) *Prog. Photovolt. Res. Appl.*, **15**, 51–68.

16 Shah, A., Torres, P., Tscharner, R., Wyrsch, N., and Keppner, H. (1999) *Science*, **285**, 692–698.

17 Dimmler, B. and Schock, H.W. (1996) *Prog. Photovolt. Res. Appl.*, **4**, 425–433.

18 Chopra, K.L., Paulson, P.D., and Dutta, V. (2004) *Prog. Photovolt. Res. Appl.*, **12**, 69–92.

19 Brabec, C.J., Sariciftci, N.S., and Hummelen, J.C. (2001) *Adv. Funct. Mater.*, **11**, 15–26.

20 Grätzel, M. (2001) *Nature*, **414**, 338–344.

21 Grätzel, M. (2001) *Nature*, **353**, 737–739.

22 Hoppe, H. and Sariciftcei, N.S. (2004) *J. Mater. Res.*, **19**, 1924–1945.

23 Zhao, J., Wang, A., Green, M.A., and Ferrazza, F. (1998) *Appl. Phys. Lett.*, **73**, 1991–1993.

24 Kerschaver, E.V. and Beaucarne, G. (2006) *Prog. Photovolt. Res. Appl.*, **14**, 107–123.

25 Zhao, J., Wang, A., and Green, M.A. (1999) *Prog. Photovolt. Res. Appl.*, **7**, 471–474.

26 Wu, X. (2004) *Sol. Energy*, **77**, 803–814.

27 Nelson, J. (2002) *Curr. Opin. Solid State Mater. Sci.*, **6**, 87–95.

28 Desertec Foundation (2009) Desertec: Red Paper, Desertec Foundation.

29 Stanford University Global Climate & Energy Project (2006). An Assessment of Solar Energy Conversion Technologies and Research Opportunities, Stanford University Global Climate & Energy Project.

30 Lewis, N.S. and Nocera, D.G. (2006) *Proc. Natl. Acad. Sci. USA*, **103**, 15729–15735.

31 Gust, D., Moore, T.A., and Moore, A.L. (2000) *Acc. Chem. Res.*, **34**, 40–48.

32 Wasielewski, M.R. (1992) *Chem. Rev.*, **92**, 435–461.

33 Gust, D., Moore, T.A., and Moore, A.L. (2009) *Acc. Chem. Res.*, **42**, 1890–1898.

34 Wasielewski, M.R. (2006) *J. Org. Chem.*, **71**, 5051–5066.

35 de la Garza, L., Jeong, G., Liddell, P.A., Sotomura, T., Moore, T.A., Mo, A.L., and Gust, D. (2004) *Nanotechnology and the Environment;*, American Chemical Society, Washington, DC, pp. 361–367.

36 Kamat, P.V. (2007) *J. Phys. Chem. C*, **111**, 2834–2860.

37 Kroto, H.W., Heath, J.R., O'Brien, S.C., Curl, R.F., and Smalley, R.E. (1985) *Nature*, **318**, 162–163.

38 Curl, R.F. (1997) *Angew. Chem.*, **109**, 1636–1647.

39 Kroto, H.W. (1997) *Angew. Chem.*, **109**, 1648–1664.

40 Smalley, R.E. (1997) *Angew. Chem.*, **109**, 1666–1673.

41 Kratschmer, W., Lamb, L.D., Fostiropoulos, K., and Huffman, D.R. (1990) *Nature*, **347**, 354–358.

42 Hebard, A.F., Rosseinsky, M.J., Haddon, R.C., Murphy, D.W., Glarum, S.H., Palstra, T.T.M., Ramirez, A.P., and Kortan, A.R. (1991) *Nature*, **350**, 600–601.

43 Friedman, S.H., DeCamp, D.L., Sijbesma, R.P., Srdanov, G., Wudl, F., and Kenyon, G.L. (1993) *J. Am. Chem. Soc.*, **115**, 6506–6509.

44 Echegoyen, L. and Echegoyen, L.E. (1998) *Acc. Chem. Res.*, **31**, 593–601.

45 Martin, N., Sanchez, L., Illescas, B., and Perez, I. (1998) *Chem. Rev.*, **98**, 2527–2548.

46 Imahori, H. and Sakata, Y. (1999) *Eur. J. Org. Chem.*, **1999**, 2445–2457.

47 Guldi, D.M. (2000) *Chem. Comm.*, 321–327.

48 Kadish, K.M and Ruoff, R.S. (eds) (2000) *Fullerenes: Chemistry, Physics, and Technology*, John Wiley-VHC Verlag GmbH, Weinheim.

49 Guldi, D.M. (2002) *Fullerenes: From Synthesis to Optoelectronic Properties* (eds D.M. Guldi and N. Martín), Kluwer Academic, Norwell, MA, pp. 237–265.

50 Boyd, P.D.W and Reed, C.A. (2005) *Acc. Chem. Res.*, **38**, 235–242.
51 Guldi, D.M. and Martín, N. (2002) *J. Mater. Chem.*, **12**, 1978–1992.
52 Hiroshi, I., Kiyoshi, H., Tsuyoshi, A., Masanori, A., Seiji, T., Tadashi, O., Masahiro, S., and Yoshiteru, S. (1996) *Chem. Phys. Lett.*, **263**, 545–550.
53 Imahori, H. and Sakata, Y. (1997) *Adv. Mater.*, **9**, 537–546.
54 Marcus, R.A. (1956) *J. Chem. Phys.*, **24**, 966–978.
55 Marcus, R.A. and Sutin, N. (1985) *Biophys. Acta*, **811**, 265.
56 Marcus, R.A. (1993) *Angew. Chem. Int. Ed.*, **32**, 1111–1121.
57 Schneider, H.-J. and Yatsimirsky, A.K. (2008) *Chem. Soc. Rev.*, **37**, 263–277.
58 Steed, J.W. and Atwood, J.L. (2009) *Supramolecular Chemistry*, 2nd ed., John Wiley & Sons, Inc., Chichester.
59 Guldi, D.M. and Prato, M. (2004) *Chem. Commun.*, 2517–2525.
60 Balbinot, D., Atalick, S., Guldi, D.M., Hatzimarinaki, M., Hirsch, A., and Jux, N. (2003) *J. Phys. Chem. B*, **107**, 13273–13279.
61 Sarova, G.H., Hartnagel, U., Balbinot, D., Sali, S., Jux, N., Hirsch, A., and Guldi, D.M. (2008) *Chem. Eur. J.*, **14**, 3137–3145.
62 Braun, M., Atalick, S., Guldi, D.M., Lanig, H., Brettreich, M., Burghardt, S., Hatzimarinaki, M., Ravanelli, E., Prato, M., Eldik, R.V., and Hirsch, A. (2003). *Chem. Eur. J.*, **9**, 3867–3875.
63 Kral, V., Springs, S.L., and Sessler, J.L. (1995) *J. Am. Chem. Soc.*, **117**, 8881–8882.
64 Springs, S.L., Gosztola, D., Wasielewski, M.R., Kral, V., Andrievsky, A., and Sessler, J.L. (1999) *J. Am. Chem. Soc.*, **121**, 2281–2289.
65 Grimm, B., Karnas, E., Brettreich, M., Ohta, K., Hirsch, A., Guldi, D.M., Torres, T., and Sessler, J.L. (2009) *J. Phys. Chem. B*, **114**, 14134–14139.
66 Tashiro, K., Aida, T., Zheng, J.-Y., Kinbara, K., Saigo, K., Sakamoto, S., and Yamaguchi, K. (1999) *J. Am. Chem. Soc.*, **121**, 9477–9478.
67 Liao, M.-S., Watts, J.D., and Huang, M.-J. (2007) *J. Phys. Chem. B*, **111**, 4374–4382.
68 Sun, D., Tham, F.S., Reed, C.A., Chaker, L., Burgess, M., and Boyd, P.D.W. (2000) *J. Am. Chem. Soc.*, **122**, 10704–10705.
69 Boyd, P.D.W. and Reed, C.A. (2004) *Acc. Chem. Res.*, **38**, 235–242.
70 Schmittel, M., He, B., and Mal, P. (2008) *Org. Lett.*, **10**, 2513–2516.
71 Tashiro, K. and Aida, T. (2007) *Chem. Soc. Rev.*, **36**, 189.
72 Yanagisawa, M., Tashiro, K., Yamasaki, M., and Aida, T. (2007) *J. Am. Chem. Soc.*, **129**, 11912–11913.
73 Ishii, T., Aizawa, N., Yamashita, M., Matsuzaka, H., Kodama, T., Kikuchi, K., Ikemoto, I., and Iwasa, Y. (2000) *J. Chem. Soc. Dalton Trans.*, 4407.
74 Hosseini, A., Taylor, S., Accorsi, G., Armaroli, N., Reed, C.A., and Boyd, P.D.W. (2006) *J. Am. Chem. Soc.*, **128**, 15903–15913.
75 Grimm, B., Schornbaum, J., Cardona, C.M., van Paauwe, J., Boyd, P., and Guldi, D.M. (2011) *Chem. Sci.*, **2**, 1530–1537.
76 Yamaguchi, T., Ishii, N., Tashiro, K., and Aida, T. (2003) *J. Am. Chem. Soc.*, **125**, 13934–13935.
77 Nobukuni, H., Shimazaki, Y., Tani, F., and Naruta, Y. (2007) *Angew. Chem. Int. Ed.*, **46**, 8975–8978.
78 Huerta, E., Isla, H., Perez, E.M., Bo, C., Martin, N., and Mendoza, J.D. (2010) *J. Am. Chem. Soc.*, **132**, 5351–5353.
79 Pérez, E.M., Sierra, M., Sánchez, L., Torres, M.R., Viruela, R., Viruela, P.M., Ortí, E., and Martín, N. (2007) *Angew. Chem. Int. Ed.*, **46**, 1847–1851.
80 Perez, E.M. and Martin, N. (2008) *Chem. Soc. Rev.*, **37**, 1512–1519.
81 Grimm, B., Santos, J., Illescas, B.M., Munoz, A., Guldi, D.M., and Martin, N. (2010) *J. Am. Chem. Soc.*, **132**, 17387–17389.
82 Effing, J., Jonas, U., Jullien, L., Plesnivy, T., Ringsdorf, H., Diederich, F., Thilgen, C., and Weinstein, D. (1992) *Angew. Chem. Int. Ed.*, **31**, 1599–1602.
83 Bhattacharya, S., Sharma, A., Nayak, S.K., Chattopadhyay, S., and Mukherjee, A.K. (2003) *J. Phys. Chem. B*, **107**, 4213–4217.
84 Saha, A., Nayak, S.K., Chottopadhyay, S., and Mukherjee, A.K. (2003) *J. Phys. Chem. B*, **107**, 11889–11892.
85 Datta, K., Banerjee, M., and Mukherjee, A.K. (2004) *J. Phys. Chem. B*, **108**, 16100–16106.

86 Sessler, J.L. and Jayawickramarajah, J. (2005) *Chem. Commun.*, 1939–1949.

87 Sessler, J.L., Jayawickramarajah, J., Gouloumis, A., Torres, T., Guldi, D.M., Maldonado, S., and Stevenson, K.J. (2005) *Chem. Commun.*, 1892–1894.

88 Torres, T., Gouloumis, A., Sanchez-Garcia, D., Jayawickramarajah, J., Seitz, W., Guldi, D.M., and Sessler, J.L. (2007) *Chem. Commun.*, 292–294.

89 Sanchez, L., Sierra, M., Martín, N., Myles, A.J., Dale, T.J., Rebek, J., Seitz, W., and Guldi, D.M. (2006) *Angew. Chem. Int. Ed.*, **45**, 4637–4641.

90 Sánchez, L., Martín, N., and Guldi, D.M. (2005) *Angew. Chem. Int. Ed.*, **44**, 5374–5382.

91 Hahn, U., Elhabiri, M., Trabolsi, A., Herschbach, H., Leize, E., Van Dorsselaer, A., Albrecht-Gary, A.-M., and Nierengarten, J.-F. (2005) *Angew. Chem. Int. Ed.*, **44**, 5338–5341.

92 McClenaghan, N.D., Grote, Z., Darriet, K., Zimine, M., Williams, R.M., De Cola, L., and Bassani, D.M. (2005) *Org. Lett.*, **7**, 807–810.

93 Seitz, W., Jimenez, A.J., Carbonell, E., Grimm, B., Rodriguez-Morgade, M.S., Guldi, D.M., and Torres, T. (1010) *Chem. Commun.*, 127–129.

94 Dirksen, A., Hahn, U., Schwanke, F., Nieger, M., Reek, J.N.H., Vögtle, F., and Cola, L.D. (2004) *Chem. Eur. J.*, **10**, 2036–2047.

95 Wessendorf, F., Gnichwitz, J.-F., Sarova, G.H., Hager, K., Hartnagel, U., Guldi, D.M., and Hirsch, A. (2007) *J. Am. Chem. Soc.*, **129**, 16057–16071.

96 Maurer, K., Grimm, B., Wessendorf, F., Hartnagel, K., Guldi, D.M., and Hirsch, A. (2010) *Eur. J. Org. Chem.*, **2010**, 5010–5029.

97 Redmore, N.P., Rubtsov, I.V., and Therien, M.J. (2003) *J. Am. Chem. Soc.*, **125**, 8769–8778.

98 Screen, T.E.O., Thorne, J.R.G., Denning, R.G., Bucknall, D.G., and Anderson, H.L. (2003) *J. Mater. Chem.*, **13**, 2796–2808.

99 Thomas, K.G., Biju, V., Guldi, D.M., Kamat, P.V., and George, M.V. (1999) *J. Phys. Chem. B*, **103**, 8864–8869.

100 Guldi, D.M., Swartz, A., Luo, C., Gómez, R., Segura, J.L., and Martín, N. (2002) *J. Am. Chem. Soc.*, **124**, 10875–10886.

101 Dong, T.-Y., Chang, S.-W., Lin, S.-F., Lin, M.-C., Wen, Y.-S., and Lee, L. (2006) *Organometallics*, **25**, 2018–2024.

102 Goldsmith, R.H., Sinks, L.E., Kelley, R.F., Betzen, L.J., Liu, W., Weiss, E.A., Ratner, M.A., and Wasielewski, M.R. (2005) *Proc. Natl. Acad. Sci. USA*, **102**, 3540–3545.

103 Weiss, E.A., Tauber, M.J., Kelley, R.F., Ahrens, M.J., Ratner, M.A., and Wasielewski, M.R. (2005) *J. Am. Chem. Soc.*, **127**, 11842–11850.

104 Tauber, M.J., Kelley, R.F., Giaimo, J.M., Rybtchinski, B., and Wasielewski, M.R. (2006) *J. Am. Chem. Soc.*, **128**, 1782–1783.

105 Goldsmith, R.H., Wasielewski, M.R., and Ratner, M.A. (2006) *J. Phys. Chem. A*, **110**, 20258–20262.

106 Müllen, K. and Wegner, G. (1998) *Electronic Materials: The Oligomer Approach;*, John Wiley-VHC Verlag GmbH, Weinheim.

107 Wessendorf, F., Grimm, B., Guldi, D.M., and Hirsch, A. (2010) *J. Am. Chem. Soc.*, **132**, 10786–10795.

108 Ashton, P.R., Campbell, P.J., Glink, P.T., Philp, D., Spencer, N., Stoddart, J.F., Chrystal, E.J.T., Menzer, S., Williams, D.J., and Tasker, P.A. (1995) *Angew. Chem. Int. Ed.*, **34**, 1865–1869.

109 D'Souza, F., Chitta, R., Gadde, S., Zandler, M.E., Sandanayaka, A.S.D., Araki, Y., and Ito, O. (2005) *Chem. Commun.*, 1279–1281.

110 Sandanayaka, A.S.D., Araki, Y., Ito, O., Chitta, R., Gadde, S., and D'Souza, F. (2006) *Chem. Commun.*, 4327–4329.

111 Gutiérrez-Nava, M., Nierengarten, H., Masson, P., Van Dorsselaer, A., and Nierengarten, J.-F. (2003) *Tetrahedron Lett.*, **44**, 3043–3046.

112 Solladie, N., Walther, M.E., Gross, M., Figueira Duarte, T.M., Bourgogne, C., and Nierengarten, J.-F. (2003) *Chem. Commun.*, 2412–2413.

113 Guldi, D.M., Ramey, J., Martinez-Diaz, M.V., Escosura, A.D.L., Torres, T.,

Da Ros, T., and Prato, M. (2002) *Chem. Commun.*, 2774–2775.

114 Martinez-Diaz, M.V., Fender, N.S., Rodriguez-Morgade, M.S., Gomez-Lopez, M., Diederich, F., Echegoyen, L., Stoddart, J.F., and Torres, T. (2002) *J. Mater.Chem.*, **12**, 2095–2099.

115 Pérez, E.M., Sanchez, L., Fernandez, G., and Martin, N. (2006) *J. Am. Chem. Soc.*, **128**, 7172–7173.

116 Perez, E.M., Capodilupo, A.L., Fernandez, G., Sanchez, L., Viruela, P.M., Viruela, R., Orti, E., Bietti, M., and Martin, N. (2008) *Chem. Commun.*, 4567–4569.

117 Gayathri, S.S., Wielopolski, M., Pérez, E.M., Fernandez, G., Sanchez, L., Viruela, R., Orti, E., Guldi, D.M., and Martin, N. (2009) *Angew. Chem. Int. Ed.*, **48**, 815–819.

118 Pérez, E.M. and Martín, N. (2010) *Pure Appl. Chem.*, **82**, 523–533.

119 Isla, H., Gallego, M., Perez, E.M., Viruela, R., Orti, E., and Martin, N. (2010) *J. Am. Chem. Soc.*, **132**, 1772–1773.

120 D'Souza, F., Maligaspe, E., Ohkubo, K., Zandler, M.E., Subbaiyan, N.K., and Fukuzumi, S. (2009) *J. Am. Chem. Soc.*, **131**, 8787–8797.

121 D'Souza, F., Chitta, R., Gadde, S., Rogers, L.M., Karr, P.A., Zandler, M.E., Sandanayaka, A.S.D., Araki, Y., and Ito, O. (2007) *Chem. Eur. J.*, **13**, 916–922.

122 Wilson, S.R., MacMahon, S., Tat, F.T., Jarowski, P.D., and Schuster, D.I. (2003) *Chem. Commun.*, 226–227.

123 Rio, Y., Seitz, W., Gouloumis, A., Vázquez, P., Sessler, J.L., Guldi, D.M., and Torres, T. (2010) *Chem. Eur. J.*, **16**, 1929–1940.

124 Ballesteros, B., de la Torre, G., Shearer, A., Hausmann, A., Herranz, M.Á., Guldi, D.M., and Torres, T. (2010) *Chem. Eur. J.*, **16**, 114–125.

125 Rodríguez-Morgade, M.S., Plonska-Brzezinska, M.E., Athans, A.J., Carbonell, E., de Miguel, G., Guldi, D.M., Echegoyen, L., and Torres, T. (2009) *J. Am. Chem. Soc.*, **131**, 10484–10496.

7
Supramolecular Chemistry of Fullerene-Containing Micelles and Gels

Hongguang Li, Sukumaran Santhosh Babu, and Takashi Nakanishi

7.1
Introduction

The biological aspects of unique nanocarbon allotropes such as fullerene (C_{60}) are an interesting research topic. However, lack of solubility of C_{60} in water [1] becomes a serious barrier to this perspective. Based on the successful preparation of several amphiphilic C_{60} derivatives, exciting biological activities of C_{60} were revealed in 1993 as reported by the research groups from Japan and the United States of America [2, 3]. The initial discoveries of the biological activities of C_{60} include inhibition of HIV-1 protease and DNA cleavage. In the former case, the unique geometry and highly hydrophobic character of C_{60} was utilized where the spherical C_{60} molecule was found to sit inside the cleft-like active site of HIV-1 protease driven by the hydrophobic/hydrophobic interaction. Triggered by these initial discoveries, more amphiphilic C_{60} derivatives soluble in aqueous media have been synthesized and a variety of new biological activities were found, including neuroprotective properties, antiapoptotic activity, antibacterial activity, and gene transfection [4].

A question, in connection with the research on the biological activities of amphiphilic C_{60} derivatives, immediately coming into mind is whether they stay as individual molecules or aggregates in an aqueous environment. This issue is very important because the aggregate formation can significantly influence the biological activity of these amphiphilic C_{60} derivatives by altering the lifetime of the triplet excited state of the C_{60} [4, 5]. Despite this disadvantage in biological aspect, the aggregation behavior of amphiphilic C_{60} derivatives is interesting from a viewpoint of colloid and interface science since the hydrophobic part in this case is a spherical C_{60} moiety instead of flexible alkyl chains as in traditional surfactants. Deeper investigations showed that, like traditional surfactants, amphiphilic C_{60} derivatives can form a variety of self-assembled supramolecular structures in water including micelles, vesicles, and tubules. Micelles and vesicles formed by surfactants or lipids are good ways to introduce C_{60} into aqueous media. In this case, normally C_{60} and detergent molecules are mixed first in a good solvent for C_{60} such as toluene. The organic solvent is then evaporated and the dry mixture is redissolved in water.

Supramolecular Chemistry of Fullerenes and Carbon Nanotubes, First Edition. Edited by Nazario Martin and Jean-Francois Nierengarten.

In this chapter, a brief review will be given on the investigations of C_{60} in water, with special emphasis on C_{60}-containing supramolecular self-assemblies. Instead of thoroughly going through the progress achieved during the past two decades, only highlights will be given on the development of water-soluble C_{60} either by covalent or by noncovalent methods. C_{60}-containing supramolecular self-assemblies formed in organic solvents will also be mentioned where necessary for comparison. In addition to these fullerene assemblies in solution, gelation properties of fullerenes are also described in this chapter. In most of the fullerene gels, organic solvents are used for the gel formation with the help of π–π interactions of C_{60} molecules or with other π organogelators and often with hydrogen bonding moieties.

7.2
Solubilization of Pristine C_{60} in Surfactant Assemblies

Surfactants are a class of organic molecules consisting of both hydrophilic and hydrophobic moieties. The hydrophilic part can be ionic (such as quaternary ammonium cations or carboxylic anions) or nonionic (such as polyethylene glycol) and the hydrophobic part is typically a long alkyl chain. In some cases, the hydrophilic part can be multifunctional and the hydrophobic part can have more than one alkyl chain. Owing to their amphiphilic character, surfactants are surface active and can reduce the surface tension of water by adsorbing onto the air/water interface. In aqueous solutions, above a threshold concentration called critical micellar concentration (CMC), they self-assemble into a variety of supramolecular structures including micelles of different shapes, uni- and multilamellar vesicles, lytropic liquid crystals with a lamellar, hexagonal, or cubic arrangement as well as bicontinuous architectures [6]. Some of them are schematically shown in Figure 7.1. In these surfactant assemblies, the hydrophilic part of a surfactant molecule is in contact with the continuous aqueous phase, while the hydrophobic part is hidden inside to minimize the total free energy.

The hydrophobic cavity created by surfactant organization provides ideal micro-environment to host hydrophobic guest molecules such as C_{60}. Till now various surfactant systems have been evaluated for their ability to enhance the solubility of C_{60} in water and properties of C_{60} molecules in surfactant assemblies and their biological activities have been investigated.

7.2.1
Solubilizaiton in Micelles

The cores of surfactant micelles, composed of hydrophobic alkyl chains, have been used as host environments for hydrophobic guest molecules including dyes such as pyrene. After they are solubilized into the hydrophobic micellar core, properties of the guest molecules (pyrene) such as the fluorescence spectra can change, which can give important information on micelle formation including the CMC, aggregation number, and microviscosity of the hydrophobic micellar core [7, 8]. As another type of

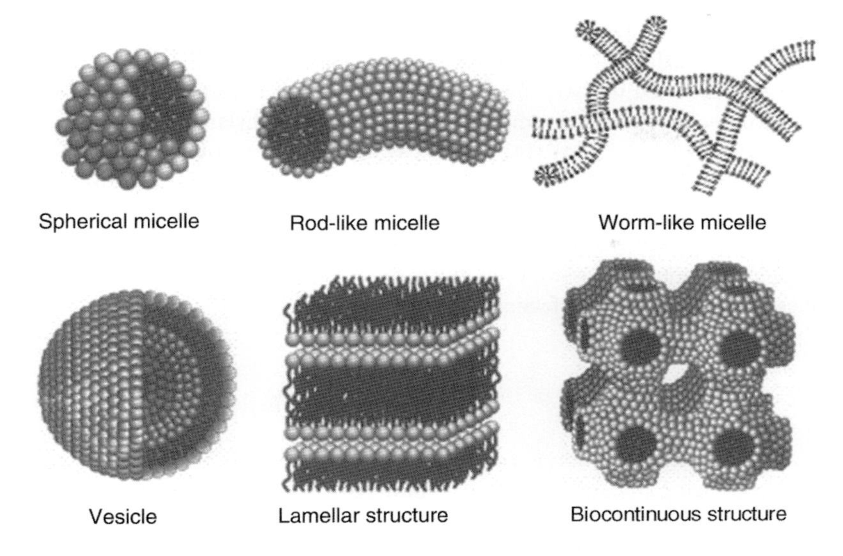

Figure 7.1 Typical surfactant self-assemblies in aqueous solutions.

hydrophobic guest molecules, solubilization of C_{60} generally obeys the same rule as for other guest molecules. However, obvious differences have also been found due to the unique geometrical and chemical properties of C_{60}. Unlike the smaller hydrophobic guest molecules that can be successfully solubilized into the micelles under sonication or even vigorous handshaking, direct solubilizing solid C_{60} into aqueous micellar solutions has been proven to be unsuccessful. In addition to the bigger size, the failure of direct solubilization also arises from the formation of C_{60} clusters in solid state induced by the π–π interactions between adjacent C_{60} molecules.

To successfully solubilize C_{60} molecules into micelles, C_{60} clusters must be dissociated first. This is usually done by using a good solvent for C_{60} during mixing. A typical procedure adopted so far is dissolving C_{60} and surfactant molecules in toluene or chloroform to make them mix totally on a molecular level. The organic solvent is then evaporated under reduced pressure or by raising temperature. The solid mixture containing C_{60} and surfactant molecules is then collected and redissolved in water. Utilizing such a strategy, various surfactant micellar systems have been investigated to solubilize C_{60} including those formed by nonionic surfactants, ionic ones, and amphiphilic macromolecules [9–13]. Eastoe *et al.* found that micelles formed by nonionic surfactants usually can solubilize a higher amount of C_{60} compared to those formed by ionic ones [10]. Photophysical investigations on C_{60} solubilized in micellar solutions of nonionic surfactants revealed that the lifetime of the excited triplet state of C_{60} has been increased by a factor of 3 compared to that in toluene [12]. Formation of C_{60} anion with an unusual long lifetime of 16 ms was also observed, which is approximately 270 times longer than that formed in C_{60}/γ-cyclodextrin complexes [12].

Besides micelles formed by single component of surfactant as mentioned above, those formed by surfactant mixtures can also be applied for C_{60} solubilization. Hao and coworker reported the solubilization of C_{60} in micelles formed by a cationic/anionic ("catanionic") surfactant mixture of tetradecyltrimethylammonium hydroxide (TTAOH)/lauric acid (LA) [14]. In this system, the type of micelles can be manipulated from spherical micelles to rod-like ones and finally to worm-like ones via addition of solid LA to an aqueous solution of TTAOH. Micellar solutions in different regions were examined for C_{60} solubilization. It was found that all the investigated solutions successfully solubilize C_{60}, with spherical micelles and rod-like ones exhibiting better performances.

7.2.2
Solubilization in Vesicles

As far as biological application is concerned, vesicle-based solubilization of C_{60} is especially promising since artificial vesicles have been regarded as model systems to mimic biological membranes. Using a similar method adopted in micellar solubilization of C_{60}, C_{60} has been solubilized in vesicular solutions formed by several artificial or natural double-tailed amphiphiles including positively charged dioctadecyldimethylammonium bromide (DODAB), negatively charged dihexadecyl hydrogen phosphate (DHP), and lecithin as reported by Guldi and coworkers in 1993 [15]. These membrane-embedded C_{60} molecules were then subjected to investigations of reduction by *iso*-propanol radicals. It was found that the yields of C_{60} reduction were highest for the positively charged DODAB vesicular system and lowest for the negatively charged DHP vesicular system due to their ion pair affinity.

Nearly at the same time, solubilization of C_{60} into aqueous vesicular solution formed by phosphatidyl choline as well as several aqueous micellar solutions was reported by Leach and coworkers [9]. UV–vis measurements were selected as a main tool to characterize the systems. By careful comparison between the spectroscopic characteristics of C_{60} solubilized in membranes and those dissolved in a good solvent like chloroform or toluene as well as those in thin solid films, the authors have assigned the spectroscopic changes in C_{60} before and after solubilization mainly to the formation of C_{60} aggregates inside the membranes.

After these two pioneering works, vesicular solubilization of C_{60} has been absent for a while in literatures until 2006 when Hao and coworkers reported the solubilization of C_{60} as well as a C_{60}/C_{70} mixture in a vesicular system formed by salt-free cationic/anionic (catanionic) surfactant mixtures [14, 16]. Owing to the unique properties of salt-free catanionic surfactant vesicular systems, several advantages of this system for C_{60} solubilization have been found including a higher amount of C_{60} solubilized and a considerably high colloidal stability (Figure 7.2). The C_{60}-solubilized vesicular solutions were subjected to rheological measurements showing that the solubilization of C_{60} induced a decrease in shear viscosity of the system.

Recently, Ikeda *et al.* reported a novel method to introduce C_{60} and its higher analogue C_{70} into lipid membranes [17, 18]. Unlike previous studies where a volatile and toxic organic solvent is involved during the solubilization process, a C_{60}

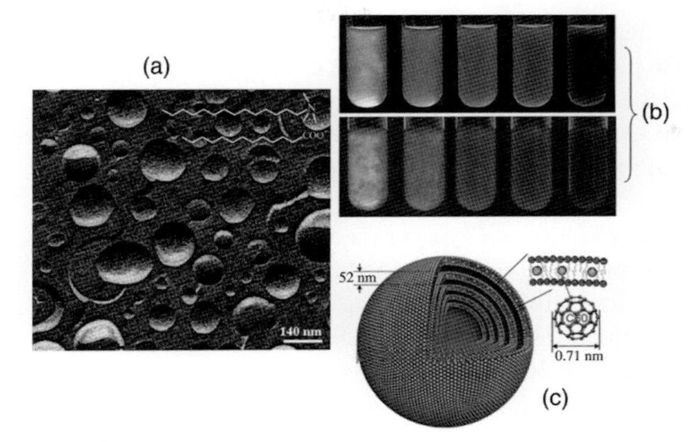

Figure 7.2 (a) Freeze-fracture TEM image of a 100 mM vesicular phase formed by a catanionic surfactant mixture tetradecyltrimethylammonium laurate (TTAL) whose molecular structure is shown inset. (b) Phase behavior of 50 mM TTAL aqueous solutions with increasing amount of C_{60} solubilized without (*top*) and with (*bottom*) crossed polarizers. The concentration of C_{60} from left to right is 0, 0.10, 0.20, 0.30, and 0.82 mM, respectively. (c) Schematic illustration of a multilamellar vesicle with C_{60} solubilized inside the hydrophobic layers. Reprinted with permission from Ref. [16]. © 2006, American Chemical Society.

exchanging method was adopted as shown in Figure 7.3. By heating a water-soluble C_{60}/γ-cyclodextrin or C_{70}/γ-cyclodextrin host–guest complex in the presence of liposomes, the C_{60} or C_{70} molecules readily transfer from the host–guest complex to liposomes. The advantage of this method is avoiding the use of toxic organic

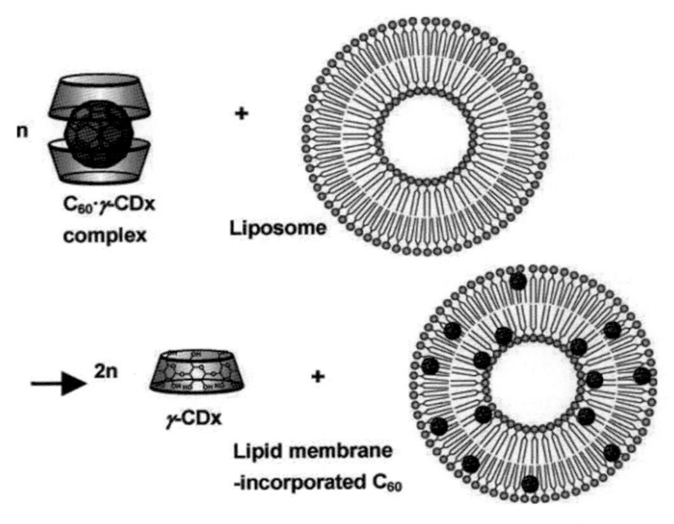

Figure 7.3 Solubilization of C_{60} into liposomes via an exchange method. C_{60} molecules are transferred from C_{60}/γ-CD complexes to the hydrophobic layers of liposomes upon heating. Reprinted with permission from Ref. [17]. © 2005, Royal Society of Chemistry.

solvent for mixing during the solubilization process. The C_{60}- and C_{70}-containing liposomes were subjected to biological tests toward DNA cleavage under irradiation of visible light. It was found the efficiency of DNA cleavage for C_{60}/liposome complex was much higher than that exhibited by C_{60}/γ-cyclodextrin host–guest complex in water. It was also found that the performance of C_{70}-containing liposomes is better than those containing C_{60}.

Although in most cases solubilization of C_{60} occurs in aqueous solutions, it should be pointed out that this concept should not be trapped here. Indeed, there are also examples showing the solubility enhancement of C_{60} in organic solvents in the presence of pseudomicellar structures. Jenekhe and Chen synthesized a poly(phenylquinoline)-block-polystyrene copolymer, which acts as an amphiphilic macromolecule in trifluoroacetic acid/dichloromethane mixture and forms a variety of spherical, vesicular, cylindrical, and lamellar aggregates, depending on the mixing ratio of the two solvents [19]. Interestingly, in the presence of C_{60} and C_{70} only spherical aggregates with aggregation numbers in excess of 10^9 were formed.

7.3 Self-Assemblies of Amphiphilic C_{60} Derivatives

Triggered by the discovery of the biological activities exhibited by C_{60}, many amphiphilic C_{60} derivatives were synthesized and some of them were found to be able to form a variety of interesting supramolecular assemblies in aqueous solutions [1, 20]. Unlike in host–guest complexes and surfactant assemblies where the existence of C_{60} is mainly due to noncovalent hydrophobic/hydrophobic interaction, in these assemblies formed by amphiphilic C_{60} derivatives, fullerenes are covalently connected to the other parts, especially coupled with hydrophilic groups.

The aggregation behavior of amphiphilic C_{60} derivatives could be described by the well-known hydrophilic–lipophilic balance (HLB) value. For example, if only one chain bearing a terminal quaternary ammonium cation is introduced to the C_{60} core (Scheme 7.1, **1**), the HLB value is not high enough to allow **1** to dissolve in aqueous condition. Instead, **1** forms a monolayer at the air–water interface [21]. After the HLB value is increased by introducing a second chain (Scheme 7.1, **2**), the water solubility becomes much better and vesicles with a fractal distribution are formed in aqueous solutions [22]. Another example comes from a dendritic C_{60} derivative bearing 18 terminal carboxyl groups (Scheme 7.1, **3**). Due to the high HLB value, **3** is highly water soluble and cannot form large aggregates that are detectable by electron microscopy observations [23, 24]. After introducing five hydrophobic alkyl chains on the opposite site [25] (Scheme 7.1, **4**) or complex with oppositely charged surfactant molecules [26], the HLB value is lowered and well-defined supramolecular assemblies are constructed.

Besides introducing hydrophilic functional group onto the C_{60} moiety via an external chain, the hydrophilic part can also be directly introduced on C_{60}. One example is a potassium salt of pentaphenyl C_{60} derivative (Scheme 7.2, **5**). With a combination of dynamic and static laser light scattering measurements, **5** was found

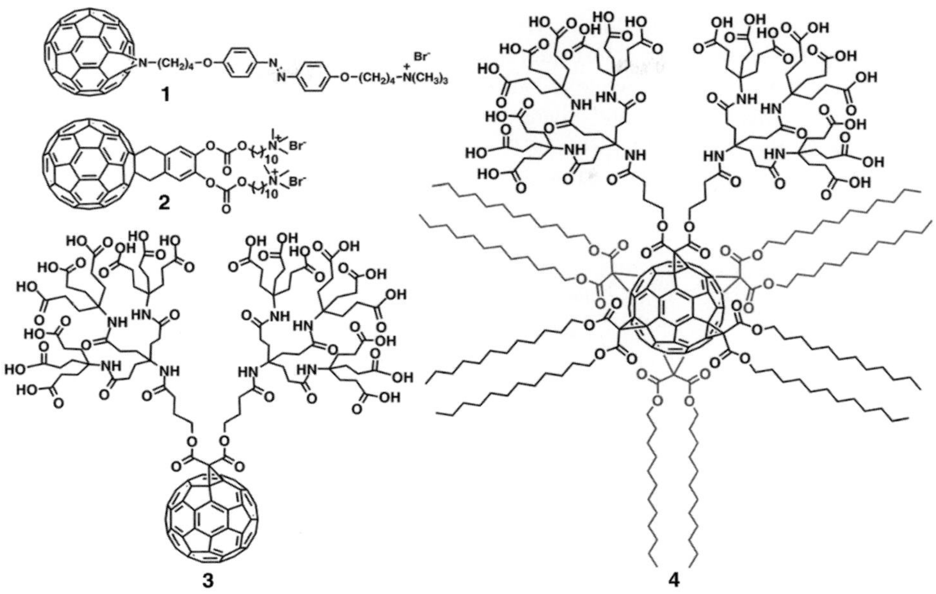

Scheme 7.1

to form vesicular supramolecular assemblies in water [27]. This is quite interesting since the molecular structure of **5** where both the hydrophilic and the hydrophobic parts are restricted on the same C_{60} is totally different from that of vesicle-forming amphiphiles such as double-tailed surfactants or lipids [28]. Similar low molecular weight C_{60}-based amphiphiles are reported by Tour and coworkers, **6** [29], and Liang and coworkers, **7** [30] (depicted in Scheme 7.2).

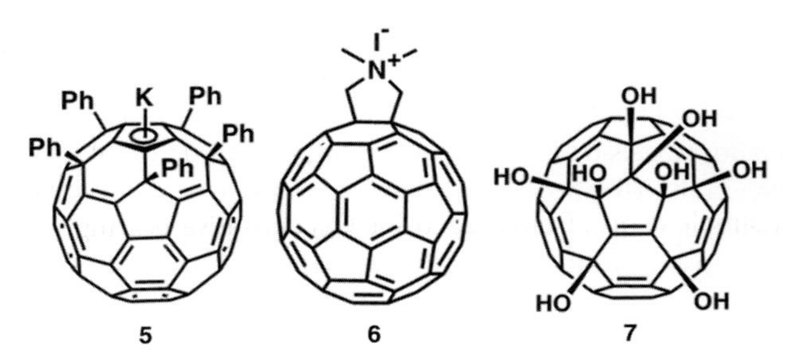

Scheme 7.2

In addition to low molecular weight C_{60} derivatives described above, amphiphilic C_{60} derivatives with poly(ethylene oxide) chain as hydrophilic parts have also been synthesized. Song *et al.* reported the synthesis of amphiphilic C_{60} derivatives carrying poly(ethylene oxide) and several C_{60} end-capped dimers with poly(ethylene oxide) spacer (Scheme 7.3, **8**,**9**) [31]. The aggregation behavior in water and THF indicated that at short poly(ethylene oxide) chain length, formation of spherical aggregates are dominated in water, while at longer chain length, networks begin to form. Sometimes, mixed hydrophilic functional groups are also introduced to the same C_{60} core as in the case of **10** (Scheme 7.3). This molecule was found to form vesicular aggregates in aqueous solutions [32].

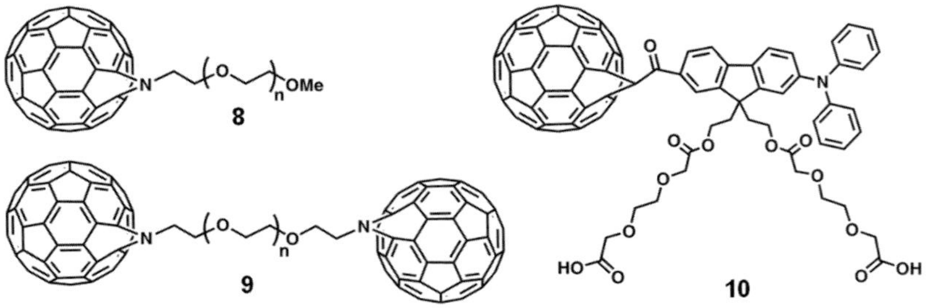

Scheme 7.3

7.4
Gels of Fullerenes

The self-assembly, organogelation of C_{60} alone and in conjugation with other photonically active units, has been an exciting research topic. The appropriate functional groups such as amide, urea, hydroxyl, and so on that provide the necessary hydrogen bonding interactions, aliphatic chains to control solubility, van der Waals interactions, and C_{60}–C_{60} spherical π–π stacking enable C_{60} molecules to organize and finally to form organogels. The recent developments have led to coassembled photoconductive gels of C_{60} and thereby extended the applications to photovoltaics.

The first C_{60}-based organogel was reported by the group of Shinkai in 1999, synthesizing a C_{60} amphiphile bearing two ammonium groups [33]. The methanol solution of **2** (Scheme 7.1) at lower concentrations produced membrane-like assemblies upon sonication and organogels upon keeping the solution (10–20 mmol/dm^3) for a few days. Accordingly, a time-dependent structural transition was observed in the TEM images as globular aggregates (90–135 nm) at lower concentrations and fibrous aggregates (10–20 nm) of organogels at higher concentrations. Interestingly, the ordered packing of **2** in the organogel fibers was established by the sharp and periodical peaks in the XRD pattern resulting from the monolayer or the tilted bilayer structure.

The chiral C_{60} assemblies in organogels were also reported by Shinkai and coworkers [34]. A cholesterol-appended C_{60} gelator **11** (Scheme 7.4) with a natural C-3 (*S*)-configuration forms transparent brown gel above a concentration of 0.0162 mol/dm^3 in dichloromethane. The xerogel of **11** in dichloromethane has fibrous structure with ~100 nm diameter, captured by SEM. In the gel state, the self-assembly of **11** imprinted a 440 nm band with a positive CD sign, having a

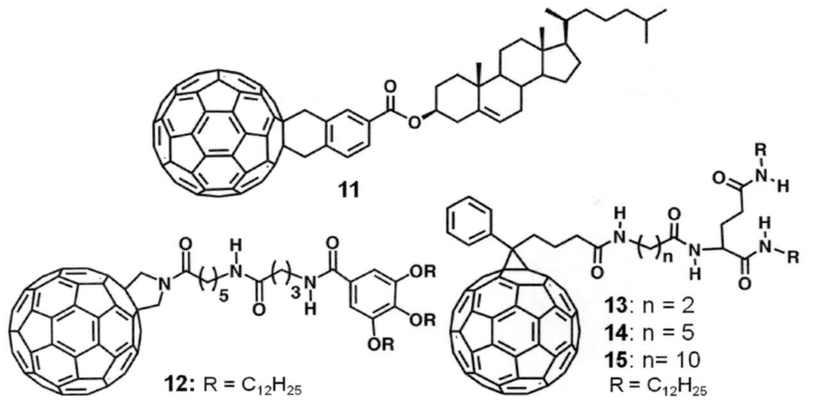

Scheme 7.4

characteristic absorption maximum at 434 nm in the UV–vis spectrum, attributed to the chiroptical contribution from the chiral aggregates. The chiral, columnar one-dimensional packing of the cholesterol moieties enabled to form a helical column decorated with chirally oriented C_{60} moieties outside.

The creation of one-dimensional nanostructures is of prime importance, especially in the case of photoconductive organic materials such as C_{60}. Nakamura and coworkers have achieved this by incorporating 3,4,5-tris(dodecyloxy)benzamide moiety to the surface of C_{60} [35]. A C_{60} derivative **12** (Scheme 7.4) formed stable organogels in chloroform and adapted a stable monolayer structure at the air/water interface. The hydrogen bonding and π–π interaction between the C_{60} parts enabled to have a monolayer fibrous structure of 1.2 nm in height, 5–10 mm in length, and 8 nm in width in the cast film.

The control over solubility of C_{60} obtained by derivatization using an L-glutamide moiety (**13–15**) (Scheme 7.4) has led to the formation of organogels, especially in mixed organic solvents [36]. The temperature-dependent phase transition between gel and sol states of **15** in cyclohexane-THF (6 : 1) mixture indicated the formation of a strong organogel. Interestingly, the coassembly of C_{60} derivatives with the corresponding porphyrin molecules also resulted in the formation of stable assemblies and gels. It was indicated by the bathochromic shift and splitting in the UV–vis as well as CD spectra in cyclohexane-THF (20 : 1) mixture. The emission intensity of porphyrin exhibited a remarkable decrease with increasing concentration of the fullerene derivatives, indicating the facile electron transfer mediated in the organogel scaffold.

The significant achievement in organogels of C_{60} is the possibility to incorporate pristine C_{60} or C_{60} derivatives into the assemblies of other electron donor units, especially porphyrins, without disturbing the self-assembly properties. Shinkai and coworkers studied the ground-state intermolecular interaction between Zn(II) porphyrin-appended cholesterol gelator **16** (Scheme 7.5) and C_{60} in the coassembled organogel [37, 38]. The sandwich complex formation between **16**/C_{60} (2 : 1) enhances

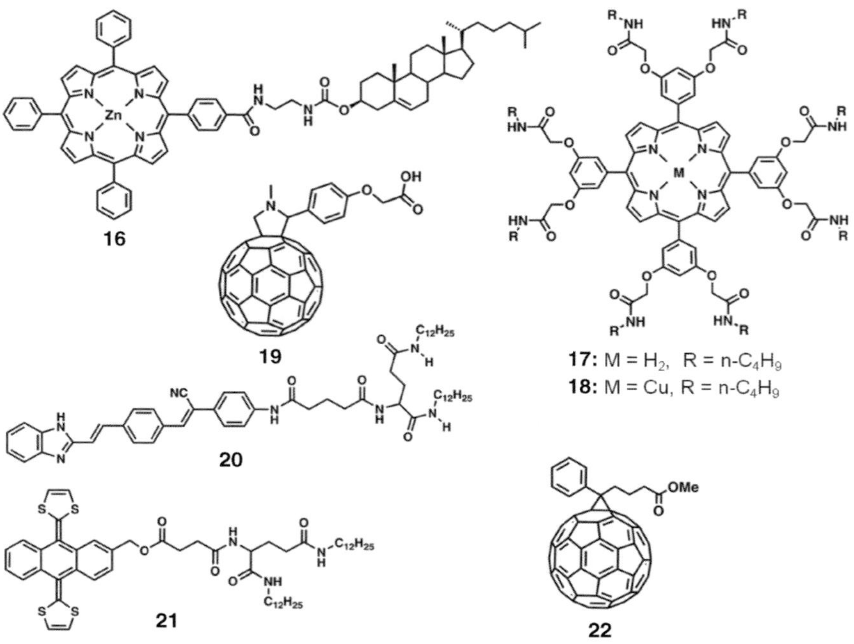

17: M = H_2, R = n-C_4H_9
18: M = Cu, R = n-C_4H_9

Scheme 7.5

the gelation ability of **16** resulting in a fibrous structure with 50–150 nm diameter. A bathochromic shift of the Soret absorption band and enhancement in the negative exciton-coupled CD signal in the presence of C_{60} established the intermolecular interactions between **16** and C_{60} in the gel phase.

The encapsulation of C_{60} molecule by the circular H-bonded array created by porphyrin gelator **17** (Scheme 7.5) in a 1 : 2 complex manner is another example [39]. The sheet-like morphology of **17** undergoes a drastic change to a fibrous network on increasing the concentration of C_{60} from 0–0.5 equivalents, due to a strong porphyrin–C_{60} interaction. The gelation of copper porphyrin **18** (Scheme 7.5) is significantly improved by addition of C_{60}, implying that **18** also exhibits versatile gelation ability comparable to that of **17** [40].

In recent years, there has been significant improvement in understanding the photoconductive and solar cell properties of C_{60}-based organogels. Xue *et al.* have integrated electron acceptors such as C_{60} and C_{60} derivative **19** into regular and ordered aggregates of a donor gelator **20** [41]. The comparison of changes in emission intensity of the gel upon addition of C_{60} and **19** (Scheme 7.5) revealed the relatively effective quenching by the later one due to the hydrogen-bonded 1 : 2 complex, thereby making the electron transfer more facile. Photocurrent obtained from the photovoltaic cell fabricated using hybrid gels of **19** and **20** (2 : 1) as the active layer exhibited stable and large photocurrent than the device containing pristine C_{60} and **20** in the same mixing ratio. Interestingly, when the ratio of **20** to **19** is increased up to 1 : 1, photocurrent becomes very weak, even less comparable with the one containing C_{60}.

Zhang and coworkers have reported the gelation and photocurrent generation from the 2 : 1 complex of ex-TTF-based organogelator **21** and C_{60} (Scheme 7.5) [42]. A steady, rapid, and reproducible cathodic photocurrent of 25 nA/cm^2 was obtained when the xerogel prepared from toluene solution of **21** (1.9 mg/mL) and phenyl-C_{61}-butyricacid methyl ester (PCBM) **22** (1.8 mg/mL) on ITO surface was exposed to white light (33 mW/cm^2). The multiple light scattering and enhanced light absorption through the porous network structure and enhanced charge separation due to the efficient contact between electrolyte and ITO through increased specific surface area of the gel structure generate steady photocurrent. Hence, organogelation is crucial in the formation of regular donor/acceptor self-assembly, making a good connection, efficient separation between donor and acceptor at the molecular level, and an efficient charge channel along the fiber direction resulting in enhanced photocurrent.

Apart from aggregation, organogelation, its effect on the development of donor/acceptor heterojunction solar cells of poly(3-hexylthiophene) (P3HT) and **22** has also been discussed in the recent past [43, 44].

7.5
Conclusions and Outlook

Apart from the research interests in superconductive materials, electronic devices such as organic solar cells, optical limiting properties, and mimicking photosynthetic solar energy transduction, the direction in biological aspects forms an indispensable part of this newly formed discipline. The success made by chemists on imparting water solubility to C_{60}, either by a noncovalent or a covalent method, has greatly accelerated the research in biological aspects and has made many dreams to realities. It should be pointed out, however, that research on water-soluble C_{60} is not limited to developing necessary candidates for biological investigations. Instead, it has met and entered several other major scientific research areas besides biology. Solubilization of C_{60} in surfactant micelles and aggregation behavior of amphiphilic C_{60} derivatives lead to new insights in colloid and interface science. The well-ordered C_{60}-containing supramolecular assemblies may also find potential applications in material science.

Despite the progress achieved so far, there is still plenty of room for novelty especially considering that the performance of self-assembled objects of C_{60} and its derivatives has not yet met the expectations. Developments in organic synthesis, supramolecular chemistry, and nanoscience have provided many new macromolecular architectures and nanocontainers that can be alternatives to host pristine C_{60} [45]. Given the infinite imagination of chemists, novel amphiphilic C_{60} derivatives are still to come. For instance, authors have developed novel hydrophobic amphiphiles based on alkylated C_{60} derivatives [46, 47]. These molecules have self-organized to various supramorphological architectures with controlled dimension under organic solvent conditions [48]. Moreover, the obtained high dimensional assemblies showed interesting materials performance such as superhydrophobicity [49], templates for nanostructured metals [50], and CNT photothermal conversion indicator [51]. These aspects together with C_{60}-containing gels have paved the way for construction of meaningful supramolecular C_{60} architectures. And of course water-soluble C_{60} assemblies would be approached for real biological applications in near future.

References

1 Nakamura, E. and Isobe, H. (2003) Functionalized fullerenes in water. The first 10 years of their chemistry, biology, and nanoscience. *Acc. Chem. Res.*, **36**, 807–815.

2 Friedman, S.H., DeCamp, D.L., Sijbesma, R.P., Srdanov, G., Wudl, F., and Kenyon, G.L. (1993) Inhibition of the HIV-1 protease by fullerene derivatives: model building studies and experimental verification. *J. Am. Chem. Soc.*, **115**, 6506–6509.

3 Tokuyama, H., Yamago, S., Nakamura, E., Shiraki, T., and Sugiura, Y. (1993) Photoinduced biochemical activity of fullerene carboxylic acid. *J. Am. Chem. Soc.*, **115**, 7918–7919.

4 Ros, T.D. and Prato, M. (1999) Medicinal chemistry with fullerenes and fullerene derivatives. *Chem. Commun.*, 663–669.

5 Guldi, D.M. and Prato, M. (2000) Excited-state properties of C_{60} fullerene derivatives. *Acc. Chem. Res.*, **33**, 695–703.

6 Israelachvili, J. (1991) *Intermolecular and Surface Forces*, Academic Press, London.

7 Feitosa, E., Brown, W., Vasilescu, M., and Swanson-Vethamuthu, M. (1996) Effect of temperature on the interaction between the nonionic surfactant $C_{12}E_5$ and poly (ethylene oxide) investigated by dynamic light scattering and fluorescence methods. *Macromolecules*, **29**, 6837–6846.

8 Panmai, S., Prud'homme, R.K., Peiffer, D.G., Jockusch, S., and Turro, N.J. (2002) Interactions between hydrophobically modified polymers and surfactants: a fluorescence study. *Langmuir*, **18**, 3860–3864.

9 Bensasson, R.V., Bienvenue, E., Dellinger, M., Leach, S., and Seta, P. (1994) C_{60} in model biological systems: a visible–UV absorption study of solvent-dependent parameters and solute aggregation. *J. Phys. Chem.*, **98**, 3492–3500.

10 Beeby, A., Eastoe, J., and Heenan, R.K. (1994) Solubilisation of C_{60} in aqueous micellar solution. *Chem. Commun.*, 173–175.

11 Yamakoshi, Y.N., Yagami, T., Fukuhara, K., Sueyoshi, S., and Miyata, N. (1994) Solubilization of fullerenes into water with polyvinylpyrrolidone applicable to biological tests. *Chem. Commun.*, 517–518.

12 Eastoe, J., Crooks, E.R., Beeby, A., and Heenan, R.K. (1995) Structure and photophysics in C_{60}-micellar solutions. *Chem. Phys. Lett.*, **245**, 571–577.

13 Lai, D.T., Neumann, M.A., Matsumoto, M., and Sunamoto, J. (2000) Complexation of C_{60} fullerene with cholesteryl group-bearing pullulan in aqueous medium. *Chem. Lett.*, 64–65.

14 Li, H. and Hao, J. (2007) Phase behavior of salt-free catanionic surfactant aqueous solutions with fullerene C_{60} solubilized. *J. Phys. Chem. B*, **111**, 7719–7724.

15 Hungerbühler, H., Guldi, D.M., and Asmus, K.D. (1993) Incorporation of C_{60} into artificial lipid membranes. *J. Am. Chem. Soc.*, **115**, 3386–3387.

16 Li, H., Jia, X., Li, Y., Shi, X., and Hao, J. (2006) A salt-free zero-charged aqueous onion-phase enhances the solubility of fullerene C_{60} in water. *J. Phys. Chem. B*, **110**, 68–74.

17 Ikeda, A., Sato, T., Kitamura, K., Nishiguchi, K., Sasaki, Y., Kikuchi, J., Ogawa, T., Yogo, K., and Takeya, T. (2005)

Efficient photocleavage of DNA utilising water-soluble lipid membrane-incorporated [60]fullerenes prepared using a [60]fullerene exchange method. *Org. Biomol. Chem.*, **3**, 2907–2909.

18 Ikeda, A., Doi, Y., Hashizume, M., Kikuchi, J., and Konish, T. (2007) An extremely effective DNA photocleavage utilizing functionalized liposomes with a fullerene-enriched lipid bilayer. *J. Am. Chem. Soc.*, **129**, 4140–4141.

19 Jenekhe, S.A. and Chen, X.L. (1998) Self-assembled aggregates of rod-coil block copolymers and their solubilization and encapsulation of fullerenes. *Science*, **279**, 1903–1907.

20 Guldi, D.M., Zerbetto, F., Georgakilas, V., and Prato, M. (2005) Ordering fullerene materials at nanometer dimensions. *Acc. Chem. Res.*, **38**, 38–43.

21 Oh-ishi, K., Okamura, J., Ishi-i, T., Sano, M., and Shinkai, S. (1999) Large monolayer domain formed by C_{60}-azobenzene derivative. *Langmuir*, **15**, 2224–2226.

22 Sano, M., Oishi, K., Ishi-i, T., and Shinkai, S. (2000) Vesicle formation and its fractal distribution by bola-amphiphilic [60]fullerene. *Langmuir*, **16**, 3773–3776.

23 Brettreich, M. and Hirsch, A. (1998) A highly water-soluble dendro[60] fullerene. *Tetrahedron Lett.*, **39**, 2731–2734.

24 Quaranta, A., McGarvey, D.J., Land, E.J., Brettreich, M., Burghardt, S., Schönberger, H., Hirsch, A., Gharbi, N., Moussa, F., Leach, S., Göttingere, H., and Bensasson, R.V. (2003) Photophysical properties of a dendritic methano[60] fullerene octadeca acid and its *tert*-butyl ester: evidence for aggregation of the acid form in water. *Phys. Chem. Chem. Phys.*, **5**, 843–848.

25 Brettreich, M., Burghardt, S., Böttcher, C., Bayerl, T., Bayerl, S., and Hirsch, A. (2000) Globular amphiphiles: membrane-forming hexaadducts of C_{60}. *Angew. Chem. Int. Ed.*, **39**, 1845–1848.

26 Hao, J., Li, H., Liu, W., and Hirsch, A. (2004) Well-defined self-assembling supramolecular structures in water containing a small amount of C_{60}. *Chem. Commun.*, 602–603.

27 Zhou, S., Burger, C., Chu, B., Sawamura, M., Nagahama, N., Toganoh, M., Hackler, U.E., Isobe, H., and Nakamura, E. (2001) Spherical bilayer vesicles of fullerene-based surfactants in water: a laser light scattering study. *Science*, **291**, 1944–1947.

28 Burger, C., Hao, J., Ying, Q., Isobe, H., Sawamura, M., Nakamura, E., and Chu, B. (2004) Multilayer vesicles and vesicle clusters formed by the fullerene-based surfactant $C_{60}(CH_3)_5K$. *J. Colloid Interface Sci.*, **275**, 632–641.

29 Cassell, A.M., Asplund, C.L., and Tour, J.M. (1999) Self-assembling supramolecular nanostructures from a C_{60} derivative: nanorods and vesicles. *Angew. Chem. Int. Ed.*, **38**, 2403–2405.

30 Zhang, G., Liu, Y., Liang, D., Gan, L., and Li, Y. (2010) Facile synthesis of isomerically pure fullerenols and formation of spherical aggregates from $C_{60}(OH)_8$. *Angew. Chem. Int. Ed.*, **49**, 5293–5295.

31 Song, T., Dai, S., Tam, K.C., Lee, S.Y., and Goh, S.H. (2003) Aggregation behavior of C_{60}-end-capped poly(ethylene oxide)s. *Langmuir*, **19**, 4798–4803.

32 Verma, S., Hauck, T., El-Khouly, M.E., Padmawar, P.A., Canteenwala, T., Pritzker, K., Ito, O., and Chiang, L.Y. (2005) Self-assembled photoresponsive amphiphilic diphenylaminofluorene-C_{60} conjugate vesicles in aqueous solution. *Langmuir*, **21**, 3267–3272.

33 Oishi, K., Ishi-i, T., Sano, M., and Shinkai, S. (1999) Unexpected discovery of a novel organic gel system comprised of [60]fullerene containing amphiphiles. *Chem. Lett.*, **28**, 1089–1090.

34 Ishi-i, T., Ono, Y., and Shinkai, S. (2000) Chirally-ordered fullerene assemblies found in organic gel systems of cholesterol-appended [60]fullerenes. *Chem. Lett.*, 808–809.

35 Tsunashima, R., Noro, S.-i., Akutagawa, T., Nakamura, T., Kawakami, H., and Toma, K. (2008) Fullerene nanowires: self-assembled structures of a low-molecular-weight organogelator fabricated by the Langmuir–Blodgett method. *Chem. Eur. J.*, **14**, 8169–8176.

36 Watanabe, N., Jintoku, H., Sagawa, T., Takafuji, M., Sawada, T., and Ihara, H. (2009) Self-assembling fullerene derivatives for energy transfer in molecular gel system. *J. Phys. Conf. Ser.*, **159**, 012016.

37 Ishi-i, T., Jung, J.H., and Shinkai, S. (2000) Intermolecular porphyrin–fullerene interaction can reinforce the organogel structure of a porphyrin-appended cholesterol derivative. *J. Mater. Chem.*, **10**, 2238–2240.

38 Ishi-i, T., Iguchi, R., Snip, E., Ikeda, M., and Shinkai, S. (2001) [60]Fullerene can reinforce the organogel structure of porphyrin-appended cholesterol derivatives: novel odd–even effect of the $(CH_2)n$ spacer on the organogel stability. *Langmuir*, **17**, 5825–5833.

39 Shirakawa, M., Fujita, N., and Shinkai, S. (2003) [60]Fullerene-motivated organogel formation in a porphyrin derivative bearing programmed hydrogen-bonding sites. *J. Am. Chem. Soc.*, **125**, 9902–9903.

40 Shirakawa, M., Fujita, N., Shimakoshi, H., Hisaeda, Y., and Shinkai, S. (2006) Molecular programming of organogelators which can accept [60] fullerene by encapsulation. *Tetrahedron*, **62**, 2016–2024.

41 Xue, P., Lu, R., Zhao, L., Xu, D., Zhang, X., Li, K., Song, Z., Yang, X., Takafuji, M., and Ihara, H. (2010) Hybrid self-assembly of a π gelator and fullerene derivative with photoinduced electron transfer for photocurrent generation. *Langmuir*, **26**, 6669–6675.

42 Yang, X., Zhang, G., Zhang, D., and Zhu, D. (2010) A new ex-TTF-based organogelator: formation of organogels and tuning with fullerene. *Langmuir*, **26**, 11720–11725.

43 Huang, W.Y., Huang, P.T., Han, Y.K., Lee, C.C., Hsieh, T.L., and Chang, M.Y. (2008) Aggregation and gelation effects on the performance of poly(3-hexylthiophene)/fullerene solar cells. *Macromolecules*, **41**, 7485–7489.

44 Koppe, M., Brabec, C.J., Heiml, S., Schausberger, A., Duffy, W., Heeney, M., and McCulloch, I. (2009) Influence of molecular weight distribution on the gelation of P3HT and its impact on the photovoltaic performance. *Macromolecules*, **42**, 4661–4666.

45 Inokuma, Y., Arai, T., and Fujita, M. (2010) Networked molecular cages as crystalline sponges for fullerenes and other guests. *Nature Chem.*, **2**, 780–783.

46 Nakanishi, T. (2010) Supramolecular soft and hard materials based on self-assembly algorithms of alkyl-conjugated fullerenes. *Chem. Commun.*, 3425–3436.

47 Babu, S.S., Möhwald, H., and Nakanishi, T. (2010) Recent progress in morphology control of supramolecular fullerene assemblies and its applications. *Chem. Soc. Rev.*, **39**, 4021–4035.

48 Nakanishi, T., Schmitt, W., Michinobu, T., Kurth, D.G., and Ariga, K. (2005) Hierarchical supramolecular fullerene architectures with controlled dimensionality. *Chem. Commun.*, 5982–5984.

49 Nakanishi, T., Michinobu, T., Yoshida, K., Shirahata, N., Ariga, K., Möhwald, H., and Kurth, D.G. (2008) Nanocarbon superhydrophobic surfaces created from fullerene-based hierarchical supramolecular assemblies. *Adv. Mater.*, **20**, 443–446.

50 Shen, Y., Wang, J., Kuhlmann, U., Hildebrandt, P., Ariga, K., Möhwald, H., Kurth, D.G., and Nakanishi, T. (2009) Supramolecular templates for nanoflake-metal surfaces. *Chem. Eur. J.*, **15**, 2763–2767.

51 Shen, Y., Skirtach, A.G., Seki, T., Yagai, S., Li, H., Möhwald, H., and Nakanishi, T. (2010) Assembly of fullerene–carbon nanotubes: temperature indicator for photothermal conversion. *J. Am. Chem. Soc.*, **132**, 8566–8568.

8
Fullerene-Containing Supramolecular Polymers and Dendrimers

Takeharu Haino and Toshiaki Ikeda

8.1
Introduction

The discovery of fullerenes in 1985 [1] has created a new research field in carbon clusters. The major breakthrough in fullerene science was achieved by Krätschmer and Huffman who produced practical amounts of [60]fullerene [2]. This pushed fullerene science forward, and many chemists have started developing ways to synthesize numerous variations of unique carbon-based compounds. Outstanding magnetic [3], superconducting [4], electrochemical [5], and photophysical [6] properties are possible due to the unique three-dimensional geometry [7] of [60]fullerene. These properties have also attracted great interest in the fields of chemistry and materials science. In recent years, fullerene chemistry has collaborated with macromolecular science. This synergetic collaboration gives rise to an opportunity to generate a new fullerene-based polymer science with a great number of applications [8]. The synthesis of fullerene-containing polymeric materials has been initially attempted via direct polymerization of a [60]fullerene molecule through cycloaddition reactions [9]. However, the products contain serious problems with regard to their structures and properties. The highly reactive 30 double bonds of [60]fullerene are chemically equivalent; thus, one-, two-, and three-dimensionally extended polymeric architectures formed will lose the stereo- and regioregularity, and are less processable than conventional polymers. The direct polymerization of [60] fullerene is difficult to use. Therefore, recent efforts have shifted to incorporating [60]fullerene onto macromolecular cores to create [60]fullerene-containing polymers [10] and dendrimers [11]. They have proved to be innovative materials, functioning as organic solar cells, magnets, photonics, optical-limiting devices, and semiconductors. These attractive functions originate from the precisely controlled nanostructures of fullerene. There are intense efforts underway to develop regulable materials, the functions of which can be tuned by regulating the fullerene nanostructures.

Supramolecular chemistry offers an easy access to the construction of well-organized nanostructures. The formation of a defined supramolecular structure stabilized by

Supramolecular Chemistry of Fullerenes and Carbon Nanotubes, First Edition. Edited by Nazario Martin and Jean-Francois Nierengarten.

noncovalent forces is a thermodynamically driven process [12]. Therefore, the structure of supramolecular macromolecules can be predicted in terms of the thermodynamic minima in an equilibrium process. Supramolecular assemblies can enable their structures and regulate their functions. For instance, the class of supramolecular macromolecules includes micelles, colloids, liquid crystals, gels, aromatic stacks, and hydrogen-bonded polymers and dendrimers. One of their characteristic features is a reversible change in their structures at the supramolecular level upon external stimuli [13]. This ability offers a new way to engineer functional organizations at the nanoscale [14]. This chapter deals with the construction of fullerene-containing supramolecular polymers and dendrimers via self-assembly of multiple components using a variety of noncovalent interactions [15].

8.2
Fabrication of [60]Fullerene Polymeric Array

In general, the control of noncovalent forces to define the size and shape of resulting supramolecular [60]fullerene ensembles in relation to their functions is a major task for a variety of phases, including the solid-state phase, the mesophase, and the solution phase. To engineer the multimolecular ensembles of [60]fullerene on the nanometer scale, noncovalent forces can be manipulated, including ion–ion, ion–dipole, hydrogen bonding, dipole–dipole, π–π stacking, and van der Waals interactions. These weak forces always compete with the solvation of the functional group, which is responsible for the fullerene association in common organic solvents. In contrast, the weak forces appear to be stronger in the solid state than in solution due to the absence of a competitive process of solvation.

Flat aromatic rings can stack on the smooth and dense surface of the [60]fullerene molecule to create attractive π–π stacking interactions. Thus, aromatic stacking can be used effectively to regulate the structures of [60]fullerene nanoarrays in the solid state. Crystallizing [60]fullerene from a solution of aromatic solvents produces one-dimensional nanoarrays containing [60]fullerene molecules interspersed with solvent molecules [16]. The aromatic solvents fill up the free spaces formed by the ordering of the [60]fullerene molecules in the crystals. This prevents the [60]fullerene molecules from randomly aggregating in the solid state. The characteristics of a solvent for cocrystallization of [60]fullerene are crucial for the resulting nanoarrays.

Common aromatic solvents and a variety of aromatic molecules can crystallize with [60]fullerene. Cocrystallization of the [60]fullerene molecules and *p*-bromocalix[4]arene propyl ether **1** forms preorganized linear columns [17] (Figure 8.1). The aggregation of the globular [60]fullerene molecules results in residual spaces among them; these spaces are filled by calix[4]arene **1**. The bromo, phenyl, and propyl groups form effective van der Waals contacts with the smooth surface of the [60]fullerene molecules, which likely drive the formation of the columnar array of [60]fullerene in the solid state. It is worth noting that the well-organized columnar array of [60]fullerene is completely separated by calix[4]arene **1**, which prevents the intercolumnar communication of the [60]fullerene molecules. This allows the directional

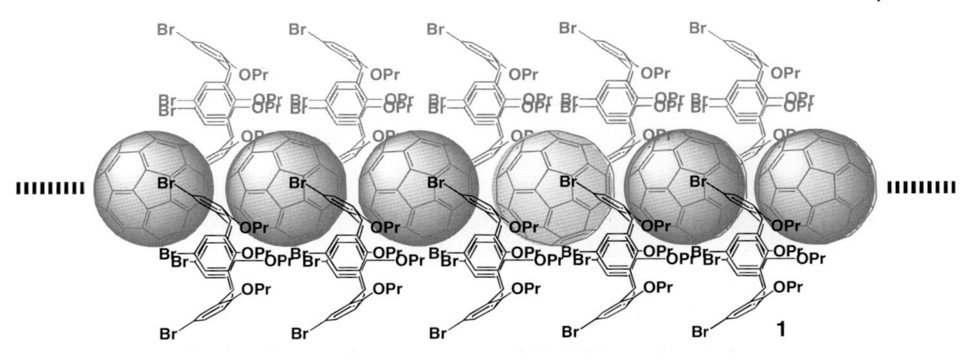

Figure 8.1 Cocrystallization of the [60]fullerene with *p*-bromocalix[4]arene propyl ether **1**.

polymerization of the [60]fullerene molecules. In fact, heat and pressure on the crystals produce the noncross-linked linear [2 + 2] addition polymer [18].

Balch *et al.* have introduced the supramolecular concept into the rational design of [60]fullerene nanoarray [19]. Their [60]fullerene–iridium complex **2** has two phenylethylbenzene components that create the π basic cavity surrounded by the four aromatic rings. The smooth and globular surface of the adjacent [60]fullerene unit is accommodated within the complementary cavity, giving rise to effective π–π stacking interactions. Therefore, the [60]fullerene–iridium complex **2** iteratively forms the host–guest complex in a head-to-tail manner, resulting in a supramolecular polymer in the solid state (Figure 8.2). This complementary interaction works reasonably well to form the [60]fullerene nanoarray.

Haino, Fukazawa, and coworkers have discovered that calix[5]arenes encapsulate groups of fullerenes into their cone-shaped cavity, complementary to the exterior of [60]fullerene. Their extensive efforts to synthesize a variety of calix[5]arene-based host molecules have revealed that bis-calix[5]arene structures capture both higher fullerenes and [60]fullerene [20]. Atwood *et al.* have reported that calix[5]arene **3**, toluene, and the [60]fullerene molecule form cocrystals; one has a 1 : 1 : 1 composition with the [60]fullerene molecules arranged in a supramolecular zigzag array, while the other has a 4 : 2 : 5 composition, which gives rise to the Z-array of the [60] fullerene molecules (Figure 8.3) [21].

A helical array of the [60]fullerene molecules has also resulted from using calix[5] arene **4** instead of **3** [22]. *p-tert*-Butyl calix[5]arene **4** complexes with the [60]fullerene

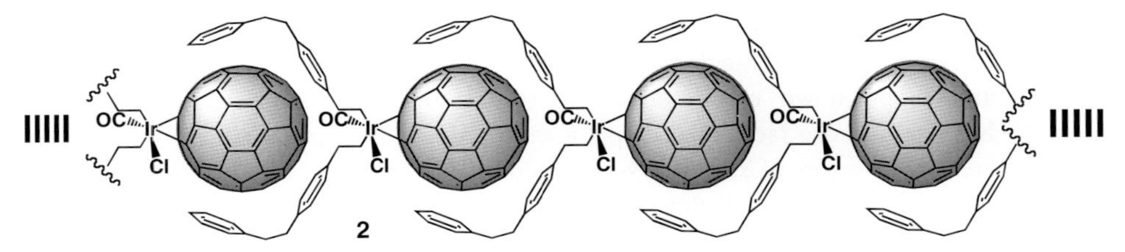

Figure 8.2 Supramolecular polymer of $(\eta^2\text{-}C_{60})Ir(CO)Cl$ **2** in the solid state.

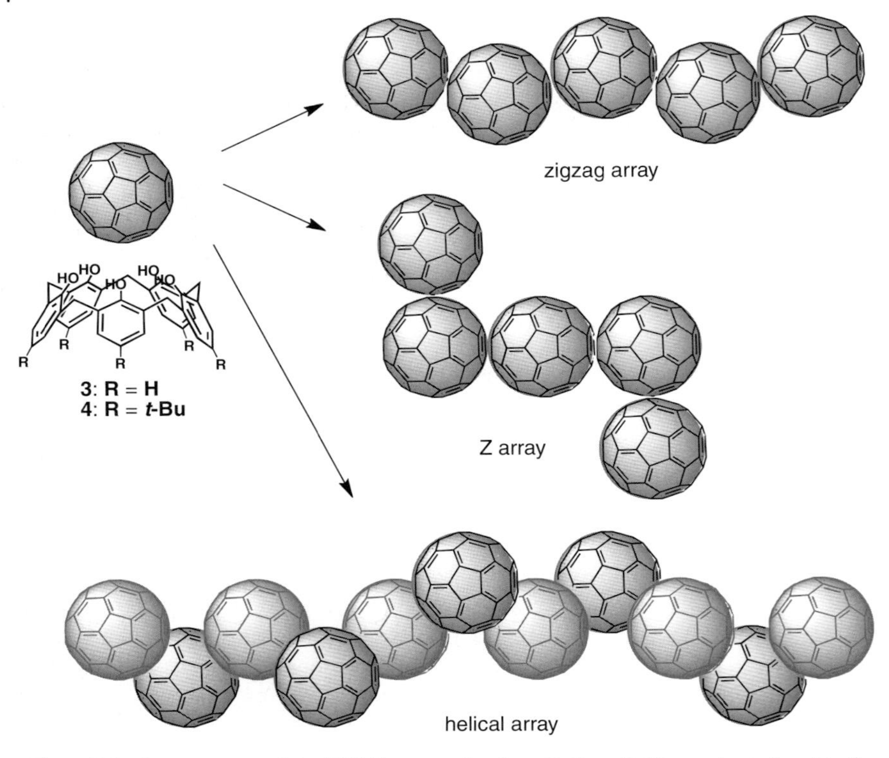

Figure 8.3 Arrangements of the [60]fullerene molecules with the calix[5]arene hosts **3** or **4** in the solid state. The calix[5]arene molecules are omitted in the representation of the assemblies for clarity.

molecules form a 1 : 1 host–guest complex and linear striations of [60]fullerenes within bundled nanofibers. The [60]fullerene molecules adopt the form of a one-dimensional, helical array in the solid state. The [60]fullerene array grows along its principal axis to form the strands of [60]fullerene fibers. *p-tert*-Butyl groups occupy the residual space produced by the [60]fullerene assemblies, allowing effective van der Waals interactions with the smooth and dense aromatic surface of the [60] fullerene molecules. This van der Waals interaction likely drives the dramatic change in the [60]fullerene nanoarray.

Double-concave graphene and permethoxylated hexa-peri-hexabenzocoronene (HBC) **5** can be cocrystallized with [60]fullerene to form a linear array of [60]fullerene molecules (Figure 8.4) [23]. Permethoxylated HBC **5** is not planar due to steric congestion in the bay region: the outer aromatic rings alternate between "flipped up" and "flipped down" configuration with respect to the inner ring. As a result, HBC **5** has two concave faces and adopts a centrosymmetric conformation. The double-concave structure of **5** has two surfaces complementary to the globular [60]fullerene molecule. The cocrystallization of **5** and [60]fullerene gives rise to a columnar packing arrangement in which the two [60]fullerene molecules are positioned on each side of HBC **5** through π–π stacking and van der Waals interactions. This is a remarkable

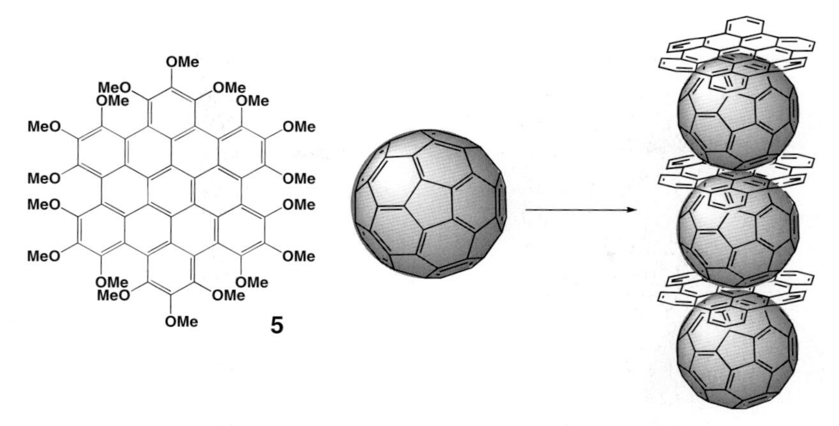

Figure 8.4 Permethoxylated hexa-peri-hexabenzocoronene **5**, [60]fullerene, and the crystal structure of the host–guest complex.

example of the construction of the linear nanoarray of the [60]fullerene molecule in the solid state.

Template-assisted nanoordering of [60]fullerene molecules is an intriguing area of research. The [60]fullerene molecules are encapsulated in the single-wall carbon nanotube. Some of the molecules self-assemble into a supramolecular polymer chain with nearly uniform center-to-center distances, which resemble a nanoscale fullerene peapod [24]. The encapsulated fullerene molecules coalesce in the carbon nanotube, likely as a result of strong π–π interactions between the exterior of the [60]fullerene and the interior of the single-wall carbon nanotube.

An artificial helical nanotube formed by self-assembly of amino acid functionalized naphthalenediimide (NDI) (**6**) can encapsulate [60]fullerene molecules in solution (Figure 8.5) [25]. The crystal structure of the nanotube reveals a helix pitch of 9 Å and three NDI units per turn [26]. The C_{60}–C_{60} interaction is observed in the nanotube, and the chirality information of the helical nanotube is transferred to the encapsulated [60]fullerene to produce the induced circular dichroism in the visible region attributed to electronic transitions of [60]fullerene.

Kawauchi *et al.* have demonstrated that syndiotactic poly(methyl methacylate) (st-PMMA) (**7**) encapsulates [60]fullerene molecules in its tubular cavity, which appeared in a helical conformation (Figure 8.6) [27]. The encapsulated [60]fullerene molecules are aligned to produce the robust peapod-like 1D array in which approximately 86% of the helical hollow spaces are filled with [60]fullerene molecules. Interestingly, the chirality of the helical supramolecular structure is induced by the addition of phenylethylalcohol during its formation. The chirality is memorized even after the chiral phenethylalcohol is removed.

These characteristic examples of constructing [60]fullerene nanoarrays using weak intermolecular forces show that controlling intermolecular interactions is effective in constructing selectively organized polymeric nanoarrays of [60]fullerene molecules in the solid state.

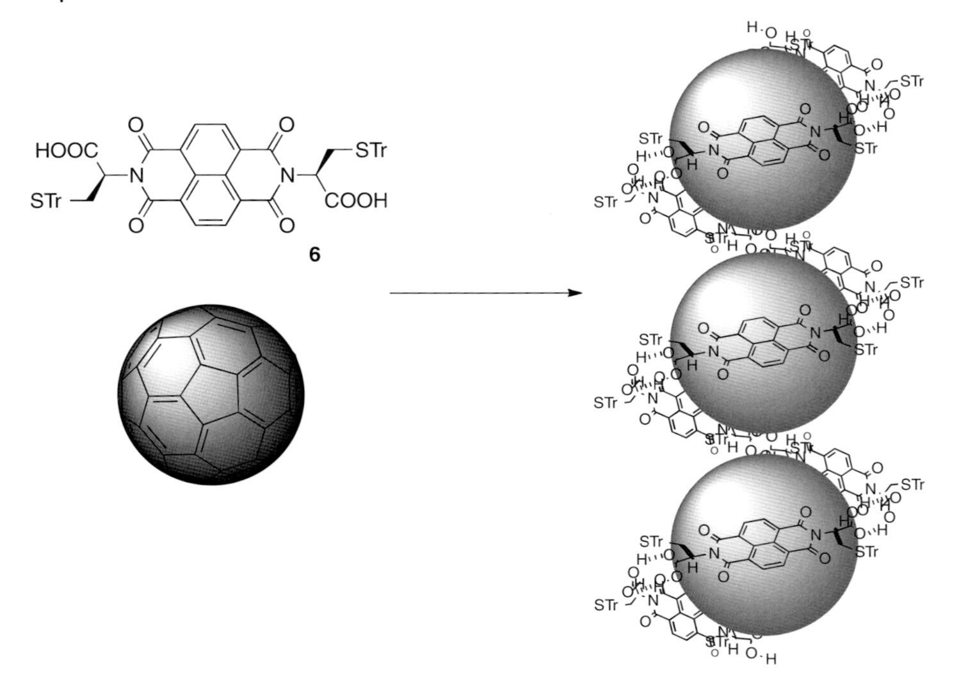

Figure 8.5 Guest-induced helical assembly of naphthalenediimide (NDI) **6** with [60]fullerene molecules.

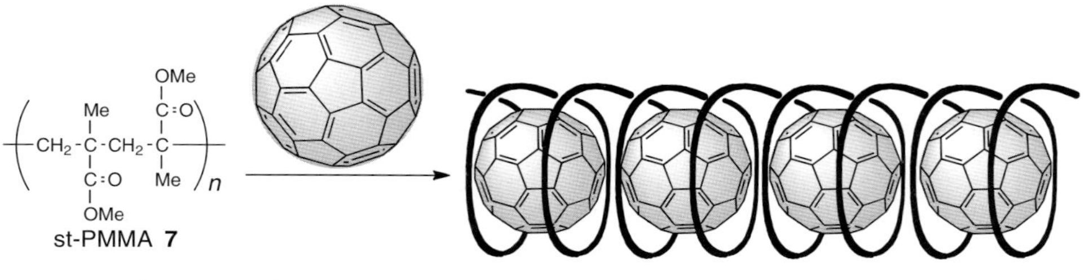

st-PMMA **7**

Figure 8.6 Schematic representation of the encapsulation of [60]fullerene in the st-PMMA helical cavity.

8.3
Supramolecular Polymerization of Functionalized [60]Fullerene

In this section, the scope of [60]fullerene-based supramolecular polymer chemistry is discussed. The approach to synthesize fullerene-containing polymers in a supramolecular manner may allow the development of unprecedented architectures in the high degree of the organization. The previous sections have primarily discussed the self-assembly of the [60]fullerene and its derivatives. The introduction of functionalized

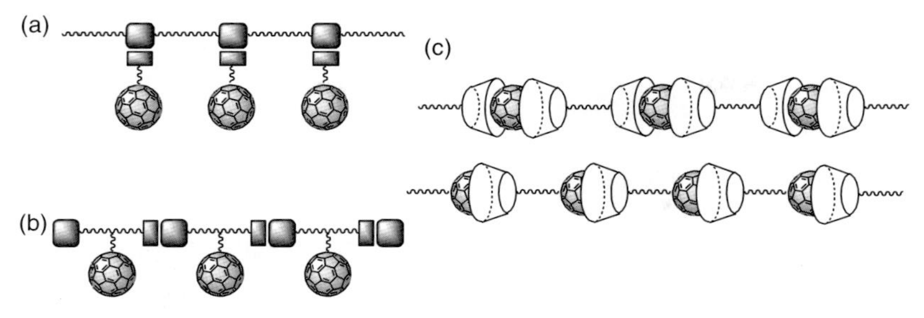

Figure 8.7 Strategies for the synthesis of [60]fullerene-based supramolecular polymers.

[60]fullerene derivatives into polymer chains through noncovalent interactions is an appealing approach, which can increase the synthetic flexibility of the [60]fullerene-based polymers in terms of their structures and functions [10b]. The construction of fullerene-containing supramolecular polymers in organic solution is presented in this section. One could envision introducing [60]fullerene into the side chain of conventional polymers via supramolecular interactions (Figure 8.7a). Binding sites can be preinstalled into a polymer backbone. The [60]fullerene possessing the binding counterpart finds its complement on the polymer chain to form the supramolecular complex. Then, the [60]fullerenes are aligned on the polymer chain. Self-assembling of [60]fullerene-attached ditopic subunits containing complementary pairs at each end offers a quick way to produce supramolecular polymer with [60]fullerene attached as a side chain (Figure 8.7b). Any supramolecular motif can be used for supramolecular polymerization. The molecular association at each site occurs in a selective manner; thus, the resulting supramolecular polymeric architectures can be designed by programming supramolecular interactions into a monomer. Figure 8.7c shows supramolecular polymerization directed by molecular recognition of [60]fullerene. This fascinating strategy produces [60]fullerene main-chain polymers in which 1D [60]fullerene array can be organized.

The combination of the characteristic properties of [60]fullerene and those of polymers suggest many potential applications for [60]fullerene–polymer hybrids. Simple mixing of [60]fullerene with polymers leads to the formation of large [60] fullerene agglomerates in polymer matrix. When the [60]fullerene molecule possesses functional groups capable of creating noncovalent interactions with those of a polymer, the noncovalent interaction between the functional groups enables the [60] fullerene molecules to be connected to the polymer.

8.3.1
Ionic Interaction

Ionic interaction between charged polymers and fullerene derivatives produces fullerene nanoarrays with a polymer template [28]. Schanze and coworkers reported that the layer-by-layer self-assembly approach was used to fabricate photovoltaic cells with the active layer material consisting of PPE-EDOT-SO_3^- **8** and [60]fullerene derivatives **9** [29]. Multiple layer-by-layer deposition of the polymers and **9** produced

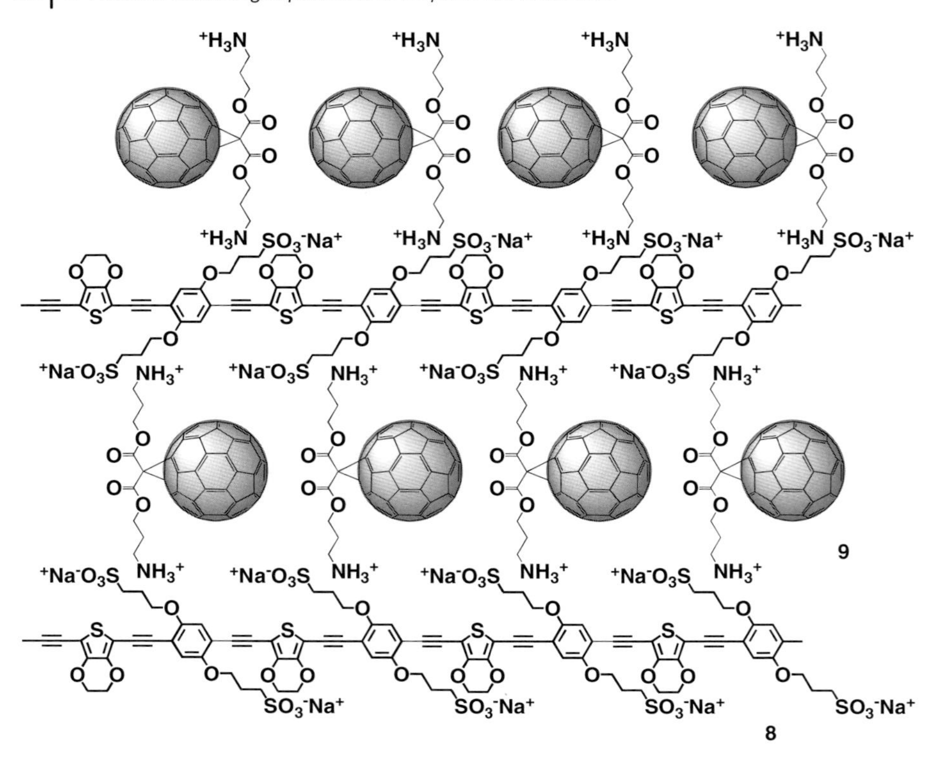

Figure 8.8 Schematic representation of the layer-by-layer structure of the film formed from **8** and **9**.

50 layers on an ITO electrode, and uniform films are obtained as confirmed by AFM studies. The device exhibited good photovoltaic response in this type of self-assembly approach. The ideal layer-by-layer structure is shown in Figure 8.8. It was reported that an integration of the cationic and anionic layers occurred during the layer-by-layer assembly process. An interpenetration of the polymer chains within the layers leads to a more disordered structure. Disorder within the film leads to intimate mixing of the donor and acceptor components. This creates the possibility of forming a bulk heterojunction, which separates and transports charge effectively through the photovoltaic cell.

Yashima and coworkers have demonstrated that phosphonate-functionalized polyphenylacetylene gathers the chiral cationic [60]fullerene molecules around its periphery via ionic interaction [30]. The macromolecular helicity of the polymer dynamically equilibrates between P- and M-forms. The negatively charged phosphonate groups remain outside of the helical polyacetylene. Optically active C_{60}-bissadduct **11** can access the negative charged periphery via ionic interactions. The dynamic helicity of the polymer is sensitive to a chiral environment; the small bias produced by the complexation of the chiral **11** greatly increases the macromolecular helicity (Figure 8.9).

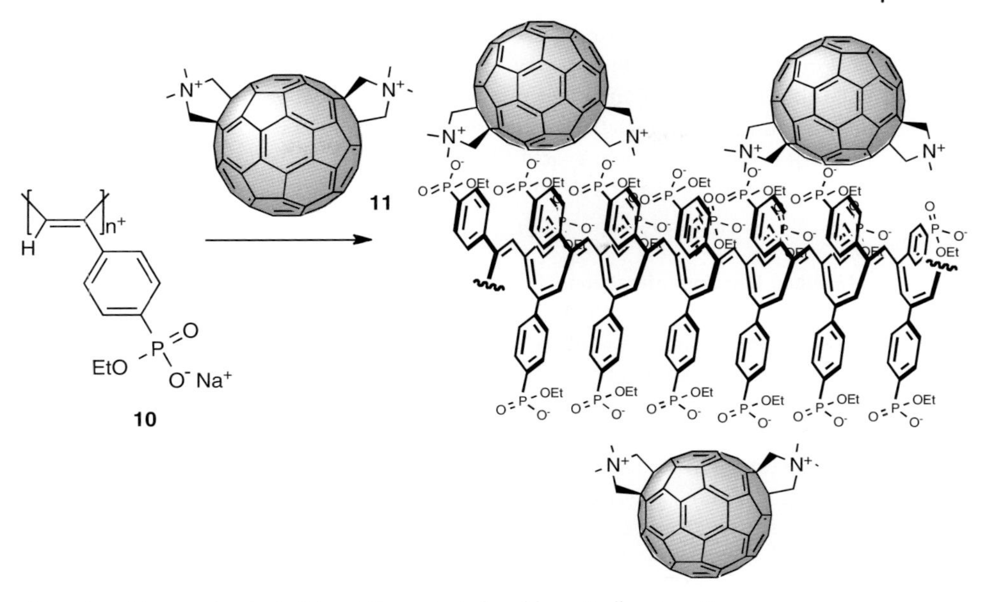

Figure 8.9 Macromolecular helicity of **10** can be induced by optically active **11**.

Fullerene-containing supramolecular polymers based on DNA templation was reported by Chu and coworkers [31]. Cationic [60]fullerene derivatives bind to the DNA main chain through Coulombic interactions and modify the structure by binding the [60]fullerenes in terms of fractal dimensions. Shinkai and coworkers report photocurrent generation in a supramolecular fullerene polymer [32]. A [60]fullerene/porphyrin/DNA ternary complex is deposited on the ITO electrode by oxidative polymerization of 3,4-ethylenedioxythiophene (EDOT). The photocurrent can be effectively generated because it is observed when the porphyrin is excited by light.

Multifunctionalized [60]fullerenes can interact with a properly functionalized polymer. Dai *et al.* utilized Coulombic interaction to introduce hydrogensulfonated fullerenol **13** to PANI-EB polymer **12** (Figure 8.10). The supramolecular polymer

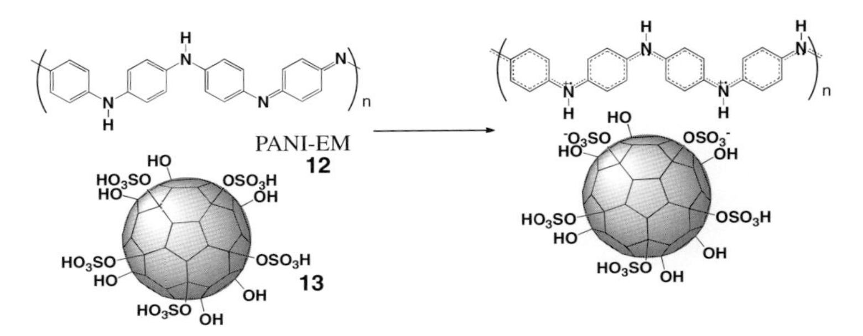

Figure 8.10 Polyaniline emeraldine base (PANI-EB) doped with hydrogensulfonated fullerenol.

showed outstanding electroconductivity, as high as 100 S/cm – about six orders of magnitude higher than the typical value for fullerene-doped conductive polymer [33].

Multifunctional [60]fullerenes possessing eight pyridyl or morpholine moieties create strong ionic interaction with acidic polymers, such as polystylenesulfonic acid (PSSA), polyvinylphosphonic acid (PVPA), PAA, and PMA [34]. The formed composites give rise to very stable films due to the ionic interactions.

8.3.2
Hydrogen Bonding Interaction

Complementary noncovalent interaction-driven synthesis of fullerene-containing supramolecular polymers has been studied intensively in recent years. The degree of polymerization of the supramolecular polymer formed through noncovalent interactions can be controlled by external stimuli; thus, the stimuli-responsive functional polymer can be created. One of the most useful noncovalent interactions is hydrogen bonding. An initial attempt to introduce the [60]fullerene molecules into a polymer chain has been reported by Goh [35]. Poly(stylene-*co*-4-vinylpyridine) (PSVPy) (**14**) possesses the pyridine moiety and a proton accepting group that forms hydrogen bond to methanofullerene carboxylic acid **15**. This salt bridge increases the loading amount of [60]fullerene into the polymer backbone, and the [60]fullerene is well dispersed in the polymer matrix. Filler **15** reinforces the material properties of the polymer. The storage modulus of the polymer is improved by fabricating the supramolecular [60]fullerene-containing polymer (Figure 8.11).

Shinkai took advantage of the diblock copolymer PS-*b*-P4VPy and [60]fullerene carboxylic acid **16** to prepare a fullerene-containing supramolecular polymer. The polymer provides distinct block structures. Site-selective complexation of **16** occurs to the pyridine moieties, where the sterically bulky **16** is gathered at the one end of the polymer. The flexible polymer changes its conformation from a random coil to a rigid rod-like conformation upon the complexation of **16**. The poor solubility of the [60] fullerene-bound end and the high solubility of the polystylene block result in microphase separation that gives rise to the micelle in which the more soluble polystylene portion remains outside the micelle (Figure 8.11).

Multiple hydrogen bonding between complementary pairs creates a strong and selective supramolecular interaction. Complementary arrays found in nucleic acids achieve high selective molecular recognition. This complementary pairing is used for grafting functionalized [60]fullerene **18**. A uracil and 2,6-diacylaminopyridine form a

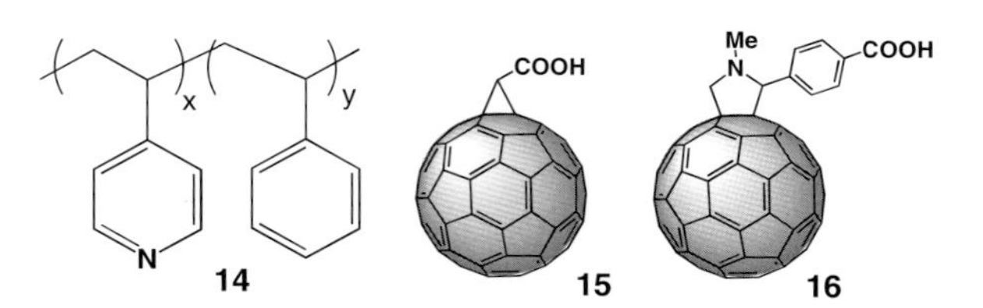

Figure 8.11 [60]Fullerene carboxylic acids and PSP4VPy polymer.

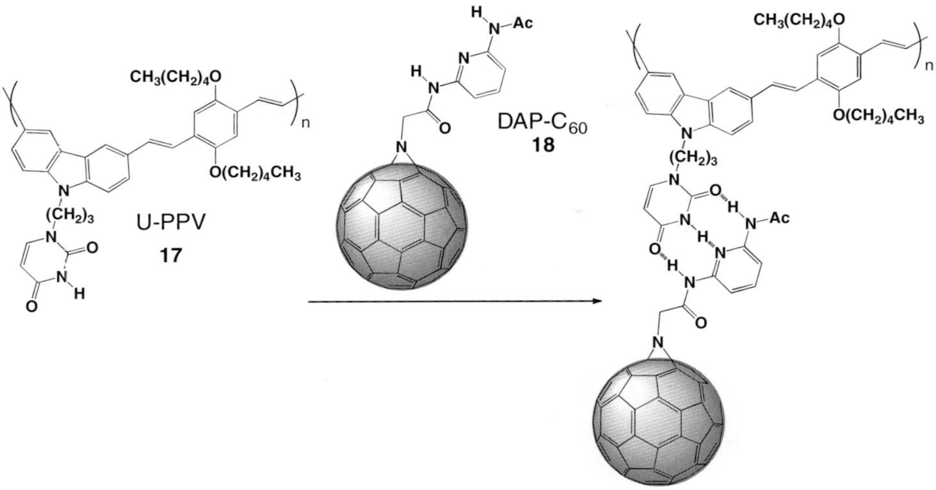

Figure 8.12 Self-complementary hydrogen-bonded supramolecular polymer.

heterodimer through complementary threefold hydrogen bonding. [60]Fullerene possessing the 2,6-diacylaminopyridine group finds its complements in uracil-functionalized poly-*p*-phenylenevinylenecarbazole **17** to form [60]fullerene-grafted supramolecular polymer via hydrogen bonds (Figure 8.12) [36]. The complementary pairing forces the [60]fullerene molecules to remain around the polymer chain, enhancing the quenching of the polymer fluorescence.

Bassani and coworkers have described the creation of a hydrogen-bonded supramolecular polymer from the [60]fullerene barbituric acid **19** and pentathienylmelamine **20** (Figure 8.13) [37]. The [60]fullerenes and the oligothiophene moieties are aligned with the aid of the complementary hydrogen bonding interactions. The hydrogen-bonded polymeric architecture is processed into a film for photovoltaic devices. The supramolecular polymers are presumably well organized by the complementary hydrogen bonding networks in which [60]fullerene and the oligothiophene moieties are aligned separately. This results in the improvement of efficiency in light harvesting in the visible region by the oligothiophene moiety and the photovoltaic response.

Hummelen and coworkers present a fascinating [60]fullerene-containing polymer built up by hydrogen bonding interactions [38]. 2-Ureido-4-pyrimidone has a donor–donor–acceptor–acceptor (DDAA) hydrogen bonding motif, which gives rise to a very stable hydrogen bonding dimer. [60]Fullerene derivative **21** having two 2-ureido-4-pyrimidone components forms self-complementary hydrogen-bonded polymer in an organic solution (Figure 8.14).

A tetraphenylporphyrin with eight amide groups forms hydrogen-bonded dimer with a [60]fullerene molecule; moreover, one-dimensional aggregation gives rise to a supramolecular fullerene polymer in which the encapsulated [60]fullerene molecules (Figure 8.15) are arranged in the complementary space formed between two

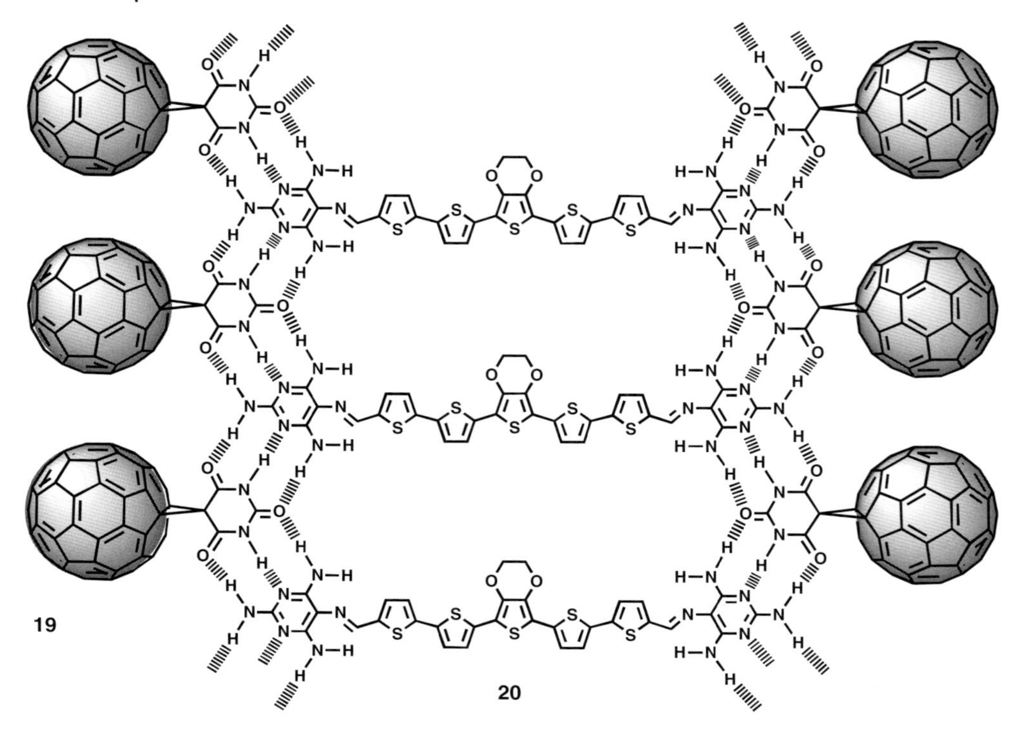

Figure 8.13 Hydrogen-bonded oligothiophene–fullerene polymer.

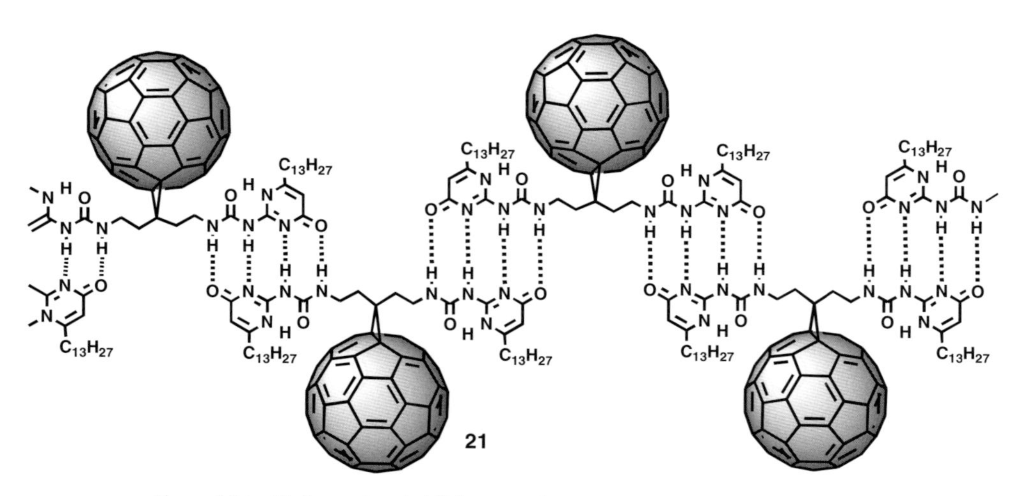

Figure 8.14 Hydrogen-bonded fullerene polymer.

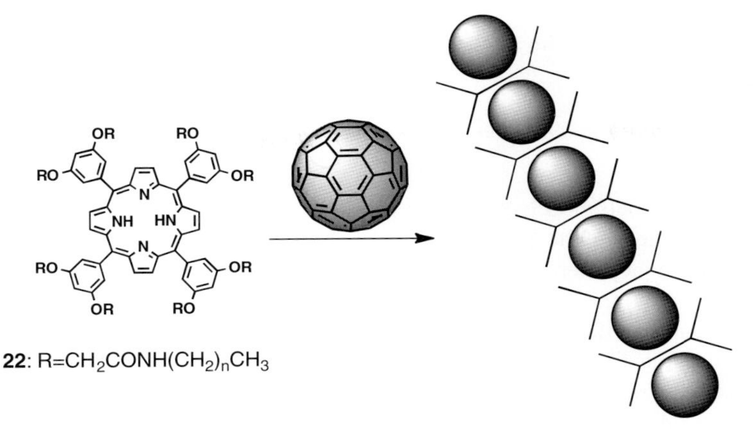

Figure 8.15 Supramolecular polymer formation motivated by [60]fullerene complexation.

porphyrin molecules [39]. The supramolecular polymer acts as a gelator for aromatic solvents. SEM observation of the xerogels confirmed the characteristic morphology in the gel state. This is an interesting example of the fullerene-assisted morphology regulation of the supramolecular polymer.

8.3.3
Host–Guest Interaction

Hydrophobic interaction is one of the most reliable forces to produce a host–guest complex. Of course, [60]fullerene is extremely hydrophobic; thus, some of the initial attempts to capture a [60]fullerene molecule are carried out in water. [60]Fullerene is encapsulated within the cavity of the cyclodextrin molecule to form a 2 : 1 complex. Liu *et al.* synthesized the water-soluble double-CD **23**. End-to-end intermolecular inclusion complexation of [60]fullerene with double-CD **23** forms supramolecular polymer in water (Figure 8.16) [40]. An STM image of the polymer shows the regular linear arrangement of the [60]fullerene nanoarray on HOPG. TEM observation of the supramolecular polymer displays the presence of a linear structure with a length in the range of 150–250 nm, suggesting that the polymer is constituted with 60–80 units of the minimum component.

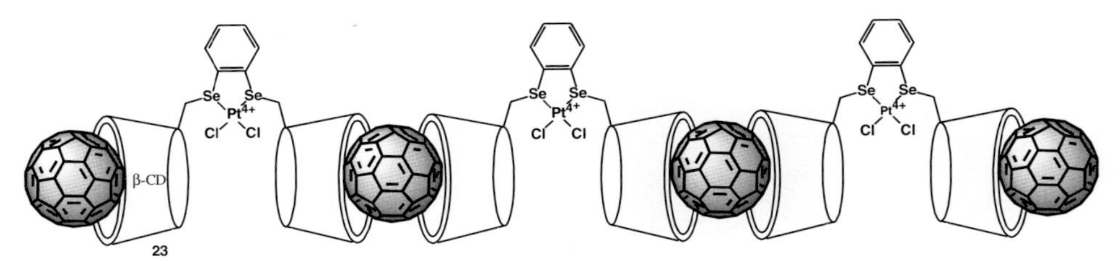

Figure 8.16 A supramolecular polymer whose structure is determined by complementary host–guest interactions.

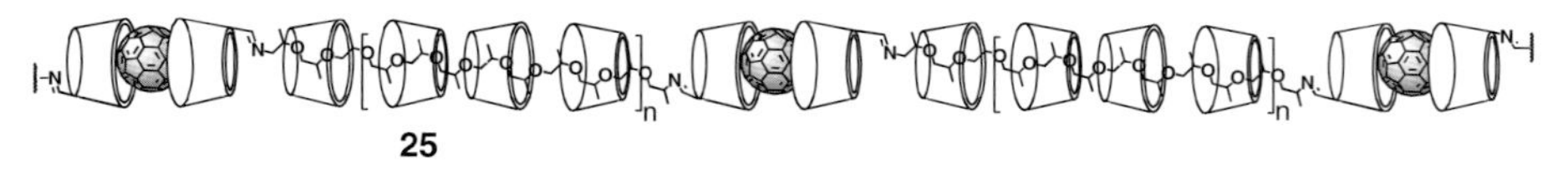

Figure 8.17 Supramolecular polymer formed by assembly of polyrotaxane **25** and [60]fullerene.

A CD-based polyrotaxane **25** end-capped with β-CD units is employed to create polymeric aggregates in the presence of [60]fullerene (Figure 8.17) [41]. TEM observation of the [60]fullerene-containing polymer shows the presence of many linear arrays of different lengths with the longest polymer in the range of 600–700 nm. The detailed structure of this polymer is confirmed by STM measurements, which reveal a fine fibrous structure. The estimated length (14.5 nm) of **25** is consistent with the measured length, and the width and height profiles of the fiber match the outer diameter of β-CD. According to the GPC analysis, the supramolecular polymer connected by host–guest interactions shows narrow polydispersity and a high M_w (293 000).

Supramolecular organization of organic/inorganic hybrid is produced by using gold nanoparticles and [60]fullerene [42]. Kaifer and coworkers prepared γ-cyclodextrin-capped gold nanoparticles (3.2 nm diameter) that grew to form water-soluble network aggregates with a diameter of about 300 nm in the presence of [60]fullerene. A 2 : 1 complexation between the γ-cyclodextrins on the particle and a [60]fullerene molecule creates supramolecular networks through hydrophobic interaction and enhances the particle aggregation.

Another type of design for the fullerene-based polymeric molecules, which are shaped like a shuttlecock, is developed by Nakamura and coworkers (Figure 8.18) [43]. The attachment of the five aromatic groups to one pentagon of a [60]fullerene molecule gives rise to a deeply conical molecule **25** that has the complementary cavity at the exterior of the globular surface of a [60]fullerene molecule. The stacking in a head-to-tail manner is driven by the attractive noncovalent interactions between the spherical fullerene moiety and the hollow cone formed by the five aromatic side groups of an adjacent molecule in the same column, which creates the infinite one-dimensional array in hexagonal columnar liquid crystalline phase.

Finally, the complementary molecular affinity between a [60]fullerene component and a complementary host in an organic solution has been used to construct supramolecular polymeric nanoarrays of [60]fullerene. Haino and coworkers have

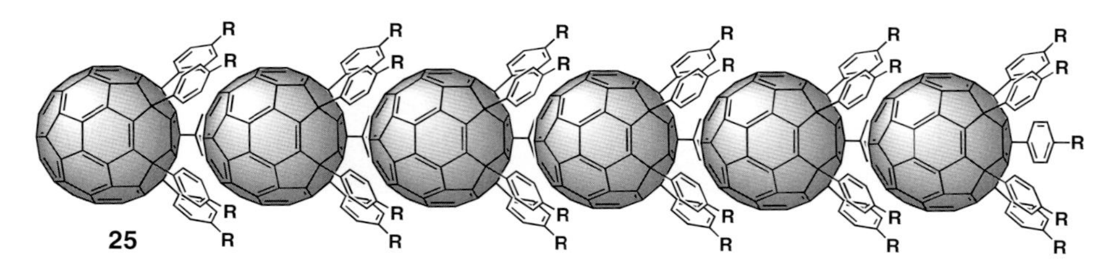

Figure 8.18 Linear array of the shuttlecock-like [60]fullerene molecules.

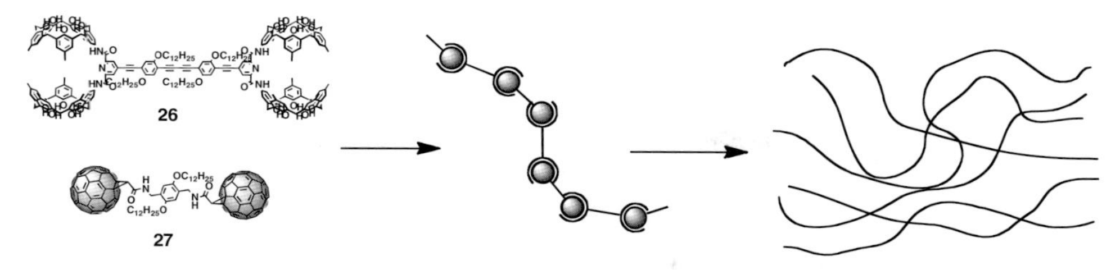

Figure 8.19 Supramolecular polymer formation via the complementary interactions.

developed supramolecular fullerene polymer through the iterative complexation of ditopic calix[5]arene **26** and dumbbell [60]fullerene **27** (Figure 8.19) [44]. Although the trimeric supramolecular copolymer forms in solution, according to the pulsed field gradient NMR studies, the higher degree of polymerization occurs in the solid state. SEM and AFM observations of the supramolecular polymer networks confirm the formation of the entwined fiber with a length of 100 nm and a diameter of 250–500 nm (Figure 8.20).

Ditopic host **26** selectively encapsulates C_{60} moieties grafted onto poly(phenylacetylene) (Figure 8.20) [45]. This process creates a remarkably stable cross-linkage, leading to an increase in molecular weight and unique morphological changes to the polymer in its solid state. The stability of the supramolecular cross-linkage is influenced by the solvent properties; in fact, the molecular weight of the polymer is increased only in toluene upon the addition of **26**. Thus, the supramolecular cross-linkage of the polymer can be regulated by the solvent system. The macroscopic solid-state morphologies of the polymer are highly influenced by supramolecular cross-linking. Nanoparticle-like morphologies are likely favored by the immiscible nature of the C_{60} moiety. The supramolecular cross-linking of polymer **28** gives rise to the dramatic change in its morphology, and the polymer bundles together to form uniform fibrils. The phase profile indicates the formed fibrils have a diameter of 19 nm. The fibrils, in turn, align into a well-oriented 2D array on a HOPG surface.

Martín and coworkers have demonstrated that the double-armed fullerene host possessing π-extended TTF analogue, 2-[9-(1,3-dithiol-2-ylidene)anthracen-10(9*H*)-ylidene]-1,3-dithiole (exTTF), shows good affinity for fullerene [46], and they recently created a head-to-tail donor–acceptor hybrid that iteratively associates to form

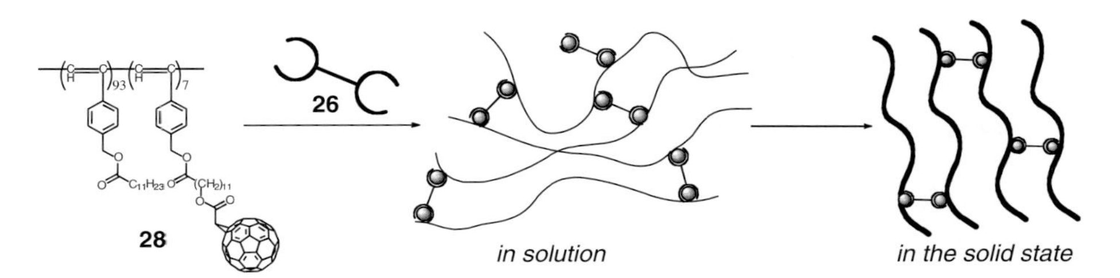

Figure 8.20 Supramolecular cross-linking of fullerene-grafted polyacetylene.

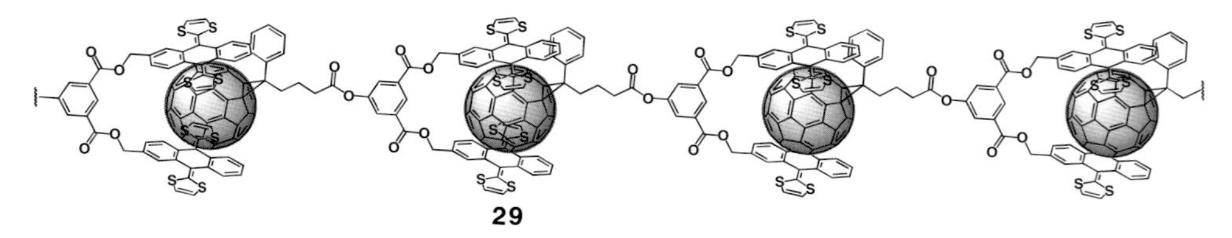

Figure 8.21 Supramolecular polymer driven by head-to-tail donor–acceptor interactions.

supramolecular fullerene-containing polymer (Figure 8.21) [47]. Self-association study of **29** using dynamic light scattering and NMR measurements confirm sizeable aggregates of more than 400 mer formed in solution. Atomic force microscopy measurements reveal the formation of the supramolecular polymer networks in solid state. The images show 15–300 nm winding, necklace-like fragments with height in the range of 1.7–2.5 nm. These dimensions are consistent with those of assemblies of **29**.

8.4
Supramolecular [60]Fullerene Dendrimer

Dendrimers are globular macromolecules with uniform size. All the bonds emerge from a core with regular branching patterns, and each repeating unit contributes a branch point. As a result of their three-dimensional structures and finely controlled size, dendritic architectures represent useful building blocks for creating functional materials. Dendritic wedges can be combined with [60]fullerene, producing electro- and photochemically interesting functions [48]; thus, fullerene-functionalized dendrimers have generated significant research activities.

Both divergent and convergent approaches have been used for the preparation of fullerene-functionalized dendrimers. The former approach of growth starts at the core and carries on outward toward the periphery of the dendrimer. The latter approach starts around the periphery of the dendrimer and proceeds inward. The choice of the approach is usually determined by the structural features of desired dendrimers. Although a number of dendrimers have been constructed with covalent bonds to date, fullerene-functionalized dendrimers built by self-assembling supramolecules provide another quick and convenient option to build noncovalent dendrimers.

8.4.1
Dendrimers with Peripheral Fullerene

Introduction of fullerenes at the periphery of a dendritic core produces fullerene-terminated dendrimers. [60]Fullerene molecules are covalently attached at the periphery of half the globular structure that shows metal coordination-driven self-assembly to give rise to the perfect globular shape surrounded by 16 fullerene molecules (Figure 8.22). The core of the dendrimer **30** is not accessible to highly condensed surface. This steric requirement prevents the copper center to approach the electrode surface, and electrode oxidation does not occur. The copper core is buried in the dendritic box made up of fullerene molecules; light energy that reaches the central core is brought back to the exterior by energy transfer. This unique photophysical feature obviously comes from the characteristics of the fullerene-functionalized dendrimer structure [49].

[60]Fullerene is concentrated on the globular surface of the dendrimers via electrostatic interactions (Figure 8.22). Ionic interactions are responsible for the self-assembly of fullerene molecules at the periphery of the dendrimer **31** [50]. 64-$[Fe^{II}(\eta^5\text{-}C_5H_5)(\eta^6\text{-}C_6Me_6)]$ is successfully introduced at the periphery of the dendritic polyamine. 64-Fe^{II} cores can be reduced to Fe^{I} by the 19-electron complex. The 19-electron Fe^{I} dendrimer shows deep blue green color and quickly reduces the 64 [60]fullerene molecules to produce the black precipitate when a toluene solution of [60]fullerene is added. The Mössbauer spectrum of this black solid indicates the presence of the Fe^{II} sandwich complex. The resulting 64 cationic centers at the periphery of the dendrimer and then gathers the 64 [60]fullerene monoanions, which are most likely located at the dendrimer periphery with tight ion pairs. However, the size of the dendrimer formed and the number of the fullerene monoanions at the periphery are unknown. This dendrimer may be a strong candidate for molecular batteries [51].

Aida and coworkers have reported the successful integration of fullerene molecules at the dendritic periphery via metal–ligation interactions (Figure 8.23) [52]. Dendritic molecule **32** is surrounded by 24-zinc porphyrin units at which [60]fullerene derivative **33** possessing pyridyl moieties bind to create the apical coordination. The addition of **33** gives rise to a large spectral change in the Soret and Q-bands, characteristic of the axial coordination of zinc porphyrins. A saturation profile observed during titration shows a molar ratio of 1 : 12 for **32** and **33** with exceedingly large association constant. The given stoichiometry and association constant suggest the simultaneous coordination of two zinc porphyrin units of **3** with two pyridine moieties of **4**. Thus, 36-[60]fullerene molecules are located at the periphery of the dendrimer. This supramolecular assembly is a photochemical functional material. Donor and acceptor arrays are spatially segregated in the dendritic structure. The effective electron transfer occurs from zinc porphyrin cores to fullerene moieties to generate a charge separation with the largest ratio of the charge separation to charge recombination rate constants (3400), which is one order of magnitude greater than those of the preceding examples.

2-[9-(1,3-dithiol-2-ylidene)anthracen-10(9*H*)-ylidene]1,3-dithiole (exTTF) is a well-known π-extended analogue of tetrathiafulvalene, and it shows concave–convex complementarity between its curved aromatic surface and the exterior of the fullerene (Figure 8.24) [53]. Four exTTF units are introduced in the dendron wedge that is connected to [60]fullerene [54]. The dendron has two fullerene binding sites, each of which is composed of two exTTF units. The self-assembly of dendron **34** divergently occurs to produce an intriguing supramolecular dendrimer. The generation and size of the supramolecular dendrimer is influenced by the temperature and concentration changes; thus, the determination of its size and shape seems difficult.

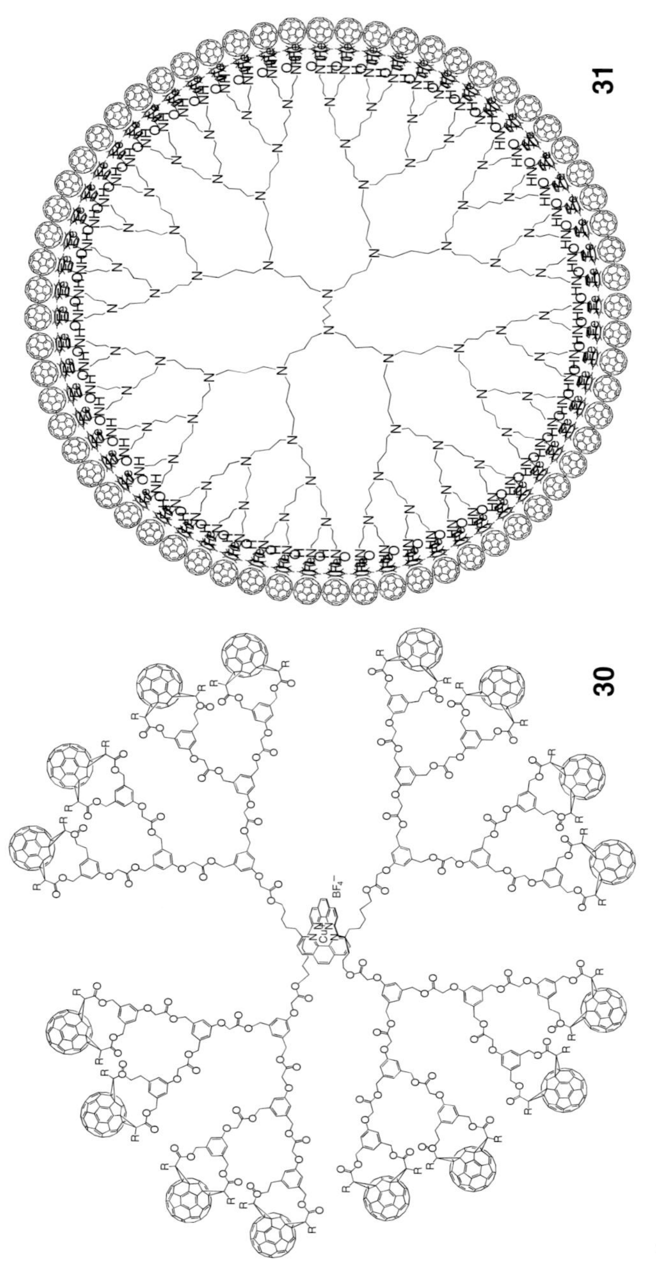

Figure 8.22 Supramolecular dendrimer with fullerene molecules at the periphery.

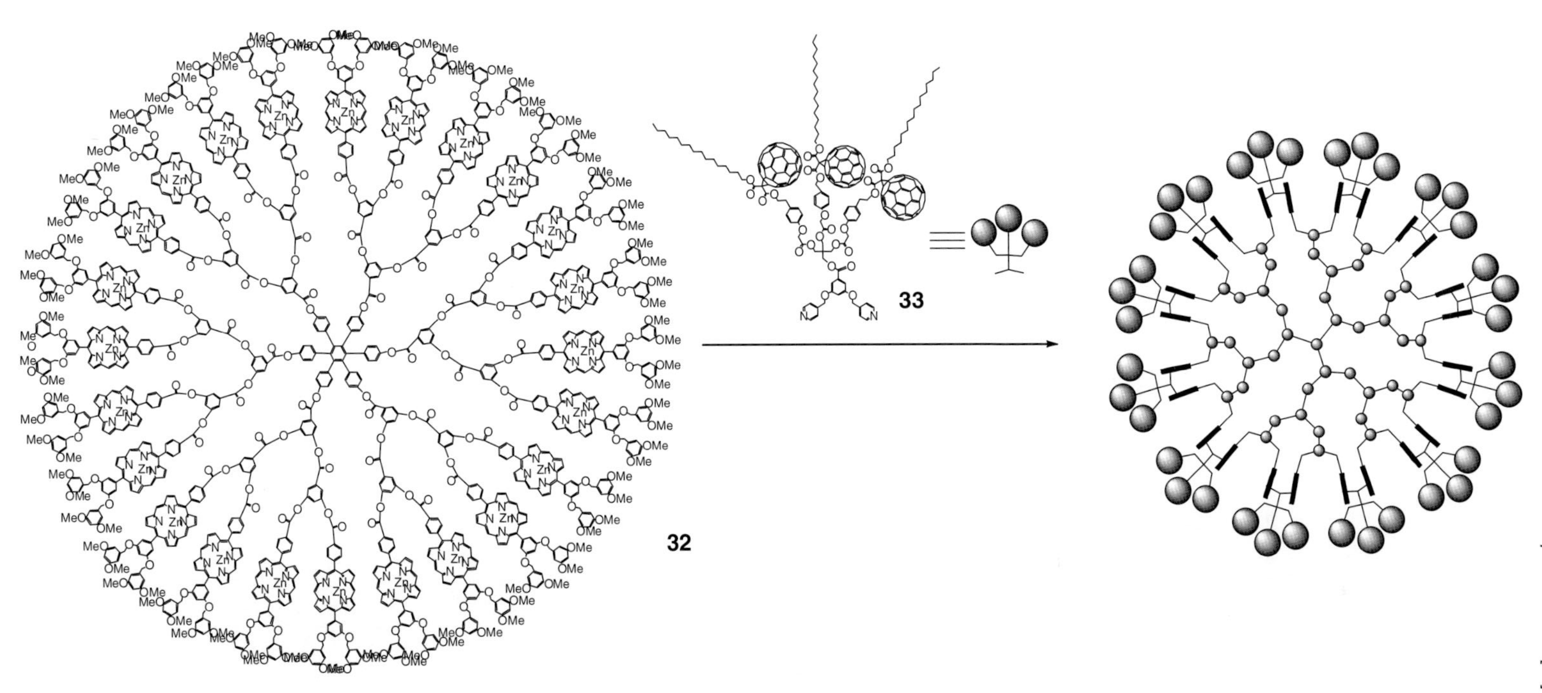

Figure 8.23 Ligation-induced multimolecular assembly of fullerene derivatives at the periphery of the porphyrin-functionalized dendrimer.

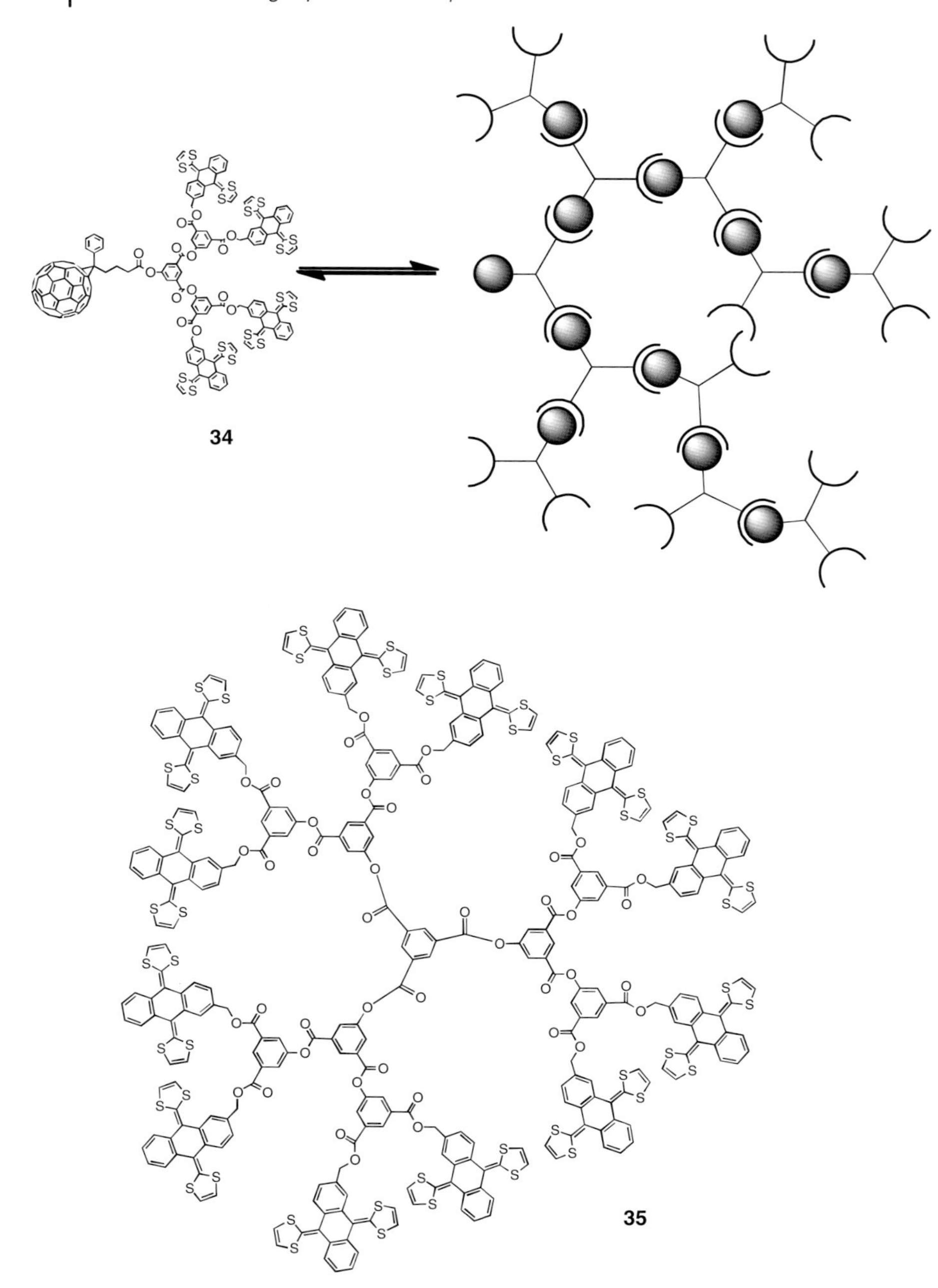

Figure 8.24 Charge transfer-driven supramolecular assembly of fullerene-functionalized dendrimers.

However, mass spectrometry and atomic force microscopy indicate the huge size of the supramolecular dendrimers that form via host–guest interactions.

This host–guest motif can be functionalized to gather fullerene molecules at the periphery of the dendrimer. The 24-ExTTF units are introduced at the periphery of the dendrimer **35** [55]. Theoretically, it has 12 fullerene binding sites. Upon the addition of the [60] fullerene molecules, the binding sites capture many fullerene molecules at the periphery to create the fullerene-condensed surface around the dendrimer.

Hydrogen bonding interaction is one of the most useful supramolecular interactions, and it can be utilized to form fullerene-functionalized dendrimers (Figure 8.25). Nierengarten and coworkers have demonstrated that strong self-assembled dimers based on 2-ureido-4-[1H]pyrimidione can be formed efficiently even from highly crowded fullerene-functionalized dendron **36** [56]. This supramolecular hydrogen-bonding motif provides an extremely large dimerization constant ($K_d > 10^7$ l/mol); thus, the motif is incorporated into the final stage of the dendrimer synthesis. Even though five peripheral [60]fullerene molecules are attached on the dendron, its dimer is detected by MALDI-TOF mass spectroscopy.

Complementary hydrogen bonding pair is also a valuable supramolecular motif to create supramolecular fullerene-functionalized dendrimer (Figure 8.25). Hirsch has taken advantage of the Hamilton-type supramolecular motif. Cyanuric acid-attached [60]fullerene **38** binds to homotritopic Hamilton-type receptor **37** to form the supramolecular fullerene-functionalized dendrimer in a ratio of 3 : 1 [57]. The titration curves between the dendrons and the receptor show a sigmoidal shape, indicating a positive cooperative effect for the subsequent binding of guest molecules. The degree of the cooperation depends on the shape, size, and rigidity of the dendritic wedges of the fullerodendrons.

Ammonium ion forms strong host–guest complexes with 18-crown-6. This supramolecular motif combines the fullerene-functionalized dendritic branch **39** possessing an ammonium moiety with fluorescent biscrown ether **40** (Figure 8.26) [58]. Binding of eight fullerenes efficiently quenches its fluorescence, and the 1 : 2 host–guest complex is confirmed by ESI mass spectrometry.

8.4.2
Dendrimers with Inner Fullerene

In this type of dendrimer, the inner core and shell of the dendrimer is surrounded by the sterically crowded, close-packed periphery. This structural feature provides a unique guest binding environment; in fact, metal cluster and small molecules are entrapped inside individual dendrimers via noncovalent interactions [59]. In this section, supramolecular encapsulation of [60]fullerene into the inner shell of dendrimers will be described.

Nierengarten, Shinkai, and Armaroli have revealed that Fréchet-type dendrimers [60] capture [60]fullerene molecules to form 1 : 1 host–guest complexes (Figure 8.27) [61]. The increase in their binding constants (**41** : 5, **42** : 12, and **43** : 68 l/mol) is in parallel with the numbers of molecules generated [62]. Fullerene

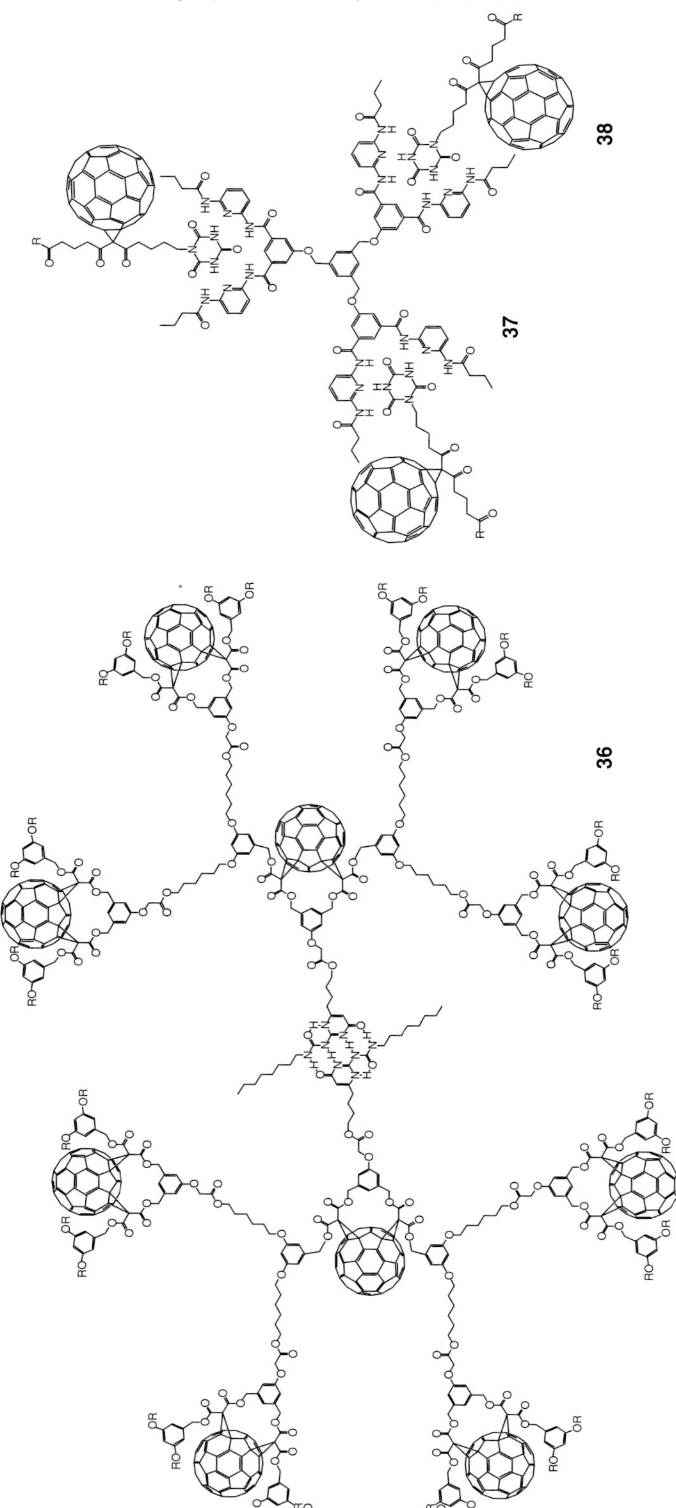

Figure 8.25 Hydrogen bonding-driven self-assembly of supramolecular fullerene-functionalized dendrimers.

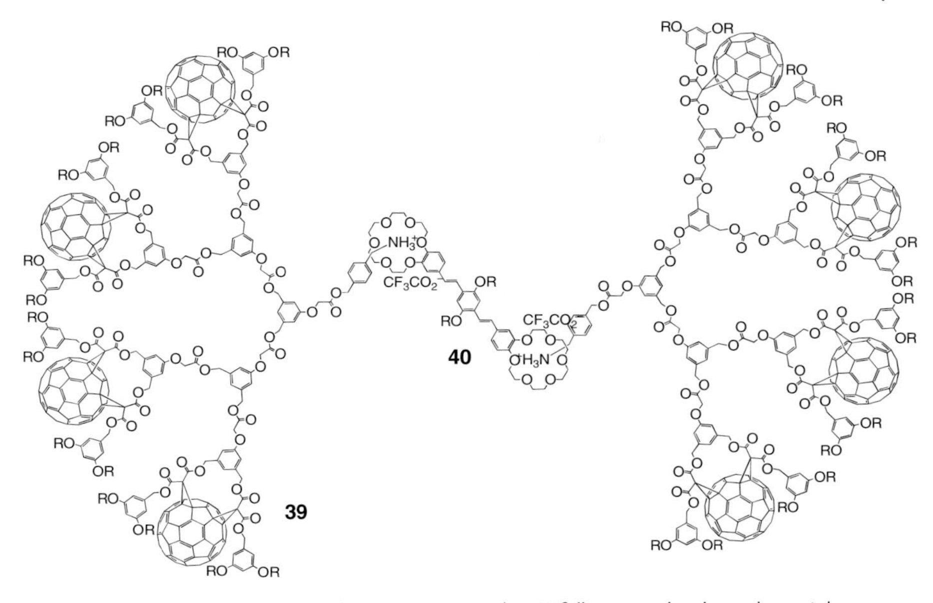

Figure 8.26 Hydrogen-bonded dendrimer possessing the [60]fullerene molecules at the periphery.

spheres with more aromatic units of the dendrimers receive more attractive π–π interactions in the dendritic shell, the interior space of which prefers the inclusion of [60]fullerene. Thus, the internal cavity within the dendrimers becomes more suitably defined as the dendrimer gets larger. Additional ^{13}C NMR studies of the [60]fullerene complexes with the dendrimers conclude that the interior region of the branching shell located close to the central core of the dendrimers is capable of producing the cavity size solely for [60]fullerene inclusion.

Cyclotriveratrylene (CTV) forms weak supramolecular complex of fullerenes (Figure 8.28) in methylene chloride solution. The introduction of dendritic wedges

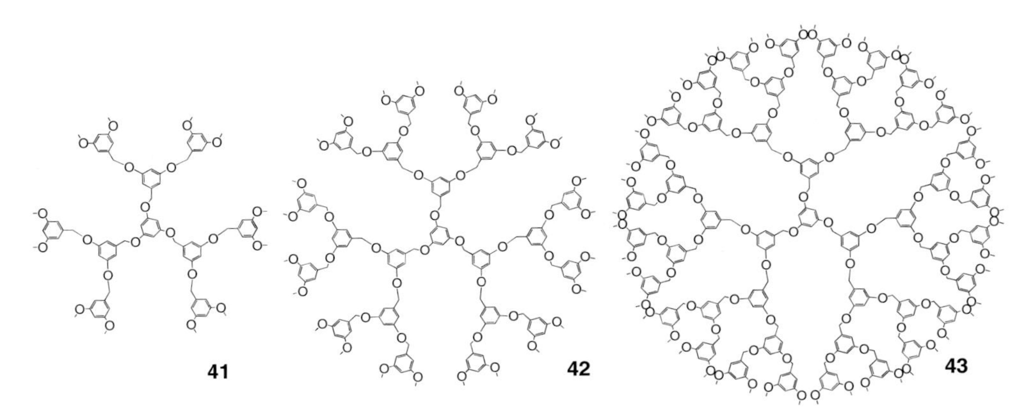

Figure 8.27 Fréchet-type dendrimers.

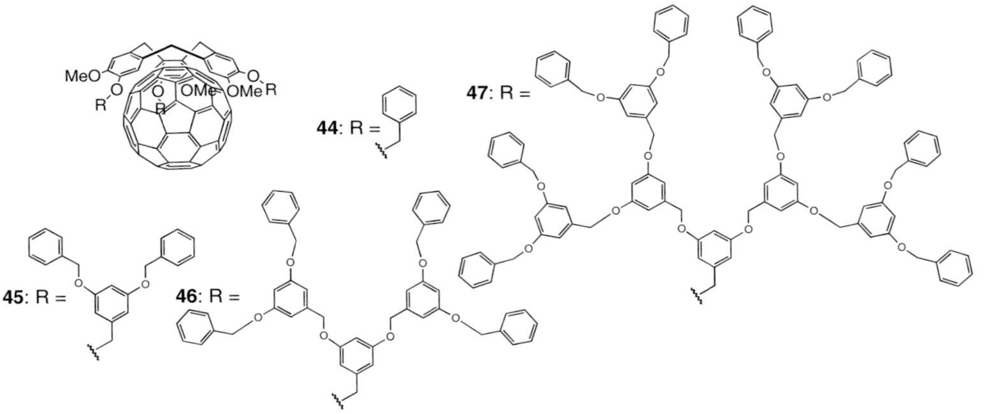

Figure 8.28 Cyclotriveratrylene-based dendrimers.

at the three hydroxyl groups provides the increased stability of the supramolecular complexes [61, 63]. The binding constant significantly increases as the generation of the dendritic substituents of the CTV core increases. Thus, the surrounding dendritic branches appear to improve the inclusion ability of the CTV central core for fullerenes. The electrostatic interactions with the encapsulated fullerene molecule are produced by the aromatic rings of the dendrons, but the interactions are not the only contribution. Most likely, the fullerene molecules can stabilize the interior space surrounded by the CTV and the dendritic wedges of the complex.

The coordination of [60]fullerene to transition metal complexes (e.g., Vaska's trans-Ir(CO)Cl(PPh_3)$_2$) is well-known (Figure 8.29). Catalano has reported that the new dendrimer **48** containing an iridium compound shows reversible binding to [60]fullerene [64]. A diphenyl phosphine ligand possessing Fréchet-type dendritic wedge complexes with iridium complex can form an analogue of Vaska's molecule. The complex binds [60]fullerene in a ratio of 1 : 1 to produce the supramolecular [60]fullerene-functionalized metallodendrimer. The complexation of the [60]fullerene molecule is reversible in solution; the free energy ΔG^0_{265} is approximately −3 kcal/

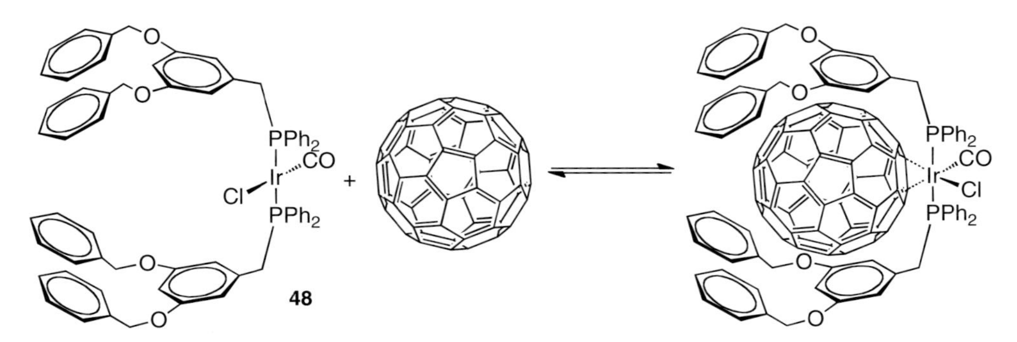

Figure 8.29 [60]Fullerene complexation with dendrimer containing iridium compound **48**.

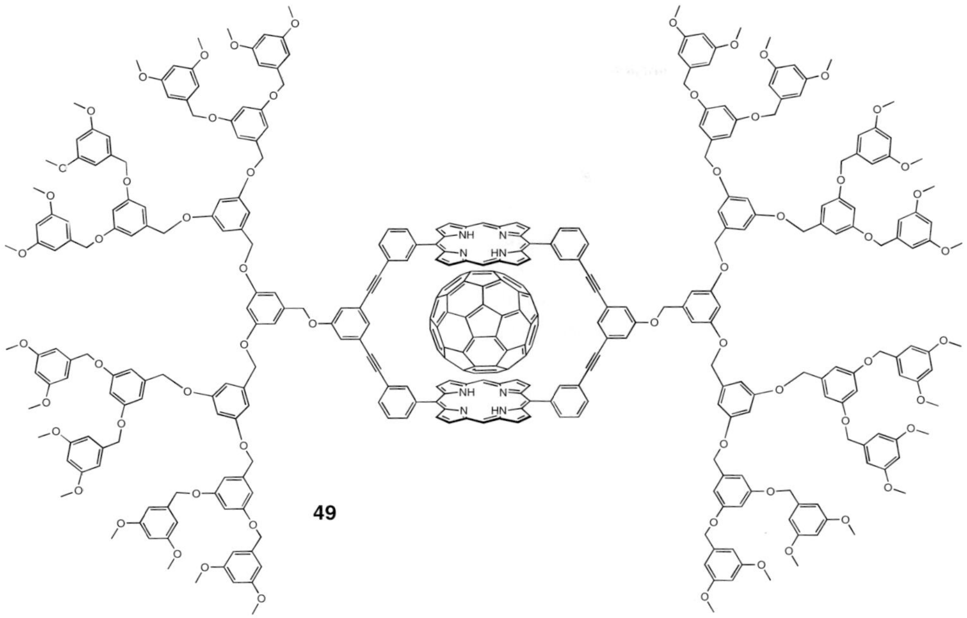

49

Figure 8.30 Dendritic host with [60]fullerene.

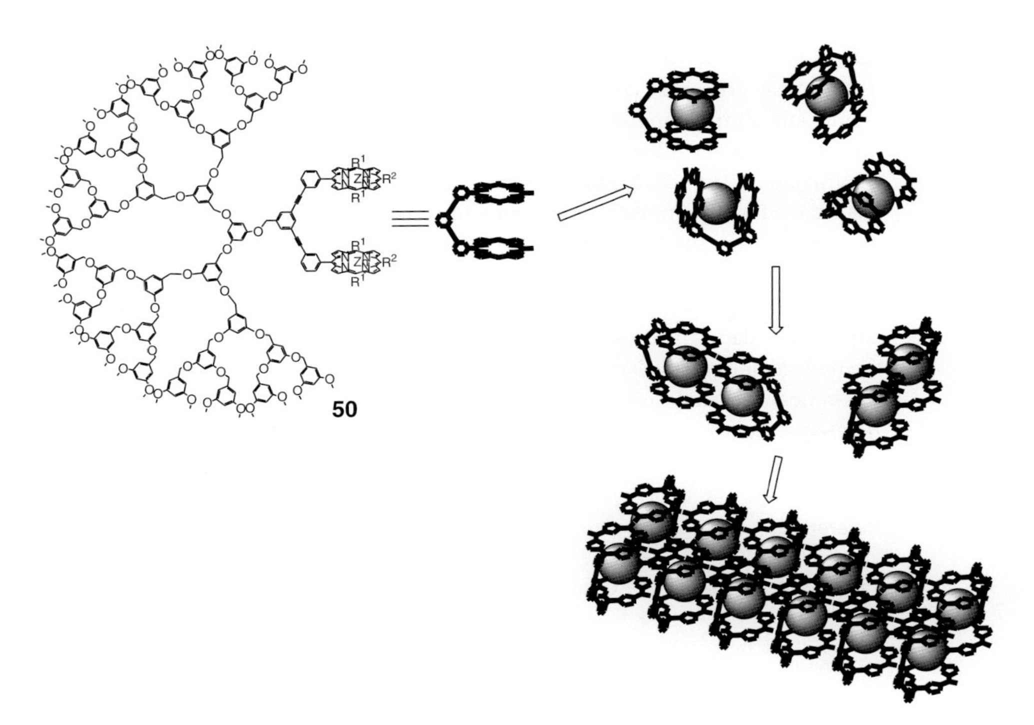

Figure 8.31 Self-assembly of the dendron with [60]fullerene.

mol. This methodology presents a way to introduce the [60]fullerene molecule into the core of the dendrimer.

Bisporphyrin macrocycle **49** is known to be a good fullerene host molecule [65]. This fullerene host is introduced into the dendritic core and enclosed by the Fréchet's dendritic wedge that provides site-isolated structure (Figure 8.30). The dendritic host is a very good carrier of [60]fullerene because the entrapped [60]fullerenes are easily transferred into a polymer film without any damage to its geometry.

Bisporphyrin cleft molecule **50**, which has eight carboxylic acids as a hydrogen bonding group, assembles with the [60]fullerene molecule to form an artificial peapod (Figure 8.31) [66]. [60]Fullerene is encapsulated within the cleft provided by the two porphyrin components. Consequently, bisporphyrin cleft loses its conformational flexibility, and the two porphyrin units adopt a parallel arrangement, as seen when [60]fullerene is encapsulated. This preorganized structure is assembled via hydrogen bonding interactions to produce a parallel array of the fullerene peapods. A fibrous morphology with a uniform diameter of 15 nm is observed in microscopy measurements.

8.5 Conclusions

In this chapter, we have provided an overview of the efforts to synthesize supramolecular fullerene-functionalized polymers and dendrimers. Fullerene and its derivatives have attracted great attention of many chemists due to their unique physicochemical properties and potential applications. Chemists have already made significant advances in fullerene chemistry by generating new functional materials. Further progress of fullerene science requires a breakthrough technology of fullerene manipulation. The collaboration of supramolecular and macromolecular chemistries has resulted in supramolecular fullerene-functionalized polymers and dendrimers as well as new supramolecular fullerene materials with enormous potential in chemistry, nanotechnology, and related fields. Their functionality is derived from the supramolecular level, and is unique to the assembled state. Thus, controlling the supramolecular structures of fullerene-functionalized polymers and dendrimers at the molecular level involves a "bottom-up" approach that requires accessing complex fullerene-based molecular architectures. This work will accelerate the progress of fullerene science.

References

1 Kroto, H.W., Heath, J.R., O'Brien, S.C., Curl, R.F., and Smalley, R.E. (1985) *Nature*, **318**, 162–163.

2 Krätschmer, W., Lamb, L.D., Fostiropoulos, K., and Huffman, D.R. (1990) *Nature*, **347**, 354–358.

3 (a) Lee, G.H., Huh, S.H., Jeong, J.W., and Ri, H.-C. (2002) *J. Magn. Magn. Mater.*, **246**, 404–411; (b) Coffey, D. and Trugman, S.A. (1992) *Phys. Rev. Lett.*, **69**, 176–179; (c) Politis, C., Buntar, V.,

Krauss, W., and Gurevich, A. (1992) *Europhys. Lett.*, **17**, 175–179.

4 (a) Jérome, D. (1991) *Science*, **252**, 1509–1514; (b) Tanigaki, K., Ebbesen, T.W., Saito, S., Mizuki, J., Tsai, J.S., Kubo, Y., and Kuroshima, S. (1991) *Nature*, **352**, 222–223; (c) Wang, H.H., Kini, A.M., Savall, B.M., Carlson, K.D., Williams, J.M., Lathrop, M.W., Lykke, K.R., Parker, D.H., Wurz, P. *et al.* (1991) *Inorg. Chem.*, **30**, 2962–2963; (d) Margadonna, S. and Prassides, K. (2002) *J. Solid State Chem.*, **168**, 639–652.

5 (a) Shinohara, H. (2000) *Rep. Prog. Phys.*, **63**, 843–892; (b) Roncali, J. (2005) *Chem. Soc. Rev.*, **34**, 483–495; (c) Kronholm, D. and Hummelen, J.C. (2007) *Mater. Matters*, **2**, 16–19.

6 (a) Sariciftci, N.S. (1999) *Curr. Opin. Solid State Mater. Sci.*, **4**, 373–378; (b) Imahori, H. and Fukuzumi, S. (2004) *Adv. Funct. Mater.*, **14**, 525–536; (c) Konishi, T., Ikeda, A., and Shinkai, S. (2005) *Tetrahedron*, **61**, 4881–4899.

7 (a) Lamb, L.D., Huffman, D.R., Workman, R.K., Howells, S., Chen, T., Sarid, D., and Ziolo, R.F. (1992) *Science*, **255**, 1413–1416; (b) Krätschmer, W. (1995) *Synth. Met.*, **70**, 1309–1312.

8 (a) Sariciftci, N.S. and Heeger, A.J. (1994) *Mol. Cryst. Liq. Cryst. Sci. Technol. Sect. A*, **256**, 317–326; (b) Sariciftci, N.S. and Heeger, A.J. (1995) *Proc. SPIE Int. Soc. Opt. Eng.*, **2530**, 76–86; (c) Sariciftci, N.S. and Heeger, A.J. (1995) *Synth. Met.*, **70**, 1349–1352; (d) Wudl, F. (2002) *J. Mater. Chem.*, **12**, 1959–1963.

9 (a) Rao, A.M., Zhou, P., Wang, K.-A., Hager, G.T., Holden, J.M., Wang, Y., Lee, W.-T., Bi, X.-X., Eklund, P.C., Cornett, D.S., Duncan, M.A., and Amster, I.J. (1993) *Science*, **259**, 955–957; (b) Wang, Y., Holden, J.M., Dong, Z.-H., Bi, X.-X., and Eklund, P.C. (1993) *Chem. Phys. Lett.*, **211**, 341–345; (c) Zhou, P., Dong, Z.-H., Rao, A.M., and Eklund, P.C. (1993) *Chem. Phys. Lett.*, **211**, 337–340; (d) Iwasa, Y., Arima, T., Fleming, R.M., Siegrist, T., Zhou, O., Haddon, R.C., Rothberg, L.J., Lyons, K.B., Carter, H.L., Hebard, A.F., Tycko, R., Dabbagh, G., Krajewski, J.J., Thomas, G.A., and Yagi, T. (1994) *Science*, **264**, 1570–1572; (e) Gügel, A., Belik, P., Walter, M., Kraus, A., Harth, E., Wagner, M., Spickermann, J., and Müllen, K. (1996) *Tetrahedron*, **52**, 5007–5014; (f) Blank, V.D., Buga, S.G.B., Dubitsky, G.A., Serebryanaya, N.R., Popov, M.Y., and Sundqvist, B. (1998) *Carbon*, **36**, 319–343; (g) Sundqvist, B. (1999) *Adv. Phys.*, **48**, 1–134.

10 (a) Geckeler, K.E. and Samal, S. (1999) *Polym. Int.*, **48**, 743–757; (b) Giacalone, F. and Martín, N. (2006) *Chem. Rev.*, **106**, 5136–5190; (c) Wang, C., Guo, Z.-X., Fu, S., Wu, W., and Zhu, D. (2004) *Prog. Polym. Sci.*, **29**, 1079–1141; (d) Dai, L. (1999) *Polym. Adv. Technol.*, **10**, 357–420; (e) Segura, J.L., Martín, N., and Guldi, D.M. (2005) *Chem. Soc. Rev.*, **34**, 31–47; (f) Zhou, N., Merschrod, E.F., and Zhao, S.Y. (2005) *J. Am. Chem. Soc.*, **127**, 14154–14155; (g) Nishimura, T., Takatani, K., Sakurai, S.I., Maeda, K., and Yashima, E. (2002) *Angew. Chem. Int. Ed.*, **41**, 3602–3604; (h) Gutiérrez Nava, M., Setayesh, S., Rameau, A., Masson, P., and Nierengarten, J.-F. (2002) *New J. Chem.*, **26**, 1584–1589.

11 (a) Hahn, U., Nierengarten, J.-F., Vögtle, F., Listorti, A., Monti, F., and Armaroli, N. (2009) *New J. Chem.*, **33**, 337–344; (b) Hosomizu, K., Imahori, H., Hahn, U., Nierengarten, J.-F., Listorti, A., Armaroli, N., Nemoto, T., and Isoda, S. (2007) *J. Phys. Chem. C*, **111**, 2777–2786; (c) El-Khouly, M.E., Kang, E.S., Kay, K.-Y., Choi, C.S., Aaraki, Y., and Ito, O. (2007) *Chem. Eur. J.*, **13**, 2854–2863; (d) Herschbach, H., Hosomizu, K., Hahn, U., Leize, E., Dorsselaer, A., Imahori, H., and Nierengarten, J.-F. (2006) *Anal. Bioanal. Chem.*, **386**, 46–51; (e) Hahn, U., Hosomizu, K., Imahori, H., and Nierengarten, J.-F. (2005) *Eur. J. Org. Chem.*, 85–91; (f) Hirsch, A. and Vostrowsky, O. (2001) *Top. Curr. Chem.*, **217**, 51–93.

12 Ciferri, A. (ed.) (2005) *Supramolecular Polymers*, 2nd edn, Taylor & Francis, New York.

13 (a) Castellano, R.K., Clark, R., Craig, S.L., Nuckolls, C., and Rebek, J. (2000) *Proc. Natl. Acad. Sci. U.S.A.*, **97**, 12418–12421; (b) Castellano, R.K.,

Rudkevich, D.M., and Rebek, J. (1997) *Proc. Natl. Acad. Sci. U.S.A.*, **94**, 7132–7137.

14 Rudkevich, D.M. (2007) *Eur. J. Org. Chem.*, 3255–3270.

15 (a) Brunsveld, L., Folmer, B.J.B., Meijer, E.W., and Sijbesma, R.P. (2001) *Chem. Rev.*, **101**, 4071–4097; (b) Sijbesma, R.P. and Meijer, E.W. (2003) *Chem. Commun.*, 5–16; (c) Sijbesma, R.P., Beijer, F.H., Brunsveld, L., Folmer, B.J.B., Hirschberg, J.H.K.K., Lange, R.F.M., Lowe, J.K.L., and Meijer, E.W. (1997) *Science*, **278**, 1601–1604.

16 (a) Ermer, O. (1991) *Helv. Chim. Acta*, **74**, 1339–1351; (b) Crane, J.D., Hitchcock, P.B., Kroto, H.W., Taylor, R., and Walton, D.R.M. (1992) *J. Chem. Soc. Chem. Commun.*, 1764–1765; (c) Meidine, M.F., Hitchcock, P.B., Kroto, H.W., Taylor, R., and Walton, D.R.M. (1992) *J. Chem. Soc. Chem. Commun.*, 1534–1537; (d) Birkett, P.R., Christides, C., Hitchcock, P.B., Kroto, H.W., Prassides, K., Taylor, R., and Walton, D.R.M. (1993) *J. Chem. Soc., Perkin Trans. 2*, 1407–1408; (e) Izuoka, A., Tachikawa, T., Sugawara, T., Suzuki, Y., Konno, M., Saito, Y., and Shinohara, H. (1992) *J. Chem. Soc. Chem. Commun.*, 1472–1473; (f) Izuoka, A., Tachikawa, T., Sugawara, T., Saito, Y., and Shinohara, H. (1992) *Chem. Lett.*, 1049–1052.

17 Barbour, L.J., Orr, G.W., and Atwood, J.L. (1998) *Chem. Commun.*, 1901–1902.

18 Sun, D. and Reed, C.A. (2000) *Chem. Commun.*, 2391–2392.

19 Balch, A.L., Catalano, V.J., Lee, J.W., and Olmstead, M.M. (1992) *J. Am. Chem. Soc.*, **114**, 5455–5457.

20 (a) Haino, T., Fukunaga, C., and Fukazawa, Y. (2007) *J. Nanosci. Nanotechnol.*, **7**, 1386–1388; (b) Haino, T., Seyama, J., Fukunaga, C., Murata, Y., Komatsu, K., and Fukazawa, Y. (2005) *Bull. Chem. Soc. Jpn.*, **78**, 768–770; (c) Haino, T., Yanase, M., and Fukazawa, Y. (1997) *Angew. Chem. Int. Ed. Engl.*, **36**, 259–260; (d) Haino, T., Yanase, M., and Fukazawa, Y. (1997) *Tetrahedron Lett.*, **38**, 3739–3742; (e) Haino, T., Yanase, M., and Fukazawa, Y. (1998) *Angew. Chem. Int. Ed.*, **37**, 997–998; (f) Haino, T., Araki, H., Fujiwara, Y., Tanimoto, Y., and Fukazawa, Y. (2002) *Chem. Commun.*, 2148–2149; (g) Haino, T., Yamanaka, Y., Araki, H., and Fukazawa, Y. (2002) *Chem. Commun.*, 402–403; (h) Haino, T., Yanase, M., Fukunaga, C., and Fukazawa, Y. (2006) *Tetrahedron*, **62**, 2025–2035; (i) Haino, T., Fukunaga, C., and Fukazawa, Y. (2006) *Org. Lett.*, **8**, 3545–3548; (j) Yanase, M., Haino, T., and Fukazawa, Y. (1999) *Tetrahedron Lett.*, **40**, 2781–2784; (k) Haino, T., Yanase, M., and Fukazawa, Y. (2005) *Tetrahedron Lett.*, **46**, 1411–1414.

21 (a) Atwood, J.L., Barbour, L.J., Heaven, M.W., and Raston, C.L. (2003) *Angew. Chem. Int. Ed.*, **42**, 3254–3257; (b) Atwood, J.L., Barbour, L.J., and Raston, C.L. (2002) *Cryst. Growth Des.*, **2**, 3–6.

22 Hubble, L.J. and Raston, C.L. (2007) *Chem. Eur. J.*, **13**, 6755–6760.

23 Wang, Z., Dötz, F., Enkelmann, V., and Müllen, K. (2005) *Angew. Chem. Int. Ed.*, **44**, 1247–1250.

24 (a) Smith, B.W., Monthioux, M., and Luzzi, D.E. (1998) *Nature*, **396**, 323–324; (b) Smith, B.W., Monthioux, M., and Luzzi, D.E. (1999) *Chem. Phys. Lett.*, **315**, 31–36; (c) Burteaux, B., Claye, A., Smith, B.W., Monthioux, M., Luzzi, D.E., and Fischer, J.E. (1999) *Chem. Phys. Lett.*, **310**, 21–24.

25 Pantoş, G.D., Wietor, J.-L., and Sanders, J.K.M. (2007) *Angew. Chem. Int. Ed.*, **46**, 2238–2240.

26 Pantoş, G.D., Pengo, P., and Sanders, J.K.M. (2007) *Angew. Chem. Int. Ed.*, **46**, 194–197.

27 Kawauchi, T., Kumaki, J., Kitaura, A., Okoshi, K., Kusanagi, H., Kobayashi, K., Sugai, T., Shinohara, H., and Yashima, E. (2008) *Angew. Chem. Int. Ed.*, **47**, 515–519.

28 Lu, F., Li, Y., Liu, H., Zhuang, J., Gan, L., and Zhu, D. (2005) *Synth. Met.*, **153**, 317–320.

29 Mwaura, J.K., Pinto, M.R., Witker, D., Ananthakrishnan, N., Schanze, K.S., and Reynolds, J.R. (2005) *Langmuir*, **21**, 10119–10126.

30 Nishimura, T., Tsuchiya, K., Ohsawa, S., Maeda, K., Yashima, E., Nakamura, Y., and Nishimura, J. (2004) *J. Am. Chem. Soc.*, **126**, 11711–11717.

31 Ying, Q., Zhang, J., Liang, D., Nakanishi, W., Isobe, H., Nakamura, E., and Chu, B. (2005) *Langmuir*, **21**, 9824–9831.

32 Bae, A.-H., Hatano, T., Sugiyasu, K., Kishida, T., Takeuchi, M., and Shinkai, S. (2005) *Tetrahedron Lett.*, **46**, 3169–3173.

33 (a) Dai, L., Lu, J., Matthews, B., and Mau, A.W.H. (1998) *J. Phys. Chem. B*, **102**, 4049–4053; (b) Lu, J., Dai, L., and Mau, A.W.H. (1998) *Acta Polym.*, **49**, 371–375.

34 (a) Lu, Z., Goh, S.H., and Lee, S.Y. (1999) *Macromol. Chem. Phys.*, **200**, 1515–1522; (b) Goh, S.H., Lee, S.Y., Lu, Z.H., and Huan, C.H.A. (2000) *Macromol. Chem. Phys.*, **201**, 1037–1047.

35 Ouyang, J., Goh, S.H., and Li, Y. (2001) *Chem. Phys. Lett.*, **347**, 344–348.

36 (a) Fang, H., Wang, S., Xiao, S., Yang, J., Li, Y., Shi, Z., Li, H., Liu, H., and Zhu, D. (2003) *Chem. Mater.*, **15**, 1593–1597; (b) Fang, H., Shi, Z., Li, Y., Xiao, S., Li, H., Liu, H., and Zhu, D. (2003) *Synth. Met.*, **135–136**, 843–844.

37 Huang, C.-H., McClenaghan, N.D., Kuhn, A., Hofstraat, J.W., and Bassani, D.M. (2005) *Org. Lett.*, **7**, 3409–3412.

38 Sánchez, L., Rispens, M.T., and Hummelen, J.C. (2002) *Angew. Chem. Int. Ed.*, **41**, 838–840.

39 Shirakawa, M., Fujita, N., and Shinkai, S. (2003) *J. Am. Chem. Soc.*, **125**, 9902–9903.

40 Liu, Y., Wang, H., Liang, P., and Zhang, H.-Y. (2004) *Angew. Chem. Int. Ed.*, **43**, 2690–2694.

41 Liu, Y., Yang, Y.-W., Chen, Y., and Zou, H.-X. (2005) *Macromolecules*, **38**, 5838–5840.

42 Liu, J., Alvarez, J., Ong, W., and Kaifer, A.E. (2001) *Nano Lett.*, **1**, 57–60.

43 (a) Sawamura, M., Kawai, K., Matsuo, Y., Kanie, K., Kato, T., and Nakamura, E. (2002) *Nature*, **419**, 702–705; (b) Matsuo, Y., Muramatsu, A., Hamasaki, R., Mizoshita, N., Kato, T., and Nakamura, E. (2004) *J. Am. Chem. Soc.*, **126**, 432–433; (c) Zhong, Y.-W., Matsuo, Y., and Nakamura, E. (2007) *J. Am. Chem. Soc.*, **129**, 3052–3053.

44 Haino, T., Matsumoto, Y., and Fukazawa, Y. (2005) *J. Am. Chem. Soc.*, **127**, 8936–8937.

45 Haino, T., Hirai, E., Fujiwara, Y., and Kashihara, K. (2010) *Angew. Chem. Int. Ed.*, **49**, 7899–7903.

46 Pérez, E.M., Sánchez, L., Fernández, G., and Martín, N. (2006) *J. Am. Chem. Soc.*, **128**, 7172–7173.

47 Fernández, G., Pérez, E.M., Sánchez, L., and Martín, N. (2008) *Angew. Chem. Int. Ed.*, **47**, 1094–1097.

48 (a) Nierengarten, J.-F., Armaroli, N., Accorsi, G., Rio, Y., and Eckert, J.-F. (2003) *Chem. Eur. J.*, **9**, 36–41; (b) Nierengarten, J.-F. (2003) *Top. Curr. Chem.*, **228**, 87–110; (c) Rio, Y., Accorsi, G., Nierengarten, H., Bourgogne, C., Strub, J.-M., Van, D.A., Armaroli, N., and Nierengarten, J.-F. (2003) *Tetrahedron*, **59**, 3833–3844; (d) Pollak, K.W., Leon, J.W., Fréchet, J.M.J., Maskus, M., and Abruña, H.D. (1998) *Chem. Mater.*, **10**, 30–38; (e) Cardullo, F., Diederich, F., Echegoyen, L., Habicher, T., Jayaraman, N., Leblanc, R.M., Stoddart, J.F., and Wang, S. (1998) *Langmuir*, **14**, 1955–1959.

49 (a) Nierengarten, J.-F. (2000) *Chem. Eur. J.*, **6**, 3667–3670; (b) Armaroli, N., Boudon, C., Felder, D., Gisselbrecht, J.-P., Gross, M., Marconi, G., Nicoud, J.-F., Nierengarten, J.-F., and Vicinelli, V. (1999) *Angew. Chem. Int. Ed.*, **38**, 3730–3733.

50 van de Coevering, R., Kreiter, R., Cardinali, F., van Koten, G., Nierengarten, J.-F., and Gebbink, R.J.M.K. (2005) *Tetrahedron Lett.*, **46**, 3353–3356.

51 Ruiz, J., Pradet, C., Varret, F., and Astruc, D. (2002) *Chem. Commun.*, 1108–1109.

52 Li, W.-S., Kim, K.S., Jiang, D.-L., Tanaka, H., Kawai, T., Kwon, J.H., Kim, D., and Aida, T. (2006) *J. Am. Chem. Soc.*, **128**, 10527–10532.

53 Pérez, E.M., Sánchez, L., Fernández, G., and Martín, N. (2006) *J. Am. Chem. Soc.*, **128**, 7172–7173.

54 Fernandez, G., Perez, E.M., Sanchez, L., and Martin, N. (2008) *J. Am. Chem. Soc.*, **130**, 2410–2411.

55 Fernandez, G., Sanchez, L., Perez, E.M., and Martin, N. (2008) *J. Am. Chem. Soc.*, **130**, 10674–10683.

56 Hahn, U., Gonzalez Juan, J., Huerta, E., Segura, M., Eckert, J.-F., Cardinali, F., de Mendoza, J., and Nierengarten, J.-F. (2005) *Chem. Eur. J.*, **11**, 6666–6672.

57 Hager, K., Hartnagel, U., and Hirsch, A. (2007) *Eur. J. Org. Chem.*, 1942–1956.

58 Nierengarten, J.-F., Hahn, U., Trabolsi, A., Herschbach, H., Cardinali, F., Elhabiri, M., Leize, E., Van Dorsselaer, A., and Albrecht-Gary, A.-M. (2006) *Chem. Eur. J.*, **12**, 3365–3373.

59 (a) Jansen, J.F.G.A., de Brabander-van den Berg, E.M.M., and Meijer, E.W. (1994) *Science*, **266**, 1226–1229; (b) Zhao, M.Q. and Crooks, R.M. (1999) *Adv. Mater.*, **11**, 217–220; (c) Archut, A., Azzellini, G.C., Balzani, V., De Cola, L., and Vögtle, F. (1998) *J. Am. Chem. Soc.*, **120**, 12187–12191; (d) Jansen, J.F.G.A., Meijer, E.W., and Debrabandervandenberg, E.M.M. (1995) *J. Am. Chem. Soc.*, **117**, 4417–4418; (e) Vögtle, F., Gestermann, S., Kauffmann, C., Ceroni, P., Vicinelli, V., and Balzani, V. (2000) *J. Am. Chem. Soc.*, **122**, 10398–10404; (f) Balzani, V., Ceroni, P., Gestermann, S., Gorka, M., Kauffmann, C., and Vögtle, F. (2000) *J. Chem. Soc. Dalton Trans.*, 3765–3771; (g) Balzani, V., Ceroni, P., Gestermann, S., Kauffmann, C., Gorka, M., and Vögtle, F. (2000) *Chem. Commun.*, 853–854.

60 (a) Hawker, C.J. and Fréchet, J.M.J. (1990) *J. Am. Chem. Soc.*, **112**, 7638–7647; (b) Hawker, C.J. and Fréchet, J.M.J. (1990) *J. Chem. Soc. Chem. Commun.*, 1010–1013.

61 Eckert, J.-F., Byrne, D., Nicoud, J.-F., Oswald, L., Nierengarten, J.-F., Numata, M., Ikeda, A., Shinkai, S., and Armaroli, N. (2000) *New J. Chem.*, **24**, 749–758.

62 Numata, M., Ikeda, A., Fukuhara, C., and Shinkai, S. (1999) *Tetrahedron Lett.*, **40**, 6945–6948.

63 Nierengarten, J.-F., Oswald, L., Eckert, J.-F., Nicoud, J.-F., and Armaroli, N. (1999) Complexation of fullerenes with dendritic cyclotriveratrylene derivatives. *Tetrahedron Lett.*, **40**, 5681–5684.

64 Catalano, V.J. and Parodi, N. (1997) *Inorg. Chem.*, **36**, 537–541.

65 Nishioka, T., Tashiro, K., Aida, T., Zheng, J.Y., Kinbara, K., Saigo, K., Sakamoto, S., and Yamaguchi, K. (2000) *Macromolecules*, **33**, 9182–9184.

66 Yamaguchi, T., Ishii, N., Tashiro, K., and Aida, T. (2003) *J. Am. Chem. Soc.*, **125**, 13934–13935.

9
[60]Fullerene-Containing Thermotropic Liquid Crystals

Daniel Guillon, Bertrand Donnio, and Robert Deschenaux

9.1
Introduction

Liquid crystals (LCs) constitute a class of dynamic soft materials, self-assembled into low-dimensional ordered periodic structures [1]. Their ability to respond to various external stimuli makes them ideal components for display applications, that is, flat-panel devices for computers and televisions. Molecular engineering of functional liquid crystals is key to controlling the self-assembling ability and the organizing processes of complex molecules into diverse periodically ordered structures [2]. Moreover, the dynamic and mobile nature of the supramolecular ordered assemblies can considerably enhance their functional capability and stimuli responsiveness, rendering them inevitable ingredients for nonlinear optics, molecular electronics, and molecular photonics, which are future technological applications.

Fullerenes and their derivatives have shown a wide range of chemical and physical properties that make them attractive candidates for a variety of interesting features in supramolecular chemistry and materials science [3]. A requirement for potential applications of this carbon allotrope is the incorporation of [60]fullerene (C_{60}) into well-ordered structures; this may pave the way for novel molecular devices and switches showing outstanding performances and characteristics by combining the electrochemical and photophysical properties of C_{60} [4] with the self-assembling features of liquid crystals. The supramolecular "host–guest" approach can be adapted to fullerene-containing liquid crystals to produce such interesting structures. These can be obtained through the formation of inclusion complexes of C_{60} with macrocyclic derivatives, or by mixing C_{60} or C_{60} derivatives with mesogenic moieties. The functionalization of C_{60} with different organic addends such as dendrimers represents another way for the design of self-organized structures. Indeed, dendritic molecules have proved particularly versatile candidates as novel and original scaffoldings for the elaboration of new LC functional materials, and research in this area has experienced a substantial development during the past decade [5]. Thus, the incorporation of C_{60} into dendritic architectures leads to self-organizing supermolecular systems that display various mesophases, including chiral nematic, smectic A,

Supramolecular Chemistry of Fullerenes and Carbon Nanotubes, First Edition. Edited by Nazario Martin and Jean-Francois Nierengarten.

smectic B, and columnar phases [6]. The direct regioselective polyaddition of a discrete number of mesogens or protomesogenic groups to C_{60} also represents another original strategy to produce multifunctional liquid–crystalline C_{60} derivatives. The selective polyaddition of side-connected malonates on C_{60} based on stepwise or simultaneous cyclopropanation reactions is the most frequently used method. Alternatively, the covalent attachment of aryl- or alkyl-containing derivatives around one pentagon of C_{60} to build pentakisfullerene derivatives is also an attractive method to create original molecular structures, and versatile for multiple functionalization of [60]fullerenes. These different approaches are described in the following sections.

9.2
Noncovalent C_{60} Derivatives

9.2.1
The Liquid–Crystalline Supramolecular Complex of C_{60} with a Cyclotriveratrylene Derivative

Several studies have shown that the cyclotriveratrylene macrocycle (CTV) is a candidate of choice for the formation of inclusion complexes with C_{60} [7, 8]. Interestingly, when adequately derivatized CTV derivatives give rise to liquid–crystalline phases [9]. Thus, it appeared judicious to combine both aspects of CTV to form mesomorphic systems containing the unaltered C_{60} sphere. The first example of a liquid–crystalline host–guest complex of C_{60} has been reported by Nierengarten *et al.* in 2000 [10], and not surprisingly consisted of a CTV derivative substituted by 18 long diverging alkyl chains, virtually programmed for self-assembly in a discrete inclusion complex with the C_{60} sphere and for self-organization of the resulting aggregates into an extended lattice with liquid–crystalline properties.

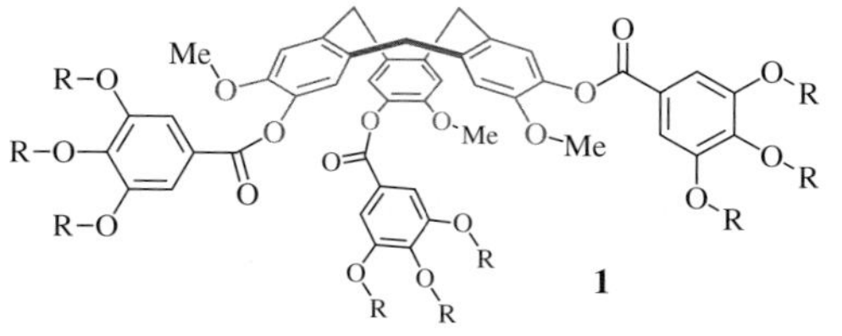

1

R = $C_{12}H_{25}$

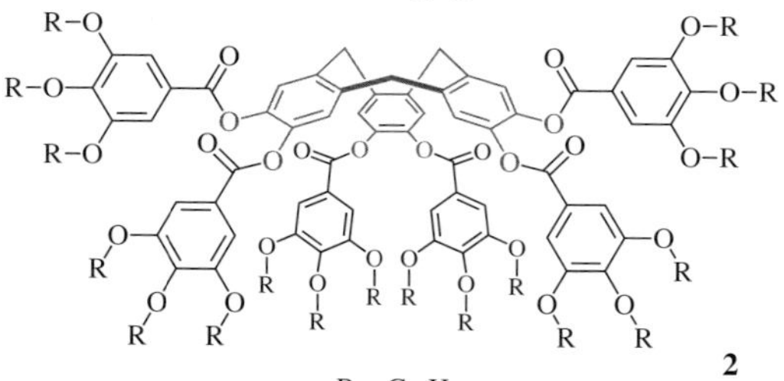

2

R = $C_{12}H_{25}$

Two CTV derivatives **1** (9 alkyl chains) and **2** (18 alkyl chains) were used for the complexation of C_{60}. The formation of the host–guest complexes in C_6H_6 solutions between C_{60} and **1** or **2** was evidenced by the continuous changes observed in the UV/Vis spectra by successive additions of the host to the C_{60} solutions. Specifically, each new addition of **1** or **2** to the C_{60} solution in C_6H_6 led to an increase in the absorption in the whole visible region with a most pronounced effect of about 430 nm. These observed spectral changes are similar to those previously described in the literature by addition of other CTV derivatives to C_{60} solutions and are characteristic of complexation.

Slow evaporation of C_6H_6 solutions of mixtures of C_{60} and **1** or **2** in various proportions afforded the corresponding supramolecular host–guest complexes. For both **1** and **2**, the 2 : 1 host–guest species have been obtained in the form of brown compounds. It should be noted that in the presence of an excess of C_{60}, the host–guest complexes were formed together with crystalline C_{60}, whereas in an excess of **1** or **2**, nonhomogeneous mixtures were obtained. In fact, two phases were present, a colorless one corresponding to pure **1** or **2**, and a brown one corresponding to the 2 : 1 host–guest complexes.

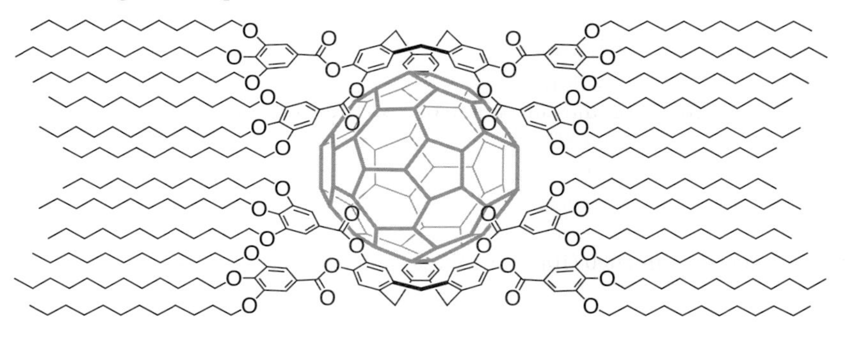

3

Macrocycle **1** and the corresponding supramolecular complex $[C_{60}\subset(\mathbf{1})_2]$ are not mesomorphic. In contrast, observation by polarized optical microscopy of both macrocycle **2** and the corresponding 2 : 1 host–guest product (directly after solvent evaporation) revealed a fluid birefringent phase at room temperature. Compound **2** clears into the isotropic liquid at 75 °C, while no clearing was detected for the complex, but a mesophase transformation instead over a narrow temperature range.

The X-ray diffraction patterns recorded below 70 °C in the liquid–crystalline phase of **2** and $[C_{60}\subset(\mathbf{2})_2]$, (**3**) are similar and contain only small- and wide-angle diffuse reflections, suggesting a low-organized phase with a nematic-like nature. These diffuse reflections indicate the presence of cybotactic groups, that is, groups of molecules locally arranged in a more ordered structure such as smectic or columnar, but not developed on a large scale [11]. However, the number and position of the diffuse reflections could not be explained by assuming only a smectic or a columnar order within the cybotactic groups, but to a lamello-columnar order [12], that is, smectic layers with the existence of columns oriented parallel to the smectic planes.

When host–guest **3** was heated above 70 °C, the birefringence of the texture under the microscope disappears and the X-ray diffraction pattern transforms into a pattern characteristic of a cubic phase. For the latter, the $I4_132$ space group was the only one fitting the observed reflections, based on the extinction rules resulting from the space group symmetry, and was considered as the most probable. It should be noted that such a space group has never been previously assigned for a liquid crystal cubic phase, despite its theoretical prediction [13]. At higher temperature, macroscopic phase segregation between C_{60} and CTV molecules occurs irreversibly, evidenced by coexistence of different phases. The dissociation of the complex was confirmed by both differential scanning calorimetry (DSC) and XRD recorded on subsequent heating–cooling cycles.

9.2.2
Supramolecular Complex Composed of Rigid Dendritic Porphyrin and Fullerene

The stability of such host–guest complexes is mainly related to the conformity of C_{60} with the organic host cavity. In this respect, Kimura *et al.* have synthesized 1,3,5-phenylene-based rigid dendritic porphyrins [14] with well-defined nanospaces formed around the porphyrin core that provide cavities for inclusion of C_{60}. In addition, the dendritic porphyrin moiety interacts strongly with C_{60} through π donor/π acceptor interactions, leading to a very good stability of the complex in toluene. In a subsequent study, Kimura *et al.* showed that compound **4** exhibits a columnar rectangular phase between 40 and 110 °C [15]. When mixed with C_{60}, compound **4** remains soluble in toluene. Titration study and quenching of fluorescence even at low concentration of C_{60} indicate the formation of an associated 1 : 1 complex. Spin-coated films of **4** in the presence of an equimolar amount of C_{60} have also been obtained; the same complex was formed. The latter exhibits two reversible transitions at 99 and 250 °C, and a needle-like texture was observed by polarized optical microscopy in the same temperature range. The authors mention a change only in the X-ray diffraction pattern of the mixture with respect to that of **4** alone, but no detail is given about the resulting structure.

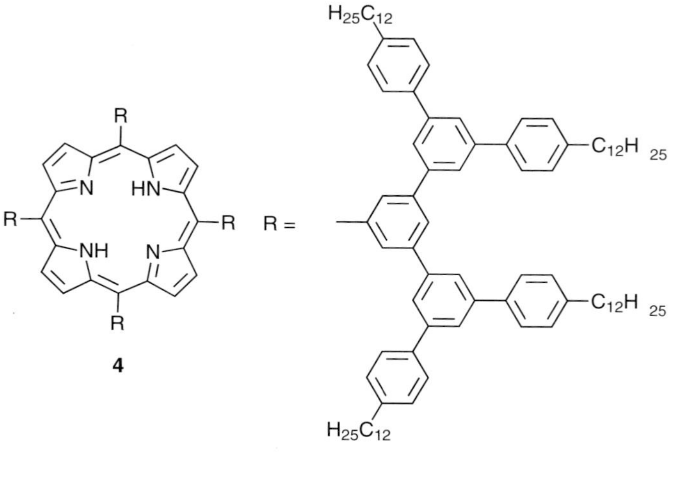

4

9.2.3
Self-Assembled Columns of C_{60}

Columnar phases based on alternating triphenylene and hexaphenyltriphenylene moieties have shown to be extremely stable [16]; the enhanced stability of the columns in the 1 : 1 mixture is, in general, reflected by the fact that the clearing temperature is higher than that of either of the two components. In addition, because of the relatively wide column–column spacing, bulky functional moieties such as crown ethers and ferrocenes can be accommodated within the alkyl chain region between the columns, without disrupting the hexagonal lattice of the columnar structure. Following this approach, Bushby *et al.* have synthesized C_{60} fullerene-containing hexaalkyloxytriphenylene **5** that was mixed with an aza derivative based on quinoxalene **6** in a 1 : 1 ratio [17]. This mixture shows a mesomorphic behavior with the presence of a hexagonal columnar phase, partly deduced from DSC and optical polarizing microscopy.

The corresponding X-ray diffraction patterns show, indeed, reflections typical of a columnar hexagonal phase, together with additional reflections with higher spacings. The authors interpret the latter as the presence of another hexagonal superlattice with a spacing $\sqrt{7}$ times larger than that of the main columnar lattice. Thus, the molecular organization results from the alternating stacking of **5** and **6** into columns, the alkyl chains forming a continuum surrounding the aromatic stacks. The presence of the supperlattice is explained by a particular ordering of the fullerenes within the liquid crystal matrix. It consists of C_{60} units wrapping around one in seven of the columns of the columnar array. This structure is stabilized by the tendency of the fullerenes for self-association that, however, is modulated by the fact that C_{60} is attached to two triphenylene cores through a spacer, the length of which should play a role too. In other words, the specific ordering results from the competition between the self-association of the C_{60} units and the self-organization of the discotic triphenylene cores to which they are linked.

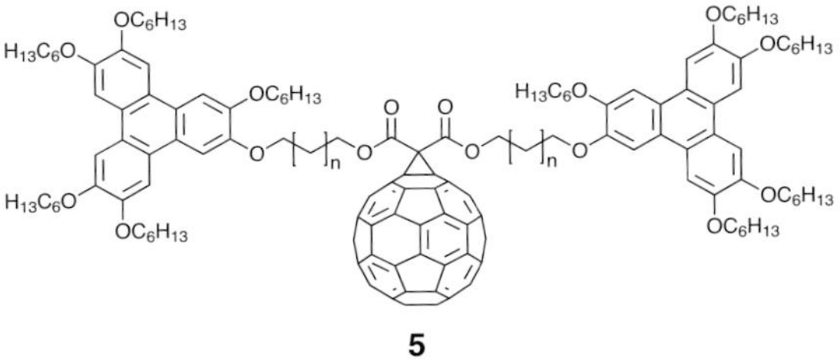

5

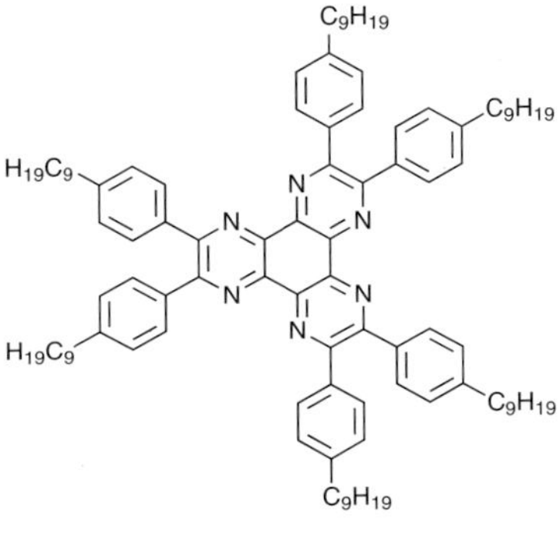

6

9.2.4
Phthalocyanine-[60]Fullerene Dyads in Liquid Crystals

Donor–acceptor dyads formed by phthalocyanine and fullerene moieties are considered as attractive photoactive systems to produce long-lived photoinduced charge-separated states and thus able to be introduced in molecular switches or photovoltaic cells [18]. On the one hand, the efficiency of organic photovoltaic cells is improved if the supramolecular organization of the molecules can be controlled on a large scale, like in liquid crystals, for example. On the other hand, a major drawback of using C_{60} in such systems is its bulkiness and therefore its difficult accommodation into a mesomorphic organization. The use of blends is one of the possibilities to obtain fullerene-containing liquid crystals. This was performed by de la Escosura *et al.* who mixed nonmesomorphic compounds **7**, **8**, or **9** with the mesomorphic discotic mesogen Zn(II)-octakis(hexadecylthio)-phthalocyanine **10** [19]. The mixtures were prepared by slow evaporation of the solvent from THF solutions and all of them showed a mesomorphic behavior over a large temperature range. The corresponding powder diffraction patterns contain the typical reflections of a columnar phase, similar to that of the well-known substituted phthalocyanines. The authors propose a model of organization in which there is a predominance of alternating stacks of dyad and phthalocyanine **10** and some domains with randomly distributed Pc–C_{60} molecules (Figure 9.1). This approach is of interest for improving the morphology of bulk heterojunctions in photovoltaic cells, resulting from an increase of the dissociation efficiency of bound electron–hole pairs at the donor–acceptor interface.

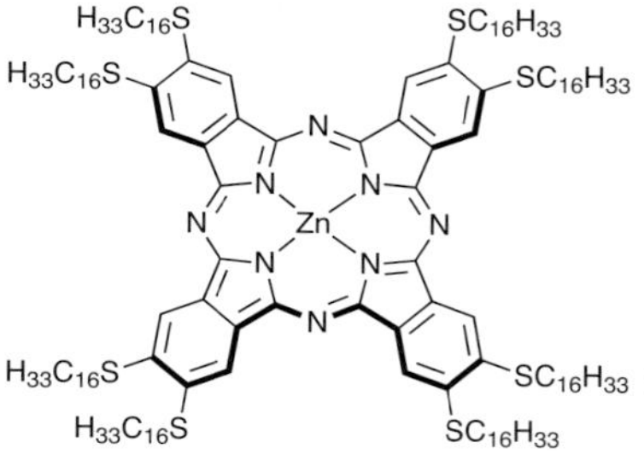

10

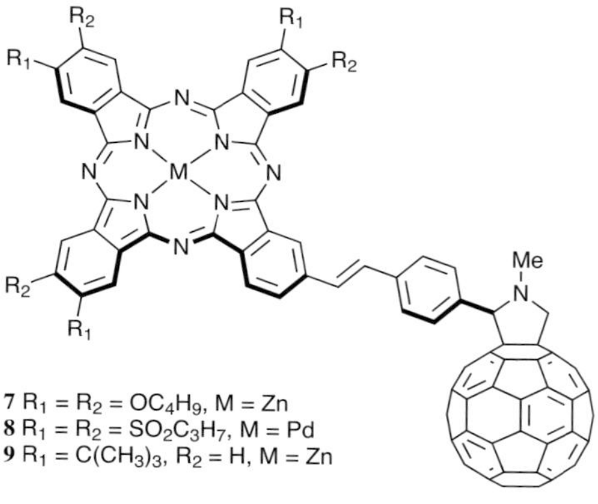

7 $R_1 = R_2 = OC_4H_9$, M = Zn
8 $R_1 = R_2 = SO_2C_3H_7$, M = Pd
9 $R_1 = C(CH_3)_3$, R_2 = H, M = Zn

9.3
Covalent C_{60} Derivatives

9.3.1
Liquid–Crystalline Methanofullerene- and Fulleropyrrolidine-Based Poly(Aryl Ester) Dendrons

Addition reaction of various generations of malonate-based dendritic addends, bearing cyanobiphenyl end groups, onto C_{60} led to methanofullerodendrimers

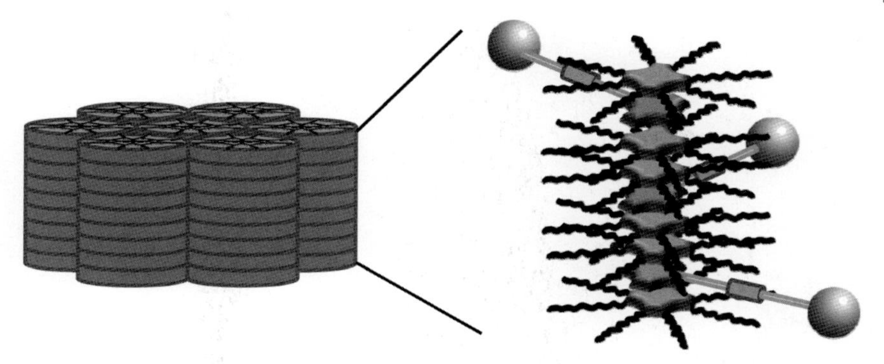

Figure 9.1 Supramolecular organization of the blends **10/7**, **10/8**, and **10/9** within the columnar phase.

11–14 [20]. All of them, **11** (first generation; T_g 48; SmA 179 I), **12** (second generation; T_g: not detected; SmA 183 N 184 I), **13** (third generation; T_g: not detected; SmA 212 I), and **14** (fourth generation; T_g: not detected; SmA 252 I), exhibit a broad SmA phase, with the mesophase stability increasing with the dendrimer generation. In all cases, periodicities are on the order of 100 Å.

11

12

13

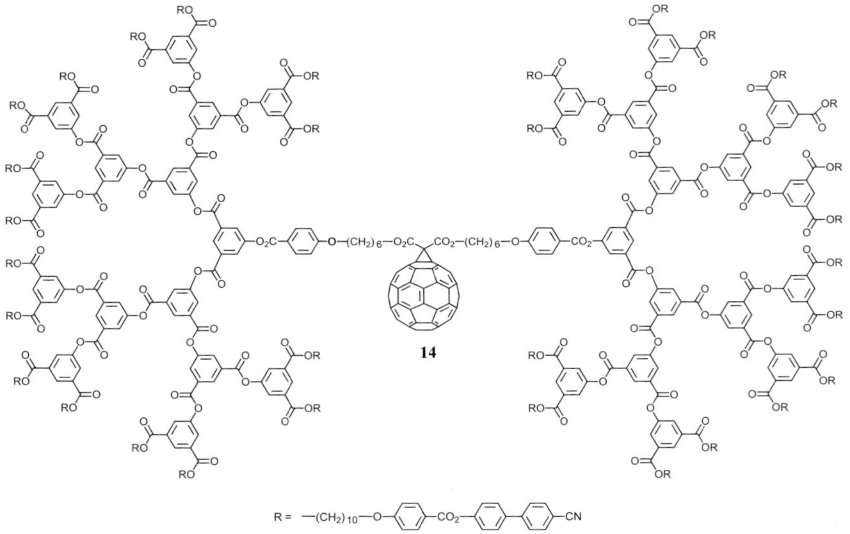

14

R = —$(CH_2)_{10}$—O—C_6H_4—CO_2—C_6H_4—C_6H_4—CN

The supramolecular organization was deduced from X-ray diffraction analysis and molecular simulation. For **11**, the molecules adopt a V-shape (constituted by pairs of mesogenic groups) and arrange in a head-to-tail fashion favored by the antiparallel packing of the polar end groups. For **12**, the branching part begins to have significant lateral extension with respect to the layer normal, and the two branches extend on both sides of C_{60}. Then, for **13** and **14** (Figure 9.2), the structure is governed by the polar cyano groups. The central part of the layer is constituted by the fullerene moiety embedded by the dendritic segments, and the layer interface is formed by partial interdigitation of the mesogenic groups. In all cases, the mesophase is stabilized by dipolar interactions (antiparallel arrangement of the cyanobiphenyl units). Thus, in **11–14**, C_{60} is buried within the dendritic branches, at least above the second generation, and as a consequence, the supramolecular organization is independent of the change of the generation number of the dendrimer and both the thermal behavior and the mesophase organization are similar to that of the corresponding mesomorphic malonate precursors (structures not shown).

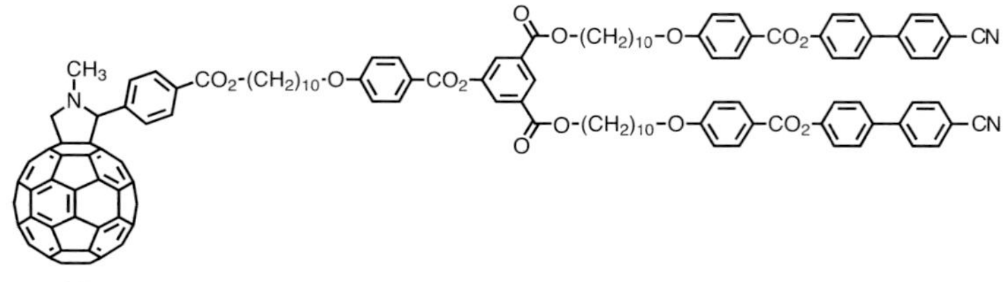

15

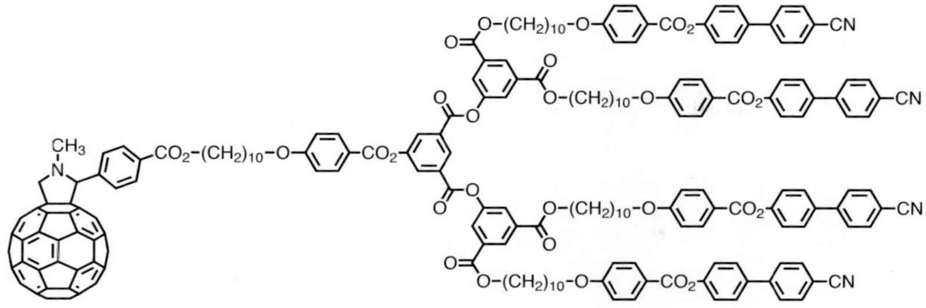

16

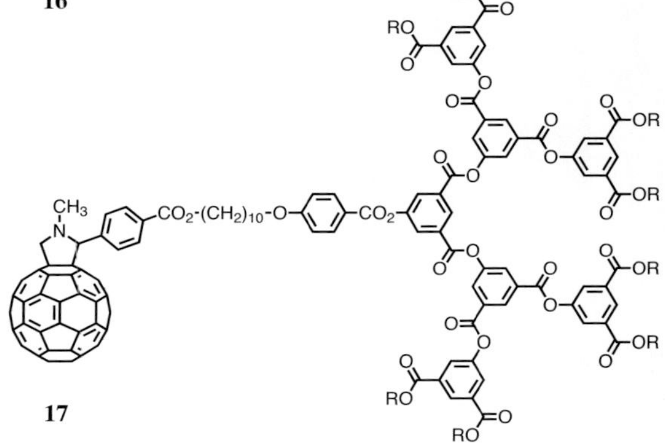

17

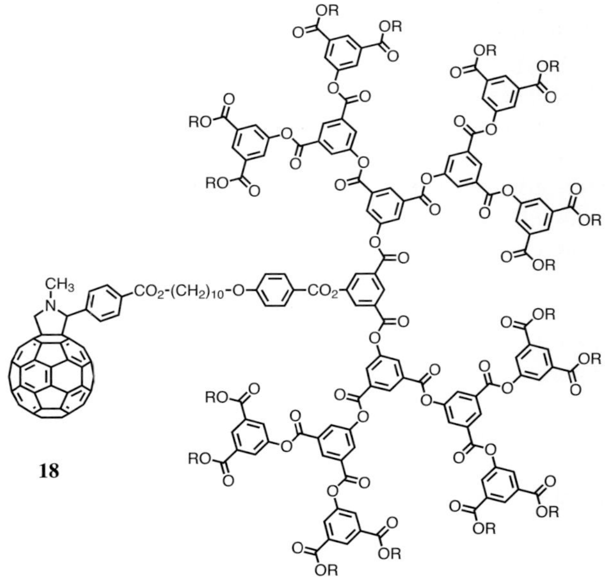

18

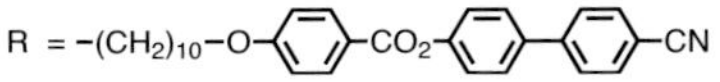

R = $-(CH_2)_{10}-O-C_6H_4-CO_2-C_6H_4-C_6H_4-CN$

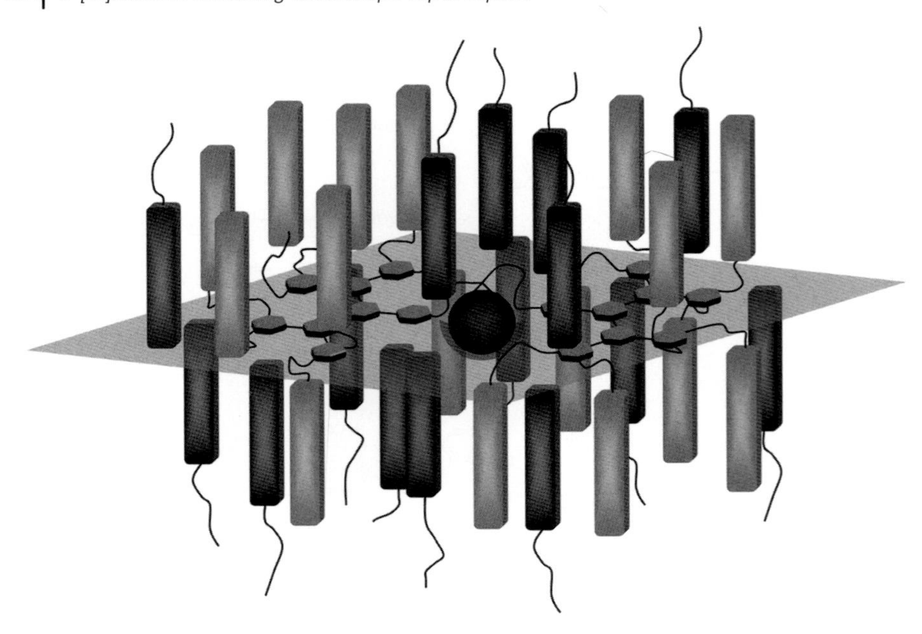

Figure 9.2 Proposed supramolecular organization of **13** within the smectic A phase. Compounds **12** and **14** show a similar supramolecular organization.

Fulleropyrrolidines are an important family of C_{60} derivatives that have the advantage over methanofullerenes to lead to stable reduced species. To promote mesomorphism in fulleropyrrolidines, C_{60} was modified with dendritic addends bearing cyanobiphenyl groups leading to four generations of fulleropyrrolidines **15–18** [21]. This appeared to be the right strategy since, with the exception of first-generation **15** that is nonmesomorphic (Cr 178 I), all other fullerene-based dendrimers give rise to a SmA phase (**16**: T_g 44, SmA 168 I; **17**: T_g 51, SmA 196 I; **18**: T_g 36, SmA 231 I), the stability of which increases with the generation number. The aldehyde precursors show a broad SmA phase, except for the one of first generation that shows a nematic phase. As for the molecular organization within the SmA phase (Figure 9.3), the second-generation molecules **16** are oriented in a head-to-tail fashion within the layers, and for each molecule the mesogenic groups point in the same direction interdigitating with mesogenic groups of adjacent layers. For the higher generations **17** and **18**, the mesogenic units are positioned above and below the dendritic core, and interdigitation occurs between layers; C_{60} is hidden in the dendritic core and has no influence on the supramolecular organization as this was the case for the methanofullerodendrimers discussed above.

9.3.2
Liquid–Crystalline Fulleropyrrolidine-Based Poly(Benzyl Ether) Dendrons

Both for fundamental aspects and for potential applications, it is of interest to control the nature and symmetry of the mesophases, with, for instance, the induction of

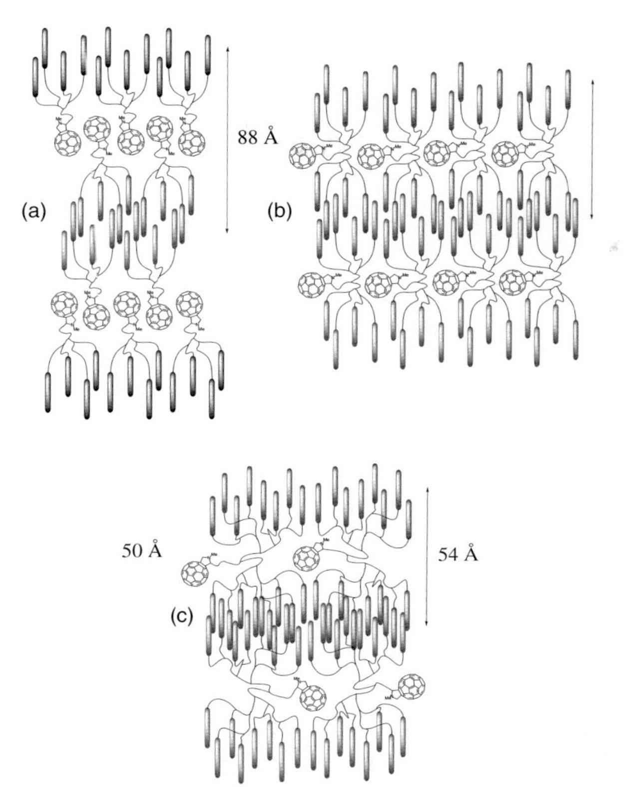

Figure 9.3 Postulated supramolecular organization of fulleropyrrolidines **16** (a), **17** (b), and **18** (c) within the smectic A phase.

mesophases with curved interfaces. Indeed, fullerene-containing liquid crystals displaying columnar phases are candidates of choice for the development of new electronic and optoelectronic devices [22]. This can be achieved by modifying the molecular structures of the dendritic addends, and poly(benzyl ether) dendrons were selected since they give rise to either spherical or cylindrical supramolecular dendrimers: due to their inherent fan, conical, or even spherical conformation, they subsequently self-organize into cubic or columnar lattices [23]. Functionalization of C_{60} with poly(benzyl ether) dendrons equipped with an aldehyde function, which exhibit columnar mesomorphism, and N-methylglycine led to fullerodendrimers **19** (T_g not detected, Col_r 80 I) and **20** (T_g 46, Col_r 74 I), which exhibit a Col_r phase with a *c2mm* symmetry, with rather large lattice parameters ($a = 128.6$ Å, $b = 86.0$ Å for **19**, and $a = 129.6$ Å, $b = 89.4$ Å for **20**) [24].

Considering a hexagonal close compact packing of the C_{60} spheres along the columnar axis, surrounded by the flexible dendritic shell, the number of molecules included in an elementary columnar slice (with a thickness of 8.7 Å) was calculated. From the values of the lattice parameters and the estimated molecular volumes

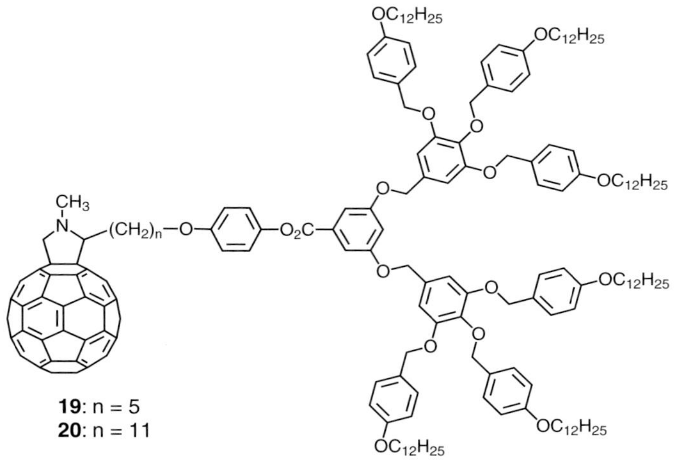

19: n = 5
20: n = 11

(4550 and 4700 Å^3 for **19** and **20**, respectively), this number turned out to be about 10 for each compound. The supramolecular organization thus results in bundles of 10 dendrimers superimposed one over the other to form an elliptic columnar core, the shape and orientation of which are in agreement with the *c2mm* symmetry. The dendritic moieties are arranged around this elliptic core to fill the intercolumnar space (Figure 9.4). This result demonstrates that C_{60} can be organized along one single direction (here, the columnar axis) provided a suitable molecular design is achieved.

To further investigate the formation of columnar phases, methanofullerenes carrying simultaneously at both extremities either two identical poly(benzyl ether) dendrons (**21**: two second-generation dendrons; **22**: two third-generation dendrons) or two poly(benzyl ether) dendrons of different generations (**23**: second- and third-generation dendrons) were synthesized [25]. All the dendritic malonate precursors exhibit a Col_h phase, with a large increase in the phase stability as the size of the dendrons increases. However, while compound **21** (Cr 52 I) does not show liquid–crystalline properties, compounds **22** (T_g 44, Col_h 93 I) and **23** (T_g 64, Col_h 74 I) display hexagonal columnar phases, clearly identified by XRD. Thus, C_{60} has no (or little) influence on the structure of the mesophases, but its impact is detrimental to the mesophase stability. The parameters of the two-dimensional hexagonal lattices (*p6mm* symmetry) are consistent with the dimensions of the basic dendritic branches, $a = 47.05$ and 46.65 Å for **22** and **23**, respectively, and are similar to that of the malonate precursors, suggesting that a similar supramolecular organization is maintained. The fan-like conformation of the dendrons forces the fullerodendrimers (and the malonates) to arrange in cylindrical columns, with the C_{60} units inside the organic shell. Depending on the combination of dendritic branches and on the geometrical parameters, two molecules of **22** and **23** self-associate into a disk in order

Figure 9.4 Postulated supramolecular organization of **19** and **20** within the rectangular columnar phase.

to pave the hexagonal 2D network. In both cases, the C_{60} units are located toward the interior of the column and loosely stacked along the columnar axis, surrounded by the dendritic part. Molecular dynamic experiments on compound **23** confirmed this supramolecular organization (Figures 9.5 and 9.6) where a lattice parameter

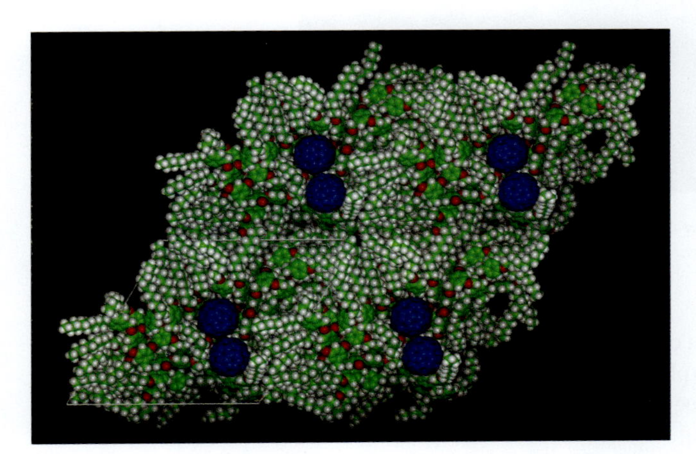

Figure 9.5 Top view of the supramolecular organization of **23** within the hexagonal columnar phase *of p6mm* symmetry.

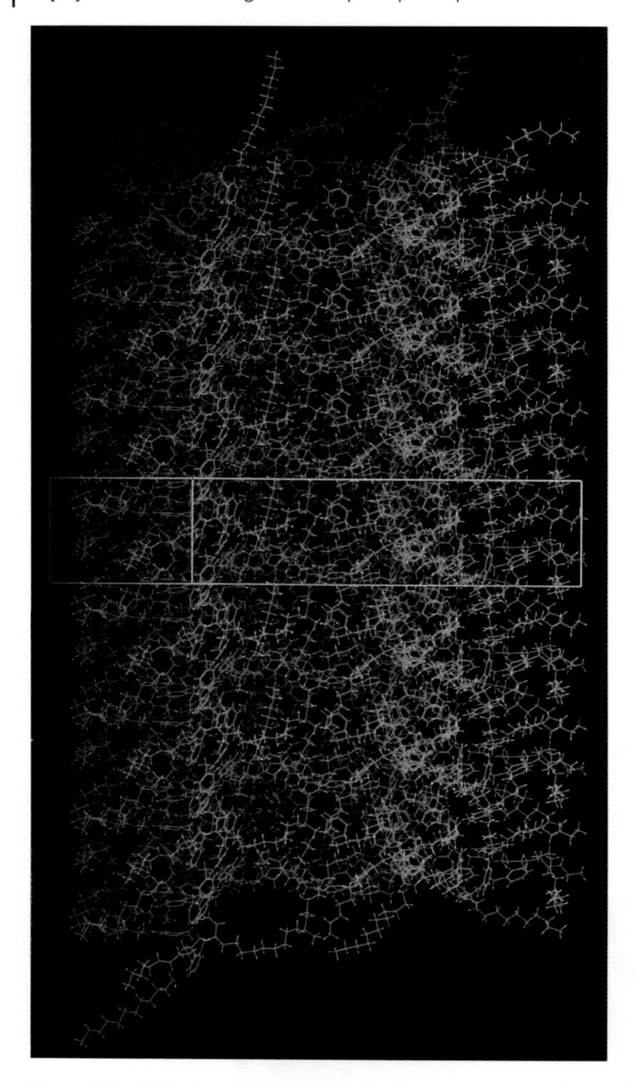

Figure 9.6 Side view of the supramolecular organization of **23** within the hexagonal columnar phase *of p6mm* symmetry.

comparable to that of the X-ray experiments and a density close to unity were found. The nonmesomorphic character of **21** indicates that mesophase induction in such bulky materials requires the connection of stronger liquid–crystalline promoters. The corresponding hemidendritic homologues, malonates and fullerodendrimers (structures not shown), also exhibit a Col_h phase, except one compound that shows a Col_r phase.

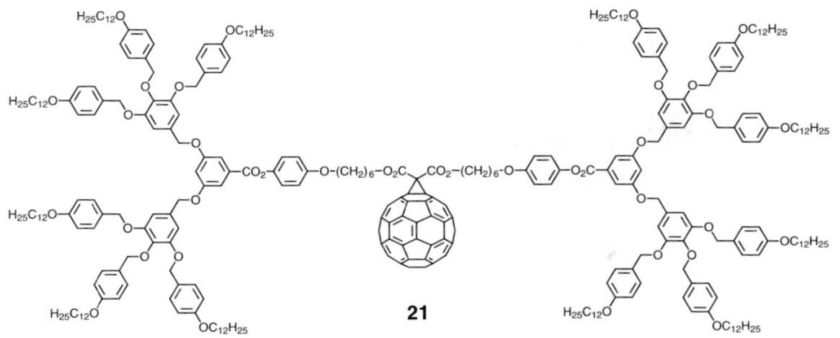

21

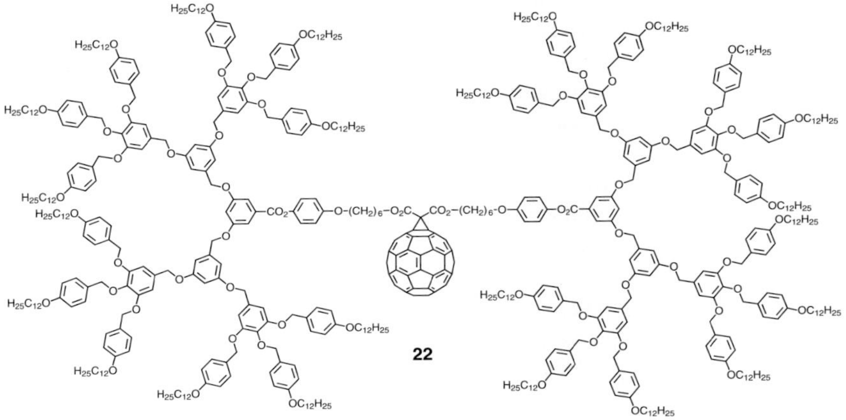

22

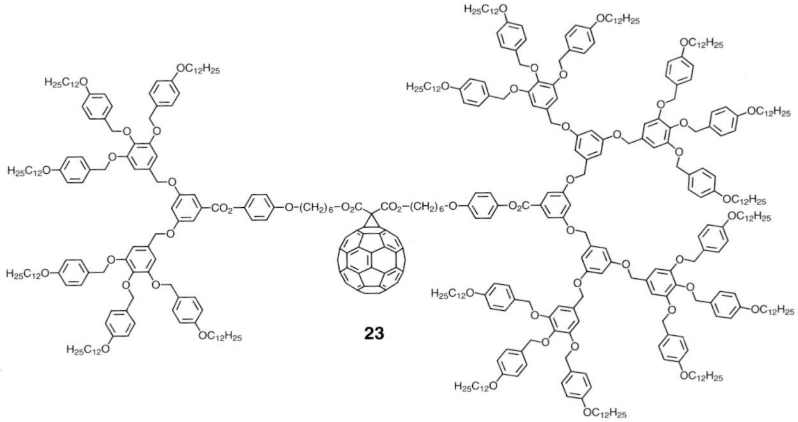

23

Thus, essentially all the malonate and fullerene derivatives equipped with such flexible poly(benzyl ether) dendrons give rise to columnar phases, and their supramolecular organization within the columnar phases is governed by the dendrimer providing that the size of the dendritic addends is large enough to compensate and circumvent the large area of the bulky C_{60}.

9.3.3
Liquid–Crystalline Fullero(Codendrimers)

As discussed above, grafting of fan-shaped poly(benzyl ether) dendrimers onto C_{60} yielded fullerodendrimers exhibiting columnar phases with either hexagonal or rectangular symmetry, while the use of dendrons bearing calamitic end groups promoted the formation of smectic phases. There are more possibilities to control the liquid–crystalline properties of codendrimers than those of homodendrimers. For example, when two different dendrons are assembled, their generation, relative proportion, and location within the molecule can be varied independently, and each modification can be used, in principle, to control the nature of the mesophases. By exploiting this modular construction, we envisioned that the liquid–crystalline properties of fullero(codendrimers) could be tuned by changing the generation and the nature of the dendrons located on C_{60}. Therefore, the assembly of poly(aryl ester) dendrons functionalized with cyanobiphenyl groups and poly(benzyl ether) dendrons carrying alkyl chains was attempted. These dendrimers were selected with the expectation that their different structural characteristics and properties would influence the overall liquid–crystalline behavior. In such voluminous structures, C_{60} is hidden in the organic matrix, and the supramolecular organization should depend only on the dendrons and should not be altered by the presence of the isotropic C_{60} hard sphere. Furthermore, owing to the different nature of the dendrons, multilevel microsegregation should be obtained, leading to long-range organization within the liquid crystal state.

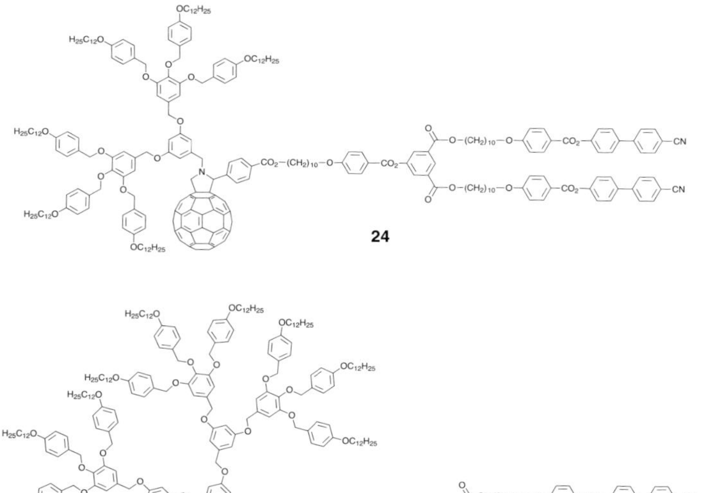

24

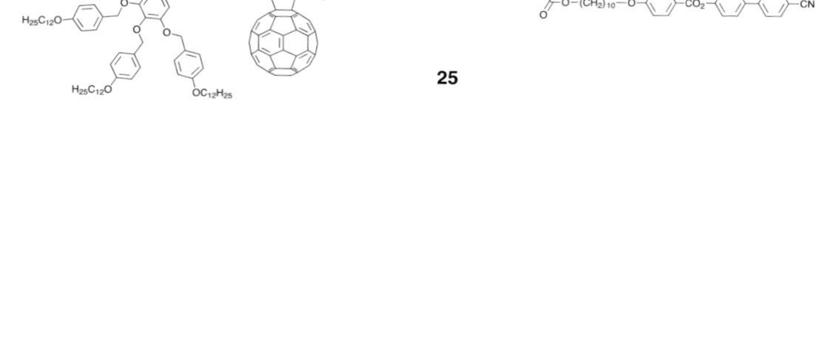

25

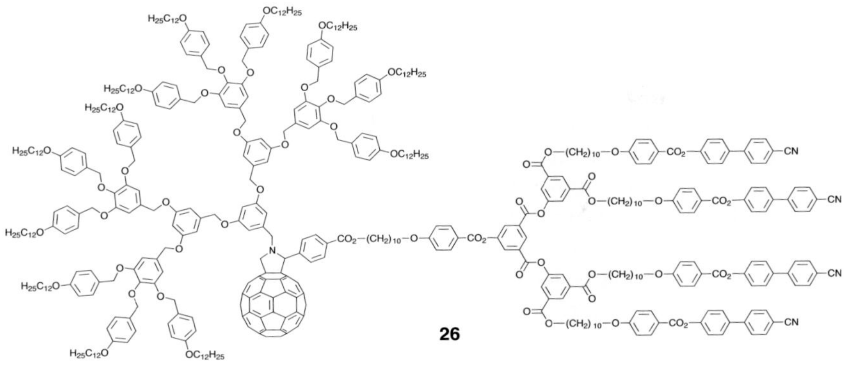

26

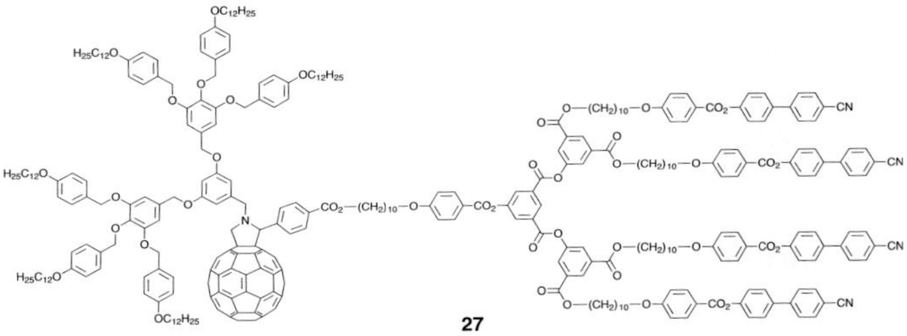

27

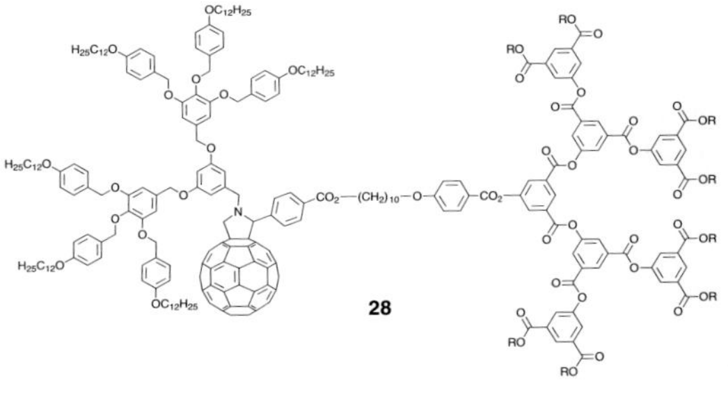

28

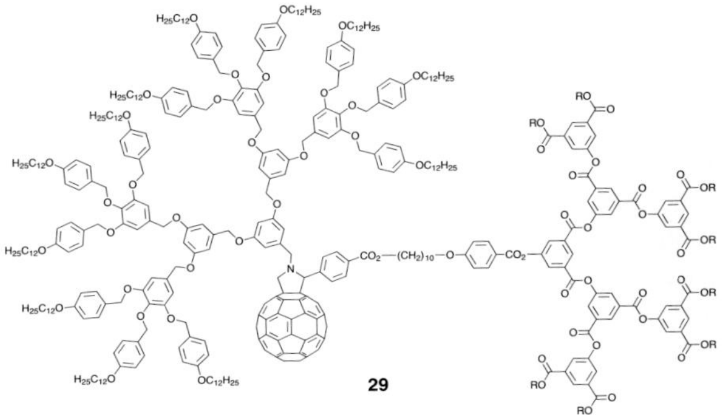

29

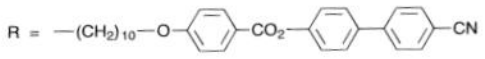

R = $-(CH_2)_{10}-O-C_6H_4-CO_2-C_6H_4-C_6H_4-CN$

Six fullero(codendrimers) **24–29** were prepared [26] and classified into two families from the point of view of their liquid–crystalline properties: the compounds that give rise to columnar phases, that is, **24** (T_g: 31; $Col_{r\text{-}c2mm}$ 105 I′ 108 I), **25** (T_g: not detected; $Col_{r\text{-}p2gg}$ 109 I), and **26** (T_g: not detected; $Col_{r\text{-}c2mm}$ 152 I), and those showing smectic phases, that is, **27** (T_g: not detected; SmC 116; SmA 155 I), **28** (T_g: not detected; SmA 210 I), and **29** (T_g: not detected; SmA 209 I).

The liquid–crystalline properties of **24–29** clearly depend on the generation of each dendron. When the generation of the poly(benzyl ether) dendron is higher than that of the poly(aryl ester) dendron, columnar mesomorphism is observed (i.e., for **24–26**); conversely, when the generation of the poly(aryl ester) dendron is higher (i.e., for **28**) than or the same as (i.e., for **27** and **29**) that of the poly(benzyl ether) dendron, smectic mesomorphism is observed. The liquid–crystalline properties of **24–29** can thus be tuned by design.

Comparison of the isotropization temperatures emphasizes the influence of the poly(aryl ester) dendrons on the thermal stability of the liquid–crystalline phases. Compounds **28** and **29**, with third-generation poly(aryl ester) dendron, show the highest isotropization temperatures (210 °C for **28** and 209 °C for **29**); in contrast, the size of the poly(benzyl ether) dendron (second generation for **28** and third generation for **29**) has no influence on the isotropization temperature. Decreasing the poly(aryl ester) dendron generation results in a decrease in clearing point (155 °C for **27** and 152 °C for **26**, both second-generation poly(aryl ester) dendron), independent of the observed mesophase. Finally, the clearing point of **24** and **25** confirms that the poly (benzyl ether) dendron has no influence on the isotropization temperature (105 °C for **24** and 109 °C for **25**).

The formation of smectic phases for **27–29** results from the antiparallel arrangement of the fullerodendrimers forming one central sublayer containing the cyanobiphenyl groups, and the poly(benzyl ether) dendrons being ejected to another sublayer. Such an intramolecular segregation occurs above the glass transition, giving enough flexibility to the poly(benzyl ether) dendron and the aliphatic spacers of the poly(aryl ester) dendron to deform in order to favor parallel arrangement of the mesogenic groups, thus producing well-developed lamellar structures.

The values of the layer periodicities, which increase only slightly with increasing molecular weight (121.5, 125.9, and 136.0 Å for **27**, **28**, and **29**, respectively) and molecular areas permit to understand the molecular organization within the layers. A bilayered structure (molecules arranged head-to-head) is thus envisaged, where the central slab of the layer is made of the cyanobiphenyl mesogenic groups arranged in an antiparallel fashion and the aliphatic chains of the poly(benzylether) dendrons pointing out at both interfaces of the layer (Figure 9.7). In this model, the area available for the aliphatic chains is compensated by that of the cyanobiphenyl mesogenic groups. In the center of the layer, the mesogenic groups are either tilted or normal to the smectic plane, depending on the aliphatic terminal chains/mesogenic groups ratio. In these layers, the C_{60} units are confined within well-defined sublayers, located on both sides of the central layer formed by the cyanobiphenyl groups; the poly(benzyl ether) dendritic portion is confined within external sublayers. The absence of X-ray signals corresponding to the C_{60} units suggests the

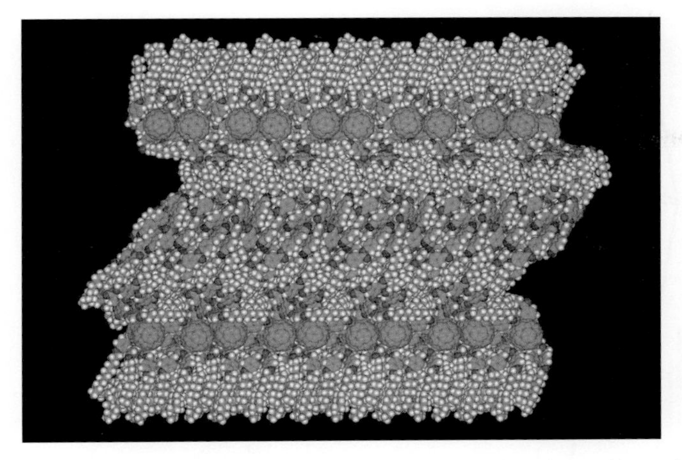

Figure 9.7 Postulated supramolecular organization of **27** obtained by molecular dynamics. Compounds **28** and **29** give similar results.

absence of long-range organization and a random disposition of fullerene within the sublayers.

As for the above-mentioned systems, the supramolecular organization of **24–26** extends over long distances despite the high molecular weight of the compounds (4.3–7.7 kDa) and in the absence of any molecular shape specificity. The values of the chains/mesogens ratio (3 for **24** and **26**, 6 for **25**) are consistent with an induced curvature in the structures (the number of aliphatic chains is larger than the number of cyanobiphenyl mesogenic groups, and thus the transverse molecular areas of both molecular moieties are significantly different, and therefore a stable lamellar structure cannot be obtained), leading to columnar phases. In addition, they develop only above certain temperatures in such a way that the conformation of the aliphatic spacers and the dendritic moieties can adapt to the most stable condensed phase.

The number of molecules per unit length along the columnar axis (about 10 Å thickness, corresponding to the diameter of C_{60}) was calculated from the rectangular lattice parameters and the estimated molecular volumes. A columnar slice contains 14 molecules of **24** and 10 molecules of the larger dendrimers **25** and **26**. The postulated model of the molecular organization is derived from that of the lamellar organization although the increasing ratio value destabilizes the layering by breaking the layers into ribbons as previously observed for lamellar-to-columnar phase transitions exhibited by polycatenar liquid crystals [27], fifth-generation carbosilane dendrimers [28], and statistical liquid–crystalline codendrimers [29]. Each columnar core is made of mesogenic groups interacting through the cyanobiphenyl groups and is surrounded by the poly(benzyl ether) dendrons including the C_{60} units. A competition occurs between the tendency of the cyanobiphenyl subunits to form layers via a head-to-head arrangement and the bulky dendrons forcing a columnar arrangement. This competition is dominated by the bulky

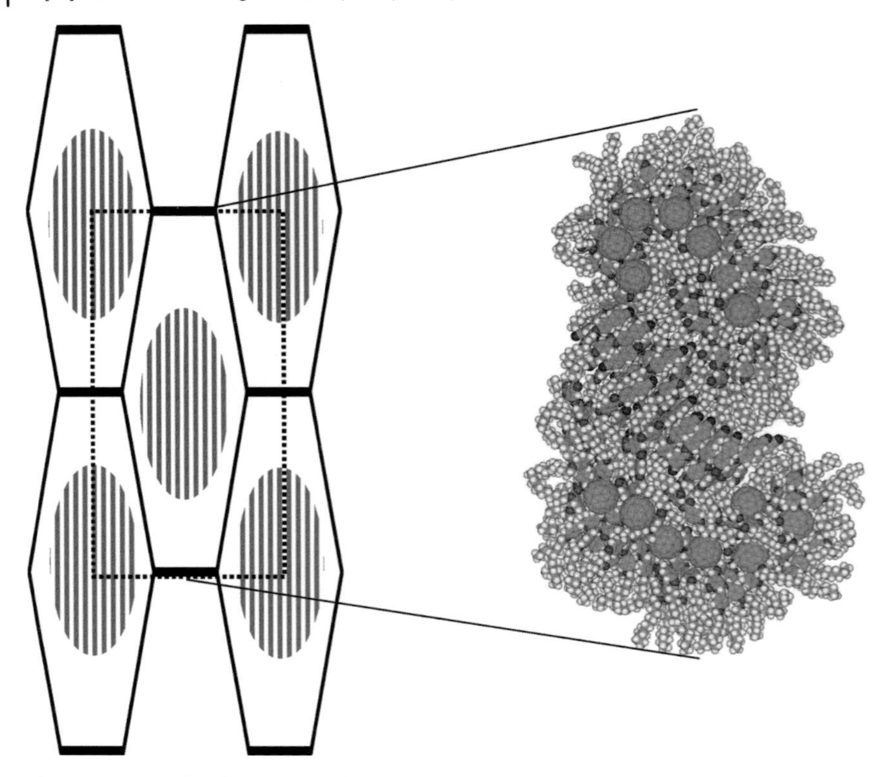

Figure 9.8 Postulated supramolecular organization of **24** within the rectangular columnar phase of *c2mm* symmetry. Compounds **25** (col_r-*p2gg*) and **26** (col_r-*c2mm*) give a similar organization.

dendrons evidenced by the presence of strong layer undulations with large amplitudes resulting in the destruction of the layers into ribbons. The symmetry of these 2D arrangements (*c2mm* or *p2gg*) is clearly associated with the chains/mesogens ratio. The absence of X-ray signal corresponding to C_{60} suggests a loose disposition of the latter. This arrangement was confirmed by molecular dynamics calculations on **24** (Figure 9.8).

The beneficial effect of codendritic architectures to generate mesomorphism for fullerodendrimers is, therefore, clearly demonstrated in these systems. Thus, XRD investigations, molecular modeling, and solution studies revealed that the supramolecular organization of compounds **24–29** is governed by (1) the "aliphatic terminal chains/mesogenic groups" ratio, (2) the effective lateral intra- and intermolecular interactions between the cyanobiphenyl mesogenic groups, (3) the effective microsegregation of the dendrons, and (4) the deformation of the dendritic core. It is worth noting that the supramolecular organization within the mesophases extends to long distances, evidenced by the presence of a large number of sharp and intense small-angle XRD peaks.

9.3.4
Polypedal [60]Fullerenes

The sixfold addition of malonates to C_{60} is an elegant means to incorporate the carbon sphere into a specific liquid–crystalline environment. The first liquid–crystalline hexaadduct, **30a**, was reported in 1999, and was equipped with malonates derived from cyanobiphenyl mesogens at the six apical positions of C_{60}. While the free malonate derivative shows only a monotropic N phase, a clear enhancement of the mesophase stability along a phase symmetry change was observed for the corresponding [6 : 0] hexaadduct with the induction of an enantiotropic SmA phase over a large temperature range (G 80 SmA 133 I) [30]. The observation of the orthogonal smectic A phase suggests that all the cyanobiphenyl units are on average oriented parallel to the layer normal, with a good adequacy of the molecular cross sections of both parts: the supermolecule adopts a cylindrical shape with the mesogenic groups being equally distributed on either side of the C_{60} sphere, in a similar manner to some classical mesomorphic dendrimers [5e] and related bulky polymetallic clusters [31]. Since this seminal report, several liquid–crystalline hexakis(methano)fullerenes have been reported (compounds **30–32**), for which the mesomorphic properties could be tailored by (i) the variation of the chemical structure of the mesogen to be grafted onto the malonate species and its topology of attachment (side-on versus end-on), (ii) the design of more intricate systems, such as heterogeneous fullerene derivatives, that is, bearing different malonate-containing mesogens, or (iii) the control of the addition patterns.

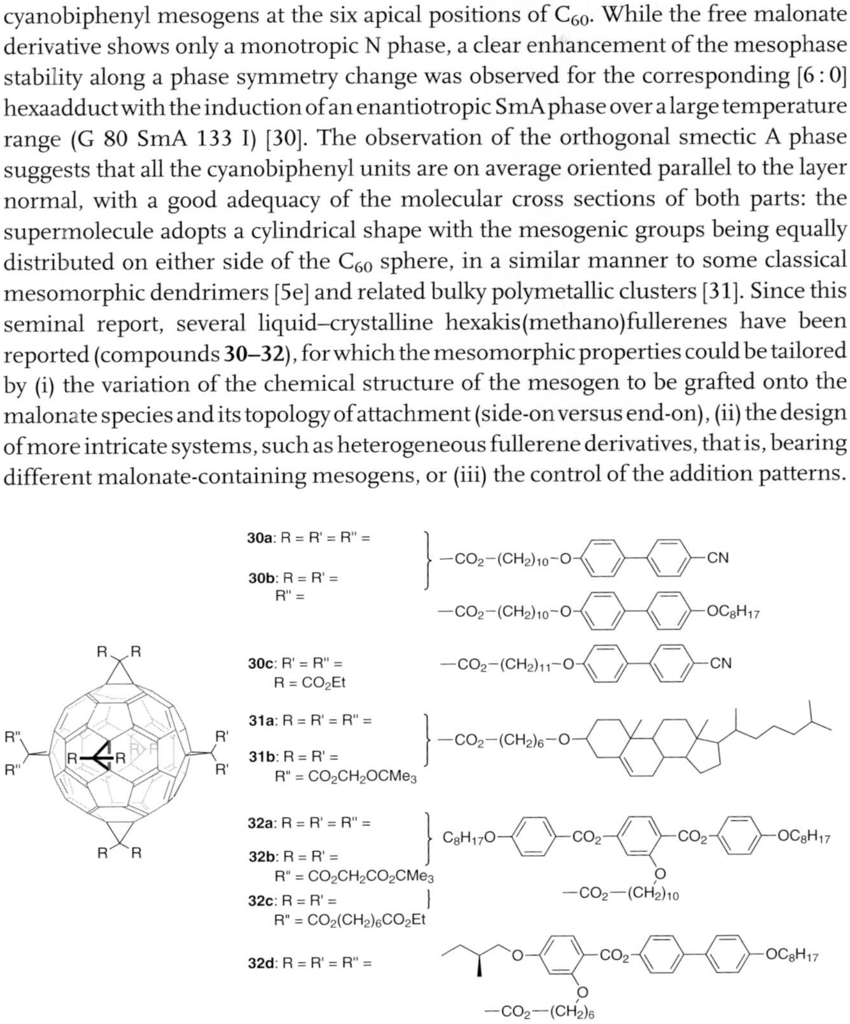

30a: R = R' = R'' =
30b: R = R' = } $-CO_2-(CH_2)_{10}-O-C_6H_4-C_6H_4-CN$
R'' = $-CO_2-(CH_2)_{10}-O-C_6H_4-C_6H_4-OC_8H_{17}$
30c: R' = R'' = $-CO_2-(CH_2)_{11}-O-C_6H_4-C_6H_4-CN$
R = CO_2Et
31a: R = R' = R'' =
31b: R = R' = } $-CO_2-(CH_2)_6-O-$cholesteryl
R'' = $CO_2CH_2OCMe_3$
32a: R = R' = R'' =
32b: R = R' =
R'' = $CO_2CH_2CO_2CMe_3$
32c: R = R' =
R'' = $CO_2(CH_2)_6CO_2Et$
$C_8H_{17}O-C_6H_4-CO_2-C_6H_3(O-CO_2-(CH_2)_{10})-CO_2-C_6H_4-OC_8H_{17}$
32d: R = R' = R'' = $-O-C_6H_3(O-CO_2-(CH_2)_6)-CO_2-C_6H_4-C_6H_4-OC_8H_{17}$

On the basis of a molecular design similar to that of compound **30a**, Felder-Flesch *et al.* reported the synthesis and mesomorphism of 2 hexaadducts of C_{60}, [6 : 0] and [5 : 1], containing either 12 mesogenic cholesterol units or 10 mesogenic cholesterol units and 2 LC-inert groups, respectively (**31a** and **31b**) [32]. The mesomorphic behavior was found quasi-similar for both supramolecular systems despite the different molecular symmetries (T_h for **31a**, C_{2v} for **31b**). Indeed, both compounds exhibit a broad SmA phase, from just above room temperature up to the transition to the isotropic liquid (**31a**: G 45 SmA 165 I; **31b**: G 40 SmA 180 I), quite different from the thermal behavior of the malonate mesogenic promoter for which a chiral nematic phase is observed, above two unidentified mesophases (M1 42, M2 67 N* 88 I). The supramolecular organization of these supermolecules within the smectic layers was revealed by small-angle X-ray diffraction and supported by molecular dynamics. It consisted of an overall smectic bilayer structure ($d = 51.1$ and 51.7 Å for **31a** and **31b**, respectively), resulting from the lateral registry of the supermolecules in cylindrical conformation, in which one of the active moieties (C_{60}) is confined in the median sublayer, sandwiched by two outer layers of cholesteryl groups; each single layer is separated by molten aliphatic chains. Moreover, and because of the quasi-spherical structure of C_{60}, the latter paves the sublayer according to a 2D hexagonal lattice, as deduced from the observation of a strong scattering corresponding to the average in-plane distance between adjacent fullerene adducts (Figure 9.9).

The heterogeneous [5 : 1] compound **30b**, bearing malonate-containing cyanobiphenyl and octyloxybiphenyl derivatives as mesomorphic promoters, obtained after a judicious design consisting in two consecutive addition steps (monoaddition of the first malonate, followed by multiaddition of the second malonate), displays a SmA phase (G 80 SmA 151 I) [33]. The mesophase is greatly stabilized with respect to the monotropic behavior of both malonate precursors (N and SmA, respectively). As reported above, the supermolecule adopts a cylindrical shape to self-assemble into a layer-like structure, with half the total number of mesogenic units on either sides of the central C_{60}, but with a random distribution of both types of mesogens. In contrast, the reduction in the number of side-mesogens as in the mixed [4 : 2] hexaadduct **30c** functionalized at the two poles of the sphere by a similar malonate mesogen was disastrous for mesomorphism, with the formation of a transient nematic phase (Cr 85 N 157 I), that is, observed only during the first heating [34]. This poor tendency for self-organization is a consequence of the mismatch between the carbonaceous sphere and the end-on mesogens cross sections (lower mesogenic content).

A nematic phase was induced in various hexaadducts bearing mesogens laterally attached to the fullerene core (compounds **32a**, **32b**, **32c**) [35]. The malonate derivative with side-on mesogens is not mesomorphic, while the corresponding hexaadducts show a room-temperature nematic phase, above the glass transition temperature (**32a**: G 13 N 60 I; **32b**: G 15 N 47 I; **32c**: G 15 N 46 I). The mixed addition [5 : 1] affects only slightly the mesophase stability, the clearing temperatures of which were found to decrease with the decreasing mesogenic character of the supermolecules ([6 : 0] $\rightarrow$ [5 : 1]). Small-angle X-ray diffraction on oriented samples revealed the presence of 3D short-range cybotactic clusters in the nematic phase, and that the same local arrangement is preserved throughout the series. The nematic organization consists of an alternation of sublayers containing the fullerene cores (two strata of C_{60}) with sublayers containing both the mesogens and the aliphatic spacers, with the aliphatic parts filling the accessible free volume between these sublayers. In this model, supported by molecular dynamics simulations, the supermolecules take the shape

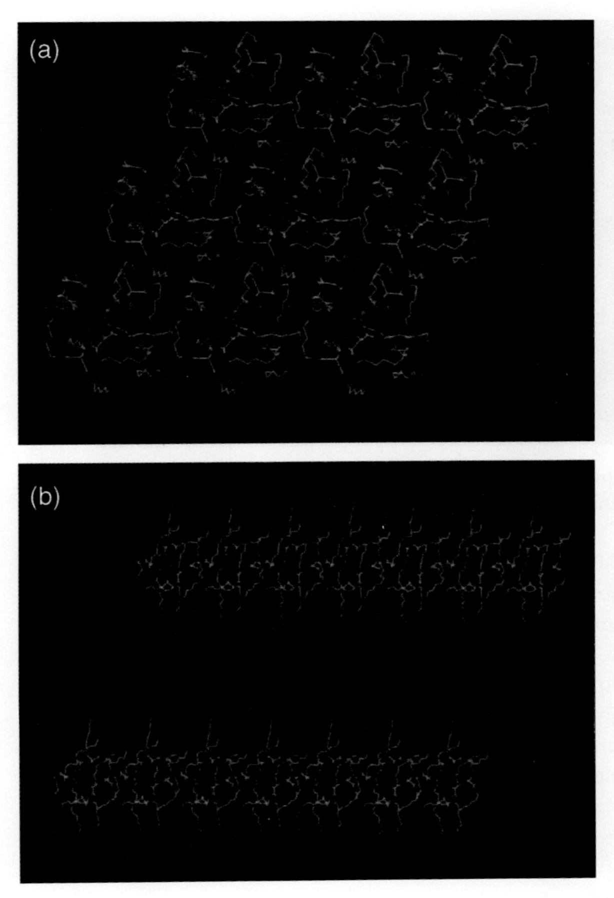

Figure 9.9 Short-range hexagonal lattice obtained from **31a** by molecular modeling: (a) top view and (b) side view.

of a "squid" with most of the side mesogens aggregating in one hemisphere (located into one monolayer), facilitating, as such, interaction of C_{60} by fragmental surface area (Figure 9.10) to stabilize this 3D edifice locally.

In order to explore the influence of polyaddition on the formation of chiral liquid–crystalline phases, a laterally branched optically active mesogenic malonate was used (G − 1 N* 133 I) to generate the corresponding chiral hexaadduct (**32d**) [36]. A highly birefringent texture that evolved to show the Grandjean planar texture and fingerprint defects characteristic of the chiral nematic phase (G 47 N* 103 I) was observed. The pitch length value of the left-handed helix ($p = 2.0\,\mu m$) is similar to that of the malonate precursor ($p = 1.9\,\mu m$) suggesting an effective shielding of the carbon sphere by the laterally attached mesogens. In the chiral nematic phase, the helical organization results in the organized packing of the supermolecules, no longer spherical. A random packing of the mesogenic units about the core is, however,

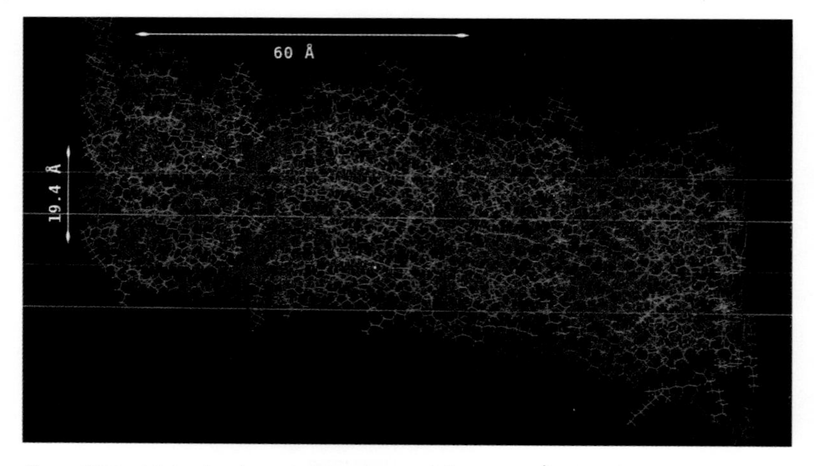

Figure 9.10 Molecular dynamics simulation of the average local packing of hexaadducts **32a–c** in the nematic phase by using the square lattice parameters.

unlikely since a packing arrangement relative to one another, both on the surface of the supermolecule and between individual supermolecules, is required. Therefore, either the mesogenic units are oriented parallel to one another and the chiral information is transmitted to the other mesogens as in the typical N* phase or, alternatively, for each supermolecule, the local director of the mesogens spirals around the C_{60} core to give poles at the top and bottom of the structure to form a molecular boojom (Figure 9.11), a likely scenario due to the short decoupling between the sphere and the mesogens. The chiral supermolecules then self-assemble together through chiral amplification and recognition processes into a helical supramolecular structure.

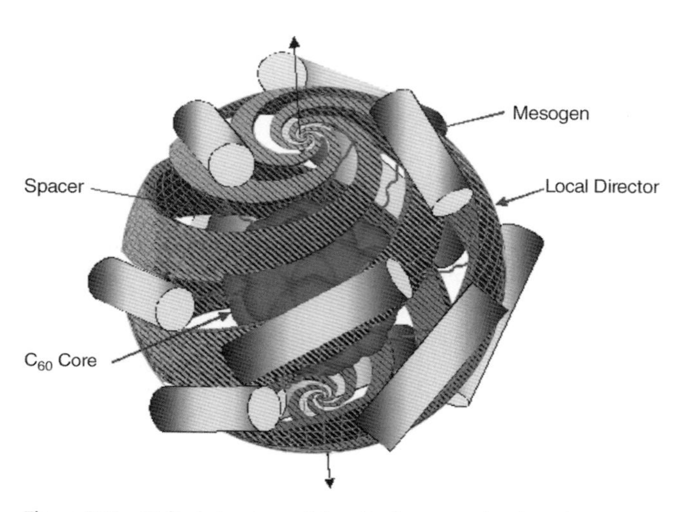

Figure 9.11 Helical structure of the chiral supermolecule **32d** in the N* phase.

9.3.5
Conical-Like "Shuttlecock" [60]Fullerenes

The discovery of the reaction between C_{60} and a Grignard compound with the general formula RMgBr (R = aryl, alkyl) in the presence of copper(I) [37] to produce penta-adducts of [60]fullerene, with a pseudo-C_{5v} symmetry, thereafter abbreviated as $[C_{60}R_5X]$ (X = H, Me, compounds **33a-c**), opened considerably the possibility to design original supramolecular materials by the introduction of a large variety of organic functionalities for further derivatization [38]. In this reaction, the five R groups are introduced selectively around one pentagon of C_{60}, converted into a cyclopentadiene ring, forming a unique cavity structure directly protruding from the fullerene core. This original design strategy for cone-shaped materials may be viewed as an attractive alternative to generate polar LC materials [39]. Thus, the Grignard compound, derived from 4-bromophenol protected with tetrahydropyranyl (THP), was reacted in the presence of a copper reagent to [60]fullerene to yield, after deprotection of the THP groups, the penta(4-hydroxyl)[60]fullerene ($[C_{60}(C_6H_4\text{-}OH)_5H]$) in high overall yields. This compound can be further derivatized by acylation reaction of the five diverging hydroxyl groups with appropriate aromatic acid chlorides to give the penta(ester) liquid–crystalline materials **33a**$_n$.

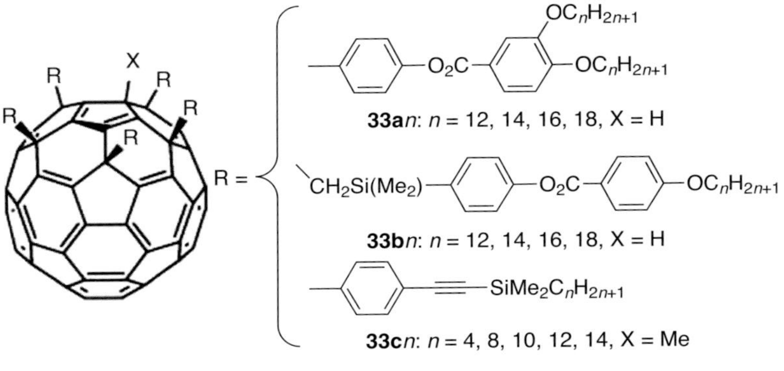

R = $-C_6H_4-O_2C-C_6H_3(OC_nH_{2n+1})_2$

33an: n = 12, 14, 16, 18, X = H

R = $CH_2Si(Me_2)-C_6H_4-O_2C-C_6H_4-OC_nH_{2n+1}$

33bn: n = 12, 14, 16, 18, X = H

R = $-C_6H_4-C{\equiv}C-SiMe_2C_nH_{2n+1}$

33cn: n = 4, 8, 10, 12, 14, X = Me

All derivatives **33a**$_n$, with n = 12, 14, 16, and 18, exhibit broad thermotropic mesophases, between the glassy state and the isotropic liquid, from −47 to −7 and from 144 to 132 °C, respectively, with the stepwise decrease in the mesophase stability temperature range as the chain length is increased [40]. These compounds additionally self-organize into lyotropic hexagonal and columnar nematic phases when mixed with various amounts of dodecane. Small-angle X-ray diffraction analysis confirmed unambiguously the formation of a columnar hexagonal phase for the four compounds, with a lattice parameter in the range of 35 Å, resulting from head-to-tail stacking of the molecules into polar columnar assemblies, in agreement with the measured stacking periodicity of about 14.5 Å: the cavity shaped by the five R substituents ensures the tight fitting of neighbored pentaadducts. Such a supramolecular polar stacking, already

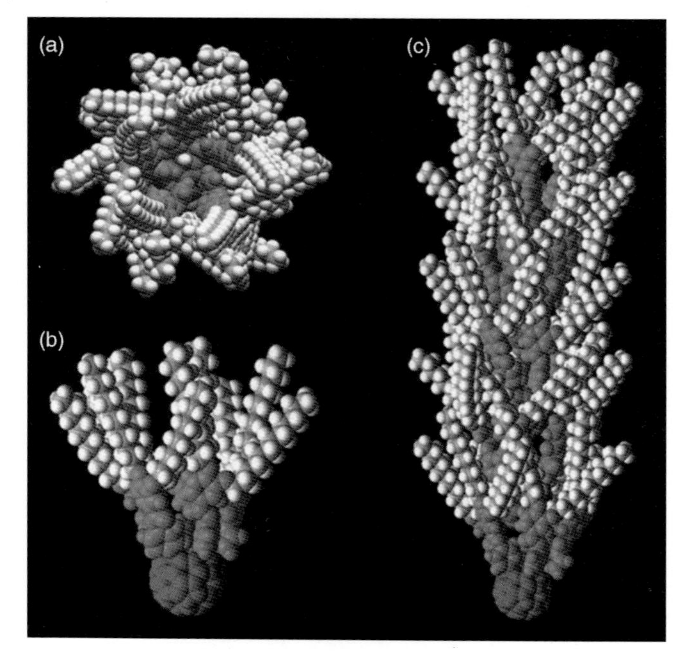

Figure 9.12 Modelized representations of conical-like [60]fullerene pentaadduct derivatives: (a) top view and (b) side view of **32a** *12*; (c) sketch of the stacking into columns as deduced by X-ray diffraction and moelcular dynamics.

present in the crystal form of a model compound, is driven by attractive interactions between fullerene moieties and the hollow cone formed by the five anisotropic groups of the neighboring molecule (of the same column); the numerous aliphatic chains wrapping the columns promote their arrangement into hexagonal arrays (Figure 9.12) and confer fluidity.

A remarkable feature of these pentaadducts is the cyclopentadienyl part that can serve as cyclopentadienyl ligand for metal complexation [41], providing that the cone solid angle is wide enough to permit its incorportaion. Implementing this concept, Nakamura and coworkers [42] developed conical columnar metallomesogens [43] integrating fullerene and ferrocene units. Thus, the liquid–crystalline buckyferrocenes $\mathbf{34}_n$ were prepared by complexation of a functionalizable pentaadduct with $[FeCp(CO)_2]_2$, followed by deprotection of the apical phenols ($Fe[C_{60}(C_6H_4\text{-}OH)_5]Cp$) and susbsequent acylation with aromatic acid chlorides. The presence of ferrocene within the conical cavity makes it shallower than in the homologous compounds $\mathbf{33a}_n$ and provides more disorder in the stacking structure (as observed in the crystal phase). This relatively weaker stacking had profound consequences for the mesomorphism as only two buckyferrocenes exhibit a columnar phase (**34** *16*: Cr −3, Col 123 I; **34** *18*: Cr 21, Col 120 I); the two shorter chain length homologues melt directly into the isotropic liquid at ~123–124 °C. X-ray diffraction experiments performed in the mesophase also confirmed a larger periodicity of the stacking, with respect to compounds **33**$_n$, with a value of 18.2–18.3 Å, and interlayer spacing increase with the chain length from 38.3 to 41.3 Å for **34** *16* and **34** *18*, respectively; the symmetry of the mesophase could however not be solved. The corresponding buckyferrocenium salt ([**34** *18*][$SbCl_6$]) also forms a mesophase between 25 and 161 °C. All these compounds exhibit interesting multielectron redox behavior, accepting or donating at least four electrons.

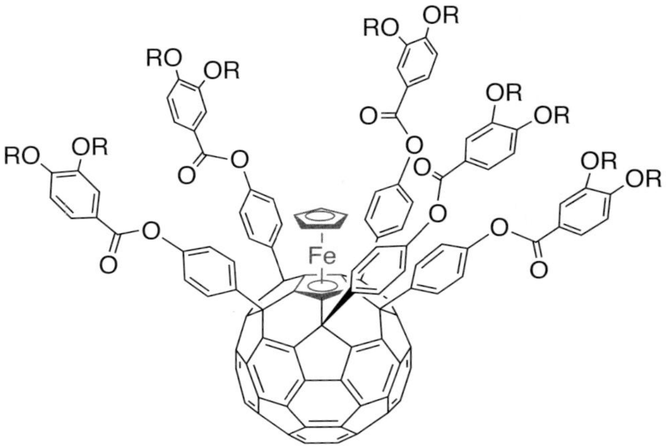

R = C_nH_{2n+1}, n = 12, 14, 16, 18

34

Enhancement of molecular flexibility and mobility within the columnar structures to improve mesophase stability and loading capacity were achieved through the insertion of a (dimethyl)silylmethylene joint between the fullerene core and the connecting mesogen units [44]. Consequently, new "tentacular" mesogens **33b**$_n$ were obtained after acylation of the hydroxylated precursory pentaadducts ([C_{60}(CH_2Si-$Me_2C_6H_4$-OH)$_5$H]) with alkoxybenzoyl chlorides. The materials, which possess a wider cavity (a cup-shaped cavity) than the **33a**$_n$ compounds, exhibit a columnar hexagonal phase (lattice parameter of about 38.5 Å) over broad temperature ranges between the glass transition temperature (rising from 2 to 35 °C) and the clearing temperature (falling in the range between 176 and 187 °C) as the carbon chain length increases. The mesophase stability was augmented by about 40 °C with respect to that of the **33a**$_n$ derivatives despite the reduction in divergent chains number. This increase in stability was attributed to the shortening of intracolumnar spacing (about 12.9 Å), corresponding to an important intermolecular interactions increase, and to the increase in the conformational freedom, with a more efficient lateral spreading of the mesogens. Both features contribute to a stronger 1D columnar stacking and to the substantial enhancement of the microphase separation between rigid and aliphatic chains, and thus to the net stabilization of the columnar phases. Increasing the number of divergent chains on the mesogenic unit parts provoked the complete destruction of mesomorphism.

Diminishing the length of the mesogenic arms to diminish the ability of the pentaadduct to embrace the neighboring fullerene and thus to decrease the tendency of the 1D stacking to avoid the induction of columnar mesophase was also attempted. Thus, a homologous series of compounds ($[C_{60}(C_6H_4\text{-}CCSiMe_2C_nH_{2n+1})_5Me]$) was prepared by the reaction of C_{60} with suitable arylmagnesium bromide derivatives containing long chains at one end (compounds **33c**$_n$) [45]. A methyl group was also installed in the cavity to prevent head-to-tail stacking. Investigation of the mesomorphic properties of compounds **33c**$_n$ revealed the formation of supramolecular lamellar assemblies, the transition temperatures of which are chain-length dependent. While the smectic phase appears between 140 and 223 °C for **33c** *8*, the mesophase exists over broader temperature ranges for the other terms of the series ($\Delta T = 115$–130 °C), with a stepwise decrease in both the glass transition (48 $\rightarrow$ 11 °C) and the clearing (178 $\rightarrow$ 126 °C) temperatures when n is raised from 10 to 14. The interlayer spacing increases monotonously with chain-length from 17.2 (**33c** *8*) to 25.8 Å (**33c** *14*) (intermediate values are $d = 20.5$ and 23.9 Å for **3c** *10* and **3c** *12*, respectively). The model proposed for the smectic organization, based on the crystal packing of the short homologues **33c** *1* and **33c** *4*, and in agreement with the stepwise variation in the lamellar periodicity, consists of the self-assembling of the C_{60}, close-packed into a sheet, arranged laterally in a probable square-like lattice, with an alternation of the pendant groups to occupy the available volume on both sides of the layer; aliphatic sublayers separate the carbonaceous sheets from each other (Figure 9.13).

Modification of the initial synthetic conditions of addition unexpectedly led to decaadduct materials $[C_{60}R_{10}X_2]$ (X = H or Me) with a pseudo-D_{5h} symmetry for the major compound. Such decaadducts present additional unique features with respect to the pentaadduct systems. First, more complex, symmetrical and non-symmetrical, molecular structures can be accessed through specifically designed synthetic procedures. Second, the presence of two cyclopentadiene parts at the two opposite poles provides a scaffold for the synthesis of dinuclear compounds. And third, the selective destruction of the spherical π electronic system of the initial C_{60} induces the formation of a hoop-shaped cyclic π electron-conjugated benzenoid system, belonging to the family of the [10]cyclophenacenes. The first series of decaadduct compounds, with the formula $[C_{60}(C_6H_4\text{-}CCSiMe_2C_nH_{2n+1})_{10}Me_2]$), was synthesized (Figure 9.14) [46]. The deca(octadecyl) compound ($n = 18$) exhibits a smectic phase between 17 and 257 °C, with a periodicity of 34.8 Å. The "bow-tie" decaadduct molecules self-assemble into layers, the main molecular axis of which is perpendicular to the layer plane: the fullerene cores are arranged in a central sheet, with some degrees of in-plane ordering, sandwiched between aliphatic outer layers. Decreasing chain length results in the formation of a 3D smectic phase ($n = 12$, Cr 114 Sm 310 I) and to crystal phases ($n = 4$ and 8). They also exhibit anisotropic luminescence that makes them useful in the design of organic light-emitting diode systems.

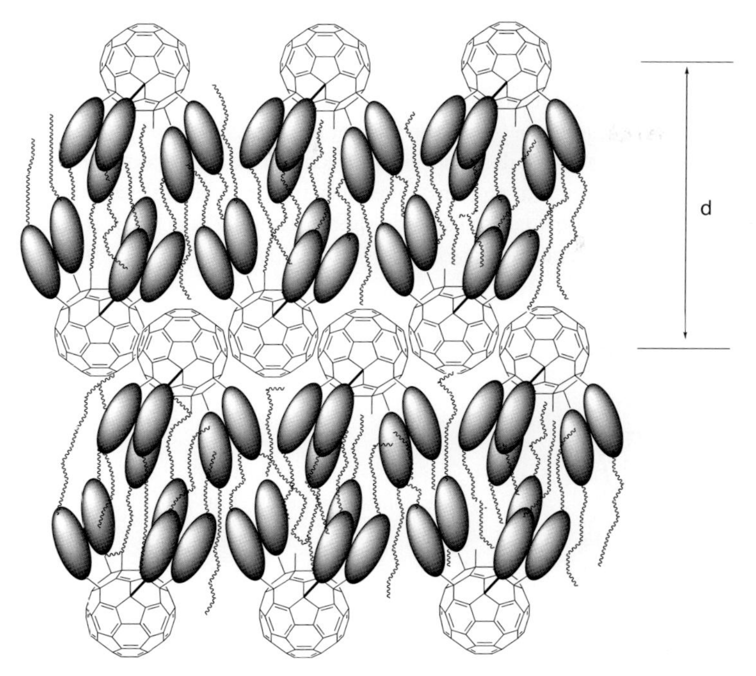

Figure 9.13 Schematic representation of the supramolecular lamellar organization of compounds **33c**$_n$.

Nonsymmetrical "bow-tie" decaadduct buckyferrocenes ($Fe[C_{60}(C_6H_4\text{-}CCSiMe_2C_nH_{2n+1})_5Me]Me_5Cp$) were also prepared by interplay of synthetic methods developed for buckyferrocene and "bow-tie" molecules, starting with the synthesis of the pentamethylated buckyferrocene, followed by the selective pentaaddition of the

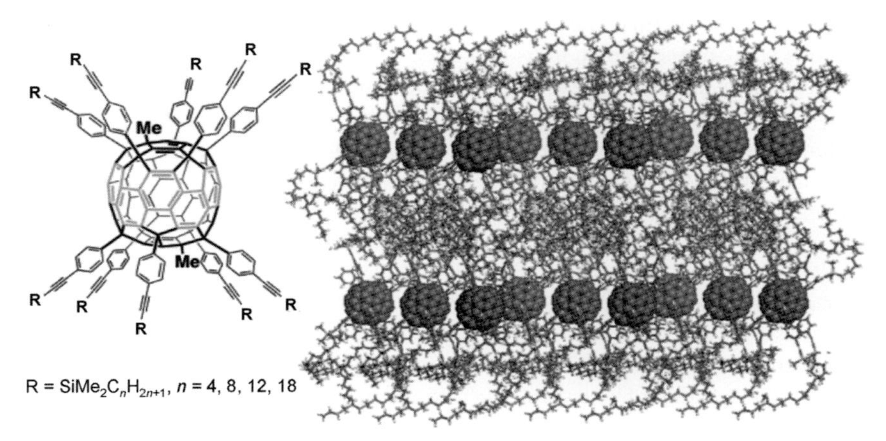

Figure 9.14 Molecular structure of the symmetrical decaadduct of C_{60}, and representation of the smectic phase organization.

protomesogens on the opposite pole [47]. These materials (**35**) combined self-assembling ability, along interesting optical and redox properties conferred by the integrated cyclophenacene and ferrocene at once. The two homologues of the series exhibit a 3D tetragonal mesophase over large temperature ranges ($n = 12$, Cr 55 LC 230 I; $n = 18$, LC 186 I). The molecules associate into tetrameric aggregate, canceling the molecular dipole, and then four aggregates self-assemble according to a centered tetragonal lattice. The tetragonal lattice expands slightly laterally ($a = 32$–32.4 Å and $c = 70.3$–70.9 Å) when the chain increases from 12 to 18.

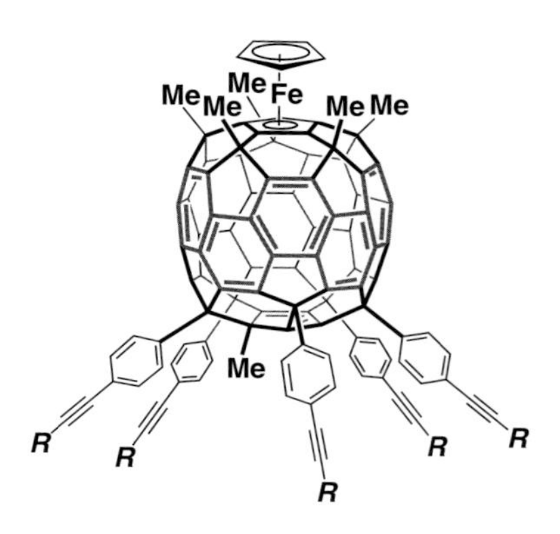

R = $SiMe_2C_nH_{2n+1}$, $n = 12$, 18

35

Thus, the above examples demonstrate that selected and judicious polyaddition of mesogenic molecules on the carbon sphere, used as a spherical template to create new 3D architectures, is a method of choice both to obtain stable anisotropic polypedal, conical and bow-tie materials and to control the C_{60} aggregation tendency. These approaches allow the preparation of a wide variety of symmetrical and nonsymmetrical [60]fullerene polyadducts with a variable and controlled number of active pendant mesogenic units and selected grafting positions. The control of these structural parameters (degree of addition and grafting positions) is essential to study the function–physical property relationships, and to govern the molecular conformation and mesogen orientations and directionalities, and thus the supramolecular organizations into various thermotropic mesophases.

9.4
Conclusions

[60]Fullerene-containing liquid crystals have reached a high degree of sophistication from the point of view of both their structure and mesomorphic properties. Owing to their unique photophysical and electrochemical characteristics, [60]fullerene-based liquid crystals are promising compounds for the development of nanotechnology by the bottom-up approach. Indeed, good studies have shown that liquid–crystalline

fullerenes can play a major role in electron transport [48] and photoconduction [49], and photoinduced electron and/or energy transfer has been achieved in donor–acceptor dyads when combined with ferrocene [50], tetrathiafulvalene (TTF) [51], or oligophenylenevinylene (OPV) [52].

Symbols

T_g	Glass transition
Cr	Crystalline or semicrystalline solid
N	Nematic phase
N*	Chiral nematic phase
SmA	Smectic A phase
SmC	Smectic C phase
Col_h	Hexagonal columnar phase
Col_r	Rectangular columnar phase
$Col_{r\text{-}c2mm}$	Rectangular columnar phase of *c2mm* symmetry
$Col_{r\text{-}p2gg}$	Rectangular columnar phase of *p2gg* symmetry
I	Isotropic liquid

Temperatures are given in °C.

Acknowledgments

R.D. acknowledges the Swiss National Science Foundation for financial support. D. G. and B.D. thank the University of Strasbourg and CNRS for their support.

References

1 Demus, D., Goodby, J.W., Gray, G.W., Spiess, H.-W., and Vill, V. (eds) (1998) *Handbook of Liquid Crystals*, Wiley-VCH Verlag GmbH, Weinheim.

2 (a) Kato, T., Mizoshita, N., and Kishimoto, K. (2006) *Angew. Chem. Int. Ed.*, **45**, 38; (b) Goodby, J.W., Saez, I.M., Cowling, S.J., Görtz, V., Draper, M., Hall, A.W., Sia, S., Cosquer, G., Lee, S.-E., and Raynes, E.P. (2008) *Angew. Chem. Int. Ed.*, **47**, 2754; (c) Goodby, J.W., Saez, I.M., Cowling, S.J., Gasowska, J.S., MacDonal, R.A., Sia, S., Watson, P., Toyne, K.J., Hird, M., Lewis, R.A., Lee, S.-E., and Vaschenko, V. (2009) *Liq. Cryst.*, **36**, 567.

3 (a) Martín, N., Sanchez, L., Illescas, B., and Pérez, I. (1998) *Chem. Rev.*, **98**, 2527; (b) Echegoyen, L. and Echegoyen., L.E. (1998) *Acc. Chem. Res.*, **31**, 593; (c) Diederich, F. and Gómez-López, M. (1999) *Chem. Soc. Rev.*, **28**, 263.

4 Guldi, D.M. (2000) *Chem. Commun.*, 321.

5 (a) Guillon, D. and Deschenaux, R. (2002) *Curr. Opin. Solid State Mater. Sci.*, **6**, 515; (b) Marcos, M. and Omenat and Serrano, J.-L. (2003) *C. R. Chim.*, **6**, 947; (c) Barberá, J., Donnio, B., Gehringer, L., Guillon, D., Marcos, M., Omenat, A., and Serrano, J.-L. (2005) *J. Mater. Chem.*, **15**, 4093; (d) Donnio, B. and Guillon, D. (2006) *Adv. Polym. Sci.*, **201**, 45;

(e) Donnio, B., Buathong, S., Bury, I., and Guillon, D. (2007) *Chem. Soc. Rev.*, **36**, 1495; (f) Marcos, M., Martín-Rapún, R., Omenat, A., and Serrano, J.-L. (2007) *Chem. Soc. Rev.*, **36**, 1889; (g) Buathong, S., Gehringer, L., Donnio, B., and Guillon, D. (2009) *C.R. Chim.*, **12**, 138.

6 (a) Deschenaux, R., Donnio, B., and Guillon, D. (2007) *New J. Chem.*, **31**, 1064; (b) Guillon, D., Donnio, B., and Deschenaux, R. (2009) in *Fullerene Polymers: Synthesis, Properties and Applications* (eds N. Martín and F. Giacalone), Wiley-VCH Verlag GmbH, Weinheim, pp. 247–270.

7 Hardie, M.J. and Raston, C.L. (1999) *Chem. Commun.*, 1153.

8 (a) Atwood, J.L., Barnes, M.J., Gardiner, M.G., and Raston, C.L. (1996) *Chem. Commun.*, 1449; (b) Matsubara, H., Hasegawa, A., Shiwaku, K., Asano, K., Uno, M., Takahashe, S., and Yamamoto, K. (1998) *Chem. Lett.*, 923; (c) Nierengarten, J.-F., Oswald, L., Eckert, J.-F., Nicoud, J.-F., and Armaroli, N. (1999) *Tetrahedron Lett.*, **40**, 5681.

9 (a) Malthête, J. and Collet, A. (1987) *J. Am. Chem. Soc.*, **109**, 7544; (b) Poupko, R., Luz, Z., Spielberg, N., and Zimmermann, H. (1989) *J. Am. Chem. Soc.*, **111**, 6094; (c) Lunkwitz, R., Tschierske, C., and Diele, S. (1997) *J. Mater. Chem.*, **7**, 2001.

10 Felder, D., Heinrich, B., Guillon, D., Nicoud, J.-F., and Nierengarten, J.-F. (2000) *Chem. Eur. J.*, **6**, 3501.

11 De Vries, A. (1970) *Mol. Cryst. Liq. Cryst.*, **10**, 219.

12 (a) Davidson, P., Levelut, A.-M., Strzelecka, H., and Gionis, V. (1983) *J. Phys. Lett.*, **44**, 823; (b) El-ghayoury, A., Douce, L., Skoulios, A., and Ziessel, R. (1998) *Angew. Chem. Int. Ed.*, **37**, 1255.

13 (a) Koch, E. and Fischer, W. (1993) *Acta Cryst. A*, **49**, 209; (b) Fischer, W. and Koch, E. (1996) *Phil. Trans R. Soc. Lond. A*, **354**, 2105; (c) Schwarz, U.S. and Gompper, G. (1999) *Phys. Rev. E*, **59**, 5528.

14 Kimura, M., Shiba, T., Yamazaki, M., Hanabusa, K., Shirai, H., and Kobayashi, N. (2001) *J. Am. Chem. Soc.*, **123**, 5636.

15 Kimura, M., Saito, Y., Ohta, K., Hanabusa, K., Shirai, H., and Kobayashi, N. (2002) *J. Am. Chem. Soc.*, **124**, 5274.

16 Arikainen, E.O., Boden, N., Bushby, R.J., Lozman, O.R., Vinter, J.G., and Wood, A. (2000) *Angew. Chem. Int. Ed.*, **39**, 2333.

17 Bushby, R.J., Hamley, I.W., Liu, Q., Lozman, O.R., and Lydon, J.E. (2005) *J. Mater. Chem.*, **15**, 4429.

18 (a) de la Torre, G., Vázquez, P., Agulló-López, F., and Torres, T. (2004) *Chem. Rev.*, **104**, 3723; (b) Gouloumis, A., Liu, S.-G., Sastre, A., Vázquez, P., Echegoyen, L., and Torres, T. (2000) *Chem. Eur. J.*, **6**, 3600.

19 de la Escosura, A., Martínez-Díaz, M.V., Barberá, J., and Torres, T. (2008) *J. Org. Chem.*, **73**, 1475.

20 Dardel, B., Guillon, D., Heinrich, B., and Deschenaux, R. (2001) *J. Mater. Chem.*, **11**, 2814.

21 Campidelli, S., Lenoble, J., Barberá, J., Paolucci, F., Marcaccio, M., Paolucci, D., and Deschenaux, R. (2005) *Macromolecules*, **38**, 7915.

22 Laschat, S., Baro, A., Steinke, N., Giesselmann, F., Hägele, C., Scalia, G., Judele, R., Kapatsina, E., Sauer, S., Schreivogel, A., and Tosoni, M. (2007) *Angew. Chem. Int. Ed.*, **46**, 4832.

23 Rosen, B.M., Wilson, C.J., Wilson, D.A., Peterca, M., Imam, M.R., and Percec, V. (2009) *Chem. Rev.*, **109**, 6275.

24 Lenoble, J., Maringa, N., Campidelli, S., Donnio, B., Guillon, D., and Deschenaux, R. (2006) *Org. Lett.*, **8**, 1851.

25 Maringa, N., Lenoble, J., Donnio, B., Guillon, D., and Deschenaux, R. (2008) *J. Mater. Chem.*, **18**, 1524.

26 Lenoble, J., Campidelli, S., Maringa., N., Donnio, B., Guillon, D., Yevlampieva, N., and Deschenaux, R. (2007) *J. Am. Chem. Soc.*, **129**, 9941.

27 Nguyen, H.-T., Destrade, C., and Malthête, J. (1997) *Adv. Mater.*, **9**, 375.

28 Richardson, R.M., Ponomarenko, S.A., Boiko, N.I., and Shibaev, V.P. (1999) *Liq. Cryst.*, **26**, 101.

29 Rueff, J.-M., Barberá, J., Donnio, B., Guillon, D., Marcos, M., and Serrano, J.-L. (2003) *Macromolecules*, **36**, 8368.

30 Chuard, T., Deschenaux, R., Hirsch, A., and Schönberger, H. (1999) *Chem. Commun.*, 2103.

31 (a) Terazzi, E., Bourgogne, C., Welter, R., Gallani, J.-L., Guillon, D., Rogez, G., and Donnio, B. (2008) *Angew. Chem. Int. Ed.*, **47**, 490; (b) Molard, Y., Dorson, F., Cîrcu, V., Roisnel, T., Artzner, F., and Cordier, S. (2010) *Angew. Chem. Int. Ed.*, **49**, 3351.

32 Felder-Flesch, D., Rupnicki, L., Bourgogne, C., Donnio, B., and Guillon, D. (2006) *J. Mater. Chem.*, **16**, 304.

33 Gottis, S., Kopp, C., Allard, E., and Deschenaux, R. (2007) *Helv. Chim. Acta.*, **90**, 957.

34 Tirelli, N., Cardullo, F., Habicher, T., Suter, U.W., and Diederich, F. (2000) *J. Chem. Soc. Perkin Trans. 2*, 193.

35 (a) Mamlouk, H., Heinrich, B., Bourgogne, C., Donnio, B., Guillon, D., and Felder-Flesch, D. (2007) *J. Mater. Chem.*, **17**, 2199; (b) Mamlouk-Chaouachi, H., Heinrich, B., Bourgogne, C., Guillon, D., Donnio, B., and Felder-Flesch, D. (2011) *J. Mater. Chem.*, **21**, 9121.

36 Campidelli, S., Brandmüller, T., Hirsch, A., Saez, I.M., Goodby, J.W., and Deschenaux, R. (2006) *Chem. Commun.*, 4282.

37 Nakamura, E. (2004) *J. Organomet. Chem.*, **689**, 4630.

38 Matsuo, Y. and Nakamura, E. (2008) *Chem. Rev.*, **108**, 3016.

39 (a) Xu, B. and Swager, T.M. (1993) *J. Am. Chem. Soc.*, **115**, 1159; (b) Serrano, J.-L. and Sierra, T. (2003) *Coord. Chem. Rev.*, **242**, 73; (c) Gorecka, E., Pociecha, D., Mieczkowski, J., Matraszek, J., Guillon, D., and Donnio, B. (2004) *J. Am. Chem. Soc.*, **126**, 15946; (d) Kishikawa, K., Nakahara, S., Nishikawa, Y., Kohmoto, S., and Yamamoto, M. (2005) *J. Am. Chem. Soc.*, **127**, 2565; (e) Takezoe, H., Kishikawa, K., and Gorecka, E. (2006) *J. Mater. Chem.*, **16**, 2412.

40 Sawamura, M., Kawai, K., Matsuo, Y., Kanie, K., Kato, T., and Nakamura, E. (2002) *Nature*, **419**, 702.

41 Sawamura, M., Iikura, H., and Nakamura, E. (1996) *J. Am. Chem. Soc.*, **118**, 12850.

42 Matsuo, Y., Muramatsu, A., Kamikawa, Y., Kato, T., and Nakamura, E. (2006) *J. Am. Chem. Soc.*, **128**, 9586.

43 Donnio, B., Guillon, D., Bruce, D.W., and Deschenaux, R. (2006) in *"Metallomesogens" Comprehensive Organometallic Chemistry III: From Fundamentals to Applications – Applications III: Functional Materials, Environmental and Biological Applications*, Elsevier, Oxford, UK, p. 195.

44 Matsuo, Y., Muramatsu, A., Hamasaki, R., Mizoshita, N., Kato, T., and Nakamura, E. (2004) *J. Am. Chem. Soc.*, **126**, 432.

45 Zhong, Y.-W., Matsuo, Y., and Nakamura, E. (2007) *J. Am. Chem. Soc.*, **129**, 3052.

46 Li, C.-Z., Matsuo, Y., and Nakamura, E. (2009) *J. Am. Chem. Soc.*, **131**, 17058.

47 Li, C.-Z., Matsuo, Y., and Nakamura, E. (2010) *J. Am. Chem. Soc.*, **132**, 15514.

48 Nakanishi, T., Shen, Y., Wang, J., Yagai, S., Funahashi, M., Kato, T., Fernandes, P., Möhwald, H., and Kurth, D.G. (2008) *J. Am. Chem. Soc.*, **130**, 9236.

49 Li, W.-S., Yamamoto, Y., Fukushima, T., Saeki, A., Seki, S., Tagawa, S., Masunaga, H., Sasaki, S., Takata, M., and Aida, T. (2008) *J. Am. Chem. Soc.*, **130**, 8886.

50 (a) Even, M., Heinrich, B., Guillon, D., Guldi, D.M., Prato, M., and Deschenaux, R. (2001) *Chem. Eur. J.*, **7**, 2595; (b) Campidelli, S., Vázquez, E., Milic, D., Prato, M., Barberá, J., Guldi, D.M., Marcaccio, M., Paolucci, D., Paolucci, F., and Deschenaux, R. (2004) *J. Mater. Chem.*, **14**, 1266; (c) Campidelli, S., Séverac, M., Scanu, D., Deschenaux, R., Vázquez, E., Milic, D., Prato, M., Carano, M., Marcaccio, M., Paolucci, F., Aminur Rahman, G.M., and Guldi, D.M. (2008) *J. Mater. Chem.*, **18**, 1504; (d) Dvinskikh, S.V., Yamamoto, K., Scanu, D., Deschenaux, R., and Ramamoorthy, A. (2008) *J. Phys. Chem. B*, **112**, 12347.

51 Allard, E., Oswald, F., Donnio, B., Guillon, D., Delgado, J.L., Langa, F., and Deschenaux, R. (2005) *Org. Lett.*, **7**, 383.

52 Campidelli, S., Deschenaux, R., Eckert, J.-F., Guillon, D., and Nierengarten, J.-F. (2002) *Chem. Commun.*, 656.

10
Supramolecular Chemistry of Fullerenes on Solid Surfaces

Roberto Otero, José María Gallego, Nazario Martín, and Rodolfo Miranda

10.1
Introduction

The fascinating chemical, electrochemical, and optical properties of the spherical molecules known as fullerenes have motivated many investigations during the past 20 years [1], resulting in a very complete understanding of the covalent chemistry of these carbon allotropes [2]. Supramolecular chemistry of fullerenes has also been developed to a great extent and, because of their singular geometry, these spherical molecules have been used both as hosts and guests in a wide variety of supramolecular ensembles [3].

The interesting properties of fullerenes make them ideal candidates for a large number of device applications. However, for the functioning of most of these devices, fullerenes must be in contact with solid supports and electrodes, raising the question of the role the solid support plays in the structural, electronic, and chemical properties of the deposited fullerenes [4]. The surface science of adsorbed fullerenes has thus also suffered a tremendous burst, aided by powerful, new microscopy techniques such as scanning tunneling microscopy (STM) or atomic force microscopy (AFM), and the full battery of time-honored surface science techniques. Obviously, the 2D arrangements of fullerenes and their derivatives deposited on solid surfaces arise from the competition between weak noncovalent intermolecular forces (such as van der Waals or dispersive forces) and molecule–substrate interactions, with the crystalline symmetry of the surface often playing an important role [4]. However, only recently our understanding of these processes has allowed us to engineer new supramolecular structures made of fullerenes exploiting not only the nonbonding interactions between the fullerenes themselves but also the site-selective interactions between fullerenes and nanoscale patterns on solid surfaces.

In this chapter, we present an overview of the state of the art in the 2D supramolecular chemistry of fullerenes on solid surfaces, with special emphasis on the new opportunities offered by the templating of solid surfaces at the

Supramolecular Chemistry of Fullerenes and Carbon Nanotubes, First Edition. Edited by Nazario Martín and Jean-Francois Nierengarten.

nanoscale. Such templates might arise naturally on surfaces through the phenomenon of *surface reconstruction*, that is, the rearrangement of the atomic structure at the surface of a solid with respect to the ideal bulk termination of the crystal. For many surfaces such as Si(111) or Au(111), such reconstructions occur spontaneously after surface recrystallization, whereas in some other cases they can be induced by the chemisorption of small molecules such as O_2, N_2, or CO. A different possibility to fabricate nanoscale templates on solid surfaces is the formation of self-assembled molecular networks. Both methods have been used to fabricate monodisperse fullerene clusters, chains, and islands, demonstrating the power of templating for the fabrication of nanoscale supramolecular fullerene structures.

10.2 Fullerenes on Nonpatterned Metal Surfaces

10.2.1 Nature and Strength of Fullerene–Metal Interactions

Before going into the specifics of site-selective adsorption on nanotemplates, it is worth to say a few words about the nature and strength of the interaction between fullerenes and metal surfaces. Besides the fundamental importance of this knowledge for the topic at hand, metal/fullerene interfaces are at the heart of any foreseeable fullerene-based electronic or optoelectronic device since charge transfer processes across the interface may control charge injection in the active device material. We have recently shown that these charge transfer processes in strong acceptors might lead even to new substrate-mediated interactions that dominate the supramolecular chemistry of the adsorbed molecules [5].

The nature of the interaction between fullerenes and metal single-crystal surfaces has been the subject of a number of recent studies. On Cu(111), high-resolution angle-resolved photoemission spectra experiments have found a dispersing photoemission peak with energies lying in the HOMO–LUMO gap, which has been interpreted as a signature of a strong hybridization between the copper and the carbon electronic states [6]. Apart from the hybridization between the d-states of the metal surface and the π orbitals at the C_{60} cage, a significant charge transfer from the substrate to C_{60} is usually observed, with reported values ranging from 0.7 to 2 e/C_{60} [7]. A covalent interaction between C_{60} molecules and metallic surfaces is also supported by STM studies, showing a significant influence of the underlying metal on the electronic structure of the absorbed C_{60} [8, 9], and DFT calculations [10].

Hybridization and charge transfer indicate a strong molecule–substrate interaction that might lead to the rearrangement of surface atoms (surface reconstruction induced by C_{60} adsorption). For example, adsorption of fullerenes on the transition metal Pd(110) [11] and Ni(110) [12] promotes the creation of long-range ordered arrays of vacancy islands. The driving force for such a dramatic surface reconstruction is the closer, and therefore stronger, interaction between the metal surface and the lateral π electrons of the C_{60} cage. There is even ample evidence supporting surface reconstruction for the much less reactive noble metal surfaces [13–16], including close-packed ones [17–19].

10.2.2
Translational and Orientational Order of Fullerene Layers on Flat Metal Surfaces

The translational order shown by fullerene monolayers on flat metal surfaces is usually hexagonal close-packed, as expected due to the spherical shape of the molecule and its nonpolarity, which leaves only isotropic dispersive interactions as the possible source for intermolecular attraction. Fullerenes, however, having a 3D structure, can adopt a number of different adsorption geometries, depending on which part of the molecule (a hexagonal ring, a pentagonal ring, or a double bond in between) is in contact with the substrate. The ability of STM to image the molecular orbitals [20] has been exploited to distinguish the orientations of adsorbed fullerene molecules (see Figure 10.1). Most of these studies reveal random orientation [21] of the molecules forming the assembled clusters or monolayers, and only in few cases does the strong interaction between the C_{60} units and the solid surface result in simple orientational order with a uniform orientation [22] or ordered alternation of two orientations [23]. As we will discuss later, adsorption of fullerenes on ordered molecular layers might also lead to self-assembled monolayers in which all the C_{60} molecules show the same orientation (see Figure 10.1) [24].

In this regard, Berndt's group has recently found a long-range orientational order in a C_{60} monolayer on Au(111) by using STM at low temperature [25]. Remarkably, a unit cell formed by 49 C_{60} molecules adopts eleven different orientations (Figure 10.2). This unit cell can be divided into a faulted and an unfaulted half similar to the (7×7) reconstruction of Si(111). In this singular case, intermolecular interactions play a leading role in stabilizing the observed superstructure.

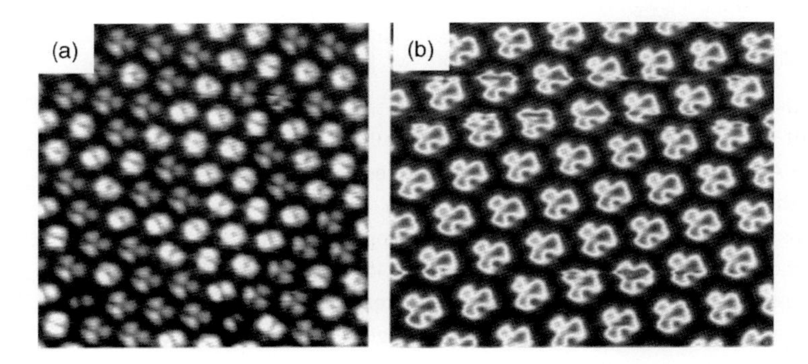

Figure 10.1 (a) STM image of an orientationally disordered C_{60} monolayer on Ag(111): $10 \times 10\,nm^2$, $V_{tip} = -2.0\,V$; (b) the $8 \times 7\,nm^2$ STM image of an orientationally ordered C_{60} deposited on top of a "pure face-on" monolayer of p-sexithiophene (6P) grown on Ag(111) ($V_{tip} = 0.5\,V$). Reproduced with permission of the American Chemical Society from Ref. [24].

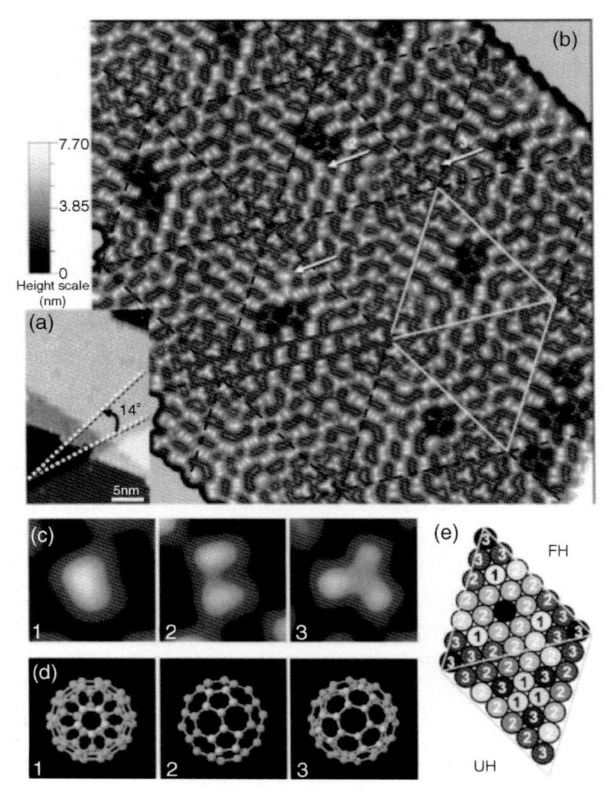

Figure 10.2 STM images ($I = 0.1$ nA, $V = 1.5$ V) of C_{60} islands deposited at room temperature on Au(111) (a) $32 \times 20\,nm^2$ and (b) pseudo-3D $20 \times 20\,nm^2$. The height scale corresponds to the highlighted rhombus (a) and (b) defines a unit cell. Added dashed lines indicate equivalent cells. (c) STM images of adsorbed C_{60} with one-, two-, or threefold symmetry axis ($1\,nm^2$), their corresponding top view models (d), and proposed C_{60} arrangement in the superlattice (e). Reproduced with permission of the American Physical Society from Ref. [25].

10.2.3
Conventional Approaches to 2D Fullerene Supramolecular Chemistry: Fullerene Functionalization

As described above, with a flat metallic substrate the spherical shape of fullerene molecules precludes the formation of more interesting translationally ordered phases other than the standard close-packed arrangement. A very similar approach to that used in solution chemistry can be used for steering the organization of fullerenes into more interesting lamella phases. Such method consist in the functionalization of the fullerene molecules with other groups that can either mediate new interactions (such as hydrogen bonding) or hinder the attractive dispersive interactions in some directions by rendering the molecular shape anisotropic.

We have recently investigated one example of the first approach, using the well-known PCBM (phenyl-C_{61}-butiric acid methyl ester) fullerene derivative, one of the most commonly used organic acceptors in heterojunction solar cells. This fullerene is provided with a butyric acid methyl ester chain capable of forming weak hydrogen bonds with the chain of neighboring molecules. While for low coverage the molecular arrangement is determined by substrate-specific adsorption (and will be discussed later) and dispersive interactions among the bulky C_{60} moieties, for high coverage the molecular tails readily form hydrogen bonds, creating a lamella structure, as depicted in Figure 10.3 [26].

Density functional theory calculations predict that two PCBM molecules are connected through two weak H bonds between the tails, leading to an energy gain of 2.19 kcal/mol in comparison with two isolated molecules. A further calculation of the tetramer by considering the presence of two additional H bonds between adjacent dimers resulted in an excellent agreement with the experimental findings (Figure 10.3). Interestingly, weak H bonds of type C−H···O have been a matter of discussion in the past 25 years, being recognized as a key aspect in many supramolecular organizations [27]. In our case, the weak H bonds between the methyl ester groups and between the phenyl rings and the carbonyl functionalities are responsible for the observed crossover site selectivity, confirming that they could be valuable tools in engineering molecular nanostructures at surfaces.

Similar lamella structures can also be found in the absence of H bonding. For example, fullerenes with long alkyl chains (compounds **1a-e** in Figure 10.4) deposited

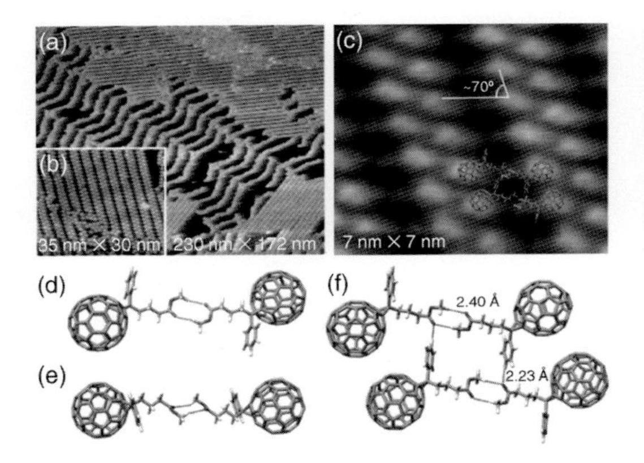

Figure 10.3 Large-scale STM image of the Au(111) surface after depositing ~0.6 ML of PCBM, showing the coexistence of two different phases: (a) the nanoscale spiderweb (created by the templating effect of the substrate surface) and (b) the sets of parallel double rows connected by an array of weak hydrogen bonds. (c) Close-up of PCBM double rows produced on Au(111) by the takeover of the intermolecular interactions. (d) Top and (e) side views of the optimized (calculated) structure for a PCBM dimer. (f) Optimized structure of a PCBM tetramer. The dotted lines mark the weak hydrogen bonds responsible for this conformation. Reproduced with permission from Ref. [26]. Copyright Wiley-VCH Verlag GmbH.

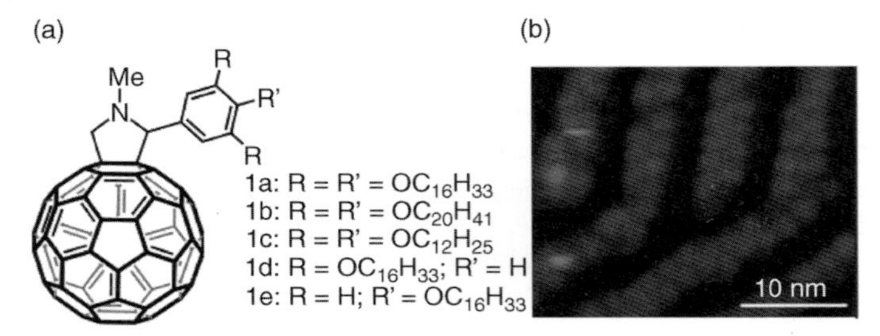

Figure 10.4 (a) Chemical structure of some fullerene derivatives **1**. (b) High-resolution STM image of **1a** on HOPG (scan range $30 \times 30\,nm^2$; $I_s = 40$ pA; $V_{bias} = +3.0$ V). Reproduced with permission of the American Chemical Society from Ref. [28].

on HOPG surfaces by spin coating from chloroform solutions also form lamellar structures, possibly by interdigitation of the alkyl chains [28]. The spacing between lamellae is determined by the alkyl chain length.

A similar experiment has been recently reported, with fullerenes functionalized with alkyl chains deposited on metal surfaces [29]. In this case, the experiments were carried out under ultrahigh vacuum conditions. Here, the role of the surface determines the adsorption configuration with a V-shape of the alkyl moieties (see Figure 10.5). The interdigitation of such moieties with the V-configuration does not lead to lamella formation, but rather to a porous network of dimers, as shown in the STM images of Figure 10.5.

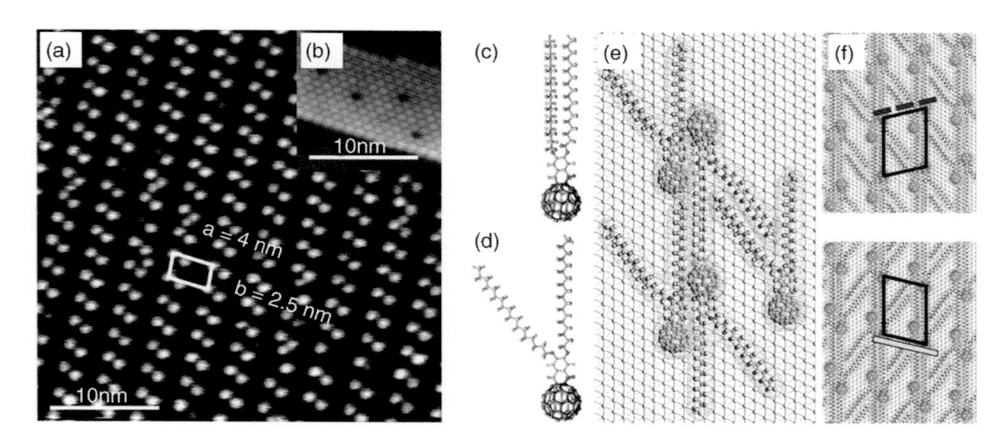

Figure 10.5 (a) STM image of a F-C_{60} self-assembled monolayer on Ag(111) (295 K, −1 V, 0.4 nA) also showing the geometry of the unit cell of the molecular structure. (b) An island of pristine C_{60} on Ag(111) imaged under identical conditions (295 K, −1 V, 0.4 nA). (c) Ground-state structure of an isolated F-C_{60} molecule containing 2.2 nm long alkyl chains with 18 C atoms each, bonded covalently to a C_{60} cage (I-shape configuration). (d) V-shape configuration of an adsorbed F-C_{60} molecule. (e) Details of the F-C_{60} assembling in the optimum geometry. (f) Adsorption geometries of the F-C_{60} SAM resulting from the proposed model. Reproduced with permission of the American Physical Society from Ref. [29].

The interactions described above reveal the important role that supramolecular chemistry can play in the construction of molecular nanostructures on solid surfaces. The arsenal of available noncovalent interactions, where alkyl chains and functional groups can play a leading role, to achieve a rational order of the molecules on the substrate is being applied on different solid surfaces and must lead to amazing results in the next coming years.

10.3 Surface Templates for Fullerene Adsorption

In this section, we show several examples demonstrating the power that templating the surface has for tailoring new supramolecular fullerene structures, from 0D single molecules or molecular clusters to 1D chains and 2D porous nanomeshes.

10.3.1 0D Point Defects and Single-Molecule Arrays

Surfaces showing well-ordered, uniform arrays of anchoring sites are excellent templates for the preparation of ordered 0D molecular nanostructures by site-selective nucleation and possibly further fullerene aggregation, yielding arrays of molecular clusters showing in many cases a very good monodispersity in size and shape. There are a variety of surface defects that can act as special nucleation sites, such as dislocation networks, stepped surfaces, or reconstructed surfaces.

Dislocation networks appear on metallic surfaces when there is a strain on the topmost layer due to lattice mismatch with the underlying bulk layers. They appear naturally by deposition of metal layers on metal surfaces with different lattice parameters. However, there are some cases such as Au(111), for which the dislocation networks appear on the surface as a reconstruction, driven by the lack of coordination of the surface atoms. It was known for many years that dislocation sites and, especially, the crossing points of the different dislocation domains, such as the elbow sites in the Au(111) herringbone reconstruction [30], act as nucleation sites for the growth of inorganic nanostructures.

For C_{60} fullerene adsorption at room temperature, no specificity has been found for the adsorption site of the molecules in the different areas of the dislocation pattern. Instead, intermolecular interactions seem to dominate the layer morphology by driving the formation of close-packed hexagonal islands extending over the different areas of the reconstruction [31]. Some specificity has, however, been recently found following low-temperature deposition (46 K), where individual C_{60} molecules appear preferentially nucleated at the elbows of the reconstruction [32]. We have also found that C_{60} functionalization can lead to preferential adsorption on the Au(111) reconstruction elbows. For example, the previously mentioned PCBM molecules are found in a regular array of single molecules at the elbows of the reconstruction after room-temperature deposition (Figure 10.6) [26]. Two effects can explain this behavior: first, the functionality seems to interact more selectively with

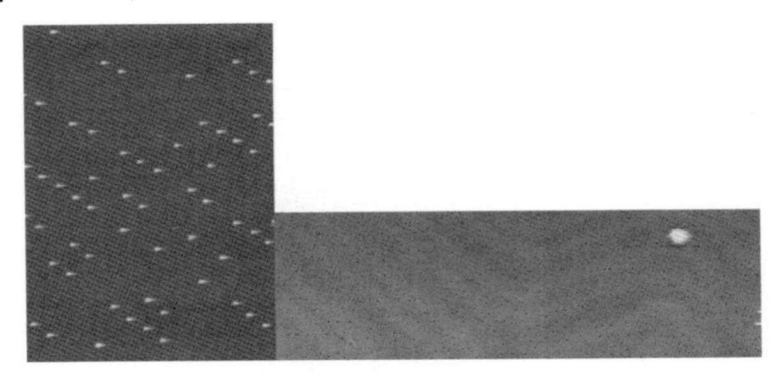

Figure 10.6 STM image ($92 \times 127\,nm^2$) of a highly regular 2D array of the C_{60} derivative PCBM molecules on the reconstructed Au(111) template surface. The inset shows a high-resolution image of a single PCBM molecule adsorbed at the elbow of the herringbone reconstruction.

the substrate than the C_{60} moiety, as evidenced by the 100% selectivity found for fcc versus hcp adsorption even at higher coverages, close to half a monolayer; second, the functionality prevents close packing of the C_{60} heads and thus destabilizes island formation at room temperature [26].

Preferential nucleation has also been observed when depositing C_{60} on top of the dislocation network created after depositing titanyl phthalocyanine (TiOPc) on Ag (111) [33]. For high deposition rates (>0.4 ML/min), TiOPc molecules form a triangular network of domains with hexagonal symmetry (Figure 10.7). The release of the compressive stress created by the relative alignment of these domains with respect to the surface symmetry directions is responsible for the formation of this network of misfit dislocations [34]. Deposition of small amounts of C_{60} on top of this network results in the formation of an ordered hexagonal arrangement of C_{60} clusters, nucleated at the intersections of six TiOPc domains, and separated ~14.0 nm (Figure 10.7).

Other kinds of surface reconstructions, not based on dislocation networks, might also provide specific adsorption sites for the adsorption of fullerenes. For example, $SrTiO_3(001)$ surfaces usually show a $c(4 \times 2)$ reconstruction, on which endohedral $Er_3N@C_{80}$ fullerenes form close-packed arrangements. However, different preparation procedures lead to the formation of a new 6×8 reconstruction that readily acts as a template for the adsorption of the endohedral fullerenes. The size of the reconstruction fits two fullerene molecules side by side, and as a result an ordered array of fullerene dimers is obtained [35]. The importance of the molecule's size and shape is underlined by the different behavior shown by C_{70} fullerene. For this smaller and nonspherical fullerene, 1D chain formation is observed on the $c(4 \times 2)$ reconstruction, alongside other island morphologies that, in this case, seem to observe a registry with the substrate atomic arrangement (Figure 10.8) [35].

10.3.2
1D Line Defects: Molecular Chains

The fast diffusion of fullerene molecules on close-packed noble metal surfaces such as the Au(111) $22 \times \sqrt{3}$, herringbone-reconstructed surface at room temperature,

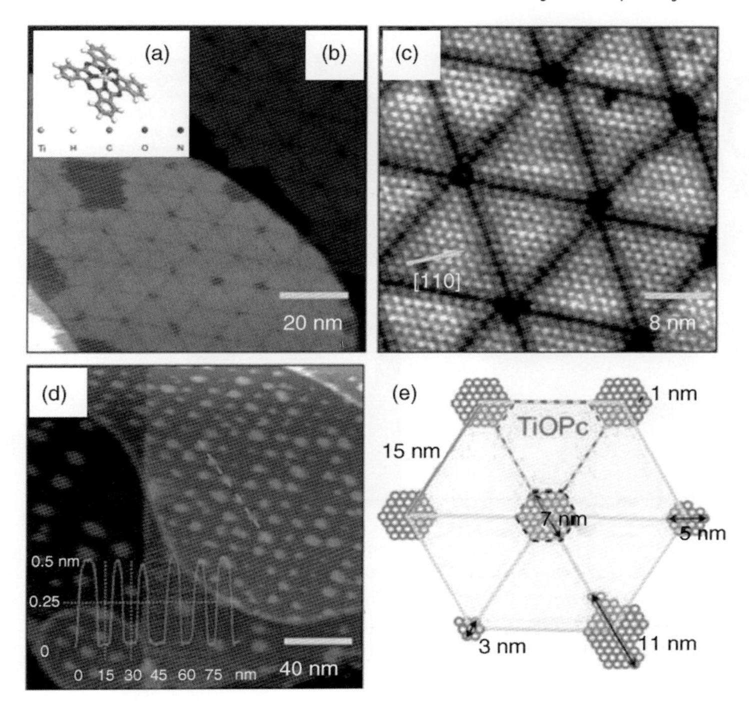

Figure 10.7 (a) Chemical structure of a TiOPc molecule. (b) STM image of TiOPc monolayer prepared at high (0.4 ML/min) flux showing a regular network of triangular domains. (c) Molecularly resolved STM image of the triangular network. (d) Large-scale STM image of a C_{60} cluster array on the TiOPc triangular dislocation networks. (e) Schematic illustration of the C_{60} cluster distribution. Reproduced with permission of the American Chemical Society from Ref. [33].

leads to preferential adsorption at step edges [37]. This observation suggested the possibility to employ vicinal Au or Cu surfaces with a periodic arrangement of monoatomic steps as a template to obtain ordered arrays of fullerene molecules [38]. In particular, the vicinal Au(11 12 12) surface still shows the dislocation network described above on the flat (111) facets, with the direction of the dislocation lines perpendicular to the step edges. Fasel and coworkers have exploited such surfaces for preparing C_{60} 1D chains with long-range order and uniform size [39]. The length of the chains is very uniform, mostly four or five C_{60} units, as determined by the fact that C_{60} molecules sit preferentially at step edges and fcc areas of the dislocation network (Figure 10.9). Only after complete decoration of the step edges, close-packed C_{60} islands grow. At higher coverages, the C_{60} layer grows out from the chains and across the terraces, which is energetically favored over the closing of the nanochain segments into extended 1D molecular chains.

In some cases, however, the templating effect of step edges must be taken with care. It was mentioned above that C_{60} adsorption might induce mass transport and surface reconstructions on certain metal surfaces. For example, Au(433) vicinal

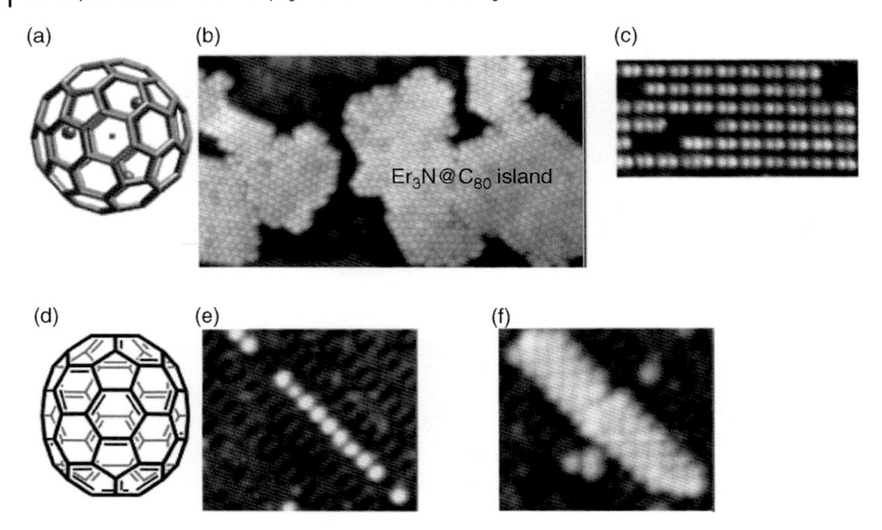

Figure 10.8 Chemical structure of $Er_3N@C_{80}$ (a) and STM images of molecular islands of it on $SrTiO_3(001)$-c(4 × 2) (55.4 × 31.3 nm^2; $V_s = +1.6$ V; $I_t = 0.10$ nA) (b); 2D endohedral fullerene array organization (31 × 16 nm^2, $V_s = +3.0$ V; $I_t = 0.1$ nA) on $SrTiO_3(001)$-(6 × 8) (c). STM images of the growth of epitaxial C_{70} (d) islands onto $SrTiO_3(001)$ showing a one-molecule-wide chain (e) and four-molecule-wide island of C_{70} (f). Imaging conditions: $V_s = +2.6$ V and $I_t = 0.10$ nA for image (e) and $V_s = +2.3$ V and $I_t = 0.10$ nA for image (*f*). Reproduced with permission of the American Chemical Society and the Royal Society of Chemistry from Refs [35, 36], respectively.

surfaces, with (111) terraces separated by {100} step edges, usually display a small faceting periodicity, with 4 nm terraces separated by bunches of about 5 or 6 steps, spanning 1.4 nm. However, deposition of C_{60} plus a subsequent annealing at 500 K leads to a much longer faceting periodicity, with wider (111) terraces separated by step bunches corresponding to a (533) facet, in such a way that each step in the bunch accommodates one single fullerene chain. As a result, a regular array 1D fullerene stripes, about six molecules wide, appear on the surface (Figure 10.10) [40].

10.3.3
2D Nanomeshes

Figure 10.9 shows how the crossing of linear defects might lead to arrays of specific adsorption sites at the centers of the squares left by the defect lines. It is, in principle, possible that the adsorption sites are not placed at the centers of the squares but at the edges of the squares, thereby steering the formation of 2D porous C_{60} materials made of intercrossing fullerene lines. Such nanomesh architecture has been suggested to be optimal for the integration of nanodevices in electronic circuits.

Such situation is readily obtained at the intermediate coverage range for PCBM on Au(111) surface. It was already discussed that PCBM adsorbs only at the elbows of the

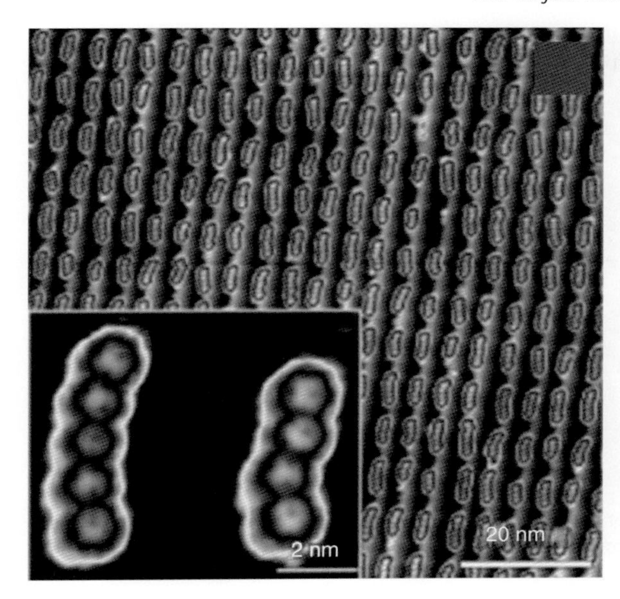

Figure 10.9 STM image of the superlattice obtained after formation of a highly regular 2D array of C_{60} nanochains on the Au(11 12 12) template surface after deposition of ~0.1 ML C_{60} taken ($T \sim 40$ K, $V_{bias} = -2$ V, $I_s = 0.03$ nA). The inset ($V_{bias} = -2.5$ V, $I_s = 0.5$ nA) shows a high-resolution image of two short chains. Reproduced with permission of the American Chemical Society from Ref. [39].

reconstruction for low surface coverage, and forms hydrogen-bonded lamella structures close to the monolayer. However, for the intermediate coverage range, PCBM molecules nucleate at the elbows of the reconstruction and continue to form 1D zigzag lines only at the fcc areas of the reconstruction, and decorate them completely.

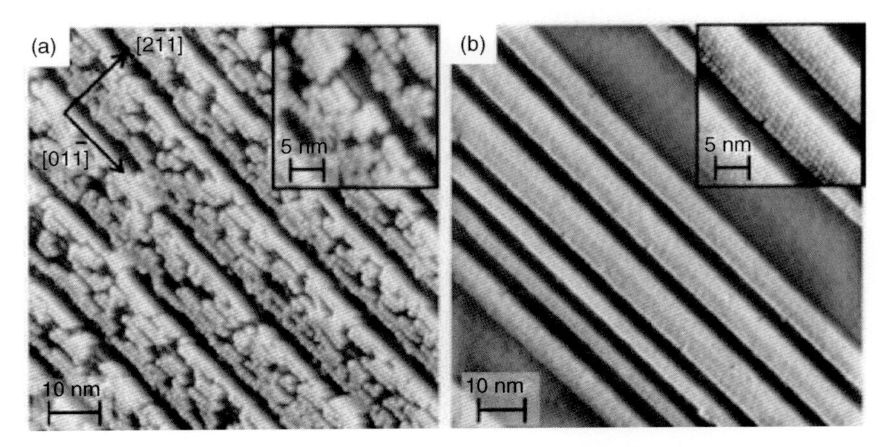

Figure 10.10 STM images of the deposition of 0.5 ML of C_{60} on Au(433) before (a) and after annealing at $T = 500$ K for 15 min (b). Insets show close-up views. Reproduced with permission of the American Chemical Society from Ref. [40].

Figure 10.11 STM images of the deposition of 0.5 ML of PCBM on Au(111). Reproduced with permission from Ref. [26]. Copyright Wiley-VCH Verlag GmbH.

Once all the fcc areas are completely decorated, the system still accommodates some molecules around the elbows of the reconstructions in the hcp areas before changing into the hydrogen-bonded network described above, for which site selectivity is lost. Thus, for coverages close to the transition, PCBM molecules are arranged in the nanomesh structure depicted in Figure 10.11 [26]. Such structure, thus, appears because of the selective decoration of the fcc areas and the elbow lines, which are, on average, perpendicular to each other.

Another square pattern that has been successfully used to create C_{60} nanomeshes is the one resulting from the partial nitridation of Cu(100) surfaces. Under proper growth conditions, nitridation of Cu(100) surfaces leads to the formation of a regular array of copper nitride islands, with a $c(2 \times 2)$ structure, separated by thin belts of pure copper. In this situation, fullerenes [41], like other organic molecules [42], prefer to adsorb on the bare metallic areas, probably because of the enhanced dispersive interaction due to the higher polarizability of the metal surface. Preference to adsorb on bare metallic areas leads to the decoration of the nitride island edge, leading to nanomesh formation (Figure 10.12).

10.4
Supramolecular Aggregation of Fullerenes and other Organic Species on Surfaces

Surface patterns like the ones presented in the previous section help organize fullerenes into well-defined structures that would otherwise be precluded by the spherical shape of the molecules. On the other hand, the flexibility of such method to produce new patterns with different symmetries and sizes is limited, and there seems to be no rational way to design new patterns. A more flexible route toward the

Figure 10.12 (a and b) STM images of the Cu-N surface after room-temperature deposition of C_{60}: (a) 0.08 ML (25 nm × 25 nm, $V_s = 0.5$ V, $I = 0.15$ nA); the adsorbed C_{60} molecules nucleate exclusively on the clean Cu area. (b) ~0.28 ML (35 nm × 35 nm, $V_s = 0.5$ V, $I = 0.15$ nA); a 2D molecular nanomesh is formed. Reproduced with permission of the American Chemical Society from Ref. [41].

rational design of surface patterns to act as templates for fullerene supramolecular aggregate formation on solid surfaces is the functionalization of the surface with a self-assembled monolayer of organic molecules (we understand self-assembled monolayer in the broader context of any ordered single layer of organic molecules on solid surfaces, not necessarily based on thiol chemistry on gold surfaces). In this section, we will review the works that have recently appeared in the literature exploring this possibility. We distinguish between host–guest systems on the one hand, in which the deposited organic layer contains nanopores that can selectively bind individual fullerenes or fullerene clusters, and mixed layers in which fullerenes and other molecular species coassemble in ordered patterns.

10.4.1
Self-Assembled Monolayers as Hosts for Fullerenes on Solid Surfaces

Some of the first attempts at using molecular monolayers adsorbed on a surface as a template for the growth of fullerene nanostructures used large macrocycles deposited from solution on HOPG. STM images of the codeposited layer of macrocycle **2** and C_{60} show a regular array of protrusions that are identified as the fullerene molecules (see Figure 10.13) [43]. Such an arrangement is not the close-packed layer usually observed on HOPG and other surfaces by depositing only the fullerenes. The position of the fullerene molecules, however, and in particular the observed dimerization at high coverages, seems not to be consistent with the fullerenes being adsorbed at the center of the macrocycle but rather on top of the bithiophene. Therefore, the driving force for this superstructure is not the interaction of the C_{60} units with the uncovered graphite surface inside the rings, but the donor–acceptor interactions between the electron-rich bithiophene fragments and the electron-poor C_{60} molecule, and strictly speaking such examples cannot be considered as hosting of the fullerene molecules. Similar results were obtained for self-assembled cyclooligothiophenes [44].

Porous networks can be constructed from other molecular building blocks, previously self-assembled on the surface. Generally speaking, porous networks are

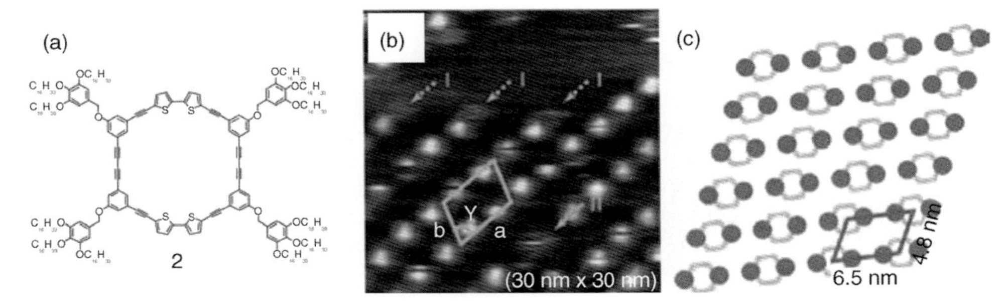

Figure 10.13 (a) Chemical structure of macrocycle **2**, (b) high-resolution STM image under ambient conditions, revealing the adsorption site of C_{60} on the macrocycle ($V_{bias} = 1.0$ V; $I_t = 0.3$ nA), and (c) proposed structural model. Reproduced with permission of the American Chemical Society from Ref. [43].

expected for strongly directional interactions such as hydrogen or coordination bonds, but in some cases they also appear on close-packed molecular layers held together by dispersive interactions, especially when the molecular shape is not very regular. For example, the star-shaped oligothiophene **3** has been used to grow extended domains (up to 200 nm long) of a 2D double-cavity open network on HOPG (Figure 10.14) [45]. Upon deposition from solution, only one of the cavities is able to trap the fullerenes, thus forming a long-range alignment of single C_{60} molecules.

A different way of building nanoporous networks is to use the directionality and strength of hydrogen or coordination bonds, whereby small molecules can form open nanoporous networks with predefined size and shape. For example, melamine and perylene tetracarboxylic diimide (PTCDI) deposited on 1 ML Ag-Si(111) $\sqrt{3} \times \sqrt{3}$R 30° surface leads to the formation of a long-range ordered honeycomb structure [46], in which melamine binds three PTCDI molecules through triple hydrogen bonds, along three axes forming an angle of 120° with each other. Further C_{60} deposition leads to the formation of C_{60} molecular clusters at the nanopores of the network (see Figure 10.15). Such clusters have a very narrow size-distribution curve, and most of them are composed of exactly seven C_{60} molecules. Such monodispersity indicates interpore diffusion of the C_{60} molecules since the intrinsically random adsorption process should have led to a rather wide pore occupancy distribution. Moreover, because of the particular size of the pore for the melamine–PTCDI network, it can accommodate up to seven C_{60} molecules and no more.

2D metal–organic coordination networks (MOCNs) offer an alternative route to the creation of nanoporous networks on metal surfaces, with the additional advantage of offering a greater thermal stability. A recent systematic STM study of the hierarchical self-assembly of Fe atoms and trimesic acid molecules (TMA, left inset in Figure 10.16a) on a Cu(100) substrate in ultrahigh vacuum has shown that the nanopores appearing in the coordination network, being formed by eight TMA molecules, can accommodate individual C_{60} molecules when they are deposited on top (Figure 10.16b). Such structures are stable up to 470 K, temperature at which

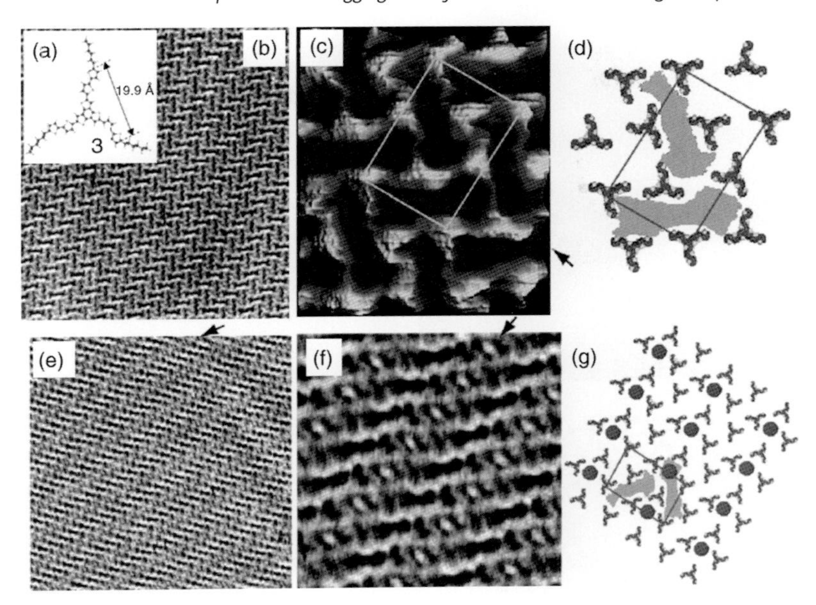

Figure 10.14 (a) Chemical structure of the star-shaped oligothiophene **3**. (b) STM image of the double-cavity self-assembly after deposition on graphite (70×70 nm^2; $V_{bias} = 0.35$ V, $I_t = 70$ pA). Vertical and horizontal cavities are clearly visible. (c) High-resolution 3D STM image (10×10 nm^2; $V_{bias} = 0.4$ V; $I_t = 20$ pA). (d) Model of the open network with the two cavities represented as gray areas. (e) STM images of the 1-C_{60} guest–host network on graphite (70×70 nm^2; $V_{bias} = 0.30$ V; $I_t = 40$ pA). All horizontal cavities remain empty while all vertical cavities are filled with a C_{60} and are no longer visible. (f) A closer look of the 1-C_{60} guest–host network (21×21 nm^2; $V_{bias} = 0.3$ V; $I_t = 40$ pA). (g) Model of the long-range alignment of C_{60} in the open network of **3**. Reproduced with permission of the American Chemical Society from Ref. [45].

release of C_{60} from the pores was observed [47]. These STM experiments reveal the reversibility of the process, which is an important issue for the fabrication of two-dimensional nanoarrays of fullerenes at solid surfaces.

In a similar way to hydrogen-bonded networks, the pore size and shape in MOCNs can be controlled by an adequate choice of the molecular species. By codepositing terphenyl or benzoic carboxylic acid molecules with Fe atoms on a Cu(100) surface, Stepanow *et al.* were able to create MOCNs with different pore sizes. Upon deposition of C_{60} on top of these networks, C_{60} clusters were formed within the pore, the cluster size being indeed controlled by the pore size [48].

10.4.2
Coassembly of Fullerenes and Other Organic Species

Fullerenes can interact via dispersive interactions with molecules other than fullerenes themselves. The different molecular geometries lead, thus, to different

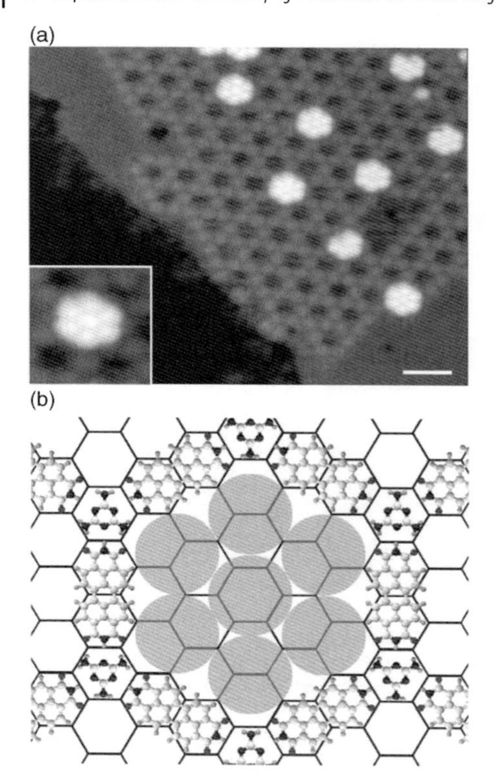

Figure 10.15 (a) STM Image of C_{60} heptamers trapped within the PTCDI–melamine network. (*Inset*) High-resolution view showing an individual cluster (scale bar 5 nm, −2 V, 0.1 nA). (b) Schematic diagram of a C_{60} heptamer.

arrangements of the fullerene molecules. This is a topic that has attracted much attention lately in relation to the interesting properties that donor–acceptor organic mixtures could have for optoelectronic and photovoltaic applications. In particular, porphyrin molecules, which can act as electron donors in certain mixtures, have been extensively used to form supramolecular ensembles by combination with C_{60} either in solution or in the solid state [49]. For example, Spillmann and coworkers have recently reported the coassembly of porphyrin derivatives **4** and **5** and C_{60} fullerene on Ag(100) under UHV conditions (see Figure 10.17) [50]. STM images at 1 ML coverage of compounds **4** and **5** in the absence of fullerenes reveal that both compounds self-organize in regular molecular rows with a high packing density.

Deposition of 0.02 ML of C_{60} onto a preadsorbed monolayer of diporphyrin **4** generates unidirectional chains composed of up to eight C_{60} molecules (~15.5 nm long, Figure 10.17a). These C_{60} units are located on top of the 3-cyanophenyl substituents, outside the porphyrin core. Under the same conditions, deposition of C_{60} on a monolayer of **5** on a Ag(1100) substrate resulted in the formation of two

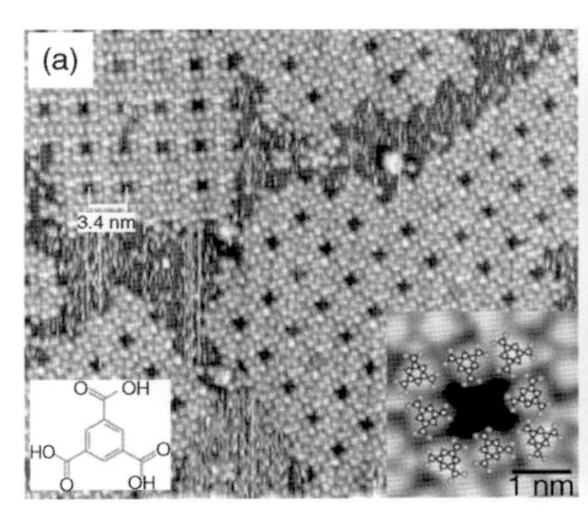

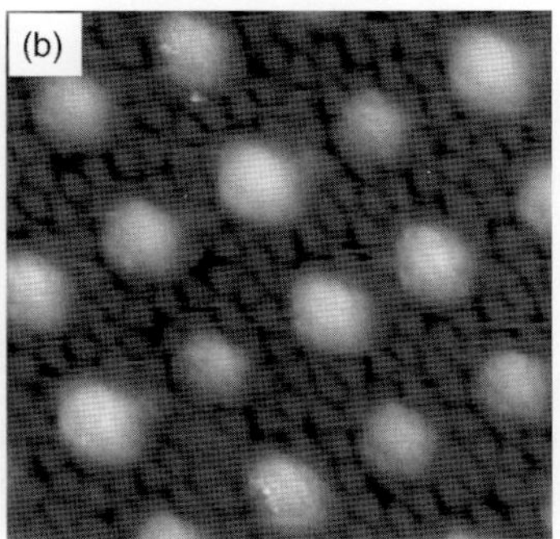

Figure 10.16 (a) STM image of the 2D metallosupramolecular nanocavities obtained on a Cu(100) substrate by hierarchical assembly of TMA and Fe atoms. The right inset shows the proposed model for a single nanocavity surrounded by eight uncoordinated carboxylate groups. (b) 14 nm × 14 nm STM images of the C_{60} binding in the receptors. The different molecular heights might reflect distinct adsorption configurations. Reproduced with permission of the Royal Society of Chemistry from Ref. [47].

segregated molecular domains: islands of C_{60} and a condensed phase composed of **5**. Thermal annealing (453 K) gave rise to the assembly shown in Figure 10.17b, with the C_{60} units regularly arranged on top of the porphyrin layer.

Nishiyama *et al.* have described the formation of a flexible porous network upon adsorption of C_{60} on a Au(111) surface covered with a compact monolayer of porphyrin **6** (Figure 10.18) bearing two carboxylic groups [51]. Although the porphyrin monolayer initially covers the entire surface in the form of long supramolecular wires densely aggregated, it is the adsorption of C_{60} molecules that leads to the formation of nanopore structures by laterally shifting the supramolecular wires, which occurs simultaneously with a partial change in the conformation of the porphyrin (Figure 10.18).

But, of course, porphyrins were not the only ones used to investigate the supramolecular chemistry of multicomponent systems in two dimensions. For example, two different ordered structures were found in the coadsorbed system C_{60}–acridine-9-carboxylic acid (ACA) on Ag(111), depending on the fullerene/ACA ratio. For local ACA coverages below 0.4 ML, a chiral structure can be found (Figure 10.19), whereas chain C_{60}:ACA structures were related to an ACA dimer phase that appears for a coverage of 0.8 ML [52]. Thus, important organization patterns, such as chirality or one-dimensional chains, can be tuned as a function of the coverage of the predeposited ACA structures.

A well-ordered organic donor/acceptor nanojunction array formed by self-assembly of fullerene C_{60} (as an electron acceptor) on the molecular nanotemplate formed by nanostripes of *p*-sexiphenyl (6P) (as an electron donor) on Ag(111) has recently been reported by Chen *et al.* [53]. In this case, STM studies reveal the formation of C_{60}/6P vertical nanojunctions with well-defined two-dimensional arrangement. Further annealing process on the molecular superstructure formed by the C_{60}/6P at 380 K leads to the insertion of C_{60} linear chains between the 6P nanostripes, thus

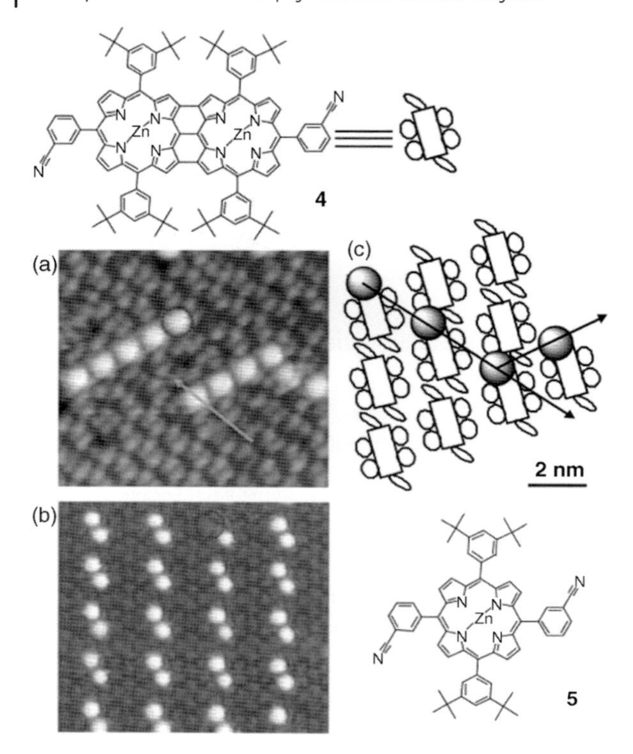

Figure 10.17 Chemical structure of porphyrins **4** and **5** and STM image of **4**·C_{60} (a) and **5**·C_{60} (b) complexes (scan range: 30 × 30 nm^2, $I_t = 16$ pA, $V_{bias} = 2.67$ V). (c) Proposed model for the assembly of **4**·C_{60} complexes, the C_{60} molecules (black spheres) are situated between the molecules of **4** approximately on top of the 3-cyanophenyl groups; the arrows indicate the major and minor growth directions. Reproduced with permission from Ref. [50]. Copyright Wiley-VCH Verlag GmbH.

resulting in the formation of a periodic lateral nanostructure constituted by C_{60} and 6P molecules (Figure 10.20).

Similar results have been reported for the C_{60}–α-sexithiophene (6T) system [54]. 6T molecules deposited on Ag(111) at room temperature grow in the form of a well-ordered nanostripe structure. After depositing C_{60} on top of a 6T bilayer and annealing, different ordered structures appear on the surface, depending on the C_{60} coverage and the annealing temperature, giving rise to different intermixed arrangements of donor and acceptor molecules on the Ag surface (Figure 10.21).

In a recent study, we carried out the growth of a self-assembled lateral superlattice, with a typical 10–20 nm, of 2-[9-(1,3-dithiol-2-ylidene)anthracen-10(9*H*)-ylidene]-1,3-dithiole (ext-TTF, **7**) as electron donor and PCBM as electron acceptor on Au(111) under UHV conditions (Figure 10.22) [55]. The aim of this work was to use the nanometer-scale pattern provided by the well-known $22 \times \sqrt{3}$ "herringbone" reconstruction of Au(111) surface as a template to steer the growth of the molecular species into one-dimensional molecular nanostructures with sizes in good agreement with

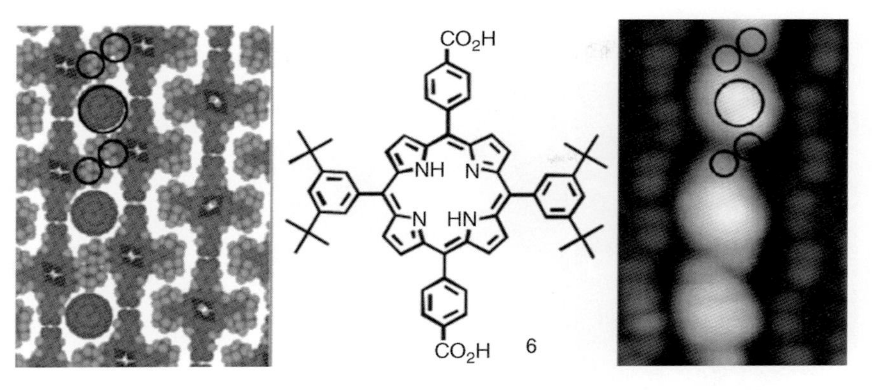

Figure 10.18 Proposed model, chemical structure, and high-resolution STM image of a linear C_{60} chain arranged on self-assembled porphyrin **6**. The circles indicate the rotation of the porphyrin aryl substituents to avoid steric hindrance. Reproduced with permission from Ref. [54]. Copyright Wiley-VCH Verlag GmbH.

the exciton diffusion lengths (10–20 nm), as required for highly efficient photovoltaic solar cells [56]. Our results reveal that the lateral segregation of donor/acceptor blends into a long-range ordered superlattice with the appropriate morphology is feasible.

PCBM and **7** were deposited in UHV from two separate glass crucibles resistively heated at 400 and 500 K, respectively, on Au(111) at room temperature. As mentioned above, UHV deposition of PCBM on Au(111) leads to the formation of a PCBM spiderweb-ordered nanostructure (see Figure 10.11) [26]. Addition of ~0.6 ML of **7** on

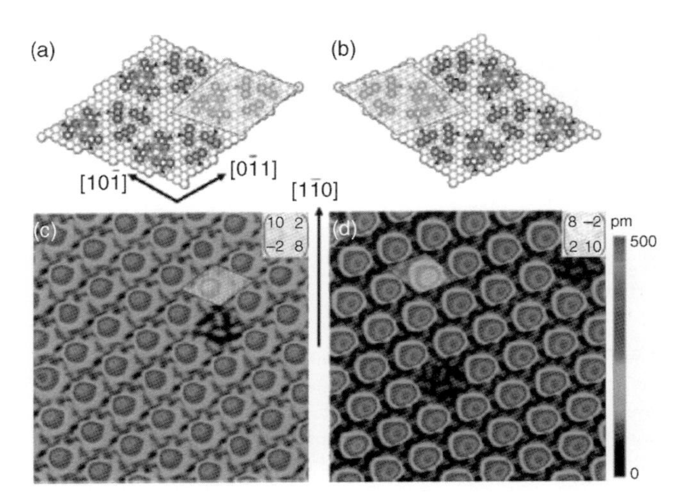

Figure 10.19 (a and b) Proposed models and STM images of the intermixed C_{60}-ACA supramolecular structures for the R and S chiral surfaces (c and d). A unit cell is indicated in each panel for reference purposes, along with the matrix notation of the domain. Reproduced with permission of the American Chemical Society from Ref. [52].

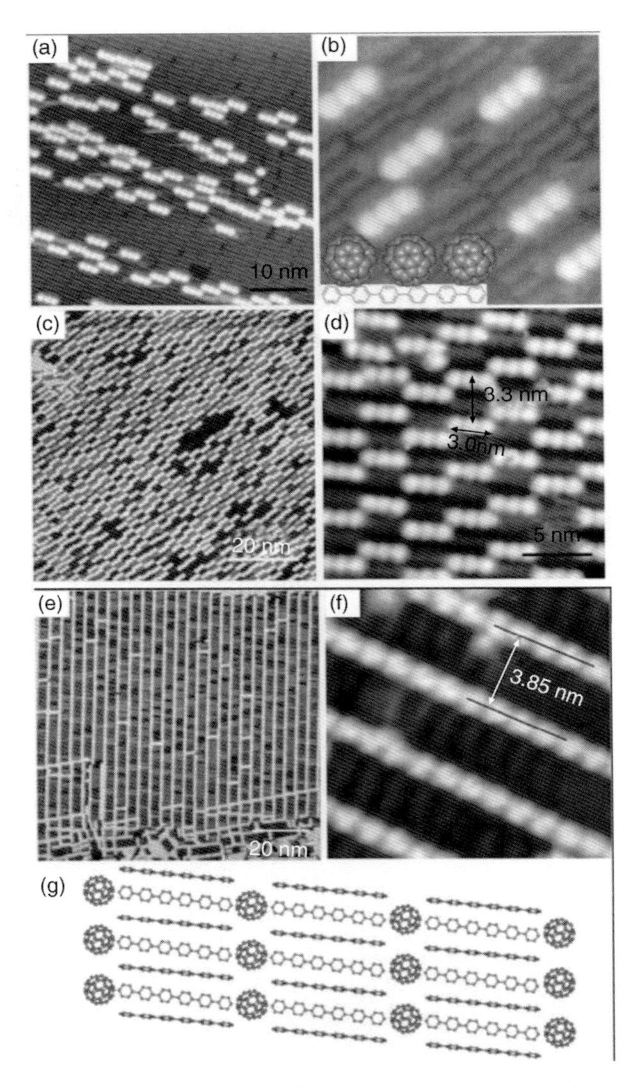

Figure 10.20 (a) $50 \times 50\,\text{nm}^2$ and (b) $15 \times 15\,\text{nm}^2$ STM images of 0.1 ML C_{60} on "face-on + edge-on" 6P layer on Ag(111) corresponding to (a) showing the preferential adsorption of C_{60} triplets atop single 6P molecule. The inset in (b) shows the proposed model. (c) $80 \times 80\,\text{nm}^2$ and (d) $20 \times 20\,\text{nm}^2$ STM images of 0.5 ML C_{60} on face-on + edge-on 6P layer on Ag(111), showing the formation of a well-ordered vertical C_{60}/6P nanojunction array. (e) $100 \times 100\,\text{nm}^2$ and (f) $13 \times 13\,\text{nm}^2$ STM images of the lateral C_{60}/6P nanojunction arrays. (g) Schematic model for the regular superlattice of alternating C_{60} and 6P linear chains. Reproduced with permission of the American Physical Society from Ref. [53].

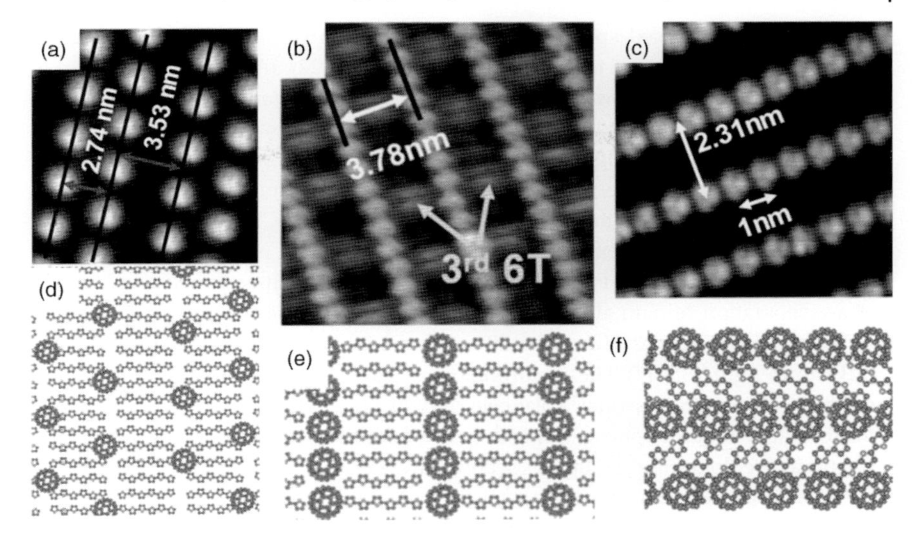

Figure 10.21 (a) High-resolution STM image of the C_{60} dot arrays obtained after depositing 0.1 ML of C_{60} on 1.5 ML 6T thin film grown on Ag (111) and annealing at 360 K for 30 min ($12.5 \times 12.5\,nm^2$, $V = -1.40$ V, $I = 70$ pA). (b) After further annealing at 360 K for 1 h, the surface is dominated by rail-like molecular arrays consisting of continuous C_{60} linear chains and interlinked 6T molecules ($15 \times 15\,nm^2$, $V = -1.44$ V, $I = 70$ pA). (c) High-resolution STM images of the perfectly ordered C_{60} single-chain arrays formed by further depositing 0.50 ML of C_{60} and subsequent annealing at 380 K ($10 \times 10\,nm^2$, $V = -1.80$ V, $I = 70$ pA). (d–f) The proposed structural models of (a), (b), and (c). Reproduced with permission from Ref. [54]. Copyright Wiley-VCH Verlag GmbH.

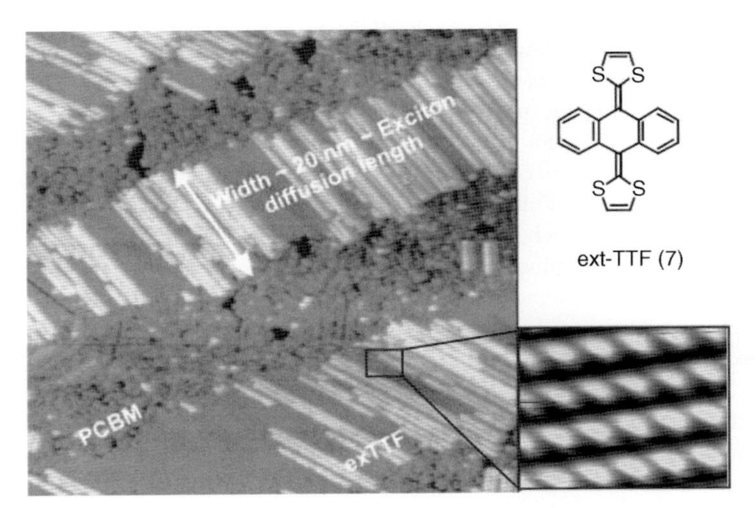

Figure 10.22 Nanoscale segregation of PCBM and ext-TTF mixtures on Au(111) ($118 \times 132\,nm^2$). The inset shows the growth morphology of the ext-TTF films at ~1 ML of coverage on Au(111). Reproduced with permission of the American Chemical Society from Ref. [55].

top of the PCBM-based spiderweb results in the formation of two distinct areas observed by STM at room temperature (see Figure 10.22): (i) a well-ordered area of ext-TTF, where the elbow-free herringbone reconstruction underneath can be recognized as a long-period corrugation of the molecular rows, and (ii) a disordered area reminiscent of the high-coverage PCBM phase.

This singular nanoscale phase segregation has been accounted for by (i) the low tendency of **7** and PCBM to mix, (ii) the high mobility of PCBM molecules at room temperature (used in the experiment), and (iii) the stronger interaction of **7** with the gold surface, in contrast to PCBM that behaves as a 2D gas. Thus, the morphology and phase separation remain almost intact upon increasing the coverage, forming ext-TTF and PCBM stripes of around 20 nm that compares well with typical exciton diffusion lengths [55, 57]. These morphological features are ideal for the construction of highly efficient photovoltaic solar cells and, in a broader sense, they are of interest for the study of other optoelectronic devices where morphology between photo- and electro-active donor and acceptor species plays a leading role.

The examples discussed in this section clearly reveal that the preparation of well-organized organic donor/acceptor nanojunction arrays is feasible as show the low-temperature STM investigations carried out on different donors (6P, 6T, and exTTFs) and C_{60} as paradigmatic case of electron acceptor molecule. These new 2D organic nanostructures represent outstanding and alternative realistic possibilities for the preparation of nanoelectronic devices.

10.5
Outlook

Although fullerenes have already been used for the direct coverage of different metal substrates from the very beginning, only recently they have been mixed with other appealing molecules that nicely complement them from a geometric and electronic point of view. As a result, organic molecules previously deposited on a substrate have been used as new templates for the formation of 2D ordered fullerene superstructures. Thus, both planar and concave host organic molecules, namely, involving macrocycles, have been successfully used to accommodate fullerenes (C_{60} and C_{70}) and fullerene derivatives in which the presence of alkyl chains or functional groups have a strong impact on the induced order.

As it has been stated, the interaction of fullerenes with the metal surface usually produces a rearrangement on the atomic structure of the surface. The different strategies discussed in this chapter clearly show that the modification of the metal surface can be avoided by decorating the solid surface with suitable organic molecules acting as templates, and exploited in the construction of hierarchical assemblies of fullerene arrays. Furthermore, the rational design of coadsorbed systems results in outstanding achievements such as the generation of chiral surfaces or the organization of intermixed layers of electron donor and acceptor components. These 2D superlattices are thus new scenarios for testing, for instance, nonconventional enantiomeric synthesis (chiral surface), or for unraveling the fundamental electron transfer process in intermixed layers for the production of optoelectronic devices at a nanometer scale.

In summary, fullerenes – probably the most studied molecules in the past 20 years – are called to play a leading role in the development of 2D supramolecular organizations, where pristine and chemically modified fullerenes can be organized at will while still preserving the outstanding chemical, electrochemical, and photophysical properties they exhibit.

Acknowledgments

Our own work has been supported by the MICINN of Spain (MAT2009-13488, FIS2010-18847, CTQ2008-00795/BTQ), Comunidad de Madrid (Nanobiomagnet S2009/MAT-1726, Madrisolar-2 S2009/PPQ-1533), CONSOLIDER-INGENIO ON MOLECULAR NANOSCIENCE (CSD2007-00010), and European Union (SMALL PITN-GA-2009-23884). R.O. thanks the Spanish Ministry for Science and Innovation for salary support under the Ramón & Cajal Program.

References

1 Martín, N. (2006) *Chem. Commun.*, 2093–2104.

2 (a) Hirsch, A. (2005) *The Chemistry of Fullerenes*, Wiley-VCH Verlag GmbH, Weinheim, Germany; (b) Guldi, D.M. and Martín, N. (2002); *Fullerenes: From Synthesis to Optoelectronic Properties*, Kluwer Academic Publishers, Dordrecht, The Netherlands; (c) Langa, F. and Nierengarten, J.-F. (2007). *Fullerenes: Principles and Applications*, RSC Publishing, Cambridge, UK.

3 Martín, N. and Nierengarten, J.-F. (eds) (2006) Special issue on Supramolecular Chemistry of Fullerenes, *Tetrahedron*, **62**, 1905–2132.

4 (a) Elemans, A.A.W., Lei, S., and De Feyter, S. (2009) *Angew. Chem. Int. Ed.*, **48**, 7298–7332; (b) Hooks, D.E., Fritz, T., and Ward, M.D. (2001) *Adv. Mater*, **13**, 227–241.

5 Tseng, T.-C., Urban, C., Wang, Y., Otero, R., Tait, S.L., Alcamí, M., Écija, D., Trelka, M., Gallego, J.M., Lin, N., Konuma, M., Starke, U., Nefedov, A., Langner, A., Wöll, C., Herranz, M.A., Martín, F., Martín, N., Kern, K., and Miranda, R. (2010) *Nat. Chem.*, **2**, 374–379.

6 Tamai, A., Seitsonen, A.P., Baumberger, F., Hengsberger, M., Shen, Z.-X., Greber, T., and Osterwalder, J. (2008) *Phys. Rev. B*, **77**, 075134.

7 Modesti, S., Cerasari, S., and Rudolf, P. (1993) *Phys. Rev. Lett.*, **71**, 2469.

8 Silien, C., Pradhan, N.A., Ho, W., and Thiry, P.A. (2004) *Phys. Rev. B*, **69**, 115434.

9 Lu, X., Grobis, M., Khoo, K.H., Louie, S.G., and Crommie, M.F. (2004) *Phys. Rev. B*, **70**, 115418.

10 (a) Wang, L.-L. and Cheng, H.-P. (2004) *Phys. Rev. B*, **69**, 045404; (b) Wang, L.-L. and Cheng, H.-P. (2004) *Phys. Rev. B*, **69**, 165417.

11 Weckesser, J., Cepek, C., Fasel, R., Barth, J.V., Baumberger, F., Greber, T., and Kern, K. (2001) *J. Chem. Phys.*, **115**, 9001.

12 Murray, P.W., Pedersen, M.Ø., Lægsgaard, E., Stensgaard, I., and Besenbacher, F. (1997) *Phys. Rev. B*, **55**, 9360.

13 Pedersen, M.Ø., Murray, P.W., Lægsgaard, E., Stensgaard, I., and Besenbacher, F. (1997) *Surf. Sci.*, **389**, 300.

14 Abel, M., Dmitriev, A., Fasel, R., Lin, N., Barth, J.V., and Kern, K. (2003) *Phys. Rev. B*, **67**, 245407.

15 Hinterstain, M., Torrelles, X., Felici, R., Rius, J., Huang, M., Fabris, S., Fuess, H., and Pedio, M. (2008) *Phys. Rev. B*, **77**, 153412.

16 Cepek, C., Fasel, R., Sancrotti, M., Greber, T., and Osterwalder, J. (2001) *Phys. Rev. B*, **63**, 125406.

17 Pai, W.W., Hsu, C.-L., Lin, M.C., Lin, K.C., and Tang, T.B. (2004) *Phys. Rev. B*, **69**, 125405.

18 Zhang, X., Yin, F., Palmer, R.E., and Guo, Q. (2008) *Surf. Sci.*, **602**, 885.

19 Altman, E.I. and Colton, R.J. (1994) *J. Vac. Sci. Technol. B*, **12**, 1906–1909.

20 (a) Hou, J.G., Jinlong, Y., Haiqian, W., Qunxiang, L., Changgan, Z., Hai, L., Wang, B., Chen, D.M., and Qingshi, Z. (1999) *Phys. Rev. Lett.*, **83**, 3001–3004; Pascual, J.I., Gómez-Herrero, J., Rogero, C., Baró, A.M., Sánchez-Portal, D., Artacho, E., Ordejón, P., and Soler, J.M. (2000) *Chem. Phys. Lett.*, **321** 78.

21 Altman, E.I. and Colton, R.J. (1993) *Phys. Rev. B*, **48**, 18244.

22 Hashizume, T., Motai, K., Wang, X.D., Shinohara, H., Saito, Y., Maruyama, Y., Ohno, K., Kawazoe, Y., Nishina, Y., Pickering, H.W., Kuk, Y., and Sakurai, T. (1993) *Phys. Rev. Lett.*, **71** 2959.

23 Gimzewski, J.K., Modesti, S., David, T., and Schlittler, R.R. (1994) *J. Vac. Sci. Technol. B*, **12**, 1942–1946.

24 Chen, W., Zhang, H., Huang, H., Chen, L., and Shen Wee, A.T. (2008) *ACS Nano*, **2**, 693–698.

25 Schull, G. and Berndt, R. (2007) *Phys. Rev. Lett.*, **99**, 226105.

26 Écija, D., Otero, R., Sánchez, L., Gallego, J.M., Wang, Y., Alcamí, M., Martín, F., Martín, N., and Miranda, R. (2007) *Angew. Chem. Int. Ed.*, **46**, 7874–7877.

27 Desiraju, G.R. (2005) *Chem. Commun.*, 2995–3001.

28 Nakanishi, T., Miyashita, N., Michinobu, T., Wakayama, Y., Tsuruoka, T., Ariga, K., and Kurth, D.G. (2006) *J. Am. Chem. Soc.*, **128**, 6328–6329.

29 Diaconescu, B., Yang, T., Berber, S., Jazdyk, M., Miller, G.P., Tománek, D., and Pohl, K. (2009) *Phys. Rev. Lett.*, **102**, 056102. .

30 Brune, H., Giovanni, M., Bromann, K., and Kern, K. (1998) *Nature*, **394**, 451–452.

31 Chambliss, D.D., Wilson, R.J., and Chiang, S. (1991) *Phys. Rev. Lett.*, **66**, 1721.

32 Tang, L., Zhang, X., Guo, Q., Wu, Y.-N., Wang, L.-L., and Cheng, H.-P. (2010) *Phys. Rev. B*, **82**, 125414.

33 Wei, Y., Robey, S.W., and Reutt-Robey, J.E. (2009) *J. Am. Chem. Soc.*, **131**, 12026–12027.

34 Wei, Y., Robey, S.W., and Reutt-Robey, J.E. (2008) *J. Phys. Chem. C*, **112**, 18537–18542.

35 Deak, D.S., Silly, F., Porfyrakis, K., and Castell, M.R. (2006) *J. Am. Chem. Soc.*, **128**, 13976–13977.

36 Deak, D.S., Porfyrakis, K., and Castell, M.R. (2007) *Chem. Commun.*, 2941–2943.

37 (a) Altman, E.I. and Colton, R.J. (1992) *Surf. Sci.*, **279**, 49–67; (b) Fujita, D., Yakabe, T., Nejoh, H., Sato, T., and Iwatsuki, M. (1996) *Surf. Sci.*, **366**, 93–98.

38 Cuberes, M.T., Schlittler, R.R., and Gimzewski, J.K. (1996) *Appl. Phys. Lett.*, **69**, 3016–3018.

39 Xiao, W., Ruffieux, P., Aït-Mansour, K., Gröning, O., Palotas, K., Hofer, W.A., Gröning, P., and Fasel, R. (2006) *J. Phys. Chem. B*, **110**, 21394–21398.

40 Néel, N., Kröger, J., and Berndt, R. (2006) *Appl. Phys. Lett.*, **88**, 163101.

41 Lu, B., Iimori, T., Sakamoto, K., Nakatsuji, K., Rosei, F., and Komori, F. (2008) *J. Phys. Chem. C*, **112**, 10187–10192.

42 Écija, D., Trelka, M., Urban, C., De Mendoza, P., Echavarren, A., Otero, R., Gallego, J.M., and Miranda, R. (2008) *Appl. Phys. Lett.*, **92**, 223117.

43 Pan, G.-B., Cheng, X.-H., Coger, S., and Freyland, W. (2006) *J. Am. Chem. Soc.*, **128**, 4218–4219.

44 (a) Mena-Osteritz, E. and Bäuerle, P. (2006) *Adv. Mater.*, **18**, 447–451; (b) Mena-Osteritz, E. (2002) *Adv. Mater.*, **14**, 609–616.

45 Piot, L., Silly, F., Tortech, L., Nicolas, Y., Blanchard, P., Roncali, J., and Fichou, D. (2009) *J. Am. Chem. Soc.*, **131**, 12864–12865.

46 (a) Theobald, J.A., Oxtoby, N.S., Phillips, M.A., Champness, N.R. and Beton, P.H. (2003) *Nature*, **424**, 1029–1031; (b) Swarbrick, J.C.,

Rogers, B.L., Champness, N.R., and Beton, P.H. (2006) *J. Phys. Chem. B.*, **110**, 6110–6114.

47 Stepanow, S., Lin, N., Barth, J.V., and Kern, K. (2006) *Chem. Commun.*, 2153–2155.

48 Stepanow, S., Lingenfelder, M., Dmitriev, A., Spillmann, H., Delvigne, E., Lin, N., Deng, X., Cai, C., Barth, J.V., and Kern, K. (2004) *Nature Mater.*, **3**, 229–233.

49 For recent examples on the complexation of C_{60} and porphyrins in solution, see (a) Nobukuni, H., Shimazaki, Y., Tani, F., and Naruta, Y. (2007) *Angew. Chem. Int. Ed.*, **46**, 8975–8978; (b) Yanagisawa, M., Tashiro, K., Yamasaki, M., and Aida, T., *J. Am. Chem. Soc.* (2007) **129**, 11912–11913.

50 Bonifazi, D., Spillmann, H., Kiebele, A., de Wild, M., Seiler, P., Cheng, F., Güntherodt, H.-J., Jung, T., and Diederich, F. (2004) *Angew. Chem. Int. Ed.*, **43**, 4759–4763.

51 Nishiyama, F., Yokohama, T., Kamikado, T., Yokoyama, S., Mashiko, S., Sakaguchi, K., and Kikuchi, K. (2007) *Adv. Mater.*, **19**, 117–120.

52 Xu, B., Tao, C., Williams, E.D., and Reutt-Robey, J.E. (2006) *J. Am. Chem. Soc.*, **128**, 8493–8499.

53 Chen, W., Zhang, H.L., Huang, H., Chen, L., and Shen Wee, A.T. (2008) *Appl. Phys. Lett.*, **92**, 193301.

54 Chen, L., Chen, W., Huang, H., Zhang, H.L., Yujara, J., and Shen Wee, A.T. (2008) *Adv. Mater.*, **20**, 484–488.

55 Otero, R., Écija, D., Fernández, G., Gallego, J.M., Sánchez, L., Martín, N., and Miranda, R. (2007) *Nano Lett.*, **7**, 2602–2607.

56 (a) Choong, V., Park, Y., Gao, Y., Wehrmeister, T., Müllen, K., Hsieh, B.R., and Tang, C.W. (1996) *Appl. Phys. Lett.*, **69**, 1492–1494; (b) Halls, J.J.M., Pichler, K., Friend, R.H., Moratti, S.C., and Holmes, A.B. (1996) *Appl. Phys. Lett*, **68**, 3120–3122.

57 For recent reviews on organic solar cells, see (a) Brabec, C.J., Sariciftci, N.S., and Hummelen, J.C. (2001) *Adv. Funct. Mater.*, **11**, 15–26; (b) Coakley, K.M. and McGehee, M.D. (2004) *Chem. Mater.*, **16**, 4533–4542; (c) Günes, S. and Neugebauer, H. (2007) *Chem. Rev.*, **107**, 1324–1338; (d) Armaroli, N. and Balzani, V. (2007) *Angew. Chem. Int. Ed.*, **46**, 52–66; (d) Thompson, B.C. and Fréchet, J.M.J. (2008) *Angew. Chem. Int. Ed.*, **47**, 58–77; (f) Balzani, V., Credi, A., and Venturi, M. (2008) *ChemSusChem*, **1**, 26–58.

11
Supramolecular Chemistry of Carbon Nanotubes

Bruno Jousselme, Arianna Filoramo, and Stéphane Campidelli

11.1
Introduction

Due to their outstanding properties, carbon nanotubes (CNTs) are a particular class of materials that are still extensively studied. However, the processing of CNTs and their integration into real applications are severely limited by a number of inherent shortcomings such as polydipersity and purity of the samples, difficulty of manipulation, and low solubility of the nanotubes. Indeed, raw carbon nanotubes also contain nonnegligible quantities of amorphous carbon and metal catalyst residues, the quantity of each depends notably on the fabrication technique used to produce the nanotubes and it can vary for the same kind of nanotubes from one batch to another. Moreover, a sample of carbon nanotubes is composed of many tubes of different lengths, diameters, and properties. For example single-wall carbon nanotubes (SWNTs) can be either metallic or semiconducting depending on their structures (i.e., on how the sheet of graphene has rolled up to form the hollow tube). Therefore, a great attention has been paid in the past 15 years to the purification, separation, and functionalization of carbon nanotubes and this research field is still very active.

The functionalization of carbon nanotubes offers the invaluable opportunity to combine the properties of the nanotubes with those of other classes of materials for many applications. To add new functions to nanotubes, two general strategies have been explored: (1) the covalent functionalization of sp^2 framework of the nanotubes or the derivatization of oxygenated functions created by oxidative treatments of the nanotubes and (2) the noncovalent functionalization through supramolecular interactions between the nanotubes and the other chemical objects.

This chapter deals with the latter functionalization technique. In particular, we will describe the principal approach explored to functionalize noncovalently the nanotubes and we will discuss some of their applications.

Supramolecular Chemistry of Fullerenes and Carbon Nanotubes, First Edition. Edited by Nazario Martin and Jean-Francois Nierengarten.

11.2 Supramolecular Carbon Nanotube Hybrids

Many synthetic protocols have been optimized for the functionalization of CNTs. Basically, either a noncovalent or a covalent approach can be used. The two methodologies differ in that the supramolecular chemistry does not interfere with the extended π electron system of CNT. Instead, an extensive organic functionalization transforms sp^2 carbons into sp^3, thus interrupting conjugation.

11.2.1 Carbon Nanotube and Surfactants

Here, we do not want to give an exhaustive list of surfactants that have been used for carbon nanotube dispersions; we rather choose to give few examples of nanotube dispersion and individualization, to discuss the organization of the molecules around the nanotubes, and to show their applications for nanotube sorting.

11.2.1.1 Suspension of Single-Wall Carbon Nanotubes (Why, How, and What for?)

The electronic properties of SWNTs can be either semiconducting or metallic depending on their atomic structure geometry (diameter and chirality). This particular feature and their nanometer-range diameter make them ideal candidates for future nanoelectronics. However, as-produced SWNT samples display a not negligible distribution, both in diameter and in chirality. In addition, they have the tendency to aggregate into bundles/ropes (due to the substantial van der Waals tube–tube interactions). These facts represent important issues for applications. Chronologically, the first problem to solve is the presence of aggregates. This bundling disturbed the study of their electronic properties and complicated attempts in separating nanotubes by size, type, or chirality, finally avoiding the possibility to consider and use them as single macromolecular species.

The first works reporting individual SWNTs used organic solvent for fabricating a solution of pristine or chemically modified nanotubes [1–4]. In these cases, the debundling was facilitated by chemical modifications. However, the chemical attacks of the nanotube sidewalls create defects in the sp^2 graphene lattice and partially destroy their outstanding electronic properties [5]. On the contrary, it is generally admitted that the use of surfactants is an effective and nonaggressive method for debundling nanotubes. The idea is that surfactants disperse nanotubes and successfully suspend them through supramolecular interactions, preserving (or very slightly modifying) their electronic properties [6]. Pioneering studies showed that anionic sodium dodecylsulfate (SDS) or nonionic triton (TX100) could form stable colloidal suspension of short single-wall nanotubes in water [7]. However, more systematic studies of the dispersion efficiency of various surfactants started only in 2003. These efforts concerned mainly laser ablation [8], HiP$_{CO}$ [9], and CoMoCat [10] SWNTs.

Islam *et al.* [8] reported on the exceptional capability of sodium dodecylbenzenesulfonate (NaDBS) surfactant to suspend high concentration of SWNTs. More in detail, the authors compared the effectiveness of various surfactants for suspending commercially purified HiP_{CO} nanotubes. To fabricate the suspensions they used surfactants in concentrations always exceeding the nominal critical micelle concentration (cmc) in a single-step process consisting in a mild bath sonication (12 W, 55 kHz, 20 h). Then, they checked the aggregation status and stability of the obtained solutions both by macroscopic observation of flocculation/precipitation in the solutions and by AFM imaging of the deposited products on silicon surfaces. It is worth to note that the AFM technique has some limitations since the imaging analysis does not allow definitively excluding the presence of small bundles of few nanotubes (due to the tip deconvolution effect) [11]. According to their report, NaDBS gives the best results, enabling to raise the SWNT concentration for stable suspensions up to 20 mg/ml, while for all the other surfactants this does not exceed the range of 0.1 mg/ml.

Moore *et al.* [9] studied the ability of suspending as-produced HiP_{CO} nanotubes by (2 wt%) aqueous anionic, cationic, or nonionic surfactants and polymers. To achieve their suspension, they used high-shear mixing and ultrasonication followed by ultracentrifugation. The obtained decant was studied by UV–Vis–NIR absorption, photoluminescence, Raman, and cryoTEM imaging. Note that the combination of these techniques permits to be more confident concerning the absence of bundles (e.g., photoluminescence is a typical signature of isolated semiconducting SWNTs [6]). Their results reported a final nanotube concentration varying in the range of 10–20 mg/l for all the tested surfactants or polymers.

In the same year, Matarredona *et al.* [10] investigated the effect of chemical pretreatment of CoMoCat carbon nanotubes on their interactions with NaDBS surfactant. Their studies were based on the electrical charge of the nanotube surface, which varied with the pH of the surrounding media (point of zero charge or PZC effect) and could be greatly affected by the nanotube purification method. They purified the as-produced CoMoCat sample by a protocol using in its final step an acid (HF) or a basic (NaOH) treatment. As expected, the purified sample presented different PZC (probably due to the presence of fluorine or sodium on the nanotube sidewalls). The idea was to study the Coulombic interactions in surfactant/nanotube systems by performing the experiments at different pH values. Their experiments demonstrated that the hydrophobic interactions dominated and that Coulomb forces could play a role only at extreme pH values.

These three works reported various hypotheses and suggestions concerning the supramolecular organization of the surfactants around the nanotube sidewalls. The consensus is that surfactants disperse SWNTs in aqueous solution mainly via hydrophobic/hydrophilic interactions. The hydrophobic part of the surfactant is adsorbed on the sidewall, while the hydrophilic head associates with water. The presence of an aromatic ring in the hydrophobic chain is supposed to reinforce the interaction with the nanotube sidewalls due to π–π stacking. For nonionic surfactants and polymers, Moore and coworkers suggested that the ability of suspending nanotubes was related to the size of the hydrophilic group (enhanced steric stabilization), while for the ionic ones this property was generally associated with the electrostatic repulsion of their charged ions. Actually, different kinds of organi-

Figure 11.1 Surfactant organization on SWNTs. Three situations are possible: (a) encapsulation in cylindrical micelles, (b) adsorption of hemispherical micelles, or (c) random organization. Reproduced with permission of the American Chemical Society from Ref. [14].

zation were proposed, and this point still remains a topic of debate. The organization scheme ranges from encapsulation in cylindrical micelles [6, 10, 12], adsorption of hemispherical micelles (also called hemimicelles) [8, 13], to random adsorption [14]. Experimentally, surfactant organization has been studied by HR-TEM [13], small-angle neutron scattering (SANS) [14], and more recently by cryo-TEM [12]. The experimental data demonstrated results consistent with situation in cylindrical micelles (a) [12], in hemimicelles (b) [13], or in random adsorption (c) [14] of Figure 11.1 depending on the experimental protocol used to prepare the suspensions.

In some cases, two different behaviors were reported for the same surfactant. In particular for SDS, Richard *et al.* [13] reported on hemimicellar configuration, while Yurekli *et al.* [14] found the signature of random adsorption. In spite of this apparently contradictory experimental situation, numerical simulations helped to understand and could elucidate the differences [15, 16]. The simulation revealed that a dynamic balance existed for the surfactant molecule between three states: (i) isolated in solution, (ii) in micelle in the solution, and (iii) adsorbed on the SWNT. Simulations suggested that this was particularly true and important for low surfactant concentrations. For example, Calvaresi *et al.* [15] reported that for all surfactant concentrations the coverage of the nanotube started by collisions between the SWNT and preformed micelles in the liquid. At very low surfactant concentration, the micelles interacted dynamically and weakly with the CNT, while the micelles started to adsorb and cover the SWNT when the surfactant concentration was increased. More in details, at low concentration, it was predicted that the micelles spread randomly on the surface and remained in a dynamical equilibrium on it. As far as the surfactant concentration increased, simulations showed that it started to self-assemble in stable superstructures (hemimicelles). Then, the hemimicelles were replaced by cylindrical micelles at higher concentration. These results converged with the concentration-dependent study by Matarredona *et al.* [10] who reported about two stable plateaus that were assigned to random face-on adsorption of the surfactants and to their ordered edge-on adsorption. It is worth to note, however, that in Calvaresi's work the nanotube was a simple isolated cylinder and chirality was not taken into account, while this parameter could influence the interaction and organization of the surfactant.

More recently, studies delved more deeply into the differences observed when varying the nanotube batches. It was pointed out that even when a standard identical procedure for dispersion was used, different results in suspensions were obtained due to differences in SWNT batches. In particular, smaller diameter HiP_{CO} presented higher solubility and dispersibility [17, 18]. In this sense, simulations predicted that small-diameter SWNTs have weaker van der Waals attraction than large-diameter ones rendering them easier to debundle and disperse [19, 20]. In addition, smaller diameter nanotubes would permit a higher packing density of the surfactant tails and less electrostatic repulsion between head groups [21].

The lesson to be learned here is that surfactant concentration is important, but an effective dispersant for a particular batch of nanotubes will not perform exactly in the same way for another one.

Biomolecules were also used for producing high-quality suspension of SWNTs and in particular a great attention was given to DNA. In 2003, Zheng *et al.* reported that single-stranded DNA (ssDNA) molecules wrapped helically around SWNTs [22]. They proposed that the aromatic part of the DNA nucleobases stacked on the nanotube sidewall via π–π interactions and that the hydrophilic charged sugar-phosphate backbone ensured the nanotube suspension in water. In addition, the flexibility of ssDNA could permit to find low-energy conformations for the molecule when the base stacking matched with the underlying SWNT lattice structure. They also noted a difference depending on the basic sequence of the strand: poly(adenine) or poly(cytosine) sequences were less efficient to suspend nanotubes compared to poly(thymine). These features suggested that the ssDNA/SWNT interactions depended on both DNA sequence and SWNT structure as it was also supported in literature by simulations [23–25]. Interesting experimental results for SWNT solubilization by ssDNA oligomers were reported by Vogel *et al.* [26]. The authors demonstrated that short $d(GT)_3/d(AC)_3$ mixtures gave better results than longer $d(GT)_n/d(AC)_n$ or isolated $d(GT)_n$ oligomers for nanotube dispersion.

The understanding and control of the interactions between the nanotubes and the surfactants are important in view of future separation of the nanotubes. Once the individualization of the nanotubes is achieved in solution, the sorting process can take place. Different approaches were reported in literature to sort SWNTs in length [27–31], diameter [29, 30, 32–38], chirality [39–44], metallic/semiconducting character [32, 35, 45–49], and handedness [50–52]. One of the most popular strategies reported in literature took advantage of the differences in surfactant organization on suspended/encapsulated individual nanotubes. The method is based on the density gradient ultracentrifugation (DGU) technique. DGU is a method widely used in biochemistry for the purification of proteins and nucleic acids. In the present context,

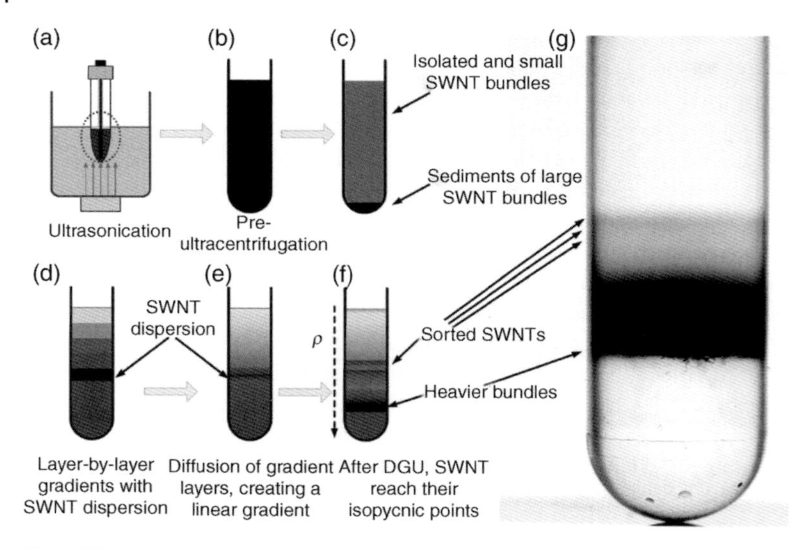

Figure 11.2 Schematic representation of SWNTs sorting by DGU (a–f) and picture of the cell tube showing the separation of SWNTs after DGU (g). Reproduced with permission of the American Chemical Society from Ref. [53].

during the ultracentrifugation process the encapsulated SWNTs migrate into the density gradient medium until they reach their corresponding isopycnic points (the points where their buoyant density equals that of the surrounding medium) [35]. Thus, objects with different buoyant density spatially separate in the gradient (overcoming the limitations of ultracentrifugation in a density constant medium). It is, therefore, easy to understand that the optimization of surfactant/nanotube interaction is crucial and that, in principle, small differences in surfactant packaging around the nanotube related to nanotube lattice structure can enable the sorting process. Figure 11.2 shows the schematic representation of the DGU process [53]. First, a SWNT/surfactant suspension was prepared by ultrasonication (Figure 11.2a), big bundles were removed by conventional ultracentrifugation (Figure 11.2b), and the supernatant was recovered (Figure 11.2c). Second, the step gradient for DGU was prepared by placing in several layers of gradient density media with decreasing concentration on top of each other. The SWNT dispersion was placed between two layers (Figure 11.2d). A linear density gradient is formed spontaneously by diffusion (Figure 11.2e). Finally, DGU led to SWNT spatial separation in the ultracentrifuge cell (as schematized in Figure 11.2f and demonstrated by a real photo of the DGU cell in Figure 11.2g).

One famous result obtained by such technique was reported by Hersam's group [35]. In their work, the group achieved a rich structure density relationship for sodium cholate (SC) encapsulated CoMoCat SWNTs that permitted their sorting by diameter and bandgap (Figure 11.3). For DGU, OptiPrep^{TM} (60% w/v iodixanol solution in water) is very often used as a gradient density medium.

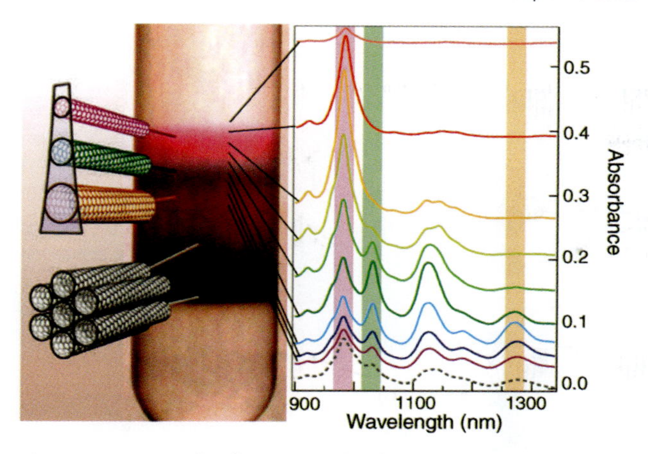

Figure 11.3 Example of SC-encapsulated CoMoCat SWNTs (0.7–1.1 nm) sorted by diameter and energy bandgap using density gradient ultracentrifugation. Adapted with permission of the Nature Publishing Group from Ref. [35].

Another strategy that has been explored to achieve nanotube sorting is based on ion exchange chromatography (IEX). Ion exchange chromatography is a technique that is commonly used to separate ions or polar molecules. It involves mainly two steps: first the molecules reversibly adsorb on oppositely charged resin and then desorption is brought about either by a change in the pH or by an increase in the salt concentration in the eluent. In this context, the design of DNA sequences for specific recognition of SWNTs is of particular importance. The idea is that SWNT–DNA hybrids are negatively charged because of the DNA phosphate backbone and that their behavior in the column depends on their linear charge density. The effective net charge of the SWNT–DNA hybrids depends mainly on the linear charge density of the phosphate backbone along the nanotube axis (i.e., how ssDNA is arranged/packed on SWNTs). Note that the effective net charge is also influenced by the electronic properties of the nanotube (charge image/polarizability effect). More in details, for a metallic nanotube the negative charges of DNA backbone induce positive screening by charge images of the nanotube and by consequence the net linear charge is reduced with respect to the value given by the wrapped DNA backbone alone. On the contrary, for a semiconducting nanotube, the lower polarizability of the nanotube, compared to the one of the surrounding water, results in an increased effective linear charge.

These properties were used to fractionate ssDNA/nanotube solutions by optical absorption, which are known to be distinct for each particular (n, m) nanotube. To achieve such a result, Zheng and coworkers [43] rationalized the exploration of ssDNA libraries and identified short DNA sequences that enabled SWNT single chirality separation as reported in Figure 11.4 (left). They observed that the successful sequences showed periodic purine–pyrimidine patterns. They suggested that these purine pyrimidine patterns could form a two-dimensional sheet by hydrogen bonding, forming well-ordered three-dimensional barrel (when folded selectively on nanotubes) (see Figure 11.4, right).

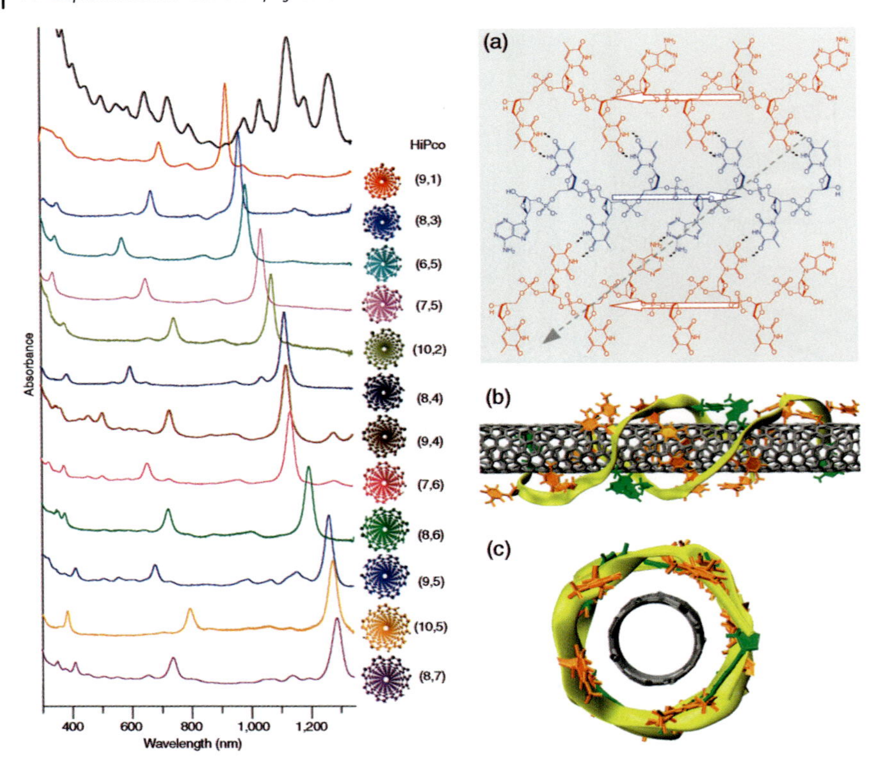

Figure 11.4 (*Left*) UV–Vis–NIR absorption spectra of semiconducting SWNTs sorted by chirality (in color) and the starting HiP_{CO} mixture (in black). (*Right*) (a) Proposed organization of a 2D DNA sheet structure formed by three antiparallel ATTTATTT strands; (b–c) schematic representation of the DNA barrel on a (8,4) nanotube formed by rolling up of the previous 2D DNA sheets. Adapted with permission of the Nature Publishing Group from reference [43].

11.2.2
π Stacking Interactions

π-conjugated molecules exhibit strong interactions with the sp^2 framework of the nanotube sidewalls. Very early, aromatic molecules bearing proper functionalization have been used to disperse carbon nanotubes in aqueous and organic media. Due to the huge possibilities offered by the organic chemistry and the versatility of the derivatization of polycyclic aromatic molecules, a wide variety of compounds have been synthesized and used to bring new functionalities to the nanotubes. In the following section, we will discuss some examples from the recent literature.

11.2.2.1 **Pyrene Derivatives**

In their initial reports, the group of Nakashima opened new avenues for nanotube functionalization. The group demonstrated that in a family of aromatic compounds

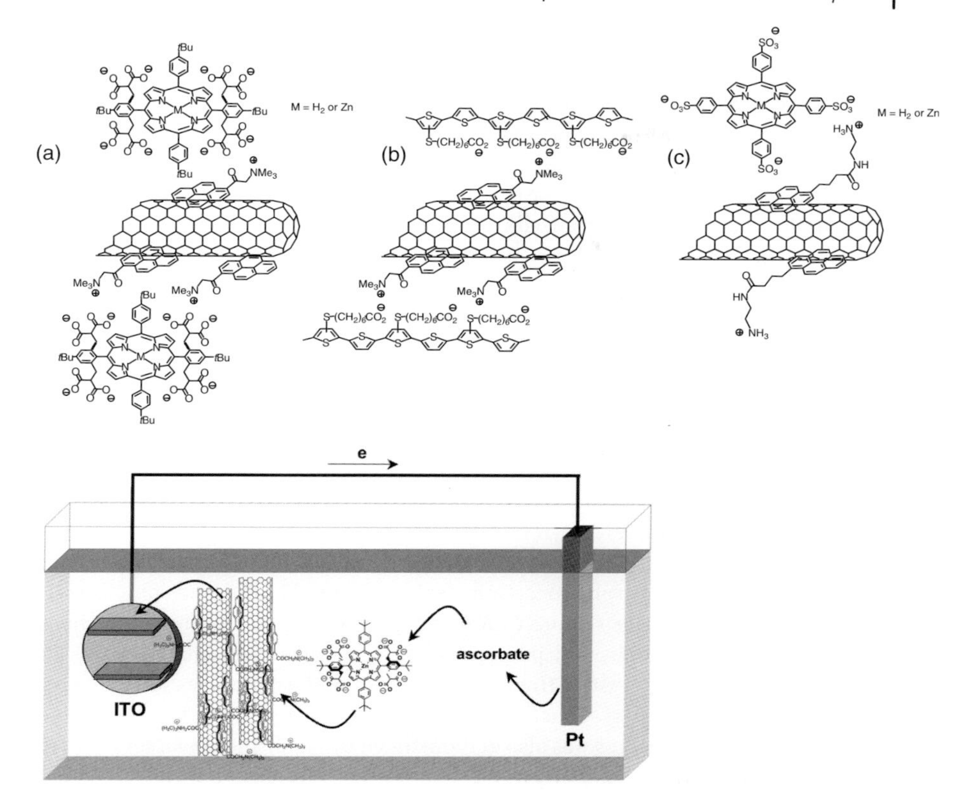

Figure 11.5 (*Top*) Examples of supramolecular donor/acceptor assemblies formed by electrostatic interactions between positively charged pyrenes and negatively charged chromophores (a–c). (*Bottom*) Schematic representation of a photoelectrochemical cell containing carbon nanotubes, positively charged pyrenes and negatively charged porphyrins.

containing a polar head, polycyclic derivatives such as phenanthrene and even better pyrene were able to give stable suspension of SWNTs in water [54, 55]. In such systems, the pyrene moiety ensured the interactions with the nanotube sidewalls, while the trimethylammonium polar head allowed the solubility in water. Rapidly 1-(trimethylammonium acetyl)pyrene bromide became popular and was used to anchor negatively charged molecules such as porphyrin or polythiophene derivatives in nanotubes (Figure 11.5a–b) [56–59]. The chromophore/nanotube hybrids presented interesting photophysical properties and they were tested as photoactive materials for current generation in electrochemical cells.

In the SWNT/pyrene^{+}/porphyrin^{8-} composite systems, fluorescence and transient absorption studies in solutions showed rapid intrahybrid electron transfer, creating intrinsically long-lived radical ion pairs. Following the initial charge separation event, the spectroscopic features of the oxidized donors disappear with time.

Through analysis at several wavelengths, it was possible to obtain lifetimes for the newly formed ion pair state of about 0.65 and 0.4 μs for H_2P^{8-} and ZnP^{8-}, respectively [57]. The favorable charge separation features that result from the combination of SWNT with porphyrins in SWNT/pyrene$^+$/MP^{8-} ($M = H_2$ or Zn) were promising for the construction of photoactive electrode surfaces.

Using electrostatically driven layer-by-layer (LBL) assembly technique, semitransparent ITO electrodes were realized from SWNT/pyrene$^+$/porphyrin^{8-} and SWNT/pyrene$^+$/polythiophene^{n-}. Photoelectrochemical cells were finally constructed using a Pt electrode connected to the modified ITO electrode. An example of a cell is given in Figure 11.5 (bottom); upon illumination, electron transfers occurred from the porphyrins or the polythiophene to the nanotubes. The electrons are then injected into the ITO layer then traveling to the Pt electrode. The oxidized electron donors were converted to their ground state through the reduction via sodium ascorbate in the electrolyte, which was used as a sacrificial electron donor. These systems gave rise to promising monochromatic internal photoconversion efficiencies (IPCE) of up to 8.5% for porphyrin systems [60] and between 1.2 and 9.3% for one and eight sandwiched layers of SWNT/pyrene$^+$/polythiophene^{n-}, respectively [59].

In a similar way, Sandanayaka *et al.* [61] described the assembly of anionic tetrasulfonatophenyl porphyrin sodium salts (Figure 11.5c) or cationic tetra-*N*-methylpyridyl porphyrins (Figure 11.6b) with positive or negative pyrene anchored in SWNTs. The photophysical properties of the hybrids were characterized and photoinduced charge transfers from the porphyrins to the nanotubes were observed in both supramolecular systems.

The inversion of charges was also demonstrated by Guldi and coworkers with the use of pyrene carbolylic acid or pyrene sulfonate derivatives and positively charged porphyrins (Figure 11.6a) [62]. The interactions between SWNT and pyrene$^-$ were investigated by absorption spectroscopy. The maxima of the pyrene$^-$ transitions in the 200–400 nm region were shifted by about 2 nm, which suggested mutually interacting π systems.

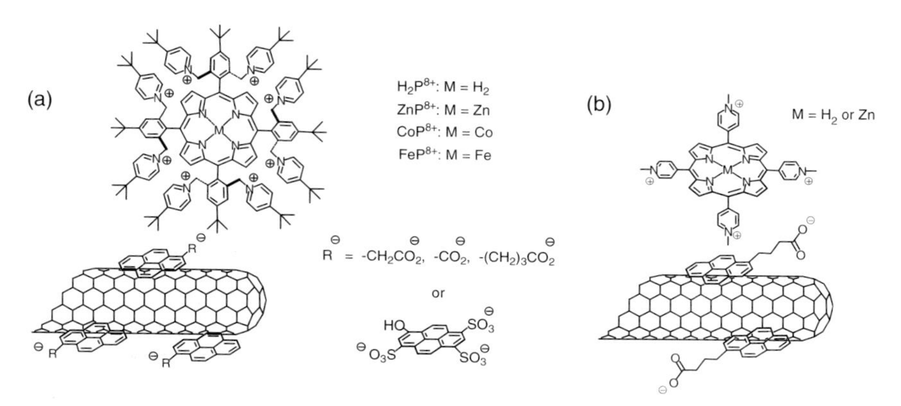

Figure 11.6 SWNT with negatively charged pyrene and positively charged porphyrins.

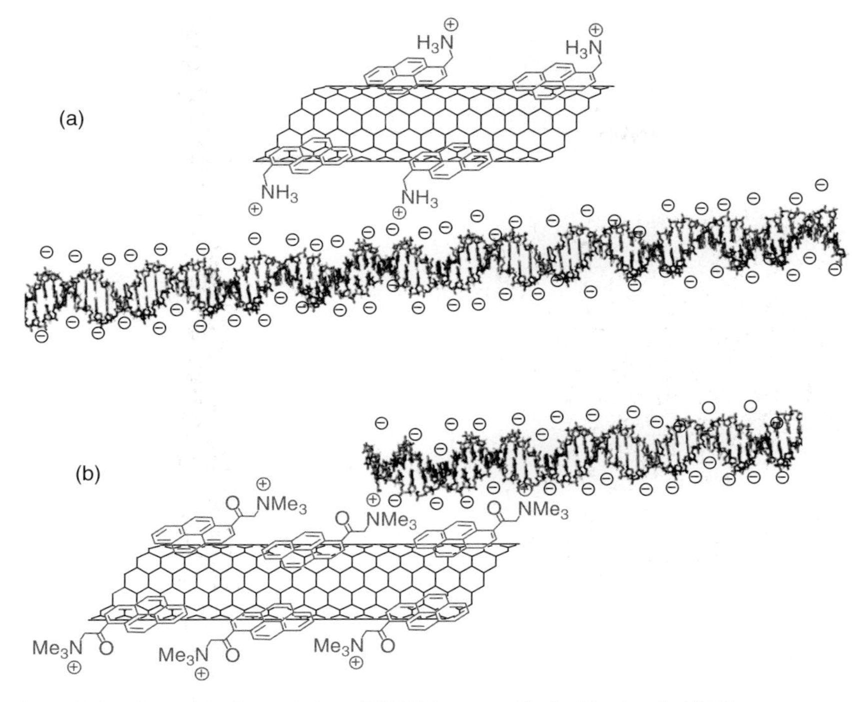

Figure 11.7 Examples of association of SWNT/pyrene with double-stranded DNA.

Positively charged pyrenes were also used to assemble DNA and carbon nanotubes. Two recent examples in the literature should be mentioned: the first in which SWNTs coated with 1-pyrenemethylamine hydrochloride were deposited on λ-DNA aligned on silicon surfaces (Figure 11.7a) [63] and the second in which SWNTs coated with 1-(trimethylammonium acetyl)pyrene bromide were used to assemble 300 nm long double-stranded DNA in solution (Figure 11.7b) [64]. In the latter case, it was demonstrated that the hybrid formation was a reversible process and that DNA can be voluntarily detached from SWNTs, by addition of a negatively charged pyrene, providing a new perspective for controlled assembly and deassembly applications.

In the previous examples, electrostatic interactions were used to build supramolecular systems adding new functions to carbon nanotubes. Different approaches based on host–guest complexation or axial complexation with transition metals were also reported.

The dispersion of SWNT with a pyrene derivative bearing an imidazole ring was achieved and the imidazoyl moiety was used for axial complexation of zinc porphyrin (ZnP) and naphthalocyanine (ZnNc) derivatives in solution (Figure 11.8) [65]. Photophysical measurements showed efficient fluorescence quenching of the ZnP and ZnNc donors in the nanohybrids and revealed that the photoexcitation of the chromophores resulted in the one-electron oxidation of the donor unit with a simultaneous one-electron reduction of the SWNT. The experiments were also conducted in the presence of electron and hole mediators (dihexyl viologen dication

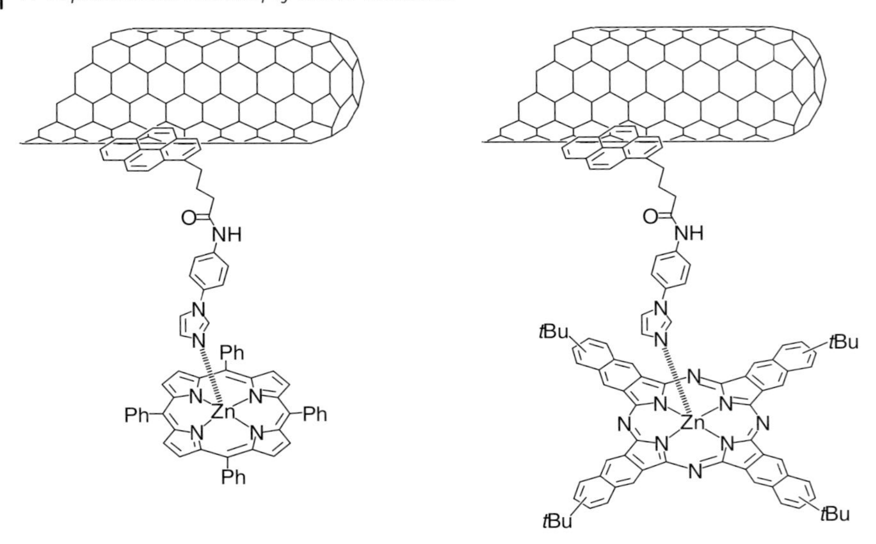

Figure 11.8 SWNT-pyrene supramolecular assemblies axially complexed with zinc-porphyrins and zinc-naphthalocyanines.

and 1-benzyl-1,4-dihydronicotinamide, respectively). Accumulation of the radical cation ($HV^{\bullet +}$) was observed in high yields, which provided additional proofs for the occurrence of photoinduced charge separation.

The same group used the specific recognition of ammonium cation with benzo-18-crown-6-ether to immobilize porphyrins and fullerene on nanotubes [66, 67]. For the realization of the nanotube/porphyrin hybrids, SWNTs were first dispersed in DMF using a pyrene derivative bearing an ammonium cation and were then complexed with porphyrins containing one or four benzo-18-crown-6-ether moieties (Figure 11.9a and b), the latter offering more complexation sites, which was expected to result in more stable ensembles owing to the cooperative binding effect. Steady-state and time-resolved emission studies revealed an efficient quenching of the singlet excited state of the porphyrins that was attributed to charge transfers between the porphyrins and the nanotubes [66]. The same strategy was used to combine SWNT-pyrene ammonium with fulleropyrrolidine bearing the 18-crown-6-ether moiety (Figure 11.9c); in this case, upon photoexcitation at 532 nm, a photoinduced electron transfer from the nanotube to the fullerene was observed [67].

Stoddart and Grüner reported the fabrication of carbon nanotube field-effect transistors (CNT-FETs) using SWNTs decorated with pyrenes modified with a β-cyclodextrin [68–70]. A SWNT can be seen as a sheet of graphene rolled up to form a hollow tube; in a SWNT all the atoms of carbon are in contact with the environment, thus making nanotubes an incredibly sensitive material and during the past decade, many works demonstrated the potential of SWNTs for the fabrication of field-effect transistors and sensors [71–74].

Taking advantage of the sensitivity of the CNT-FETs and the recognition properties between β-cyclodextrin and adamantane derivatives [75], the authors fabricated

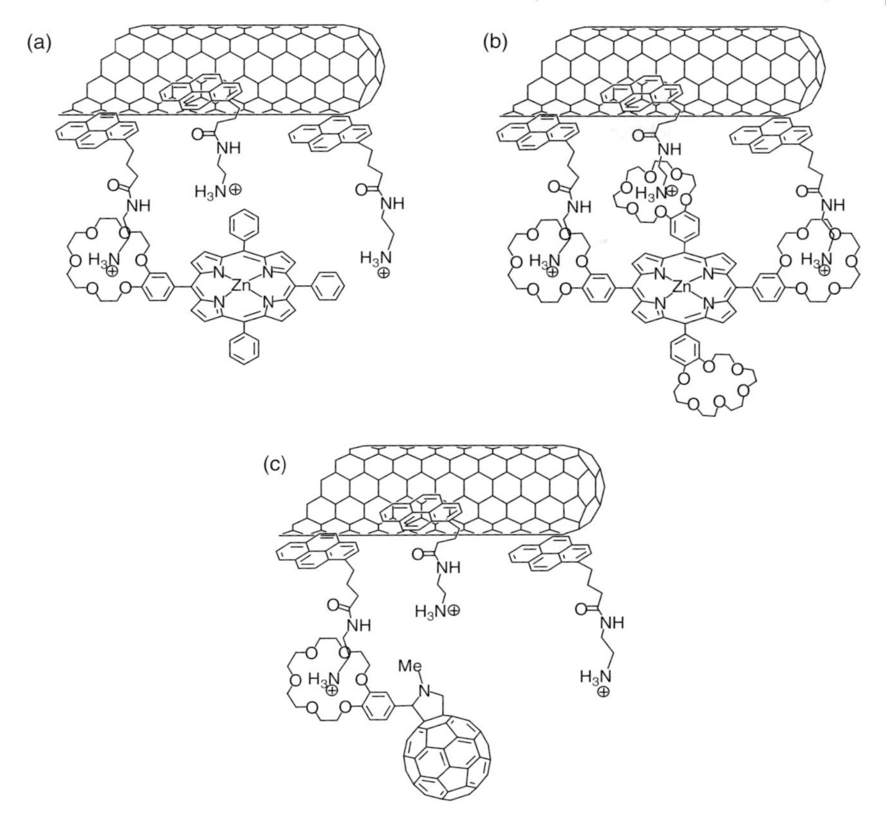

Figure 11.9 SWNT-porphyrins and SWNT-fullerene hybrids based on ammonium cation/crown ether self-assembly.

chemical detectors (Figure 11.10a) and photosensors (Figure 11.10b). In the first case, electrical response of the device allowed the detection of the complexation events: the threshold voltage of the CNT-FET shifted toward negative gate voltage and the magnitudes of the shifts depended highly on the complex formation constants (K_S) between the organic molecule and the pyrenecyclodextrin [68]. In the second case, the device was used as a tunable photosensor to sense a fluorescent adamantyl-modified ruthenium complex. When the light is on, the threshold voltage of the pyrenecyclodextrin-CNT/FET shifted toward a negative gate voltage by about 1.6 V and its resistance increased quickly, indicating a charge transfer process from the pyrenecyclodextrin/Ru complexes to the SWNTs [69].

Pyrenecyclodextrin-decorated nanotubes were also used to form SWNT hydrogels via host–guest complexations between the β-CDs of the SWNT/pyrenecyclodextrin hybrids and the dodecyl chains of a modified polyacrylic acid. The supramolecular SWNT hydrogel exhibited gel-to-sol transition by adding competitive β-CD guests (i.e., adamantyl derivatives) [76]. Very recently, the complementary approach, that is, the assembly of SWNT/adamantyl-modified pyrene with β-cyclodextrins-tethered

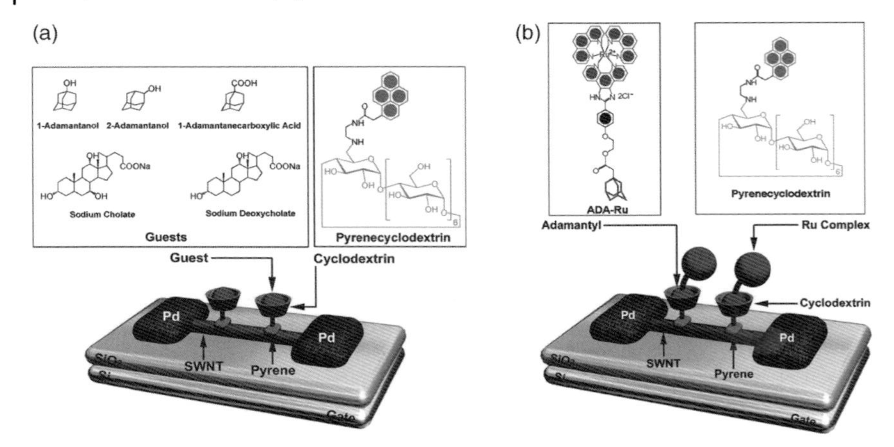

Figure 11.10 Schematic representation of the pyrenecyclodextrin-decorated CNT-FET devices: (a) chemical sensor and (b) photosensor. Adapted with permission of the American Chemical Society from Ref. [70].

ruthenium complexes, was proposed [77]. The nanotube-based supramolecular assembly was complexed with plasmid DNA and used as nonviral as a nonviral gene delivery system with the ruthenium complexes as a fluorescent probe to monitor uptake of DNA by cells.

Addition of new moieties and/or new functionalities on carbon nanotubes can be performed directly through the addition of a pyrene covalently modified previously. Many examples of pyrene modified with porphyrins [78–80], phthalocyanines [81], tetrathiafulvalenes [82, 83], thiophenes [84], ferrocenes [85], fullerenes [86, 87], azobenzene chromophores [88], spiropyrans [89], quantum dots (QDs) [90] or glycodendrons [91], and glucosamines [92] were described these past 5 years mainly for photoinduced charge transfer or sensing applications. The advantage of the preparation of robust pyrene-based derivatives is a better control of the final structures since no sequential assembly is required. The evident disadvantage is the lack of versatility: each new pyrene derivative must be prepared independently and combined with the nanotubes.

To overcome this problem, commercially available activated pyrenes (1-pyrene-butanoic acid succinimidyl ester) were used, in the first step, to disperse and/or functionalize carbon nanotubes, and in the second step, to covalently attach the molecules of interest containing amine groups to the activated carboxyl groups. This approach permitted to successfully attach biomolecules (e.g., proteins, enzymes, antigens, and DNA aptamers) [93–97] or QDs [98] to carbon nanotube sidewalls. One of the interesting features of this method is the facile immobilization of biomolecules on nanotubes for the realization of biosensors.

11.2.2.2 **Other Cyclic Aromatic Compounds**

Several examples of dispersion of carbon nanotubes using other polycyclic aromatic compounds such as anthracene [99, 100], phenathrene [55], extended pyrene [101], and coronene [102] were reported in literature. These compounds are, in general, less efficient than pyrene for nanotube functionalization. A solution for better dispersion of nanotubes and for their selection in chirality could come from the synthesis of specific aromatic bent structures that could follow the curvature of the nanotube sidewalls [103].

Few years ago, Feng *et al.* [104] described the combination of carbon nanotubes with a perylene derivative (the *N,N′*-diphenyl glyoxaline-3,4,9,10-perylene tetracarboxylic acid diacidamide). The resulting nanocomposite was tested as active film for photocurrent generation. The group of Hirsch reported the efficient individualization of SWNTs using amphiphilic perylene derivatives [105, 106]. The surfactants were based on a perylene core bearing one or two water-soluble Newkome dendrons [107] as solubilizing counterparts (Figure 11.11).

Perylene diimide-based surfactants are a particular class of compounds because they combine two important features: (1) good dispersion ability of nanotubes and (2) excellent electron accepting character. In particular, the perylene derivative bearing two dendrons of first generation were used in a series of photophysical and spectroelectrochemistry experiments to probe the charge transfers between the nanotubes and the perylene [108].

11.2.2.3 **Porphyrins and Derived Structures**

Aromatic macrocycles such as porphyrins or phthalocyanines are also suitable for nanotube dispersion. In particular, several groups demonstrated that monomeric [109–114] or polymeric porphyrins [115] and fused porphyrins [116] could stack to the nanotube surface by π stacking interactions. Some examples of monomeric phthalocyanines are also reported in the literature [117–119].

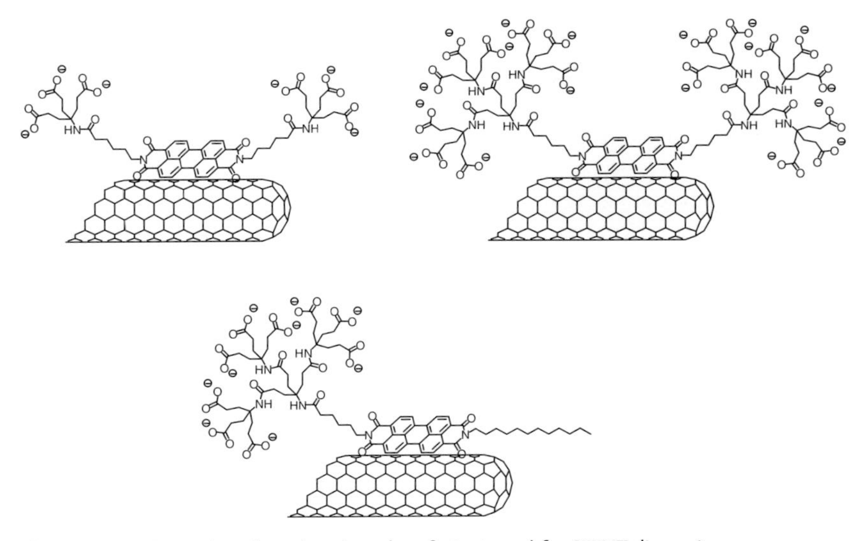

Figure 11.11 Examples of perylene-based surfactant used for SWNT dispersion.

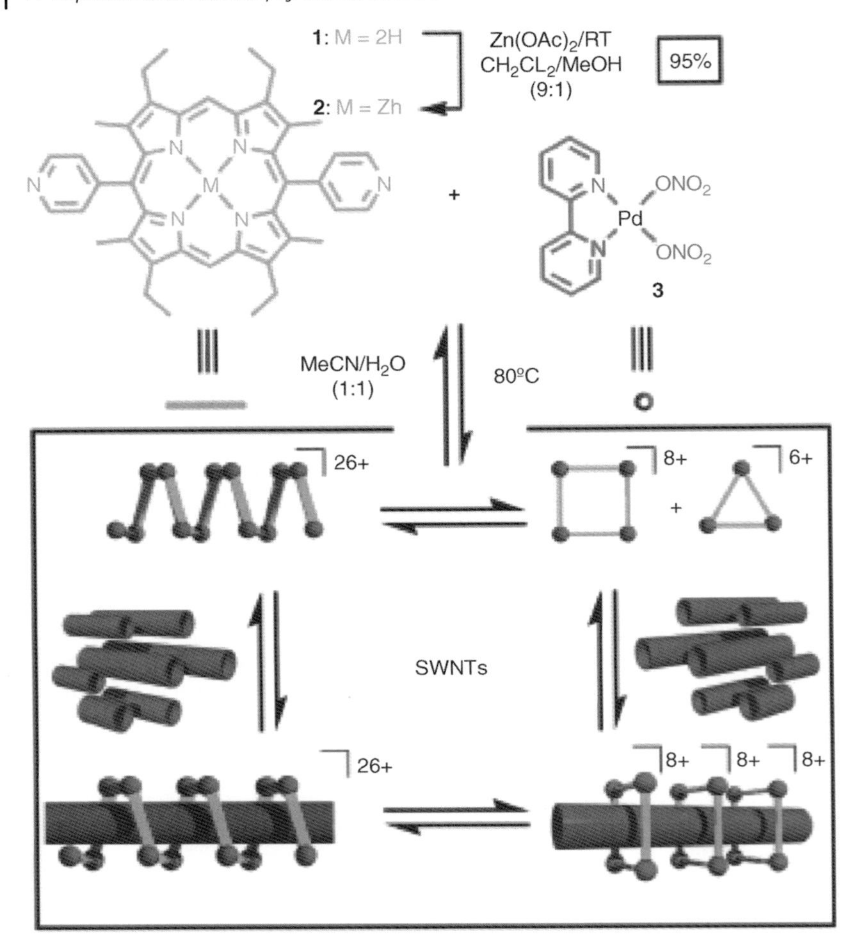

Figure 11.12 Example of supramolecular structures formation around SWNTs. Reproduced with permission of Wiley-VCH Verlag GmbH from Ref. [120].

In 2005, Chichak *et al.* [120] reported on the functionalization of SWNTs with supramolecular porphyrin polymers. The polymer formation was based on the spontaneous complexation of 5,15-bis(4-pyridyl)-porphyrin with *cis*-palladium complexes (Figure 11.12). While neither the porphyrin nor the palladium complex was able to disperse efficiently the nanotubes in water, the supramolecular polymers, in the presence of SWNTs, gave stable dispersion. The same group demonstrated later that SWNT-field effect transistor coated with the porphyrin 2 alone could act as light detector by detecting photoinduced electron transfers within the donor/acceptor system [121].

Recently, a very smart application of π stacking interactions between the porphyrins and the nanotube sidewalls was developed by Komatsu and his group [50, 51, 122]. They demonstrated the separation in chirality of SWNTs and not simply in diameter [35] using particular porphyrin derivatives. Chiral diporphyrin "tweezers" containing 1,3-phenylene, 2,6-pyridylene, or 3,6-carbazolylene bridges were found to bind left- and right-handed helical nanotube isomers with different affinities to form complexes with unequal stabilities that could be readily separated by centrifugation (Figure 11.13). Theoretical calculation showed that the difference of association enthalpies between a tweezer and the two enantiomers of a (6,5)-SWNT was 0.22 kcal/mol [51]. The diporphyrins could be liberated from the nanotubes

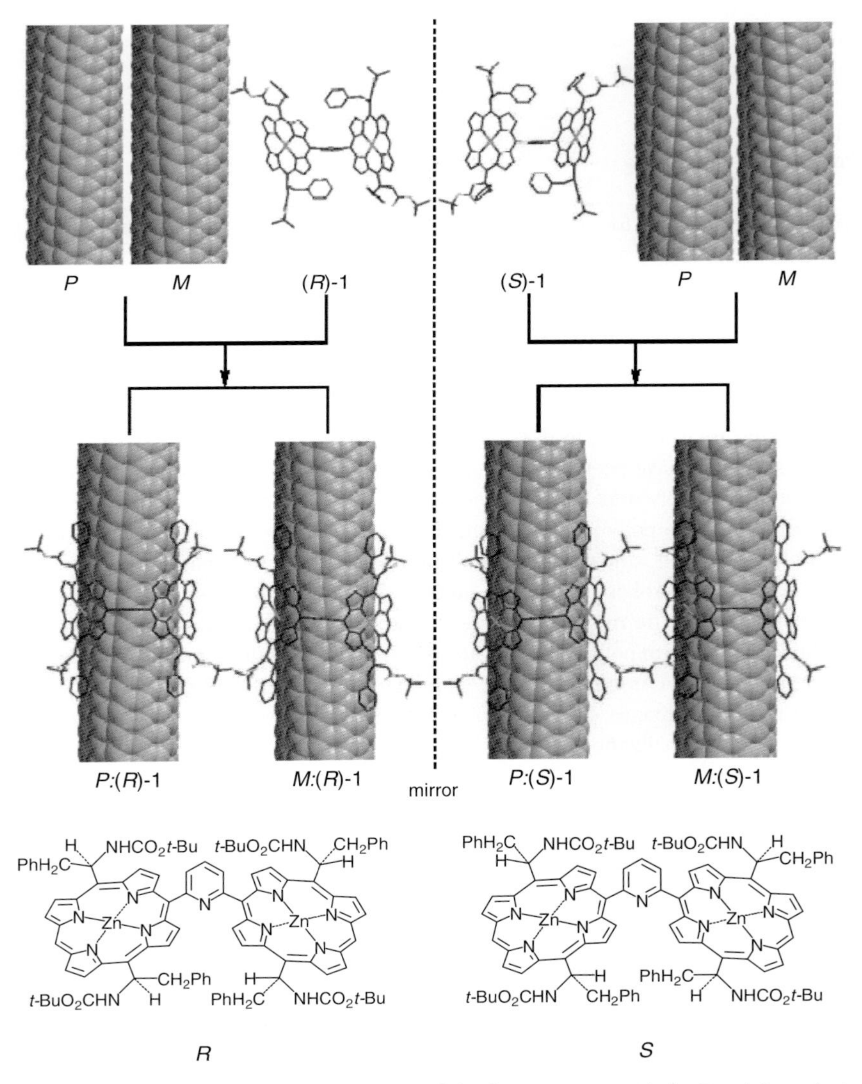

Figure 11.13 Representation of the selection of chiral (6,5)-SWNTs with the *R* and *S* porphyrin tweezers. Adapted with permission of the American Chemical Society from Ref. [51].

afterward, to provide optically enriched SWNTs. Circular dichroism (CD) of the nanotube solutions revealed the presence of Cotton effects that were not due to the tweezers and were therefore attributed to the chirality of the nanotubes.

The same group went a step further with the rationalization of the design of their chiral nanotweezers for discrimination of the handedness and diameter of SWNTs simultaneously [123]. In this example, the asymmetric porphyrin subunits were connected via a phenanthrene spacer. The dihedral angle made by two porphyrins were effective to selectively tweeze out (6,5)-SWNTs among the major components of CoMoCat-SWNTs. By CD, only (6,5)-SWNTs exhibit Cotton effects, indicating that the single enantiomer of (6,5)-SWNTs was enriched through the molecular recognition with the tweezers.

11.2.3 Polymers and Wrapping

In this section, we will focus on the association of carbon nanotubes with conducting and photo/electroactive polymers. Here, we voluntarily do not consider nanotube composites for reinforcement applications. This is a field of intense research but beyond the scope of this chapter. Note that the polymers usually employed for such reinforcement are mainly polyacrylates, polystyrenes, polyvinyls, and so on that are, in general, not appropriate for applications such as nanotube sorting or optoelectronic devices. For readers who are interested in the field of nanotube composites, we recommend them to refer to the reviews in Refs [124, 125].

Two strategies have been reported in literature for combining carbon nanotubes with conducting polymers. The first one consists in the direct ultrasonic mixing of the nanotubes and the polymers in organic solvent. The time of sonication has to be carefully controlled since under harsh conditions both the nanotubes and the polymers can be broken into small pieces. The second method is based on the polymerization of the monomer in the presence of the nanotubes. Note that this method tends to form polymer aggregates around the nanotubes and thus the carbon nanotubes/polymer assemblies are less defined than in the first case.

The main advantages of using polymers, instead of surfactant molecules, for nanotube functionalization is that the polymers reduce the entropic penalty of micelle formation and have significantly higher interactions with nanotubes than small molecules. As previously discussed for DNA and surfactants, also for polymers the chemical structure of the chains as well as their flexibility, stiffness, and hydrophobicity, combined with their electron affinity, are directly responsible for the interactions between the polymer and the nanotubes.

Poly(*meta*-phenylenevinylene) (P*m*PV) was the first π-conjugated polymer that was found to wrap around the nanotubes [126–129]. Due to the *m*-phenylene linkage and the repulsive interactions between the alkoxy solubilizing groups, P*m*PV backbone adopts a helical structure suitable for wrapping around the nanotube sidewalls. The motivation for the realization of nanotube/P*m*PV composites is the combination of the photoabsorption properties of the polymer with the conductivity and charge mobility in the nanotubes. For example, by adding multiwall carbon nanontubes

(MWNTs) to P*m*PV, the electrical conductivity of the polymer increased by eight orders of magnitude [127]. On SWNTs, it was demonstrated that P*m*PV was able to selectively pick up (11,6), (11,7), and (12,6) nanotubes from a HiP_{CO} SWNT mixture [130]. These nanotubes exhibited diameters of 1.19, 1.25, and 1.24 nm, respectively; the tubes with smaller diameters were removed gradually under centrifugation.

The backbone modification of the P*m*PV with a *m*-pyridylene instead of the *m*-phenylene gave the poly{(2,6-pyridinylenevinylene)-*co*-[(2,5-dioctyloxy-*p*-phenylene) vinylene]} (PPyPV). This polymer was also able to wrap and then disperse efficiently carbon nanotubes [131]. Field-effect transistors fabricated with SWNTs coated with P*m*PV or PPyPV demonstrated a photogating effect on *I*/*V* characteristics that could rectify or amplify current through the nanotubes. Similarly, P*m*PV derivatives functionalized with alcohol, azide, thiol, and amino acid functions in the C-5 position of the *meta*-disubstituted phenylene rings were designed and synthesized. These poly {(5-alkoxy-*m*-phenylenevinylene)-*co*-[(2,5-dioctyloxy-*p*-phenylene)-vinylene]} derivatives (PA*m*PV) were able to disperse SWNT in organic solvents; the functional groups on the polymers were used to graft pseudorotaxane molecular machines along the walls of the nanotubes to construct arrays of molecular switches and actuators [132].

The polyphenylacetylene (PPA) is structurally equivalent to the polystyrene in which the saturated $CH_2{-}CH_2$ groups are replaced with $CH{=}CH$ groups; the polyene backbone structure of the polymer allows the wrapping around nanotubes. The first example of nanotube/PPA hybrids was fabricated by *in situ* polymerization of the phenylacetylene monomer in the presence of MWNTs. This process gave stable suspension of MWNTs in common organic solvents such as chloroform, toluene, and tetrahydrofuran [133]. Later, PPAs modified either with pyrene [134] or with ferrocene [135] groups were synthesized, and absorbed on MWNTs and SWNTs. The PPA/nanotube hybrids showed good optical-limiting properties, light-emitting behavior, and efficient photoinduced charge transfer performance. Recently, it was demonstrated that donor–acceptor interactions between carbon nanotubes and poly(phenylacetylene) functionalized with carbazole or fluorene allowed to enhance the dispersion of MWNTs in organic solvents [136].

To summarize, due to their particular structure P*m*Pv, P*m*Pv derivatives and PPA polymers are able to wrap around nanotubes to give efficient dispersion in organic solvents. These nanotube/polymer hybrids open interesting perspectives for the realization of optoelectronic devices thanks to the combination of the electronic properties of the nanotubes with the optical properties of the polymers.

Recently, regioregular poly-3-alkylthiophene (PT), commonly used as a *p*-type material in organic solar cell devices, was wrapped around SWNTs [137–140]. Using scanning tunneling microscopy, Goh *et al.* [138] showed that monolayers of regioregular poly-3-hexylthiophene (P3HT) are adsorbed on SWNTs with an angle of 48° (±4°) with respect to the SWNT axis. Recently, the group of Nicholas showed that P3HT can be used to isolate individual SWNT forming a highly ordered nanohybrid structure that reduced the optical bandgap and increased carrier mobility in the polymer. They also demonstrated using photoluminescence studies that highly efficient energy transfers, rather than charge transfers, occurred from P3HT to SWNTs, explaining the relatively poor performance of P3HT-SWNT solar cells [139]. Nevertheless, the excess P3HT led to long-lived charge-separated state in a mixture P3HT with 1% of SWNTs [140].

On the contrary, more rigid polymers such as poly(*p*-phenyleneethynylene) (PPE) were found to orient their backbones along the nanotube long axis [141]. A similar behavior was reported for the assemblies between linear and rigid Zn-porphyrin conjugated polymers and SWNTs [115]. The polymer, formed by porphyrins linked in *meso*-position with acethylene bridges enabling the exfoliation of nanotube bundles. This nanotube/polyporphyrin composite presented a broad absorption from the UV to the near-IR and potential application as light-harvesting systems in photovoltaics was foreseen for this system.

The substitution of methyl end groups of the alkoxy solubilizing chains of the PPE by sodium sulfonate led to poly[2,5-bis(3-sulfonatopropoxy)-1,4-ethynylphenylene-*alt*-1,4-ethynylphenylene] sodium salt (PPES), which allowed the solubilization of SWNTs coated with phenyleneethynylene in aqueous medium. In contrast to previous studies in which the *p*-phenyleneethynylene-based polymers were shown to organize parallel to the SWNT axis, the linear and conformationally restricted conjugated PPES polymer wrapped around SWNTs forming a self-assembled helical super structure [142]. Note that no explanation was given to explain the difference between PPE and PPES. The anionic conjugated polyelectrolyte was used to assemble PPES/SWNTs on a pre-patterned, positively charged surface via electrostatic interactions. Patterned features were found to be electrically conducting with a low resistance value [143]. Recently, benzene rings of the PPES were replaced by naphthalene and the resulting polymer: the poly[2,6-{1,5-bis(3-sulfonatopropoxy)}naphthylene]ethynylene sodium salt (PNES) was used to disperse SWNTs in aqueous or organic media. The solubility of the SWNT/PNES hybrids in organic media was obtained thanks to the use of a phase transfer catalyst (i.e., the crown ether 18C6) that can complex the sodium counterions of the polymer [144]. The efficient individualization of SWNTs was confirmed by the steady-state electronic absorption spectroscopy, AFM, and TEM.

Electrochemical deposition of polypyrrole (PPy) onto an individual SWNT in field-effect transistor configuration was performed to study the electronic property of the SWNT/PPy composite [145]. The nanotube devices exhibited surprising suppressed conductance for thin and moderate PPy coatings for both metallic and semiconducting nanotubes.

Other conducting polymers such as polyaniline (PANI) [146–150] or poly-3,4-ethylenedioxythiophene:polystyrenesulfonate (PEDOT:PSS) [151–155] were mixed with carbon nanotubes in order to modify the polymer conductivities and for thermoelectricity or photovoltaic applications. To the best of our knowledge, no morphological studies were performed on these composites as it was the case for P*m*PV, PPE, PT, or polyfluorene (PFO) composites.

The last example described in this chapter is precisely the case of PFO/nanotube composites. It was demonstrated that polyfluorene-based copolymers were able to wrap selectively SWNTs with certain chiral angles or diameters depending on the chemical structures of the polymers [44, 156–159]. The incorporation of aromatic rings such as benzene, anthracene, electron-poor benzothiadiazole, or electron-rich *N,N'*-diphenyl-*N,N'*-di(*p*-butyloxyphenyl)-1,4-diaminobenzene moieties in the PFO backbone allowed a fine-tuning of the interactions between the polymers and the nanotubes. SWNTs selectively wrapped with the polymers were separated by centrifugation from the raw mixture.

11.2.4
Filling Nanotubes

Supramolecular hybrids in which the molecules are encapsulated inside the nanotubes (X@CNTs, where X is the molecule encapsulated in the nanotubes) are less developed than the externally modified nanotubes described in the previous sections. Their development is, however, increasing very fast due to their potential applications for gas storage (especially hydrogen) [160], as nanoscale reactors [161], or for theranostics [162–164].

Two main processes are used to encapsulate molecules inside CNTs: the *in situ* or *ex situ* filling, where nanotubes are, respectively, filled in one step during their growth or first synthesized and then filled in the second step [165–167]. The *in situ* process is less compatible with organic compounds since the temperatures required for the growth of carbon nanotube are superior to several hundreds of degrees. It was, however, the method used to synthesize the first example of fullerene C_{60} inside SWNTs denoted C_{60}@SWNTs [168]. Later on, researchers developed the *ex situ* process to encapsulate organic compounds inside CNTs. Most of the studies were performed on SWNTs because of their small diameter (less than 2 nm) and the fact that they possess only one carbon layer. This facilitates the determination of the arrangement of the guest molecules inside the tubes by high-resolution transmission electron microscopy (HR-TEM).

Gas phase, liquid phase, and supercritical phase routes were carried out to fill opened end carbon nanotubes [166, 167]. Each method presents some advantages and limitations described in the following section:

- Gas phase: The CNTs and the organic compounds are sealed in pyrex vessel under vacuum ($\sim 10^{-6}$ bar) and heated up slightly above the sublimation temperature of the filling material. The method led to a high ratio filling for the fullerene derivatives and has been extended to other organic compounds like metallofullerenes, strong electron donors and acceptors [169].

  Due to the high temperature, the compounds in gas phase have enough kinetic energy to enter the CNT cavity and diffuse within the nanotubes to reach high yield of filling. Unfortunately, the method is not appropriate for functionalized fullerenes and a large range of organic complexes and biomolecules that cannot be vaporized without decomposition.
- Liquid phase: This is an appropriate alternative method to insert compounds incompatible with a gas-phase filling. The method allowed insertion of various fullerenes and organic compounds [170, 171]. Three important parameters have to be considered in liquid phase: the energy of interaction of the organic molecule

with the CNT, the energy of interaction of the organic compound with solvent molecules (i.e., energy of solubilization), and the energy of the absorption of the solvent itself within the CNT. Indeed, encapsulation is unlikely to occur if the solvent interacts more efficiently with CNTs than the chosen molecule. Based on this principle, two processes called "nanoextraction" and "nanocondensation" were developed. Nanoextraction consists in mixing organic compounds and nanotubes in a solvent for which they have poor affinity but have strong affinity for each other. Nanocondensation consists in mixing nanotubes with a supersaturated solution of the guest compounds. One of the main drawbacks of these methods is that the solvent is always, at least partially, encapsulated with the organic compounds since the nanotubes are spontaneously filled with a fluid with surface tension below 100–200 mN/m [172]. Thus, the yield of encapsulation is generally lower than in the case of gas phase encapsulation. The release of the compounds is also easier due to the residue of solvent also encapsulated inside the CNT.

- Supercritical phase: This method performed mainly in sc-CO_2 under mild conditions (31 °C) allows encapsulation of fragile molecules that cannot be vaporized and leaves no encapsulated solvent in the nanotubes due to the low critical diameter of the CO_2 compared to common organic solvents [173]. Like in any liquid-phase process, the solvent–host and solvent–guest interactions are crucial factors to fill SWNTs. The low viscosity and the absence of surface tension of supercritical CO_2 combined with the low solvation effect of the guest molecules provided high diffusivity approaching that of the gas phase. Thus, this method combines the advantages of both liquid and gas phase. This technique reached 70% yield for C_{60} encapsulation into SWNTs [173].

Different classes of organic compounds were inserted into CNTs and among the wide variety of organic molecules, fullerenes and fullerene derivatives were the most studied. After the first preparation of C_{60}@SWNT peapods by the *in situ* method in 1998 (Figure 11.14) [168], the insertion of other buckminsterfullerenes (C_{70}, C_{78}, C_{80}, etc.), endohedral fullerenes, doped fullerenes, and functionalized fullerenes was carried out using gas-phase or solution-phase encapsulation methods [174, 175].

Metallocenes like ferrocenes and cobaltocenes were encapsulated in SWNTs via the gas-phase process [176–178]. The optical, electrical, and electrochemical properties of the encapsulated materials were investigated by means of photoluminescence spectroscopy, cyclic voltammetry, or in field-effect transistor configuration. The filling of MWNTs and SWNTs, respectively, with metallophthalocyanines [179, 180] and porphyrins [181] was also studied.

Strong electron donors such as tetrakis(dimethylamino)ethylene, tetramethyltetraselenafulvalene, and tetrathiafulvalene, or electron acceptor molecules such as tetracyano-*p*-quinodimethane, and tetrafluorotetracyano-*p*-quinodimethane [169] as well as azafullerene [182] were encapsulated into SWNTs using gas-phase process. These molecules were able to dope the nanotubes, and depending on their electron donor or electron acceptor characters, n- or p-type materials were obtained. Using the doping properties of the encapsulated materials, several examples of fabrication of

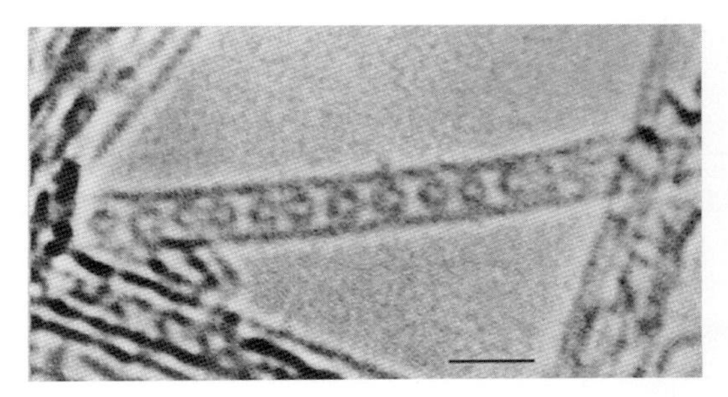

Figure 11.14 HR-TEM image of the first synthesized C_{60}@SWNTs peapods. Reproduced with permission of the Nature Publishing Group from Ref. [168].

p–n junctions were reported. p–n junctions were observed in SWNTs partially filled with iron nanoparticles produced by thermal decomposition of encapsulated ferrocene [183] or in SWNTs and DWNTs both filled with electron donor/electron acceptor couples, caesium and C_{60}, or with caesium and iodine [184, 185].

Electron-deficient polyaromatic dopant can have interesting properties. Iijima and coworkers incorporated perylene-3,4,9,10-tetracarboxylic dianhydrides (PTCDA) in opened SWNTs via gas-phase process. The high-temperature treatment of the PTCDA@SWNT provoked the coalescence of perylene in the SWNTs and the final formation of double-wall carbon nanotubes (Figure 11.15) [186]. Similar DWNT formations were also observed after thermal treatment of C_{60}@SWNTs [187] and ferrocene@SWNTs [176, 188].

Finally, it was shown that the presence of fullerene derivatives inside the nanotubes influenced notably their electrical response upon illumination. Indeed, transport characteristics of various fullerene (C_{60}, C_{70}, and C_{84}) and azafullerene ($C_{59}N$) peapods investigated in FET devices exhibited a clear photoresponse compared to those observed with empty SWNTs. This behavior was explained by charge transfer between the fullerenes and the nanotubes. The conductance of azafullerene peapods shows much higher photoresponse sensitivity than those observed in fullerene peapods, suggesting a different photoinduced charge transfer mechanism [189–193].

Encapsulation of molecules in CNTs presents several advantages compared to classical external adsorption; for example, better communication between the CNTs and the compounds encapsulated owing to the closeness of the two systems and protection of the organic chromophores from the external medium and in particular from oxygen are some of them. This last point was clearly demonstrated with the encapsulation using a solution-phase process of the β-carotene [171], which is a model system of π-conjugated molecules that shows good UV–visible absorption and large third-order optical nonlinearity. Its encapsulation inside a SWNT protected this polyene molecule from reacting with radical species such as O_2 and also suppressed its isomerization. Both β-carotene [194] and squarylium dyes [195] encapsulated in SWNTs showed significantly redshifted and broadened absorption spectra; photoluminescence measurements demonstrated an efficient energy transfer from the encapsulated compounds to the SWNTs, which paved the way for their applications in optical devices. Moreover, it was demonstrated recently that the electronic structure

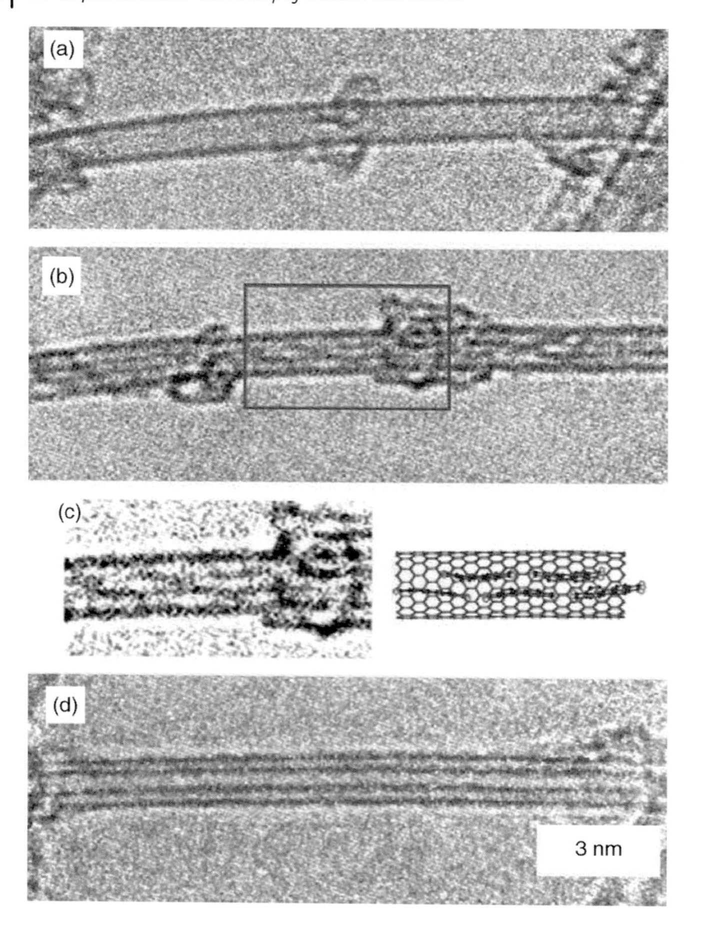

Figure 11.15 HR-TEM images taken for (a) empty-SWNTs and (b) PTCDA-doped SWNTs; (c) enhanced image of a boxed part and the corresponding schematic of PTCDA molecules arranged inside a (15,0)-SWNT; (d) DWNTs obtained by heating PTCDA-doped SWNTs at 1050 °C for 48 h. Reproduced with permission of Elsevier from Ref. [186].

of metallic or semiconducting SWNTs changes slightly the optical absorption properties of encapsulated β-carotene molecules [196].

The α-sexithiophene (6T), a conjugated molecule containing six thiophene rings and possessing interesting optical and redox properties, was encapsulated in SWNTs by sublimation (gas-phase process) [197]. Upon excitation in the UV, the nanohybrid material exhibited strong luminescence in the visible range of the spectrum and was found to be a promising source of photons for future optoelectronic devices (Figure 11.16) [197, 198].

Carbon nanotubes are more and more studied as material electrodes because they can exhibit better properties (such as transparency or conductivity) than existing materials. Recently, it was demonstrated in SWNT/Si junction for infrared energy

Figure 11.16 Cartoon of the visible light emitting 6T@SWNT peapod. Reproduced with permission of Wiley-VCH Verlag GmbH from Ref. [197].

conversion that the use C_{60}@SWNTs led to a significant increase in the open-circuit voltage compared to empty SWNTs. This effect was attributed to the adjustment of the Fermi level of SWNTs that enhance the build-in potential [199]. Similarly, in Li ion storage devices, it was demonstrated by electrochemical charge–discharge measurements that the use of SWNTs filled with 9,10-dichloroanthracene, β-carotene, or coronene could store about 2.5 times more Li ions than the empty tubes [200]. Naitoh *et al.* [201] studied the reversible resistance switching effect of a 10 nm wide C_{60}@SWNT device. They showed that the used peapod is at the origin of the particular behavior of the device: the migration of fullerene inside the SWNT changed the tunneling resistance in between the electrodes.

The effect of the confinement in SWNTs was studied by encapsulation of the 1-butyl-3-methylimidazolium hexafluorophosphate in MWNTs. This ionic liquid had a melting point of 6 °C under ambient conditions, but its confinement at the nanoscale resulted in a fully different phase transition and crystal formation involving the formation of a stable, polymorphous crystal possessing at a melting point above 200 °C [202]. Recently, HR-TEM showed that zinc-containing quaternary ammonium-based ionic liquid encapsulated in SWNT formed different morphologies such as single-chain, double-helix, zigzag tubes and random tubes depending on the diameter of the SWNTs. The ionic liquid also presented higher thermal decomposition when encapsulated than in the bulk [203].

Polymers were also incorporated into CNTs either via direct encapsulation of polymer chains or via encapsulation of the monomer and then polymerization inside the hollow tube. *In situ* polymerization was first described for polystyrene: the monomers and the initiator were encapsulated using supercritical CO_2 solvent and the polymerization was carried out into MWNTs with outer diameters of 40–50 nm [204] and also recently into SWNTs [205]. Note that other polymers such as polyacetylene [206, 207], poly(*N*-vinylcarbazole) [208], polypyrrole [208], and polyacrylonitrile [209] were synthesized in MWNTs. For the direct polymer encapsulation, a study was carried out using poly(ethylene oxide) and poly(caprolactone) with low molecular weights, the successful encapsulation was verified by TEM [210]. Single-stranded DNAs were also encapsulated using a modified electrophoresis method in SWNT [211, 212] and in DWNT [213] to modify electronic properties of carbon nanotubes.

11.3 Conclusions

Supramolecular chemistry of carbon nanotubes gathers thousands of references; it was, therefore, impossible to review all the aspects of this field. The examples discussed in this chapter were taken from the recent literature, and we tried to give an overview of research pursued on noncovalent functionalization of carbon nanotubes. Our main objective was to give useful information to the reader on different techniques of functionalization and their potential applications.

The already rich variety of CNT applications can be further improved when these carbon cylinders are functionalized. In addition, the recent development of nanotube sorting techniques, as well as the commercialization of sorted SWNTs (by NanoIntegris), represents a real breakthrough. The possibility (very soon) to obtain a specific sample of (*n*,*m*)-SWNT excites the curiosity of physicists and chemists. Indeed, for the first time we can apprehend SWNT as molecules exhibiting homogeneous properties and reactivity.

References

1 Thess, A., Lee, R., Nikolaev, P., Dai, H., Petit, P., Robert, J., Xu, C., Lee, H.Y., Kim, S.G., Rinzler, A.G., Colbert, D.T., Scuseria, G.E., Tomanek, D., Fischer, J.E., and Smalley, R.E. (1996). Crystalline ropes of metallic carbon nanotubes. *Science*, **273**, 483–487.

2 Bockrath, M., Cobden, D.H., McEuen, P.L., Chopra, N.G., Zettl, A., Thess, A., and Smalley, R.E. (1997) Single-electron transport in ropes of carbon nanotubes. *Science*, **275**, 1922–1925.

3 Martel, R., Schmidt, T., Shea, H.R., Hertel, T., and Avouris, P. (1998) Single- and multi-wall carbon nanotube field-effect transistors. *Appl. Phys. Lett.*, **73**, 2447–2449.

4 Chen, J., Hamon, M.A., Hu, H., Chen, Y., Rao, A.M., Eklund, P.C., and Haddon, R.C. (1998) Solution properties of single-walled carbon nanotubes. *Science*, **282**, 95–98.

5 Avouris, P. (2002) Molecular electronics with carbon nanotubes. *Acc. Chem. Res.*, **35**, 1026–1034.

6 O'Connell, M.J., Bachilo, S.M., Huffman, C.B., Moore, V.C., Strano, M.S., Haroz, E.H., Rialon, K.L., Boul, P.J., Noon, W.H., Kittrell, C., Ma, J., Hauge, R.H., Weisman, R.B., and Smalley, R.E. (2002). Band gap fluorescence from individual single-walled carbon nanotubes. *Science*, **297**, 593–596.

7 Liu, J., Rinzler, A.G., Dai, H., Hafner, J.H., Bradley, R.K., Boul, P.J., Lu, A., Iverson, T., Shelimov, K., Huffman, C.B., Rodriguez-Marcias, F., Shon, Y.-S., Lee, T.R., Colbert, D.T., and Smalley, R.E. (1998). Fullerene pipes. *Science*, **280**, 1253–1256.

8 Islam, M.F., Pojas, E., Bergey, D.M., Johnson, A.T., and Yodh, A.G. (2003) High weight fraction surfactant solubilization of single-wall carbon

nanotubes in water. *Nano Lett.*, **3**, 269–273.

9 Moore, V.C., Strano, M.S., Haroz, E.H., Hauge, R.H., and Smalley, R.E. (2003) Individually suspended single-walled carbon nanotubes in various surfactants. *Nano Lett.*, **3**, 1379–1382.

10 Matarredona, O., Rhoads, H., Li, Z., Harwell, J.H., Balzano, L., and Resasco, D.E. (2003) Dispersion of single-walled carbon nanotubes in aqueous solutions of the anionic surfactant NaDDBS. *J. Phys. Chem. B*, **107**, 13357–13367.

11 Markiewicz, P. and Goh, M.C. (1994) Atomic force microscopy probe tip visualization and improvement of images using a simple deconvolution procedure. *Langmuir*, **10**, 5–7.

12 Nativ-Roth, E., Regev, O., and Yerushalmi-Rozen, R. (2008) Shear-induced ordering of micellar arrays in the presence of single-walled carbon nanotubes. *Chem. Commun.*, 2037–2039.

13 Richard, C., Balavoine, F., Schultz, P., Ebbesen, T.W., and Mioskowski, C. (2003) Supramolecular self-assembly of lipid derivatives on carbon nanotubes. *Science*, **300**, 775–778.

14 Yurekli, K., Mitchell, C.A., and Krishnamoorti, R. (2004) Small-angle neutron scattering from surfactant-assisted aqueous dispersions of carbon nanotubes. *J. Am. Chem. Soc.*, **126**, 9902–9903.

15 Calvaresi, M., Dallavalle, M., and Zerbetto, F. (2009) Wrapping nanotubes with micelles, hemimicelles, and cylindrical micelles. *Small*, **5**, 2191–2198.

16 Angelikopoulos, P., Gromov, A., Leen, A., Nerushev, O., Block, H., and Campbell, E.E.B. (2010) Dispersing individual single-wall carbon nanotubes in aqueous surfactant solutions below the cmc. *J. Phys. Chem. C*, **114**, 2–9.

17 Duque, J.G., Parra-Vasquez, A.N.G., Behabtu, N., Green, M.J., Higginbotham, A.L., Price, B.K., Leonard, A.D., Schmidt, H.K., Lounis, B., Tour, J.M., Doorn, S.K., Cognet, L., and Pasquali, M. (2010) Diameter-dependent solubility of single-walled carbon nanotubes. *ACS Nano*, **4**, 3063–3072.

18 McDonald, T.J., Engttrakul, C., Jones, M., Rumbles, G., and Heben, M.J. (2006) Kinetics of PL quenching during single-walled carbon nanotube rebundling and diameter-dependent surfactant interactions. *J. Phys. Chem. B*, **110**, 25339–25346.

19 Tangney, P., Capaz, R.B., Spataru, C.D., Cohen, M.L., and Louie, S.G. (2005) Structural transformations of carbon nanotubes under hydrostatic pressure. *Nano Lett.*, **5**, 2268–2273.

20 Kamal, C., Ghanty, T.K., Banerjee, A., and Chakrabarti, A. (2009) The van der Waals coefficients between carbon nanostructures and small molecules: a time-dependent density functional theory study. *J. Chem. Phys.*, **131**, 164708.

21 Tummala, N.R. and Striolo, A. (2009) SDS surfactants on carbon nanotubes: aggregate morphology. *ACS Nano*, **3**, 595–602.

22 Zheng, M., Jagota, A., Semke, E.D., Diner, B.A., McLean, R.S., Lustig, S.R., Richardson, R.E., and Tassi, N.G. (2003) DNA-assisted dispersion and separation of carbon nanotubes. *Nat. Mater.*, **2**, 338–342.

23 Johnson, R.R., Johnson, A.T.C., and Klein, M.L. (2008) Probing the structure of DNA carbon nanotube hybrids with molecular dynamics. *Nano Lett.*, **8**, 69–75.

24 Wang, Y. (2008) Theoretical evidence for the stronger ability of thymine to disperse SWCNT than cytosine and adenine: self-stacking of DNA bases vs their cross-stacking with SWCNT. *J. Phys. Chem. C*, **112**, 14297–14305.

25 Johnson, R.R., Kohlmeyer, A., Johnson, A.T.C., and Klein, M.L. (2009) Free energy landscape of a DNA–carbon nanotube hybrid using replica exchange molecular dynamics. *Nano Lett.*, **9**, 537–541.

26 Vogel, S.R., Kappes, M.M., Hennrich, F., and Richert, C. (2007) An unexpected new optimum in the structure space of DNA solubilizing single-walled carbon nanotubes. *Chem. Eur. J.*, **13**, 1815–1820.

27 Duesberg, G.S., Muster, J., Krstic, V., Burghard, M., and Roth, S. (1998) Chromatographic size separation of

single-wall carbon nanotubes. *Appl. Phys. A*, **67**, 117–119.
28 Arnold, K., Hennrich, F., Krupke, R., Lebedkin, S., and Kappes, M.M. (2006) Length separation studies of single walled carbon nanotube dispersions. *Phys. Status Solidi B Basic Solid State Phys.*, **243**, 3073–3076.
29 Heller, D.A., Mayrhofer, R.M., Baik, S., Grinkova, Y.V., Usrey, M.L., and Strano, M.S. (2004) Concomitant length and diameter separation of single-walled carbon nanotubes. *J. Am. Chem. Soc.*, **126**, 14567–14573.
30 Vetcher, A.A., Srinivasan, S., Vetcher, I.A., Abramov, S.M., Kozlov, M., Baughman, R.H., and Levene, S.D. (2006) Fractionation of SWNT/nucleic acid complexes by agarose gel electrophoresis. *Nanotechnology*, **17**, 4263–4269.
31 Strano, M.S., Zheng, M., Jagota, A., Onoa, G.B., Heller, D.A., Barone, P.W., and Usrey, M.L. (2004) Understanding the nature of the DNA-assisted separation of single-walled carbon nanotubes using fluorescence and Raman spectroscopy. *Nano Lett.*, **4**, 543–550.
32 Zheng, M., Jagota, A., Strano, M.S., Santos, A.P., Barone, P.W., Chou, S.G., Diner, B.A., Dresselhaus, M.S., McLean, R.S., Onoa, G.B., Samsonidze, G.G., Semke, E.D., Usrey, M.L., and Walls, D.J. (2003) Structure-based carbon nanotube sorting by sequence-dependent DNA assembly. *Science*, **302**, 1545–1548.
33 Doorn, S.K., Strano, M.S., O'Connell, M.J., Haroz, E.H., Rialon, K.L., Hauge, R.H., and Smalley, R.E. (2003) Capillary electrophoresis separations of bundled and individual carbon nanotubes. *J. Phys. Chem. B*, **107**, 6063–6069.
34 Arnold, M.S., Stupp, S.I., and Hersam, M.C. (2005) Enrichment of single-walled carbon nanotubes by diameter in density gradients. *Nano Lett.*, **5**, 713–718.
35 Arnold, M.S., Grenn, A.A., Hulvat, J.F., Stupp, S.I., and Hersam, M.C. (2006) Sorting carbon nanotubes by electronic structure using density differentiation. *Nat. Nanotechnol.*, **1**, 60–65.
36 Crochet, J., Clemens, M., and Hertel, T. (2007) Quantum yield heterogeneities of aqueous single-wall carbon nanotube suspensions. *J. Am. Chem. Soc.*, **129**, 8058–8059.
37 Hennrich, F., Arnold, K., Lebedkin, S., Quintillà, A., Wenzel, W., and Kappes, M.M. (2007) Diameter sorting of carbon nanotubes by gradient centrifugation: role of endohedral water. *Phys. Status Solidi B Basic Solid State Phys.*, **244**, 3896–3900.
38 Nair, N., Kim, W.-J., Braatz, R.D., and Strano, M.S. (2008) Dynamics of surfactant-suspended single-walled carbon nanotubes in a centrifugal field. *Langmuir*, **24**, 1790–1795.
39 Wei, L., Wang, B., Goh, T.H., Li, L.-J., Yang, Y., Chan-Park, M.B., and Chen, Y. (2008) Selective enrichment of (6,5) and (8,3) single-walled carbon nanotubes via cosurfactant extraction from narrow (*n*,*m*) distribution samples. *J. Phys. Chem. B*, **112**, 2771–2774.
40 Zheng, M. and Semke, E.D. (2007) Enrichment of single chirality carbon nanotubes. *J. Am. Chem. Soc.*, **129**, 6084–6085.
41 Ju, S.-Y., Doll, J., Sharma, I., and Papadimitrakopoulos, F. (2008) Selection of carbon nanotubes with specific chiralities using helical assemblies of flavin mononucleotide. *Nat. Nanotechnol.*, **3**, 356–362.
42 Kim, S.N., Kuang, Z., Grote, J.G., Farmer, B.L., and Naik, R.R. (2008) Enrichment of (6,5) single wall carbon nanotubes using genomic DNA. *Nano Lett.*, **8**, 4415–4420.
43 Tu, X., Manohar, S., Jagota, A., and Zheng, M. (2009) DNA sequence motifs for structure-specific recognition and separation of carbon nanotubes. *Nature*, **460**, 250–253.
44 Nish, A., Hwang, J.-Y., Doig, J., and Nicholas, R.J. (2007) Highly selective dispersion of single-walled carbon nanotubes using aromatic polymers. *Nat. Nanotechnol.*, **2**, 640–646.
45 Krupke, R., Hennrich, F., von Löhneyesen, H., and Kappes, M.M. (2003) Separation of metallic from

semiconducting single-walled carbon nanotubes. *Science*, **301**, 344–347.

46 Krupke, R., Hennrich, F., Weber, H.B., Kappes, M.M., and von Löhneysen, H. (2003) Simultaneous deposition of metallic bundles of single-walled carbon nanotubes using AC-dielectrophoresis. *Nano Lett.*, **3**, 1019–1023.

47 Krupke, R., Hennrich, F., Kappes, M.M., and von Löhneyesen, H. (2004) Surface conductance induced dielectrophoresis of semiconducting single-walled carbon nanotubes. *Nano Lett.*, **4**, 1395–1399.

48 Krupke, R., Linden, S., Rapp, M., and Hennrich, F. (2006) Thin films of metallic carbon nanotubes prepared by dielectrophoresis. *Adv. Mater.*, **18**, 1468–1469.

49 Tanaka, T., Jin, H., Miyata, Y., Fuji, S., Suga, H., Naitoh, Y., Minari, T., Miyadera, T., Tsukagoshi, K., and Kataura, H. (2009) Simple and scalable gel-based separation of metallic and semiconducting carbon nanotubes. *Nano Lett.*, **9**, 1497–1500.

50 Peng, X., Komatsu, N., Bhattacharya, S., Shimawaki, T., Aonuma, S., Kimura, T., and Osuka, A. (2007) Optically active single-walled carbon nanotubes. *Nat. Nanotechnol.*, **2**, 361–365.

51 Peng, X., Komatsu, N., Kimura, T., and Osuka, A. (2007) Improved optical enrichment of SWNTs through extraction with chiral nanotweezers of 2,6-pyridylene-bridged diporphyrins. *J. Am. Chem. Soc.*, **129**, 15947–15953.

52 Gosh, S., Bachilo, S.M., and Weisman, R.B. (2010) Advanced sorting of single-walled carbon nanotubes by nonlinear density-gradient ultracentrifugation. *Nat. Nanotechnol.*, **5**, 443–450.

53 Bonaccorso, F., Hasan, T., Tan, P.H., Sciascia, C., Privitera, G., Di Marco, G., Gucciardi, P.G., and Ferrari, A.C. (2010) Density gradient ultracentrifugation of nanotubes: interplay of bundling and surfactants encapsulation. *J. Phys. Chem. C*, **114**, 17267–17285.

54 Nakashima, N., Tomonari, Y., and Murakami, H. (2002) Water-soluble single-walled carbon nanotubes via noncovalent sidewall-functionalization with a pyrene-carrying ammonium ion. *Chem. Lett.*, 638–639.

55 Tomonari, Y., Murakami, H., and Nakashima, N. (2006) Solubilization of single-walled carbon nanotubes by using polycyclic aromatic ammonium amphiphiles in water-strategy for the design of high-performance solubilizers. *Chem. Eur. J.*, **12**, 4027–4034.

56 Guldi, D.M., Rahman, G.M.A., Jux, N., Tagmatarchis, N., and Prato, M. (2004) Integrating single-wall carbon nanotubes into donor–acceptor nanohybrids. *Angew. Chem. Int. Ed.*, **43**, 5526–5530.

57 Ehli, C., Rahman, G.M.A., Jux, N., Balbinot, D., Guldi, D.M., Paolucci, F., Marcaccio, M., Paolucci, D., Melle-Franco, M., Zerbetto, F., Campidelli, S., and Prato, M. (2006) Interactions in single wall carbon nanotubes/pyrene/porphyrin nanohybrids. *J. Am. Chem. Soc.*, **128**, 11222–11231.

58 Sgobba, V., Rahman, G.M.A., Guldi, D.M., Jux, N., Campidelli, S., and Prato, M. (2006) Supramolecular assemblies of different carbon nanotubes for photoconversion processes. *Adv. Mater.*, **18**, 2264–2269.

59 Rahman, G.M.A., Guldi, D.M., Cagnoli, R., Mucci, A., Schenetti, L., Vaccari, L., and Prato, M. (2005) Combining single wall carbon nanotubes and photoactive polymers for photoconversion. *J. Am. Chem. Soc.*, **127**, 10051–10057.

60 Guldi, D.M., Rahman, G.M.A., Prato, M., Jux, N., Qin, S., and Ford, W.T. (2005) Single-wall carbon nanotubes as integrative building blocks for solar-energy conversion. *Angew. Chem. Int. Ed.*, **44**, 2015–2018.

61 Sandanayaka, A.S.D., Chitta, R., Subbaiyan, N.K., D'Souza, L., Ito, O., and D'Souza, F. (2009) Photoinduced charge separation in ion-paired porphyrin-single-wall carbon nanotube donor–acceptor hybrids. *J. Phys. Chem. C*, **113**, 13425–13432.

62 Guldi, D.M., Rahman, G.M.A., Jux, N., Balbinot, D., Hartnagel, U., Tagmatarchis, N., and Prato, M. (2005) Functional single-wall carbon nanotube nanohybrids associating SWNTs with water-soluble enzyme model systems. *J. Am. Chem. Soc.*, **127**, 9830.

63 Xin, H. and Wooley, A.T. (2003) DNA-templated nanotube localization. *J. Am. Chem. Soc.*, **125**, 8710–8711.

64 Chung, C.-L., Gautier, C., Campidelli, S., and Filoramo, A. (2010) Hierarchical functionalisation of single-wall carbon nanotubes with DNA through positively charged pyrene. *Chem. Commun.*, 6539–6541.

65 Chitta, R., Sandanayaka, A.D., Schumacher, A.L., D'Souza, L., Araki, Y., Ito, O., and D'Souza, F. (2007) Donor–acceptor nanohybrids of zinc naphthalocyanine or zinc porphyrin noncovalently linked to single-wall carbon nanotubes for photoinduced electron transfer. *J. Phys. Chem. C*, **111**, 6947–6955.

66 D'Souza, F., Chitta, R., Sandanayaka, A.S.D., Subbaiyan, N.K., D'Souza, L., Araki, Y., and Ito, O. (2007) Self-assembled single-walled carbon nanotube: zinc–porphyrin hybrids through ammonium ion–crown ether interaction: construction and electron transfer. *Chem. Eur. J.*, **13**, 8277–8284.

67 D'Souza, F., Chitta, R., Sandanayaka, A.S.D., Subbaiyan, N.K., D'Souza, L., Araki, Y., and Ito, O. (2007) Supramolecular carbon nanotube-fullerene donor–acceptor hybrids for photoinduced electron transfer. *J. Am. Chem. Soc.*, **129**, 15865–15871.

68 Zhao, Y.-L., Hu, L., Stoddart, J.F., and Grüner, G. (2008) Pyrenecyclodextrin-decorated single-walled carbon nanotube field-effect transistors as chemical sensors. *Adv. Mater.*, **20**, 1910–1915.

69 Zhao, Y.-L., Hu, L., Grüner, G., and Stoddart, J.F. (2008) A tunable photosensor. *J. Am. Chem. Soc.*, **130**, 16996–17003.

70 Zhao, Y.-L. and Stoddart, J.F. (2009) Noncovalent functionalization of single-walled carbon nanotubes. *Acc. Chem. Res.*, **42**, 1161–1171.

71 Avouris, P., Chen, Z., and Perebeinos, V. (2007) Carbon-based electronics. *Nat. Nanotechnol.*, **2**, 605–615.

72 Avouris, P., Freitag, M., and Perebeinos, V. (2008) Carbon-nanotube photonics and optoelectronics. *Nat. Photonics*, **2**, 341–350.

73 Allen, B.L., Kichambare, P.D., and Star, A. (2007) Carbon nanotube field-effect-transistor-based biosensors. *Adv. Mater.*, **19**, 1439–1451.

74 Grüner, G. (2006) Carbon nanotube transistors for biosensing applications. *Anal. Bioanal. Chem.*, **384**, 322–335.

75 Rekharsky, M.V. and Inoue, Y. (1998) Complexation thermodynamics of cyclodextrins. *Chem. Rev.*, **98**, 1875–1917.

76 Ogoshi, T., Takashima, Y., Yamaguchi, H., and Harada, A. (2007) Chemically-responsive sol–gel transition of supramolecular single-walled carbon nanotubes (SWNTs) hydrogel made by hybrids of SWNTs and cyclodextrins. *J. Am. Chem. Soc.*, **129**, 4878–4879.

77 Yu, M., Zu, S.-Z., Chen, Y., Liu, Y.-P., Han, B.-H., and Liu, Y. (2010) Spatially controllable DNA condensation by a water-soluble supramolecular hybrid of single-walled carbon nanotubes and ß-cyclodextrin-tethered ruthenium complexes. *Chem. Eur. J.*, **16**, 1168–1174.

78 Kavakka, J.S., Heikkinen, S., Kilpeläinen, I., Mattila, M., Lipsanen, H., and Helaja, J. (2007) Noncovalent attachment of *pyro*-pheophorbide *a* to a carbon nanotube. *Chem. Commun.*, 519–521.

79 Satake, A., Miyajima, Y., and Kobuke, Y. (2005) Porphyrin-carbon nanotube composites formed by noncovalent polymer wrapping. *Chem. Mater.*, **17**, 716–724.

80 Maligaspe, E., Sandanayaka, A.S.D., Hasode, T., Ito, O., and D'Souza, F. (2010) Sensitive efficiency of photoinduced electron transfer to band gaps of semiconductive single-walled carbon nanotubes with supramolecularly attached zinc porphyrin bearing pyrene glues. *J. Am. Chem. Soc.*, **132**, 8158–8164.

81 Bartelmess, J., Ballesteros, B., de la Torre, G., Kiessling, D., Campidelli, S., Prato, M., Torres, T., and Guldi, D.M. (2010) Phthalocyanine-pyrene conjugates: a powerful approach toward carbon nanotube solar cells. *J. Am. Chem. Soc.*, **132**, 16202–16211.

82 Herranz, M.A., Ehli, C., Campidelli, S., Gutiérrez, M., Hug, G.L., Ohkubo, K., Fukuzumi, S., Prato, M., Martín, N., and

Guldi, D.M. (2008) Spectroscopic characterization of photolytically generated radical ion pairs in single-wall carbon nanotubes bearing surface-immobilized tetrathiafulvalenes. *J. Am. Chem. Soc.*, **130**, 66–73.

83 Ehli, C., Guldi, D.M., Herranz, M.A., Martín, N., Campidelli, S., and Prato, M. (2008) Pyrene-tetrathiafulvalene supramolecular assembly with different types of carbon nanotubes. *J. Mater. Chem.*, **18**, 1498–1503.

84 Klare, J.E., Murray, I.P., Goldberger, J., and Stupp, S.I. (2009) Assembling p-type molecules on single wall carbon nanotubes for photovoltaic devices. *Chem. Commun.*, 3705–3707.

85 Le Goff, A., Moggia, F., Debou, N., Jégou, P., Artero, V., Fontecave, M., Jousselme, B., and Palacin, S. (2010) Facile and tunable functionalization of carbon nanotube electrodes with ferrocene by covalent coupling and p-stacking interactions and their relevance to glucose bio-sensing. *J. Electroanal. Chem.*, **641**, 57–63.

86 Guldi, D.M., Menna, E., Maggini, M., Marcaccio, M., Paolucci, D., Paolucci, F., Campidelli, S., Prato, M., Rahman, G.M.A., and Schergna, S. (2006) Supramolecular hybrids of [60]fullerene and single-wall carbon nanotubes. *Chem. Eur. J.*, **12**, 3975–3983.

87 D'Souza, F., Sandanayaka, A.S.D., and Ito, O. (2010) SWNT-based supramolecular nanoarchitectures with photosensitizing donor and acceptor molecules. *J. Phys. Chem. Lett.*, **1**, 2586–2593.

88 Zhou, X., Zifer, T., Wong, B.M., Krafcik, K.L., Léonard, F., and Vance, A.L. (2009) Color detection using chromophore-nanotube hybrid devices. *Nano Lett.*, **9**, 1028–1033.

89 Guo, X., Huang, L., O'Brien, S., Kim, P., and Nuckolls, C. (2005) Directing and sensing changes in molecular conformation on individual carbon nanotube field effect transistors. *J. Am. Chem. Soc.*, **127**, 15045–15047.

90 Hu, L., Zhao, Y.-L., Ryu, K., Zhou, C., Stoddart, J.F., and Grüner, G. (2008) Light-induced charge transfer in pyrene/CdSe-SWNT hybrids. *Adv. Mater.*, **20**, 939–946.

91 Wu, P., Chen, X., Hu, N., Tam, U.C., Blixt, O., Zettl, A., and Bertozzi, C.R. (2008) Biocompatible carbon nanotubes generated by functionalization with glycodendrimers. *Angew. Chem. Int. Ed.*, **47**, 5022–5025.

92 Sudibya, H.G., Ma, J., Dong, X., Ng, S., Li, L.-J., Liu, X.-W., and Chen, P. (2009) Interfacing glycosylated carbon-nanotube-network devices with living cells to detect dynamic secretion of biomolecules. *Angew. Chem. Int. Ed.*, **48**, 2723–2726.

93 Chen, R.J., Zhang, Y., Wang, D., and Dai, H. (2001) Noncovalent sidewall functionalization of single-walled carbon nanotubes for protein immobilization. *J. Am. Chem. Soc.*, **123**, 3838–3839.

94 Besteman, K., Lee, J.-O., Wiertz, F.G.M., Heering, H.A., and Dekker, C. (2003) Enzyme-coated carbon nanotubes as single-molecule biosensors. *Nano Lett.*, **3**, 727–730.

95 Li, C., Curreli, M., Lin, H., Lei, B., Ishikawa, F.N., Datar, R., Cote, R.J., Thompson, M.E., and Zhou, C. (2005) Complementary detection of prostate-specific antigen using In_2O_3 nanowires and carbon nanotubes. *J. Am. Chem. Soc.*, **127**, 12484–12485.

96 Maehashi, K., Katsura, T., Kerman, K., Takamura, Y., Matsumoto, K., and Tamiya, E. (2007) Label-free protein biosensor based on aptamer-modified carbon nanotube field-effect transistors. *Anal. Chem.*, **79**, 782–787.

97 Ramasamy, R.P., Luckarift, H.R., Ivnitski, D.M., Atanassov, P.B., and Johnson, G.R. (2010) High electrocatalytic activity of tethered multicopper oxidase–carbon nanotube conjugates. *Chem. Commun.*, 6045–6047.

98 Landi, B.J., Evans, C.M., Worman, J.J., Castro, S.L., Bailey, S.G., and Raffaelle, R.P. (2006) Noncovalent attachment of CdSe quantum dots to single wall carbon nanotubes. *Mater. Lett.*, **60**, 3502–3506.

99 Sandanayaka, A.S.D., Takaguchi, Y., Uchida, T., Sako, Y., Morimoto, Y., Araki, Y., and Ito, O. (2006) Light-induced electron transfer on the single wall carbon nanotube surrounded in anthracene dendron in aqueous solution. *Chem. Lett.*, **35**, 1188–1189.

100 Zhang, J., Lee, J.-K., Wu, Y., and Murray, R.W. (2003) Photoluminescence and electronic interaction of anthracene derivatives adsorbed on sidewalls of single-walled carbon nanotubes. *Nano Lett.*, **3**, 403–407.

101 Mateo-Alonso, A., Ehli, C., Chen, K.H., Guldi, D.M., and Prato, M. (2007) Dispersion of single-walled carbon nanotubes with an extended diazapentacene derivative. *J. Phys. Chem. A*, **111**, 12669–12673.

102 Voggu, R., Rao, K.V., George, S.J., and Rao, C.N.R. (2010) A simple method of separating metallic and semiconducting single-walled carbon nanotubes based on molecular charge transfer. *J. Am. Chem. Soc.*, **132**, 5560–5561.

103 Steinberg, B.D. and Scott, L.T. (2009) New strategies for synthesizing short sections of carbon nanotubes. *Angew. Chem. Int. Ed.*, **48**, 5400–5402.

104 Feng, W., Fuji, A., Ozaki, M., and Yosino, K. (2005) Perylene derivative sensitized multi-walled carbon nanotube thin film. *Carbon*, **43**, 2501–2507.

105 Backes, C., Schmidt, C.D., Hauke, F., Böttcher, C., and Hirsch, A. (2009) High population of individualized SWCNTs through the adsorption of water-soluble perylenes. *J. Am. Chem. Soc.*, **131**, 2172–2184.

106 Backes, C., Schmidt, C.D., Rosenlehner, K., Hauke, F., Coleman, J.N., and Hirsch, A. (2010) Nanotube surfactant design: the versatility of water-soluble perylene bisimides. *Adv. Mater.*, **22**, 788–802.

107 Newkome, G.R., Nayak, A., Behera, R.K., Moorefield, C.N., and Baker, G.R. (1992) Cascade polymers: synthesis and characterization of four-directional spherical dendritic macromolecules based on adamantane. *J. Org. Chem.*, **57**, 358–362.

108 Ehli, C., Oelsner, C., Guldi, D.M., Mateo-Alonso, A., Prato, M., Schmidt, C.D., Backes, C., Hauke, F., and Hirsch, A. (2009) Manipulating single-wall carbon nanotubes by chemical doping and charge transfer with perylene dyes. *Nat. Chem.*, **1**, 243–249.

109 Murakami, H., Nomura, T., and Nakashima, N. (2003) Noncovalent porphyrin-functionalized single-walled carbon nanotubes in solution and the formation of porphyrin–nanotube nanocomposites. *Chem. Phys. Lett.*, **378**, 481–485.

110 Chen, J. and Collier, C.P. (2005) Noncovalent functionalization of single-walled carbon nanotubes with water-soluble porphyrins. *J. Phys. Chem. B*, **109**, 7605–7609.

111 Rahman, G.M.A., Guldi, D.M., Campidelli, S., and Prato, M. (2006) Electronically interacting single wall carbon nanotube–porphyrin nanohybrids. *J. Mater. Chem.*, **16**, 62–65.

112 Tanaka, H., Yajima, T., Matsumoto, T., Otsuka, Y., and Ogawa, T. (2006) Porphyrin molecular nanodevices wired using single-walled carbon nanotubes. *Adv. Mater.*, **18**, 1411–1415.

113 Kauffman, D.R., Kuzmych, O., and Star, A. (2007) Interactions between single-walled carbon nanotubes and tetraphenyl metalloporphyrins: correlation between spectroscopic and FET measurements. *J. Phys. Chem. C*, **111**, 3539–3543.

114 Roquelet, C., Lauret, J.-S., Alain-Rizzo, V., Voisin, C., Fleurier, R., Delarue, M., Garrot, D., Loiseau, A., Roussignol, P., Delaire, J.A., and Deleporte, E. (2010) Pi-stacking functionalization of carbon nanotubes through micelle swelling. *ChemPhysChem*, **11**, 1667–1672.

115 Cheng, F. and Adronov, A. (2006) Noncovalent functionalization and solubilization of carbon nanotubes by using a conjugated Zn–porphyrin polymer. *Chem. Eur. J.*, **12**, 5053–5059.

116 Cheng, F., Zhang, S., Adronov, A., Echegoyen, L., and Diederich, F. (2006) Triply fused ZnII-porphyrin oligomers: synthesis, properties, and supramolecular interactions with single-walled carbon nanotubes (SWNTs). *Chem. Eur. J.*, **12**, 6062–6070.

117 Wang, X., Liu, Y., Qiu, W., and Zhu, D. (2002) Immobilization of tetra-*tert*-butylphthalocyanines on carbon nanotubes: a first step towards the development of new nanomaterials. *J. Mater. Chem.*, **12**, 1636–1639.

118 Ma, A., Lu, J., Yang, S., and Ng, K.M. (2006) Quantitative non-covalent functionalization of carbon nanotubes. *J. Clust. Sci.*, **17**, 599–608.

119 Wang, J. and Blau, W.J. (2008) Linear and nonlinear spectroscopic studies of phthalocyanine-carbon nanotube blends. *Chem. Phys. Lett.*, **465**, 265–271.

120 Chichak, K.S., Star, A., Altoé, M.V.P., and Stoddart, J.F. (2005) Single-walled carbon nanotubes under the influence of dynamic coordination and supramolecular chemistry. *Small*, **4**, 452–461.

121 Hecht, D.S., Ramirez, R.J.A., Briman, M., Artukovic, E., Chichak, K.S., Stoddart, J.F., and Grüner, G. (2006) Bioinspired detection of light using a porphyrin-sensitized single-wall nanotube field effect transistor. *Nano Lett.*, **6**, 2031–2036.

122 Peng, X., Komatsu, N., Kimura, T., and Osuka, A. (2008) Simultaneous enrichments of optical purity and (*n*,*m*) abundance of SWNTs through extraction with 3,6-carbazolylene-bridged chiral diporphyrin nanotweezers. *ACS Nano*, **2**, 2045–2050.

123 Wang, F., Matsuda, K., Rahman, A.F.M.M., Peng, X., Kimura, T., and Komatsu, N. (2010) Simultaneous discrimination of handedness and diameter of single-walled carbon nanotubes (SWNTs) with chiral diporphyrin nanotweezers leading to enrichment of a single enantiomer of (6,5)-SWNTs. *J. Am. Chem. Soc.*, **132**, 10876–10881.

124 Chae, H.G., Liu, J., and Kumar, S. (2006) Carbon nanotube-enabled materials, in *Carbon Nanotubes Properties and Applications*, CRC Taylor and Francis, pp. 213–274.

125 Spitalsky, Z., Tasis, D., Papagelis, K., and Galiotis, C. (2010) Carbon nanotube.polymer composites: chemistry, processing, mechanical and electrical properties. *Prog. Polym. Sci.*, **35**, 357–401.

126 Coleman, J.N., Dalton, A.B., Curran, S., Rubio, A., Davey, A.P., Drury, A., McCarthy, B., Lahr, B., Ajayan, P.M., Roth, S., Barklie, R.C., and Blau, W.J. (2000) Phase separation of carbon nanotubes and turbostratic graphite using a functional organic polymer. *Adv. Mater.*, **12**, 213–216.

127 Curran, S.A., Ajayan, P.M., Blau, W.J., Carroll, D.L., Coleman, J.N., Dalton, A.B., Davey, A.P., Drury, A., McCarthy, B., Maier, S., and Stevens, A. (1998) A composite from poly(*m*-phenylenevinylene-*co*-2,5-dioctoxy-*p*-phenylenevinylene) and carbon nanotubes: a novel material for molecular optoelectronics. *Adv. Mater.*, **10**, 1091–1093.

128 Star, A., Stoddart, J.F., Steuerman, D., Diehl, M., Boukai, A., Wong, E.W., Yang, X., Chung, S.-W., Choi, H., and Heath, J.R. (2001) Preparation and properties of polymer-wrapped single-walled carbon nanotubes. *Angew. Chem., Int. Ed.*, **40**, 1721–1725.

129 Curran, S., Davey, A.P., Coleman, J.N., Dalton, A.B., McCarthy, B., Maier, S., Drury, A., Gray, D., Brennan, M., Ryder, K., de la Chapelle, M.L., Journet, C., Bernier, P., Byrne, H.J., Carroll, D.L., Ajayan, P.M., Lefrant, S., and Blau, W. (1999). Evolution and evaluation of the polymer nanotube composite. *Synth. Met.*, **103**, 2559–2562.

130 Yi, W., Malkovskiy, A., Chu, Q., Sokolov, A.P., Colon, M.L., Meador, M., and Pang, Y. (2008) Wrapping of single-walled carbon nanotubes by a pi-conjugated polymer: the role of polymer conformation-controlled size selectivity. *J. Phys. Chem. B*, **112**, 12263–12269.

131 Steuerman, D.W., Star, A., Narizzano, R., Choi, H., Ries, R.S., Nicolini, C., Stoddart, J.F., and Heath, J.R. (2002) Interactions between conjugated polymers and single-walled carbon nanotubes. *J. Phys. Chem. B*, **106**, 3124–3130.

132 Star, A., Liu, Y., Grant, K., Ridvan, L., Stoddart, J.F., Steuerman, D.W., Diehl, M.R., Boukai, A., and Heath, J.R. (2003) Noncovalent side-wall functionalization of single-walled carbon nanotubes. *Macromolecules*, **36**, 553–560.

133 Tang, B.Z. and Xu, H.Y. (1999) Preparation, alignment, and optical properties of soluble poly (phenylacetylene)-wrapped carbon nanotubes. *Macromolecules*, **32**, 2569–2576.

134 Yuan, W.Z., Sun, J.Z., Dong, Y.Q., Häussler, M., Yang, F., Xu, H.P., Qin, A.J., Lam, J.W.Y., Zheng, Q., and Tang, B.Z. (2006) Wrapping carbon nanotubes in pyrene-containing poly (phenylacetylene) chains: solubility, stability, light emission, and surface photovoltaic properties. *Macromolecules*, **39**, 8011–8020.

135 Yuan, W.Z., Sun, J.Z., Liu, J.Z., Dong, Y., Li, Z., Xu, H.P., Qin, A., Häussler, M., Jin, J.K., Zheng, Q., and Tang, B.Z. (2008) Processable hybrids of ferrocene-containing poly(phenylacetylene)s and carbon nanotubes: fabrication and properties. *J. Phys. Chem. B*, **112**, 8896–8905.

136 Zhao, H., Yuan, W.Z., Mei, J., Tang, L., Liu, X.Q., Yan, M., Shen, X.Y., Sun, J.Z., Qin, A.J., and Tang, B.Z. (2009) Enhanced dispersion of nanotubes in organic solvents by donor–acceptor interaction between functionalized poly (phenylacetylene) chains and carbon nanotube walls. *J. Polym. Sci. Part A Ploym. Chem.*, **47**, 4995–5005.

137 Ikeda, A., Nobusawa, K., Hamano, T., and Kikuchi, K. (2006) Single-walled carbon nanotubes template the one-dimensional ordering of a polythiophene derivative. *Org. Lett.*, **8**, 5489–5492.

138 Goh, R.G.S., Motta, N., Bell, J.M., and Waclawik, E.R. (2006) Effects of substrate curvature on the adsorption of poly (3-hexylthiophene) on single-walled carbon nanotubes. *Appl. Phys. Lett.*, **88**, 053101.

139 Schuettfort, T., Snaith, H.J., Nish, A., and Nicholas, R.J. (2010) Synthesis and spectroscopic characterization of solution processable highly ordered polythiophene–carbon nanotube nanohybrid structures. *Nanotechnology*, **21**, 025201.

140 Stranks, S.D., Weisspfennig, C., Parkinson, P., Jonston, M.B., Herz, L.M., and Nicholas, R.J. (2011) Ultrafast charge separation at a polymer-single-walled carbon nanotube molecular junction. *Nano Lett.*, **11**, 66–72.

141 Chen, J., Liu, H.Y., Weimer, W.A., Halls, M.D., Waldeck, D.H., and Walker, G.C. (2002) Noncovalent engineering of carbon nanotube surfaces by rigid, functional conjugated polymers. *J. Am. Chem. Soc.*, **124**, 9034–9035.

142 Kang, Y.K., Lee, O.S., Deria, P., Kim, S.H., Park, T.H., Bonnell, D.A., Saven, J.G., and Therien, M.J. (2009) Helical wrapping of single-walled carbon nanotubes by water soluble poly (*p*-phenyleneethynylene). *Nano Lett.*, **9**, 1414–1418.

143 Cheng, F., Imin, P., Maunders, C., Botton, G., and Adronov, A. (2008) Soluble, discrete supramolecular complexes of single-walled carbon nanotubes with fluorene-based conjugated polymers. *Macromolecules*, **41**, 2304–2308.

144 Deria, P., Sinks, L.E., Park, T.H., Tomezsko, D.M., Brukman, M.J., Bonnell, D.A., and Therien, M.J. (2010) Phase transfer catalysts drive diverse organic solvent solubility of single-walled carbon nanotubes helically wrapped by ionic, semiconducting polymers. *Nano Lett.*, **10**, 4192–4199.

145 Liu, X.L., Ly, J., Han, S., Zhang, D.H., Requicha, A., Thompson, M.E., and Zhou, C.W. (2005) Synthesis and electronic properties of individual single-walled carbon nanotube/ polypyrrole composite nanocables. *Adv. Mater.*, **17**, 2727–2732.

146 Huang, J.E., Li, X.H., Xu, J.C., and Li, H.L. (2003) Well-dispersed single-walled carbon nanotube/ polyaniline composite films. *Carbon*, **41**, 2731–2736.

147 Jiménez, P., Maser, W.K., Castell, P., Martínez, M.T., and Benito, A.M. (2009) Nanofibrilar polyaniline: direct route to carbon nanotube water dispersions of high concentration. *Macromol. Rapid. Commun.*, **30**, 418–422.

148 Sainz, R., Small, W.R., Young, N.A., Valles, C., Benito, A.M., Maser, W.K., and in het Panhuis, M. (2006) Synthesis and properties of optically active polyaniline

carbon nanotube composites. *Macromolecules*, **39**, 7324–7332.

149 Cochet, M., Maser, W.K., Benito, A.M., Callejas, M.A., Martínez, M.T., Benoit, J.M., Schreiber, J., and Chauvet, O. (2001) Synthesis of a new polyaniline/nanotube composite: "in-situ" polymerisation and charge transfer through site-selective interaction. *Chem. Commun.*, 1451.

150 Zengin, H., Zhou, W.S., Jin, J.Y., Czerw, R., Smith, D.W., Echegoyen, L., Carroll, D.L., Foulger, S.H., and Ballato, J. (2002) Carbon nanotube doped polyaniline. *Adv. Mater.*, **14**, 1480–1483.

151 Carroll, D.L., Czerw, R., and Webster, S. (2005) Polymer–nanotube composites for transparent, conducting thin films. *Synth. Met.*, **155**, 694–697.

152 Kim, D., Kim, Y., Choi, K., Grulan, J.C., and Yu, C. (2010) Improved thermoelectric behavior of nanotube-filled polymer composites with poly (3,4-ethylenedioxythiophene) poly (styrenesulfonate). *ACS Nano*, **4**, 513–523.

153 Li, J., Liu, J.-C., and Gao, C.-J. (2010) On the mechanism of conductivity enhancement in PEDOT/PSS film doped with multi-walled carbon nanotubes. *J. Polym. Res.*, **17**, 713–718.

154 Moon, J.S., Park, J.H., Lee, T.Y., Kim, Y.W., Yoo, J.B., Park, C.Y., Kim, J.M., and Jin, K.W. (2005) Transparent conductive film based on carbon nanotubes and PEDOT composites. *Diam. Relat. Mat.*, **14**, 1882–1887.

155 Fan, B.H., Mei, X.G., Sun, K., and Ouyang, J.Y. (2008) Conducting polymer/carbon nanotube composite as counter electrode of dye-sensitized solar cells. *Appl. Phys. Lett.*, **93**, 143103.

156 Chen, F., Wang, B., Chen, Y., and Li, L.J. (2007) Toward the extraction of single species of sinale-walled carbon nanotubes using fluorene-based polymers. *Nano Lett.*, **7**, 3013–3017.

157 Hwang, J.-Y., Nish, A., Doig, J., Douven, S., Chen, C.-W., Chen, L.-C., and Nicholas, R.J. (2008) Polymer structure and solvent effects on the selective dispersion of single-walled carbon nanotubes. *J. Am. Chem. Soc.*, **130**, 3543–3553.

158 Nish, A., Hwang, J.-Y., Doig, J., and Nicholas, R.J. (2008) Direct spectroscopic evidence of energy transfer from photo-excited semiconducting polymers to single-walled carbon nanotubes. *Nanotechnology*, **19**, 095603.

159 Gao, J., Kwak, M., Wildeman, J., Herrmann, A., and Loi, M.A. (2011) Effectiveness of sorting single-walled carbon nanotubes by diameter using polyfluorene derivatives. *Carbon*, **49**, 333–338.

160 Bénard, P., Chahine, R., Chandonia, P.A., Cossement, D., Dorval-Douville, G., Lafi, L., Lachance, P., Paggiaro, R., and Poirier, E. (2007) Comparison of hydrogen adsorption on nanoporous materials. *J. Alloys Compd.*, **380**, 446–447.

161 Wang, Z., Shi, Z., and Gu, Z. (2010) Chemistry in the nanospace of carbon nanotubes. *Chem. Asian J.*, **5**, 1030–1038.

162 Pastorin, G. (2009) Crucial functionalizations of carbon nanotubes for improved drug delivery: a valuable option. *Pharm. Res.*, **26**, 746–769.

163 Xu, J., Yudasaka, M., Kouraba, S., Sekido, M., Yamamoto, Y., and Iijima, S. (2008) Single wall carbon nanohorn as a drug carrier for controlled release. *Chem. Phys. Lett.*, **461**, 189–192.

164 Ajima, K., Murakami, T., Mizoguchi, Y., Tsuchida, K., Ichihashi, T., Iijima, S., and Yudasaka, M. (2008) Enhancement of *in vivo* anticancer effects of cisplatin by incorporation inside single-wall carbon nanohorns. *ACS Nano*, **2**, 2057–2064.

165 Britz, D.A. and Khlobystov, A.N. (2006) Noncovalent interactions of molecules with single walled carbon nanotubes. *Chem. Soc. Rev.*, **35**, 637–659.

166 Monthioux, M. and Flahaut, E. (2007) Meta- and hybrid-CNTs: a clue for the future development of carbon nanotubes. *Mater. Sci. Eng. C*, **27**, 1096–1101.

167 Monthioux, M. and Flahaut, E. (2006) Hybrid carbon nanotubes: strategy, progress, and perspectives. *J. Mater. Res.*, **21**, 2774–2793.

168 Smith, B.W., Monthioux, M., and Luzzi, D.E. (1998) Encapsulated C-60 in carbon nanotubes. *Nature*, **396**, 323–324.

169 Takenobu, T., Takano, T., Shiraishi, M., Murakami, Y., Ata, M., Kataura, H., Achiba, Y., and Iwasa, Y. (2003) Stable and controlled amphoteric doping by encapsulation of organic molecules inside carbon nanotubes. *Nat. Mater.*, **2**, 683–688.

170 Simon, F., Kuzmany, H., Rauf, H., Pichler, T., Bernardi, J., Peterlik, H., Korecz, L., Fülöp, F., and Jánossy, A. (2004) Low temperature fullerene encapsulation in single wall carbon nanotubes: synthesis of N@C60@SWCNT. *Chem. Phys. Lett.*, **383**, 362–367.

171 Yanagi, K., Miyata, Y., and Kataura, H. (2006) Highly stabilized beta-carotene in carbon nanotubes. *Adv. Mater.*, **18**, 437–441.

172 Ebbesen, T.W. (1996) Wetting, filling and decorating carbon nanotubes. *J. Phys. Chem. Solids*, **57**, 951–955.

173 Khlobystov, A.N., Britz, D.A., Wang, J.W., O'Neil, S.A., Poliakoff, M., and Briggs, G.A.D. (2004) Low temperature assembly of fullerene arrays in single-walled carbon nanotubes using supercritical fluids. *J. Mater. Chem.*, **14**, 2852–2857.

174 Khlobystov, A.N., Britz, D.A., and Briggs, G.A.D. (2005) Molecules in carbon nanotubes. *Acc. Chem. Res.*, **38**, 901–909.

175 Kitaura, R. and Shinohara, H. (2006) Carbon-nanotube-based hybrid materials: nanopeapods. *Chem. Asian J.*, **1**, 646–655.

176 Guan, L., Shi, Z., Li, M., and Gu, Z. (2005) Ferrocene-filled single-walled carbon nanotubes. *Carbon*, **43**, 2780–2785.

177 Li, L.J., Khlobystov, A.N., Wiltshire, J.G., Briggs, G.A.D., and Nicholas, R.J. (2005) Diameter-selective encapsulation of metallocenes in single-walled carbon nanotubes. *Nat. Mater.*, **4**, 481–485.

178 Li, Y.F., Hatakeyama, R., Kaneko, T., Izumida, T., Okada, T., and Kato, T. (2006) Synthesis and electronic properties of ferrocene-filled double-walled carbon nanotubes. *Nanotechnology*, **17**, 4143–4147.

179 Cao, L., Chen, H.Z., Li, H.Y., Zhou, H.B., Sun, J.Z., Zang, X.B., and Wang, M. (2003) Fabrication of rare-earth biphthalocyanine encapsulated by carbon nanotubes using a capillary filling method. *Chem. Mater.*, **15**, 3247–3249.

180 Schulte, K., Swarbrick, J.C., Smith, N.A., Bondino, F., Magnano, E., and Khlobystov, A.N. (2007) Assembly of cobalt phthalocyanine stacks inside carbon nanotubes. *Adv. Mater.*, **19**, 3312–3316.

181 Kataura, H., Maniwa, Y., Abe, M., Fujiwara, A., Komada, T., Kikuchi, K., Imahori, H., Misaki, Y., Suzuki, S., and Achiba, Y. (2002) Optical properties of fullerene and non-fullerene peapods. *Appl. Phys. A Mater. Sci. Process.*, **74**, 349–354.

182 Kaneko, T., Li, Y.F., Nishigaki, S., and Hatakeyama, R. (2008) Azafullerene encapsulated single-walled carbon nanotubes with n-type electrical transport property. *J. Am. Chem. Soc.*, **130**, 2714–2715.

183 Li, Y.F., Hatakeyama, R., Shishido, J., Kato, T., and Kaneko, T. (2007) Air-stable p-n junction diodes based on single-walled carbon nanotubes encapsulating Fe nanoparticles. *Appl. Phys. Lett.*, **90**, 173127.

184 Kato, T., Hatakeyama, R., Shishido, J., Oohara, W., and Tohji, K. (2009) P-n junction with donor and acceptor encapsulated single-walled carbon nanotubes. *Appl. Phys. Lett.*, **95**, 083109.

185 Li, Y.F., Hatakeyama, R., Oohara, W., and Kaneko, T. (2009) Formation of p-n junction in double-walled carbon nanotubes based on heteromaterial encapsulation. *Appl. Phys. Express*, **2**, 095005.

186 Fujita, Y., Bandow, S., and Iijima, S. (2005) Formation of small-diameter carbon nanotubes from PTCDA arranged inside the single-wall carbon nanotubes. *Chem. Phys. Lett.*, **413**, 410–414.

187 Hernandez, E., Meunier, V., Smith, B.W., Rurali, R., Terrones, H., Nardelli, M.B., Terrones, M., Luzzi, D.E., and Charlier, J.-C. (2003) Fullerene coalescence in nanopeapods: a path to novel tubular carbon. *Nano Lett.*, **3**, 1037–1042.

188 Shiozawa, H., Pichler, T., Pfeiffer, R., Kuzmany, H., and Kataura, H. (2007) Ferrocene encapsulated in single-wall

carbon nanotubes: a precursor to secondary tubes. *Phys. Status Solidi B Basic Solid State Phys.*, **244**, 4102–4105.

189 Li, Y.F., Kaneko, T., and Hatakeyama, R. (2008) Photoinduced electron transfer in C-60 encapsulated single-walled carbon nanotube. *Appl. Phys. Lett.*, **92**, 183115.

190 Li, Y.F., Kaneko, T., and Hatakeyama, R. (2008) Electrical transport properties of fullerene peapods interacting with light. *Nanotechnology*, **19**, 415201.

191 Li, Y.F., Kaneko, T., Kong, J., and Hatakeyama, R. (2009) Photoswitching in azafullerene encapsulated single-walled carbon nanotube FET devices. *J. Am. Chem. Soc.*, **131**, 3412–3413.

192 Li, Y.F., Kaneko, T., Miyanaga, S., and Hatakeyama, R. (2010) Synthesis and property characterization of C69N azafullerene encapsulated single-walled carbon nanotubes. *ACS Nano*, **4**, 3522–3526.

193 Yongfeng, L., Kaneko, T., and Hatakeyama, R. (2009) Photoresponse of fullerene and azafullerene peapod field effect transistors. IEEE Conference on Nanotechnology, 86–89.

194 Yanagi, K., Iakoubovskii, K., Kazaoui, S., Minami, N., Maniwa, Y., Miyata, Y., and Kataura, H. (2006) Light-harvesting function of beta-carotene inside carbon nanotubes. *Phys. Rev. B*, **74**, 155420.

195 Yanagi, K., Iakoubovskii, K., Matsui, H., Matsuzaki, H., Okamoto, H., Miyata, Y., Maniwa, Y., Kazaoui, S., Minami, N., and Kataura, H. (2007) Photosensitive function of encapsulated dye in carbon nanotubes. *J. Am. Chem. Soc.*, **129**, 4992–4997.

196 Yanagi, K., Miyata, Y., Liu, Z., Suenaga, K., Okada, S., and Kataura, H. (2010) Influence of aromatic environments on the physical properties of beta-carotene. *J. Phys. Chem. C*, **114**, 2524–2530.

197 Loi, M.A., Gao, J., Cordella, F., Blondeau, P., Menna, E., Bártová, B., Hébert, C., Lazar, S., Botton, G.A., Milko, M., and Ambrosch-Draxl, C. (2010) Encapsulation of conjugated oligomers in single-walled carbon nanotubes: towards nanohybrids for photonic devices. *Adv. Mater.*, **22**, 1635–1639.

198 Yanagi, K. and Kataura, H. (2010) Carbon nanotubes breaking Kasha's rule. *Nat. Photonics*, **4**, 200–201.

199 Hatakeyama, R., Li, Y.F., Kato, T.Y., and Taneko, T. (2010) Infrared photovoltaic solar cells based on C_{60} fullerene encapsulated single-walled carbon nanotubes. *Appl. Phys. Lett.*, **97**, 013104.

200 Kawasaki, S., Iwai, Y., and Hirose, M. (2009) Electrochemical lithium ion storage properties of single-walled carbon nanotubes containing organic molecules. *Carbon*, **47**, 1081–1086.

201 Naitoh, Y., Yanagi, K., Suga, H., Horikawa, M., Tanaka, T., Kataura, H., and Shimizu, T. (2009) Non-volatile resistance switching using single-wall carbon nanotube encapsulating fullerene molecules. *Appl. Phys. Express*, **2**, 035008.

202 Chen, S., Wu, G., Sha, M., and Huang, S. (2007) Transition of ionic liquid [bmim][PF6] from liquid to high-melting-point crystal when confined in multiwalled carbon nanotubes. *J. Am. Chem. Soc.*, **129**, 2416–2417.

203 Chen, S., Kobayashi, K., Miyata, Y., Imazu, N., Saito, T., Kitaura, R., and Shinohara, H. (2009) Morphology and melting behavior of ionic liquids inside single-walled carbon nanotubes. *J. Am. Chem. Soc.*, **131**, 14850–14856.

204 Liu, Z.M., Dai, X.H., Xu, J., Han, B.X., Zhang, J.L., Wang, Y., Huang, Y., and Yang, G.Y. (2004) Encapsulation of polystyrene within carbon nanotubes with the aid of supercritical CO_2. *Carbon*, **42**, 458–460.

205 Liu, Z.F., Zhang, X.Y., Yang, X.Y., Ma, Y.F., Huang, Y., Wang, B., Chen, Y.S., and Sheng, J. (2010) Synthesis and characterization of the isolated straight polymer chain inside of single-walled carbon nanotubes. *J. Nanosci. Nanotechnol.*, **10**, 5570–5575.

206 Kim, G., Kim, Y., and Ihm, J. (2005) Encapsulation and polymerization of acetylene molecules inside a carbon nanotube. *Chem. Phys. Lett.*, **416**, 279–282.

207 Steinmetz, J., Lee, H.-J., Kwon, S., Lee, H.J., Abou-Hamad, E., Almairac, R., Goze-Bac, C., Kim, H., and Park, Y.W. (2007) Routes to the synthesis of carbon nanotube-polyacetylene composites by Ziegler-Natta polymerization of acetylene inside carbon nanotubes. *Curr. Appl. Phys.*, **7**, 39–41.

208 Steinmetz, J., Kwon, S., Lee, H.-J., Abou-Hamad, E., Almairac, R., Goze-Bac, C., Kim, H., and Park, Y.W. (2006) Polymerization of conducting polymers inside carbon nanotubes. *Chem. Phys. Lett.*, **431**, 139–144.

209 Lock, E.H., Merchan-Merchan, W., D'Arcy, J., Saveliev, A.V., and Kennedy, L.A. (2007) Coating of inner and outer carbon nanotube surfaces with polymers in supercritical CO_2. *J. Phys. Chem. C*, **111**, 13655–13658.

210 Bazilevsky, A.V., Sun, K.X., Yarin, A.L., and Megaridis, C.M. (2007) Selective intercalation of polymers in carbon nanotubes. *Langmuir*, **23**, 7451–7455.

211 Kaneko, T., Okada, T., and Hatakeyama, R. (2007) DNA encapsulation inside carbon nanotubes using micro electrolyte plasmas. *Contrib. Plasma Phys.*, **47**, 57–63.

212 Okada, T., Kaneko, T., Hatakeyama, R., and Tohji, K. (2006) Electrically triggered insertion of single-stranded DNA into single-walled carbon nanotubes. *Chem. Phys. Lett.*, **417**, 288–292.

213 Li, Y.F., Kaneko, T., and Hatakeyama, R. (2010) Tailoring the electronic structure of double-walled carbon nanotubes by encapsulating single-stranded DNA. *Small*, **6**, 729–732.

12
Supramolecular Chemistry of Fullerenes and Carbon Nanotubes at Interfaces: Toward Applications

Riccardo Marega, Davide Giust, Adrian Kremer, and Davide Bonifazi

12.1
Introduction

An interface is a surface regarded as the common boundary of two bodies, spaces, or phases. In chemical and life sciences, the properties of interfaces between different phases are of fundamental importance since they play a critical role in many technological and biological fields. For instance, the production of new materials deeply involves processes at the interfaces: thin films on surfaces are often dominated by surface effects, for example, latex films, coatings, and paints. In tribology, the science and engineering of interacting surfaces in relative motion, the friction is reduced by lubrification, which again is a surface phenomenon. Crystal growth, lubrication, catalysis, and electrochemistry all occur at solid/liquid interface. In biology, structures and dynamics of biological membranes, which surround cellular organelles and cells, play critical roles in fundamental physiological and pathological processes. For instance, the cell–cell interactions not only enable the formation of different tissues but also form the basis for the recognition of third bodies by our immune system. Furthermore, also the energy production pathways, such as photosynthesis in chloroplasts and oxidative phosphorylation in mitochondrion, occur at the interface between the membranes and the surrounding medium. These are only a few examples in which a proper understanding of the physical and chemical phenomena requires a detailed structural knowledge of the interface at a molecular or atomic scale. In general, the chemical and physical properties of an interface (e.g., composition, wettability, or tribology) can be changed by both inorganic [1] and organic chemical modifications [2]. In the past years, supramolecular chemistry [3] has played an important role in the control of the structure and the properties of organic molecules at the interfaces. The use of weak, reversible interactions has been motivated by the need to engineer smaller and smaller components in order to improve, for example, the information storage capabilities of classical silicon-based devices [4]. In this chapter, we will focus on the recent advances in the use of the supramolecular chemistry for the construction of specific fullerene and carbon nanotube (CNT) derivatives that display exploitable properties at

Supramolecular Chemistry of Fullerenes and Carbon Nanotubes, First Edition. Edited by Nazario Martin and Jean-Francois Nierengarten.

interfaces. Here, two different approaches are described. In the first, the tools of supramolecular chemistry have been used to merge CNT properties with that of organic moieties displaying useful technological or biological functions, while in the second pristine or covalently modified CNTderivatives showed postfunctionalization noncovalent interactions that revealed to be useful for specific applications. A particular focus will be given on the recent discoveries that might bring these carbon-rich nanostructures to real applications.

12.2
Fullerene Interfaces

Interface modification utilizing C_{60} and its covalent derivatives is of substantial interest, owing to the possibility of transferring the unique C_{60} properties to bulk materials by surface coating. Monolayers, containing redox centers at a fixed distance from a surface instead of freely diffusing, form an important class of new hybrid materials offering potential technological applications ranging from bioactive materials to advanced nanostructured devices for electronic applications. This chapter provides a general overview on the various approaches toward the immobilization of C_{60} and its derivatives on functional interfaces. The physical properties of C_{60}-containing films are controlled by the deposition conditions and influenced by impurities or disordered surface structures. Despite extensive efforts to form stable and well-ordered C_{60}-containing films, the strong van der Waals interactions between the carbon spheres and the resulting tendency to form aggregates still remain one of the major issues for the formation of stable and structurally ordered films at the air–solid and air–water interfaces. Therefore, an absolutely essential requirement for the control and systematic exploration of the physical properties of the carbon allotropes is the development of new methodologies for the incorporation of C_{60} into well-defined two- and three-dimensional networks. The main methods to prepare C_{60}-containing monolayers or thin films are monolayer self-assembly on solid surface, *Langmuir* and *Langmuir–Blodgett* deposition techniques. Langmuir films are prepared by spreading amphiphilic molecules on liquid surfaces, in particular water. The molecules arrange in such a way that the polar head groups are in direct contact with water, while the hydrophobic tails (i.e., the fullerene cage) stick out into the air [5, 6]. Transferring the Langmuir films onto a solid substrate (usually glass) results in the formation of Langmuir–Blodgett (LB) films that can be monolayered or multilayered depending on the number of times the substrate is dipped into the solution [5, 6]. Typically, amphiphilic fullerene compounds with a hydrophilic head group that interacts with or immerses into the aqueous subphase and a hydrophobic tail are employed for the production of LB films. In view of the vast number of recent reviews summarizing the research on C_{60}-based thin films prepared with Langmuir and Langmuir–Blodgett [7–15], this chapter will focus on the immobilization of fullerenes at interfaces via the self-assembly monolayer (SAM) methodology. In the next section, different self-assembly approaches for the construction of fullerene-containing interfaces displaying potential applications in materials science are described.

12.2.1 Fullerenes at the Liquid–Liquid and Micellar Interfaces

According to the wide number of studies carried out with fullerenes in solution, it is of great importance for fullerenes to own properties at the interface of different solvent phases. In that sense it was reported how an aqueous emulsion containing pristine C_{60} could be obtained by using pristine C_{60} dissolved in water and benzene phases under high-power sonication [16]. Recently, Ito and coworkers reported the large-scale production of C_{60}/ferrocene nanosheet crystals, exhibiting a strong charge transfer (CT) between ferrocene (Fc) and C_{60} at the interface [17]. The nanosheets were prepared by C_{60} in the presence of an excess of Fc at the toluene/isopropanol interface, followed by filtration and heating at 80 °C to remove the excess Fc. Scanning electron microscopy (SEM) analysis showed hexagonal nanosheets of C_{60}/ferrocene complexes with a size of 9.1 ± 6.2 μm and 250–550 nm of thickness. Furthermore, X-ray diffraction analysis (XRD) indicated that the nanosheets were composed of triclinic $C_{60}(Fc)_2$ units, indicating two Fc molecules per each C_{60} structure. The electronic interactions between Fc and C_{60} in the supramolecular structure of the crystal were studied and related to the formation of a smectic A phase. On that, C_{60} imposes the arrangement of the other moieties to form a partial bilayer, while Fc moieties are intercalated between the C_{60} units and the dendric core [18, 19]. Furthermore, by increasing the temperature the $C_{60}(Fc)_2$ nanosheets decomposed into face-centered cubic (fcc) C_{60} hexagonal nanosheets. Diffuse reflectance spectra showed high CT at 782 nm from which the CT transition energy was estimated as 1.59 eV in $C_{60}(Fc)_2$. This allowed to establish the driving force of the interaction between C_{60} and Fc in the crystal, which was based on π–π interactions. The no-observed CT in the crystal, after increasing temperature, also confirmed these findings. Recently, Kurth and coworkers reported that fullerene derivatives bearing three alkyloxy groups, obtained after 1,3 dipolar cycloaddition of azomethine ylides, undergoes hierarchical self-assembly furnishing various polymorphs, depending on the experimental conditions such as solvent and temperature [20, 21]. Of particular interest were the self-assembly properties of the C_{60} derivative reported in Figure 12.1, which was prepared by refluxing the corresponding benzaldehyde with *N*-methylglycine and C_{60} in dry toluene [22]. It was found to form a dark brown precipitate upon cooling a 1,4-dioxane solution thereof from 70 °C to room temperature, with close to quantitative yield, as indicated by the colorless supernatant (Figure 12.1b). Both TEM and SEM analyses revealed that the precipitates were of globular shape (1–10 μm in diameter) and exhibited wrinkled, flake-like submicrometer structures at the outer surface (Figure 12.1c–e) [23].

Fast Fourier transform (FFT) analysis of the image (Figure 12.1f, lower) resulted in a lamellar periodicity value of 4.4 nm, which corresponds to the length of a bilayer of interdigitated eicosyloxy chains since the lateral dimension of the C_{60} derivative is approximately 3.6 nm. At the nanoscopic level, the globular objects are made of interdigitated bilayers, whose stacks form sheets, which are crumpled to a discrete, spherical object with microscopic dimensions and a flake-like submicrometer structure. Thin films of these globular objects were easily engineered on different

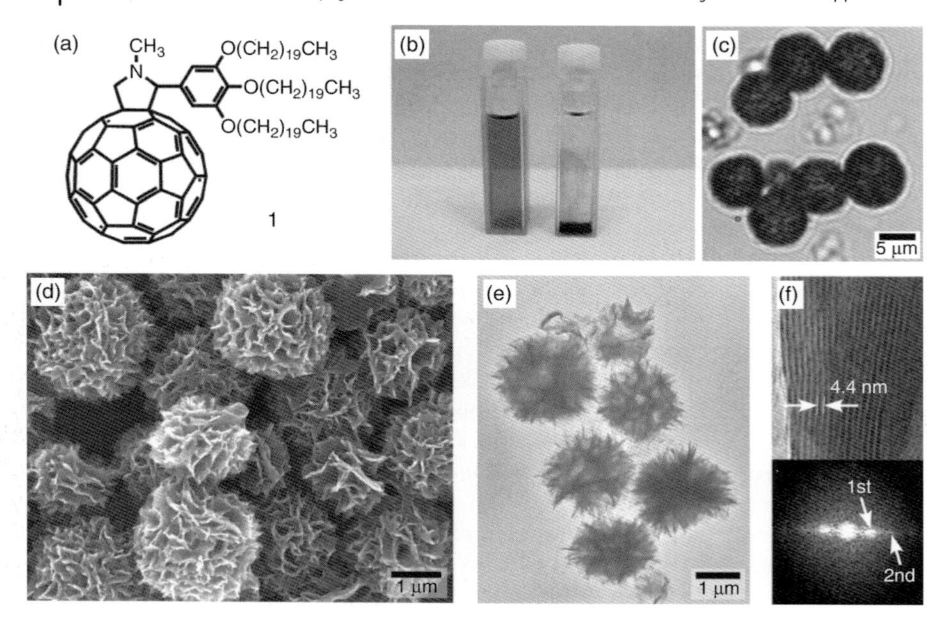

Figure 12.1 (a) Chemical structure of the C_{60} bearing three eicosyloxy aliphatic chains. (b) A photograph of the C_{60} derivative in 1,4-dioxane at 70 °C (*left*) and dark brown precipitate generated by cooling the solution to 20 °C (*right*). (c) Optical microscopy, (d) SEM, and (e) TEM images of globular objects of **1** deposited from a dilute 1,4-dioxane dispersion (various substrates). The precipitates are of globular shape, measure several micrometers in diameter, and exhibit a wrinkled, flake-like submicrometer structure at the outer surface. (f) High-resolution cryogenic TEM (HR-cryo-TEM) image (*top*) and corresponding FFT analysis of a flake edge showing the first- and second-order spots (*bottom*). The periodicity of the lamella is 4.4 nm. Reproduced with permission from Ref. [23]. Copyright 2008, Wiley-VCH Verlag GmbH.

substrates such as silicon, gold, and glass, by slow evaporation of a dilute 1,4 dioxane dispersion. Under these conditions, the globular objects formed densely packed films with a thickness of approximately 20 μm, as observed by SEM analysis. Remarkably, a thin film of the globular objects on silicon featured water-repellent superhydrophobicity with a water contact angle of 152.0°. In contrast, spin coating of the C_{60} derivative from chloroform solutions on silicon resulted in a smooth layer with a low surface roughness (5 nm, as evaluated by AFM), exhibiting a contact angle of 103.5° only. The films prepared from the globular objects showed high thermal stability and durability toward various polar solvents. In fact, upon heating to 100 °C for up to 36 h, the fractal morphology and the superhydrophobicity (152.6°) of the films were completely retained. Moreover, exposure of the surface to organic solvents such as acetone (151.5°) and ethanol (153.1°) as well as to acidic (152.1° at pH 2) or basic (151.0° at pH 12) aqueous media did not significantly change the water contact angle, highlighting the robustness of the obtained assembly. Hierarchical organization of molecular components into macroscopic objects with fractal structures provides a solution for the design and fabrication of low adhesion, low friction, and nonwetting surfaces for micro/nanoelectromechanical systems (MEMS/NEMS), microfluidics, or self-cleaning surfaces. A comprehensive discussion on the supramolecular properties of fullerene derivatives bearing aliphatic chains is presented in a recent review by Möhwald and coworkers [24]. Nakamura and coworkers recently presented a study on the behavior of an amphiphilic fluorous fullerene derivative in different solvent phases [25]. The synthesized pentakis[*p*-(perfluorooctyl)phenyl]C_{60} allows to obtain a complex, insoluble in water and toluene phases, which was soluble in perfluorooctane (C_8F_{18}). Nevertheless, the corresponding anion of pentakis[*p*-(perfluorooctyl)phenyl]fullerene, as potassium salt, was found to be highly soluble in water (up to 10 g/l at 25 °C) and to form bilayer vesicles insoluble in toluene and C_8F_{18} (Figure 12.2). The reference C_{60} derivative bearing a pentaphenyl moiety lacking the fluorous group was found to be soluble in toluene, while insoluble in water and C_8F_{18}. The fluorous vesicles in water were analyzed by static light scattering (SLS), which showed a hydrodynamic radius (R_h) of 26.7 nm and a gyration radius (R_g) of 26.7 nm, thus confirming the hollow vesicular structure in water ($R_h/R_g = 1.00$). SLS and SEM clarified the structure of water-soluble vesicles.

The fullerene cages, in the interior of the bilayer, aggregate upon their noncovalent interactions, forming the hydrophobic core of vesicles. At the same time, the fluorinated chains are exposed to the aqueous environment. The described

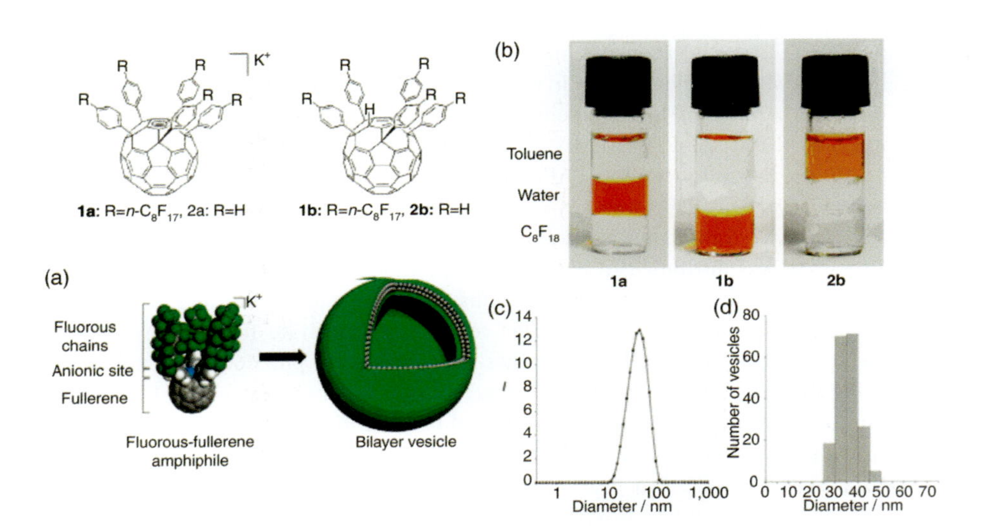

Figure 12.2 Potassium salts of water-soluble fullerene anions **1a** and **2a** and their neutral precursors **1b** and **2b** (*top*). Bilayer vesicle from fluorous fullerene amphiphile **1a** in water. (a) Drawing of **1a** and a model of its vesicle. F green, C gray, H white, and cyclopentadienide moiety blue. (b) Fluorous anion **1a** dissolves in water (middle phase). Neutral fluorous fullerene **1b** dissolves well in C_8F_{18} (*bottom phase*) and sparingly in toluene (*top phase*). Neutral phenyl fullerene **2b** dissolves in toluene. (c, d) Size distributions of vesicles of **1a** as determined by DLS (c) and by SEM (d); the two methods agree well with each other. Adapted with permission from Ref. [25]. Copyright 2010, Wiley-VCH Verlag GmbH.

water-soluble vesicles were found to be useful to form hydrophobic coating on substrates. When spin coated on indium tin oxide (ITO), the resulted material became water repellent like in poly(tetrafluoroethylene) (PTFE, Teflon) surface, as observed by contact angle analysis. The properties of these fullerene-based vesicles suggest their use for targeted delivery of organic and inorganic materials in several systems, as well as for the nanoscale modification of solid substrates. The strategies to engineer amphiphilic C_{60} derivatives have considerably increased in number, since they allow to control the aggregation of C_{60} in different solvents, toward the creation of a structure-defined organization, as well as to obtain functional materials for a wide number of applications [26, 27].

The synthesis of an amphiphilic C_{60} monomer (AF-I) was reported by Conyers and coworkers [28]. The monomer consisted in a C_{60} cage to which a Newkome-like dendrimer and five lipophilic C_{12} chains, octahedrally positioned with respect to the dendrimer, were attached. This novel fullerene-based liposome called "buckysome" was found to be water soluble and to form stable nanometer-sized vesicles. By using different characterization techniques such as cryogenic electron microscopy (Cryo-EM), transmission electron microscopy (TEM), and dynamic light scattering (DLS), the formation of bilayer membranes (~6.5 nm) as well as large multilamellar (400 nm diameter) and unilamellar (50–150 nm diameter) liposome-like vesicles, under different pH conditions in aqueous solvents, was reported. Furthermore, no toxicity was observed by testing the obtained vesicles in different cell lines, suggesting their possible biological applications. Following the aim, Vernon and coworkers proposed the synthesis of supramolecular nanocomposites based on the self-assembly of C_{60} with polymer colloids in water [29]. By using polystyrene (PS) as the core in isopropanol, C_{60} in toluene solution (0.5 mg/ml) was added to self-assembled micelles of PS_{6700}-*b*-$PDMAEMA_{9200}$, thus obtaining a solution of the C_{60}/polymer composite in a 2: 1 molar ratio. After removing the excess of toluene, the composites were analyzed by TEM and Cryo-EM, and thus structurally defined C_{60}/polymer complexes were found in water solution. The formation of controlled supramolecular structures was attributed to particle–particle interactions induced by an Ostwald ripening process in water. These results provided a good example of assembly controlled C_{60}-based supramolecular colloids, as substrate for a wide range of applications. More recently, the synthesis of isomerically pure [60]fullerenol undergoing the formation of amphiphilic aggregates in water was proposed. Despite the number of hydroxylation methods reported in order to increase the C_{60} solubility in aqueous environments [30, 31], Nakamura and coworkers obtained an octa-hydroxyl C_{60} derivative ($C_{60}(OH)_8$), bearing all the hydroxyl moieties on the same hemisphere. The location of all the hydroxyl moieties on the same hemisphere of C_{60} renders the molecule amphiphilic, with a different behavior with respect to the previously synthesized C_{60} hydroxyl derivatives [32]. Furthermore, the distribution of hydroxyl groups over the entire surface of C_{60} was demonstrated to be a source of intermolecular H bonding between molecules in aqueous solution. In fact, laser light scattering studies suggested that the assembly of $C_{60}(OH)_8$ in aqueous solution is highly concentration dependent and also that the aggregates exhibited R_h of about 100 nm. Atomic force microscopy analysis suggested that the morphology of aggregates is spherical. The height of the dried aggregates was quantified in 6 nm, much smaller than the R_h in aqueous solution, thus suggesting that the aggregate formed in solution is of low density. Finally, it was demonstrated that since the aggregate is much larger than $C_{60}(OH)_8$ it contains multidomains rich in OH groups and in C_{60} shells; thus, $C_{60}(OH)_8$ could be used as carrier for both hydrophilic and hydrophobic molecules [33].

12.2.2
Fullerenes at the Solid–Liquid Interface

The main challenge in the design of fullerene-based material with enhanced properties requires it to own electronic and physical properties by controlling the geometry during the self-assembly of molecules and even later. For example, the formation of architectures of pristine C_{60} self-assembled on clean gold surfaces from solution [34, 35] is based on the strong interactions between d-band metals and π systems of C_{60}. This usually leads to the formation of well-ordered arrangements of C_{60} on surface pattern, if regulated by the molecular orientation of the material, as in the case of Au(111) that well dictates the geometry of the resulting surfaces [36]. Nevertheless, in some cases these strong interactions between C_{60} and metal might induce an atomic modification of the substrate, leading to poorly ordered architectures and surface reconstruction, even due to the uncontrolled spatial rearrangement of unsaturated C_{60} on surface [37]. Noncovalent interactions are a source of stability upon the formation of many different fullerene-based supramolecular complexes, where the main observed van der Waals and π–π interactions could be useful to produce bulk materials with novel chemical/physical properties [38–40]. In fact, another approach to build highly organized fullerene architectures is based upon the host–guest interactions between pristine fullerene deposited from solution on a surface bearing previously self-assembled molecules. For example, calix[4]arene, which noncovalent interactions use as template for the self-assembly of fullerene on surface, allowed to obtain 2D modified material with enhanced properties [41].

More examples of engineered metal surface as nanotemplates for noncovalent hosting fullerene cages have been recently provided by using phthalocyanine and porphyrin architectures deposited on Au(100) and Au(111) electrode surfaces. These acted as 2D nanoarrays for the design of photoactive nanorods and nanowires upon π electron donating–accepting properties of the 3D supramolecular layers [42]. Several works report on the self-assembly of different monomeric units that are able to form pores suitable for C_{60} hosting on various surfaces. Heckl and coworkers proposed an interesting approach for the controlled deposition and further manipulation of isolated C_{60} molecules on graphite supports, through a room-temperature STM study at the solid/liquid interface [43]. In this STM study, they used a supramolecular 2D architecture of trimesic acid (TMA) molecules adsorbed on a graphite substrate as a host for the incorporation of C_{60} as a molecular guest. In fact, TMA molecules noncovalently self-assembled on the graphite substrate to form a chicken-wire structure with a periodic arrangement of cavities of 1.1 nm in diameter with a nearest-neighbor distance of 1.6 nm. By

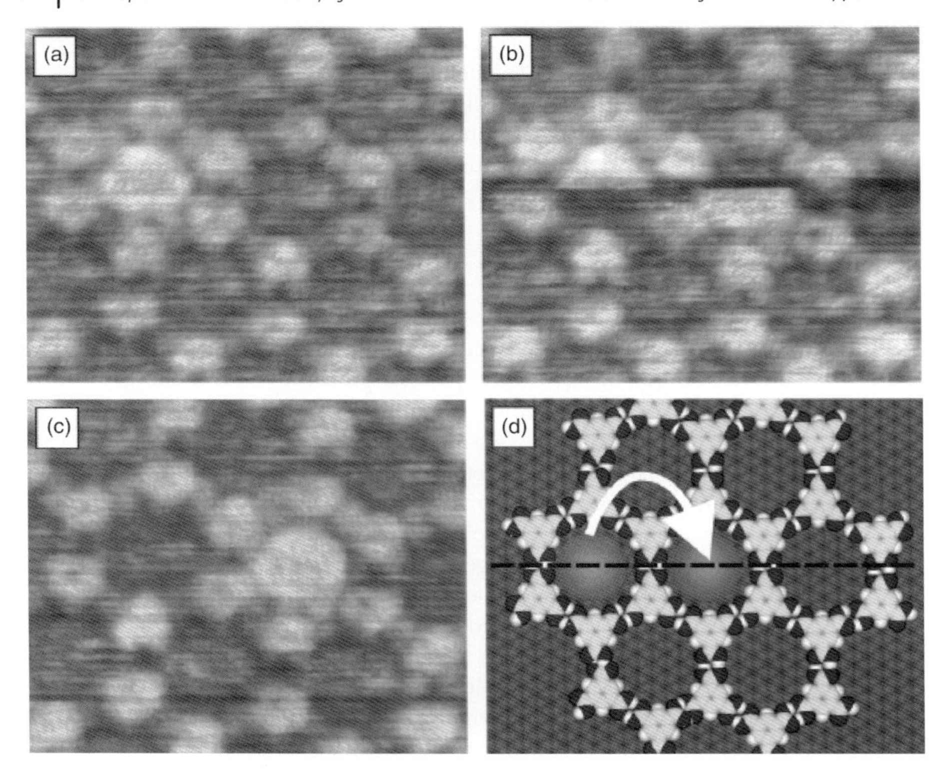

Figure 12.3 (a) STM topograph of the starting situation with a single C_{60} guest molecule inside the TMA host network ($a - c = 4.6 \times 4.6\,nm^2$). (b) Manipulation step: after half the molecule was scanned, the tunneling current was switched from imaging ($I_T \sim 70$ pA) to manipulation ($I_T \sim 150$ pA) conditions; after the molecule was transferred to the adjacent cavity, the imaging conditions were restored. (c) Final result of the lateral manipulation with the whole molecule imaged in the target cavity. (d) Illustration of the TMA network and the manipulation process; the horizontal line indicates the lines that were scanned with an increased reference current. Reproduced with permission from Ref. [43]. Copyright 2004, American Chemical Society.

choosing heptanoic acid as a solvent, which became nonconductive under their experimental conditions, it was possible to verify that self-assembly of the host–guest structure was accomplished at the solid/liquid interface. Therefore, it was possible to coadsorb C_{60} within cavities of the open TMA structure from the liquid phase, and lateral manipulation of the molecular guest by the STM tip was demonstrated at room temperature. Indeed, because of the increased tip–sample interaction as a result of lower tunneling resistance, a transfer of a C_{60} molecule from one cavity of the host structure to an adjacent one was achieved (Figure 12.3).

Among the large number of composite bulk heterojunctions tested as solar cells, fullerene/polythiophene derivatives exhibited a better performance [44]. The most frequent problem for the realization of highly efficient fullerene/polythiophene

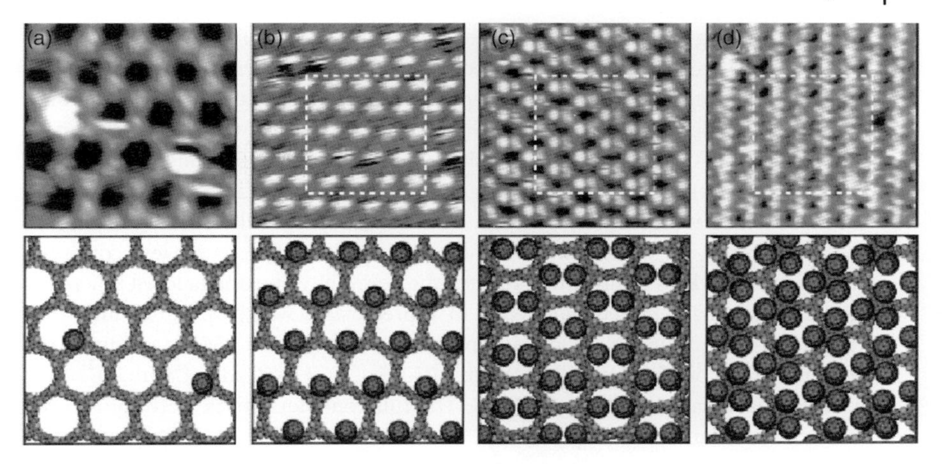

Figure 12.4 TTBTA-C_{60} host–guest architectures with sparse fullerene coverage (a) and one (b), two (c), and three (d) fullerenes per chicken wire unit cell. Tentative molecular models have been assigned to a 10.7 nm × 10.7 nm area of each image (indicated by dashed boxes in b–d). STM parameters: $V_b = -1000$ mV, $I_t = 0.1$ nA (a); $V_b = -600$ mV, $I_t = 0.1$ nA (b); $V_b = -900$ mV, $I_t = 0.03$ nA (c); $V_b = -800$ mV, $I_t = 0.3$ nA (d). Image (a) has an area of 10.7 nm × 10.7 nm; images b–d each has an area of 18 nm × 18 nm Reproduced with permission from Ref. [45]. Copyright 2009, American Chemical Society.

electroactive material was the control of the molecular assembly on the substrate, which usually led to a morphological surface disorder, which could arise from the polydispersity of thiophene derivatives and the consequent fullerene agglomeration. Nevertheless, recently Perepichka and coworkers successfully obtained hierarchically controlled 2D geometry by depositing COOH-substituted oligothiophene TTBTA (terthieno-benzenetricarboxylic acid) at solution/graphite interface, owning the H bond interactions to control the architecture of self-assembly (Figure 12.4) [45].

The relative cavities of chicken-wire structures have shown to efficiently host C_{60} molecules, forming ordered domains of one, two, or three fullerenes per cavity. The observed arrangements of fullerenes dictated by the thiophene/fullerene interactions led to a material with enhanced properties as an organic photovoltaic device. Another example of C_{60} deposition mediated by a thiophene derivative on graphite substrates was given by Fichou and coworkers [46]. By using STM at the solid/liquid interface, they showed that a self-organized nanocavity array of a star-shaped oligothiophene derivative (SSOD) could be used as a nanotemplate to produce long-range alignments of single C_{60} guest molecules. SSOD self assembles on highly oriented pyrolitic graphite (HOPG) into a long-range network made of two types of elongated cavities having different sizes and shapes. Postdeposition of C_{60} led to the formation of defect-free alignments of single C_{60} molecules by site-selective inclusion into one of the two cavities. The major achievement of this work is that the selective filling of one of the two available cavities by C_{60} will, in principle, allow the subsequent codeposition of another guest that could enhance/modulate the C_{60} properties.

By controlling the noncovalent interactions during the self-assembly, further studies on electroactive materials, based on metal surfaces bearing various scaffold molecules for unsaturated fullerene deposition, have been recently reported. Some examples consider the development of highly conductive electronic devices based on phthalocyanine–C_{60} conjugates deposited on HOPG [47] and the production of tetraacidic azobenzene scaffolds on graphite hosting unsaturated C_{60}, as templates for new molecular arrays [48]. While the 2D arrangement of fullerenes on surfaces is well established, alignment of fullerenes in 1D architectures is far less common. Kurth and coworkers recently investigated the possibility of using tailored fullerenes to produce self-assembled nanowires in a predictable way [49]. To do this, they employed the 1,3 dipolar cycloaddition of azomethine ylides [50] to produce fulleropyrrolidines bearing long alkylic chains ($C_{12}H_{25}$, $C_{16}H_{33}$, $C_{20}H_{41}$), with the aim of inducing epitaxial assembly on HOPG. The fullerene derivatives were spin coated from chloroform solutions ($c = 2 \times 10^{-5}$ M) on freshly cleaved HOPG under conditions to achieve submonolayer coverage, in order to reveal structural details of the interfacial layers. Tapping mode AFM images reveal that the C_{60} derivative with $C_{16}H_{33}$ chains formed a 1D lamellar structure on HOPG, with lamellae of about 10 nm in width and more than 100 nm in length, composed of C_{60} moieties arranged in a zigzag fashion, as ascertained by STM in ultrahigh vacuum (UHV).

Among the increasing number of interesting works on fullerene-based materials focused toward controlled self-assemblies at solid–liquid interface, the engineering of highly efficient liquid crystals opened an exciting route to the development of platforms for solar energy conversion and photoactive switches [18, 19]. Another interesting example of the self-assembly of a water-repellent material originating from a rich π conjugation was recently reported by Nakanishi and coworkers [51]. Fullerene derivatives modified with diacetylene moieties (DA) were self-assembled into flake-like microstructures that showed a bilayer structuring observed at molecular level by XRD analysis. The photocross-linking of both DA and C_{60} after UV irradiation investigated by Fourier transform infrared spectroscopy (FTIR) analysis, led to the formation of a water-repellent material with exceptional resistance properties against solvent, heat, and mechanical stress while maintaining the original lamellar organization of the self-assembly.

12.2.3
Fullerenes at the Gas–Solid Interface

Among the wide range of possible applications, fullerenes have been studied as containers for gas storage [52] and in the past decade just few studies have focused on the engineering of new fullerene-based materials as gas sensor devices [53–55] and gas separation materials [56, 57]. Actually, there is a lack of materials meeting industrial requirements for hydrogen storage in solid material, mainly because such materials should be lighter than aluminum and possess high gravimetric density (e.g., 9% w/w). Unfortunately, in these types of materials hydrogen binds strongly, as in complex light-metal hybrids, or weakly, as in carbon-based nanostructures, calthrates, and zeolites, and metal–organic frameworks (MOFs). For this reason, attention was turned to fullerenes and carbon nanotubes [58, 59], in which transition metal atoms can be uniformly distributed over the surface and bind copious H_2 amounts in a quasi-molecular form. In fact, the adsorbed H_2 donate electrons to unfilled d orbitals of the metal atoms, which in turn donate an electron in the antibonding orbital of H_2. Consequently, H_2 does not dissociate but binds quasi-molecularly with a stretched H–H bond. Furthermore, the binding energy of about 0.5 eV/H_2 molecule is in the ideal range for room-temperature application. Further investigation showed that the d–d strong interaction between the transition metal atoms leads to the formation of clustering, thus reducing the hydrogen storage capacity. It was also demonstrated that Li atoms in $C_{60}Li_{12}$ do not cluster due to the strong Li–C bond and a weak Li–Li bond, although a weak hydrogen adsorption energy and a desorption of hydrogen at low temperature were also observed (Figure 12.5) [59]. By using molecular dynamic calculations, Kawazoe and coworkers obtained encouraging predictions on Ca-decorated C_{60} [60]. In fact, Ca-decorated C_{60} exhibited no clustering and showed to bind up to 126 hydrogen atoms in their models. Finally, they observed that $C_{60}Ca_{32}$ complexes are thermodynamically stable and, with respect to the use of Li and Mg, form stable metal-coated C_{60}.

Using magnesium/lithium-decorated C_{60} within MOFs, a new model for hydrogen and methane storage into C_{60} cage was also simulated [61]. By using density functional theory (DFT) calculations Barajas-Barraza and Guirado-Lopez presented a theoretical model to study the physical and chemical interactions between H_2 molecules and the fullerene carbon cage [52]. The study performed on H_2 molecules confined to spheroidal C_{82} demonstrated how a small number of encapsulated H_2 exist in a nonbonded state within the fullerene cavity, with defined conformation, while by increasing the number of stored nitrogen an adsorption on the inner of the

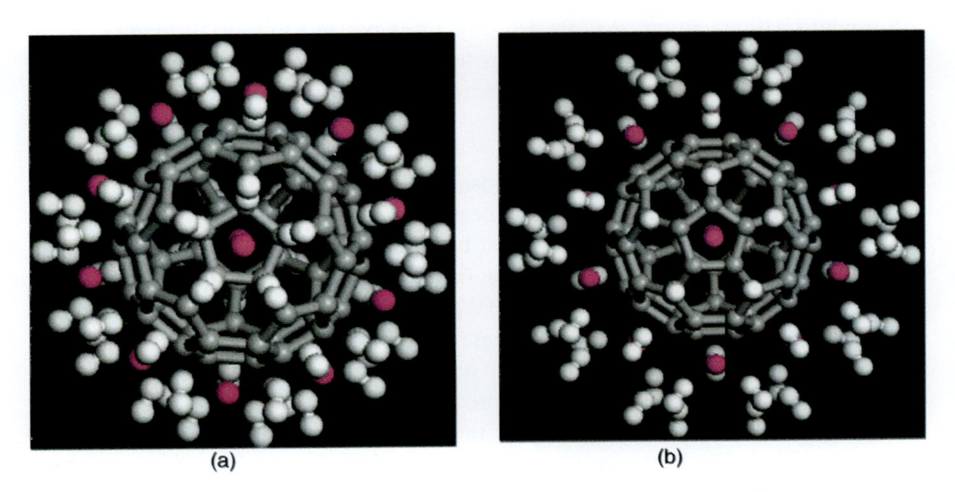

Figure 12.5 Schematic example of theoretical studies of H_2 (white) storage onto C_{60} cage decorated with Li (pink). (a) Initial and (b) optimized expected geometry of $Li_{12}C_{60}(H_2)_{60}$ Reproduced with permission from Ref. [59]. Copyright 2006, American Chemical Society.

carbon cage together with the formation of $(H_2)_m$ molecular clusters was obtained. Nevertheless, upon an increase in storage pressure, nitrogen might covalently bond to the carbon networks, thus indicating the fullerene-type material as not an ideal container for hydrogen storage. A study on a fullerene-based optical sensor for oxygen sensing at trace level was recently performed. The reported sensor is based upon the quenching of delayed fluorescence (DF) of C_{70} embedded into a highly permeable polymer membrane as organosilica (OS) and ethyl cellulose (EC) [62]. A great number of molecules are possible candidates for oxygen sensing since they can be quenched by excited state oxygen generated by molecular oxygen. Among these, fullerenes displayed a strong thermally activated E-type DF, with a unique DF quantum yield at 150 °C and triplet–triplet adsorption in the IR region sensitive to oxygen. The electronic state and transition state of fullerenes are related to the large number of π electrons. Furthermore, the fluorescence occurs between two single-excited state (Φ_F: 0.05%, $\tau \approx 650$ ps) and in the red region (650–725 nm). Thus, fullerenes were used as candidates both for oxygen sensing and for quantum yield of the triplet formation close to one. Therefore, it was demonstrated that C_{70} dissolved into OS and EC responded to oxygen concentrations (0–300 ppm in nitrogen atmosphere) with an excellent DF (< 0.1 s). Furthermore, the fullerene-based sensor displayed higher sensibility to oxygen by increasing the temperature for three reasons: (i) Φ_{DF} increases, (ii) DF lifetime decreases, and (iii) there is a higher collision rate of O_2. The fullerene-embedded matrix further exhibited a repeated use for hundreds of times and no degradation upon storage, thus allowing an effective fullerene device toward oxygen sensing down to the ppb range [62]. Another important aspect in the supramolecular chemistry of fullerenes at gas/solid interface is related to the self-assembly processes that occurs under UHV conditions [63]. Diederich and coworkers first proposed the self-assembly of pristine C_{60} on Ag(100) and Ag(111) surfaces bearing different porphyrin derivatives, which self-assembled under UHV conditions [64]. By this way it was possible to form different types of geometries (Figure 12.6) with cavities for noncovalent hosting of the pristine C_{60} cages, thus leading to a highly organized surface with enhanced properties as a photoactive device.

Among the different hosts that can supramolecularly host C_{60} derivatives, for optimum "face-to-face" contact, the host should have a complementary structure to the convex surface of C_{60}, as it is the case for the concave surface of corannulene ($C_{20}H_{10}$, COR), which is the simplest bowl-shaped fullerene fragment. Fasel and coworkers were the first to report the formation of a surface-supported COR/C_{60} host–guest system on Cu(110) in UHV [65]. *In situ* variable-temperature scanning tunneling microscopy (VT-STM) studies revealed two distinctly different states of C_{60} on the COR host lattice, with different binding energies and bowl–ball separations. The transition from a weakly bound precursor state to a strongly bound host–guest complex was found to be thermally activated. Other interesting examples of the UHV assembly of C_{60} host–guest structures are (i) 2D honeycomb-like network by UHV codeposition of 3,4,9,10-perylenetetracarboxylic diimide (PTCDI) and 2,4,6-triamino-1,3,5-triazine (melanine) on a Ag-terminated silicon surface, toward highly ordered assemblies of pristine C_{60} [66], (ii) vacuum-deposited organic solar cells

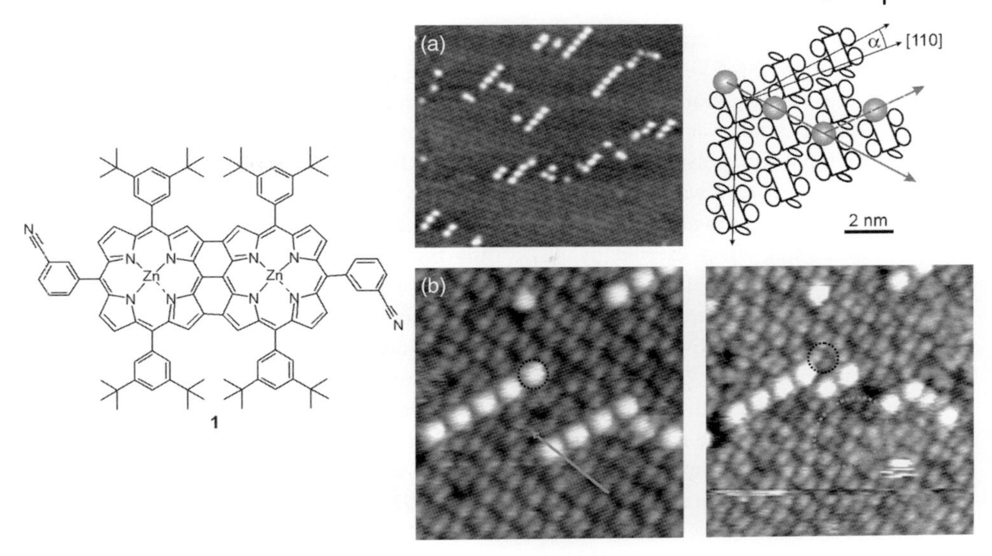

Figure 12.6 STM images of **1**–C_{60} assembly. (a) (*Left*) STM images showing the preferential direction of the chain-like assembly of C_{60} on a previously deposited monolayer of **1** (scan range: $77 \times 65\,nm^2$, $I_t = 22$ pA, $V_{bias} = 2.59$ V). (*Right*) Proposed model for the chain-like assembly ($\alpha = \sim 7.5°$); the C_{60} molecules (gray spheres) are located between the diporphyrin molecules approximately on top of the 3-cyanophenyl residues; the solid and dashed arrow indicate the major and minor growth directions, respectively. (b) Detailed view before (*left*) and after (*right*) manipulation sequence of the C_{60} molecule on a layer of **1** (scan range: $21 \times 21\,nm^2$, $I_t = 11$ pA, $V_{bias} = 3.01$ V). The arrow traces the lateral displacements of the STM tip during the manipulation (tunneling parameters: $I_t = 214$ pA, $V_{bias} = 1.43$ V, $V_{tip} = 5$ nm/s) and the ellipse indicates the intact layer of **1** after repositioning of the C_{60} molecule; the circle denotes a C_{60} molecule that vanished during the repositioning experiment. Reproduced with permission from Ref. [64]. Copyright 2004, Wiley-VCH Verlag GmbH.

based on low-bandgap oligothiophene/C_{60} derivatives [67], and (iii) sexithiophene nanostripes on Au(111) as a route to fullerene-quantum computers [68].

12.2.4
Fullerenes at the Biological Interface

Fullerene's chemical/physical behaviors have often been exploited for studying interactions at the biological interface. Nevertheless, the lack of solubility of fullerene cages in water represented an obstacle to the study of its properties in such important environment. For this reason, some strategies to engineering water-soluble fullerene derivatives have been employed, for example, by covalently attaching specific appendages [69–72] and by inclusion into cyclodextrins [73], or toward noncovalent interactions like the aforementioned host–guest calix[4]arene [41] or resorcin[4]arene [74] complexes. It was recently reported that pristine nC_{60} aggregates might interact with normal amplification of genes from DNA. The results observed by real-time polymerase chain reaction (RT-PCR) technique involving the amplification of certain DNA genes indicated that fullerene aggregates inhibited the normal PCR gene amplification, a dose-dependent inhibition. Fullerene inhibition to PCR gene amplification was related to fullerene noncovalent interactions with DNA and/or mainly with polymerase, thus preventing the transduction of gene sequences. Furthermore, it was demonstrated how the contemporary presence of large amount of bovine serum albumin proteins (BSA) in the buffer assay avoided the polymerase inhibition effect exerted by nC_{60}, thus indicating a competitive effect between polymerase and fullerene interaction toward the more favorable protein and fullerene one [75]. Among their unique photochemical and photophysical properties, fullerene derivatives have been widely employed as photocleavage agents toward target DNA [76, 77]. Recently, Gao *et al.* reported the synthesis of a water-soluble fullerene derivative as an efficient DNA photocleavage agent. A synthesized water-soluble fullerene-naphthaleneacetate derivative (NP-C_{60}) was driven to produce a noncovalent complex with Riboflavin (RF) in water solution, thus creating a hybrid that displays high affinity toward calf thymus DNA. The noncovalent formation of the NP-C_{60} derivative and RF in water is based upon π–π interactions between the aromatics of RF and C_{60} and demonstrated to be effective for increasing NP-C_{60} water solubility and for the observed photocleavage effect on DNA. These results suggest an electron transfer (ET) occurring between DNA to the excited state of RF/NP-C_{60}, following the equation, $\text{DNA} + \text{NP-C}_{60}{}^{*} \rightarrow \text{DNA}^{\bullet+} + \text{NP-C}_{60}{}^{\bullet-}$, while no photo-induced effect was observed using RF alone, even observing an electron transfer to DNA [78]. Fullerene derivatives were also employed for their hydrophobic properties and noncovalent interactions on biological substrates for gene delivery [79–81]. In that sense, a novel cationic tetra(piperazino)fullerene epoxide (TPFE) was recently tested for *in vivo* gene delivery (Figure 12.7) [81]. The cationic C_{60} derivative, covalently bearing four cationic piperazine and epoxide function on the cage, forms a supramolecular complex in the presence of plasmidic DNA encoding enhanced green fluorescent protein (EGFP). The complex formed by electrostatic interactions displayed a high adsorption on the cellular compartments of mouse fetuses, and a high ratio of gene delivery, evaluated through the total fluorescence of the codified EGFP. In

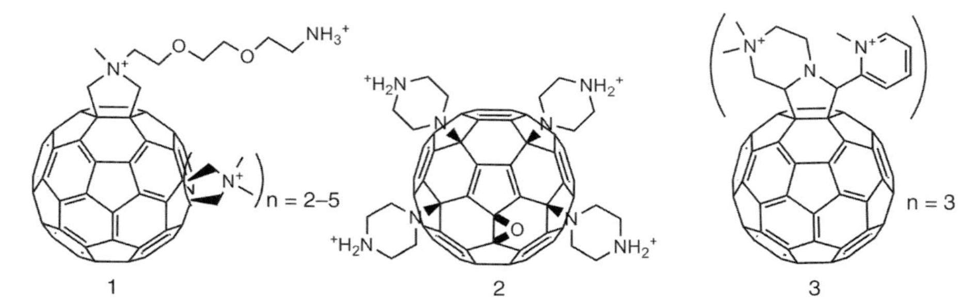

Figure 12.7 Structure of the polycationic C_{60} derivatives used as gene delivery systems (1[80] and 2[81]) and as light-activated antimicrobial photosensitizers (3[87]).

addition, no acute toxicity was observed about the complex over its multiorgan distribution, as a consequence of the selectivity upon π–π interaction of TPFE with DNA nucleobases, thus avoiding undesired interaction with lipid-rich compartments and the consequent accumulation *in vivo*. The electrostatic interactions were used to explain the anti-HIV properties of fullerene derivatives, upon the inhibition of HIV-protease by the accommodation into the hydrophobic cavity of the enzyme, and interaction with specific amino acid residues [82, 83]. TPFE derivative has also been used to deliver mouse insulin 2 gene (*Ins2*) expressing plasmid, in the same way as EGFP, thus demonstrating how the complex bearing the gene *Ins2* effectively reduced concentrations of glucose in blood, by increasing the insulin transcription [84]. Among the fullerene applications over biological substrates, fullerene derivatives have been studied as effective and selective antimicrobial photosensitizers [85], as well as photodynamic agents for killing cancer cells upon the photoinduced reactive oxygen species (ROS) formation and quenching, by photoexcited fullerene intersystem crossing from single to triplet state [86]. Recently, it was demonstrated how fullerenes bearing cationic moieties are more effective as antimicrobial agents against a wide spectrum of Gram-negative and Gram-positive bacteria, compared to more hydrophobic or unsaturated fullerene derivatives. In that sense, different species of bacteria more easily allow the uptake of positively charged compounds through membrane carrier. Thus, water-soluble cationic fullerenes, such as the sarcosine-bispyridine derivatives [87], have been shown to be effective photosensitizers because they are easily taken up into cytosolic compartment of bacteria, where they generate toxic oxygen species upon photoexcitation (Figure 12.7) [88–90].

Among the photochemical properties of fullerenes, the extended π electron delocalization over the carbon core also determines the antioxidant activity. This property allows the quenching of radical species, such as oxygen free radicals, from the fullerenes at their ground state, enabling them to act as radical scavengers in biological systems [91–93]. Recently, it was demonstrated how supramolecular complexes of nonmodified hydrated pristine fullerenes as nanoclusters (C_{60}HyFn) are effective in protecting DNA from ionizing radiation *in vitro*, upon removal of hydroxilic radicals (•OH) from solution [94]. Furthermore, the observed •OH removing efficacy of C_{60}HyFn was found to be in reverse correlation with the concentration used, thus supposing a "nonstoichiometric" mechanism for its antioxidant effect. This probably involves hydration toward free radical self-neutralization, catalyzed by the ordered structures of C_{60}HyFn in water. Thus, the same C_{60}HyFn derivative injected in mice exposed to X-ray irradiation resulted in a 15% of animal survival, confirming the antioxidant effect also *in vivo*. Furthermore, C_{60}HyFn at low concentration exhibited DNA protection from radical-induced damage, with respect to the 8-oxoguanine marker. The antioxidant properties of fullerenes were also studied on the superoxide dismutation. Tris-malonyl $C_{60}(C_3)$ was prepared for this study, and its antioxidant effect was confirmed by the observed high quenching of superoxide anions in solution [95]. Thus, it was demonstrated that the unpaired electrons of superoxide radical were transferred to C_{60} upon contact with the fullerene carbon cage. The process, converting $O_2^{\bullet-}$ species to neutral oxygen O_2, was demonstrated to be the rate-determining step of the reaction. Afterward, other superoxide radical might react with $C_{60}^{\bullet-}$ forming hydrogen peroxide, upon catching up the additional electron previously transferred. The entire process was shown to be clearly exothermic, based on steps with low activation barriers.

All these examples of the supramolecular chemistry of fullerene and fullerene derivatives at the interfaces highlight the importance of this novel carbon allotrope for both basic research advancements and possible technological applications. Several recent studies have shown that fullerenes exhibit interesting biological activities both *in vitro* and *in vivo* [72, 81, 96]. In the biological context, some elegant examples of the use of C_{60} immobilized on a surface have been described. Higashi *et al.* demonstrated that positively charged SAMs on a gold substrate can immobilize double-stranded DNA without disrupting its intrinsic higher order structure and that site-specific cleavage of the DNA is successfully achieved by covalent incorporation of C_{60} into the SAM [97]. The strategy employed in this study included the preparation, on a gold surface, of a well-ordered mixed monolayer assembly that contained (a) quaternary ammonium salts that interact electrostatically with the phosphate groups of DNA and (b) free primary amino groups for the covalent attachment of C_{60}. The authors spectroscopically observed a highly guanidine-selective DNA cleavage, consistent with the involvement of singlet oxygen generated by interaction of the photoexcited fullerene core with molecular oxygen [98].

Electrodes modified with fullerene monolayers offer exceptional electroactive interfaces capable of electronically coupling redox-active biomolecules or enzymes with electrodes and activate the respective biological function [99]. Along this line, Willner and coworkers described the use of C_{60} as an electron mediator for an electrocatalyzed biotransformation (Figure 12.8) [100]. In this work, a methano[60]

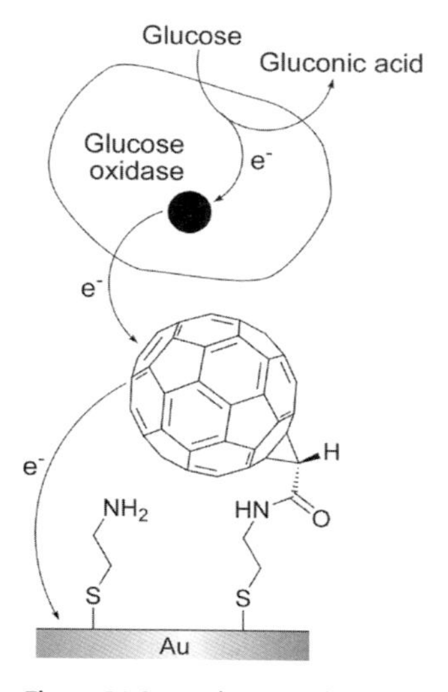

Figure 12.8 Reduction of glucose by glucose oxidase mediated by a C_{60} SAM covalently immobilized on Au as reported by Willner and coworkers. Reproduced with permission from Ref. [100]. Copyright 1998, Elsevier.

fullerene carboxylic acid derivative was covalently attached to a cysteamine monolayer preadsorbed on a gold electrode and the resulting monolayer was shown to provide an active interface for mediating the biocatalyzed oxidation of glucose to gluconic acid. The fullerene moiety acts as an electron relay in the electrical communication between the electrode and a soluble glucose oxidase with an electron transfer rate constant of $k_{ET} = 3 \times 10^4$ M/s. The authors proposed that given the size of the carbon cage, the electron mediator cannot penetrate into the protein to yield intimate contact with the FAD-site (FAD, flavin adenine dinucleotide) for electron transfer. Thus, the mediated electrical contact is assumed to proceed via long-distance tunneling.

With the aim of understanding the role of the coenzyme redox couple NAD/NADH in biological redox processes involving electron transfer from a substrate, the group of Zhou prepared a C_{60}-glutathione-modified Au electrode for the electrocatalytical oxidation of NADH [101]. In this modified electrode, C_{60} was covalently anchored to a preorganized SAM of glutathione on Au. The structural properties of the fullerene-containing surface were assessed by FTIR spectroscopy, cyclic voltammetry, and electrochemical impedance spectroscopy (EIS). Electrocatalysis studies (NAD/NADH oxidation processes occur at 1.10 V and 1.30 V (versus SCE) at a glassy carbon and Pt electrode) revealed that the NADH form is electrochemically oxidized by the fullerene radical cation, $C_{60}^{\bullet +}$, confined to the surface. CV measurements conducted in the presence of an increasing concentration of NADH showed a linear dependence of the current intensity on concentration, suggesting that as more biocompatible modified electrodes become available, these fullerene-coated surfaces can represent a new class of functional materials for the detection of redox-active biological molecules.

12.3 Carbon Nanotubes

In the second part of the chapter, we will review and discuss the most important examples of the supramolecular chemistry of CNT and CNT derivatives at the interfaces, with a particular focus on the recent advancements that might bring these carbon-rich nanostructures to real applications.

12.3.1 Carbon Nanotubes at the Liquid–Liquid Interface

Smalley and coworkers employed the common phase transfer catalyst tetraoctylammonium bromide (TOAB) to afford length-dependent extraction of SWNTs [102]. The sidewalls of the purified nanotube were functionalized with chloroaniline sulfonate using a previously described method [103], then concentrated nanotube solutions in water were prepared without the use of surfactants. The functionalized nanotubes showed a relatively high solubility in water and in other polar solvents (e.g., methanol, ethanol, etc.). As a consequence, the nanotubes were dissolved in water (0.04–0.2 mg/ml) and placed in a vial. Then, a solution of TOAB in an organic solvent (either ethyl acetate or toluene) was added to the vial at a 1: 1 volume ratio resulting in a liquid–liquid phase system. After the two-phase mixture is shaken, the system becomes a gray emulsion. By using an excess of TOAB, the emulsions settled and complete extraction of the nanotubes into the organic layer occurred. It should be pointed out that previous approaches that have involved sulfonates or other groups [104] interacting with TOAB have shown an electrostatic interaction. For this reason, it is expected that the extraction of the nanotubes occurs via stoichiometric electrostatic interactions between the anionic SO_3^- moiety on the functional groups on the nanotube sidewall and the cationic TOA^+. Length-dependent extractions were thus obtained by employing a substoichiometric ratio of TOAB. Under these TOAB-starved conditions, the nanotubes' solubility and their colloidal interactions can be efficiently controlled, allowing separations based on the length of the nanotubes. The starting material had a broad distribution of nanotube lengths with an average of 275 nm. The addition of enough TOAB to complex 30% of all SO_3^- moieties resulted in only very short nanotubes extraction into the organic layer as determined by AFM. Increased ion pairing results in longer nanotubes extraction and after 75% of the SO_3^- moieties are ion paired, the length distribution approached that of the starting material.

Ziegler and coworkers reported that SWNT bundles can be selectively removed from an aqueous dispersion containing individually suspended carbon nanotubes coated with gum arabic, via interfacial trapping [105]. After the dispersion of the nanotubes in water through the gum arabic interaction, a two-phase system was created by the addition of toluene. The absorbance spectra of the bulk aqueous phase showed high absorbance due to the presence of both individual and bundled SWNTs. After the interfacial trapping, the absorbance of the suspension was clearly decreased, while the spectral features were better resolved and blueshifted (Figure 12.9).

These changes are in agreement with the removal of nanotubes from the aqueous phase and the presence of a higher fraction of individualized SWNTs. By the comparison of the fluorescence intensities between these solution and the ones obtained in similar conditions with the use of ultracentrifugation, it was concluded that the amount of individual nanotubes significantly increased.

Niu and coworkers reported an elegant example of SWNT assembly at the water/oil interface, by preparing imidazolium-modified SWNTs (SWNT-Im) [106]. At first, SWNT-Im were dispersed in water by ultrasonication and after the addition of chloroform, they remained in the upper layer. Then, after ultrasonication for few seconds and standing overnight, SWNTs-Im spontaneously transferred to the W/O interface. By fixing the area of the W/O interface and controlling the concentration of SWNT-Im aqueous solution, films with different thicknesses could be obtained. Interfacial assembly of SWNTs was driven by minimizing the interfacial energy, ΔE. Due to the assembly of SWNTs from the aqueous solution to the W/O interface, ΔE can be given by $-\pi R^2 \gamma_{ow}(1\text{-}\cos\theta)$, where R, γ_{ow}, and θ represent the equivalent radius of SWNTs, the tension of the W/O interface, and the contact angle of SWNTs with the interface, respectively. Therefore, the utilization of the W/O interface on the one hand

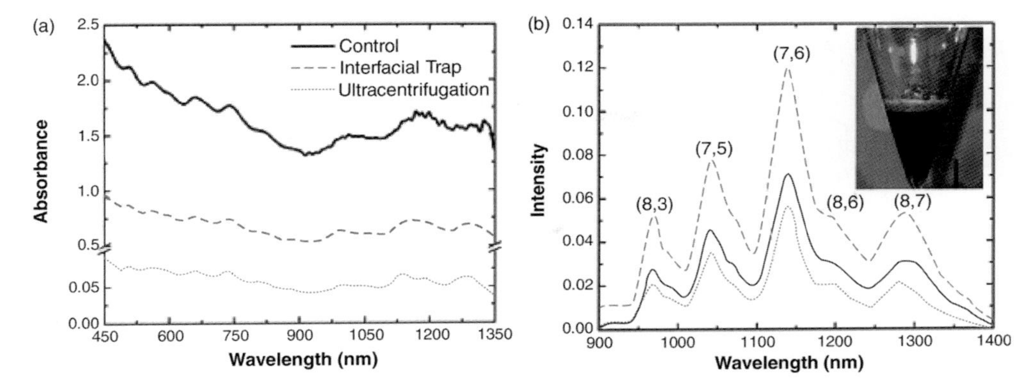

Figure 12.9 (a) Absorbance and (b) fluorescence (exc. 662 nm) spectra for the interfacial trapping process of gum arabic-suspended SWNTs using an initial SWNT mass concentration of 0.03 mg/ml. The control spectra (solid lines) are the SWNTs after homogenization and ultrasonication. The inset shows the interfacial trapping process in a separatory funnel. The control sample is then either subjected to ultracentrifugation (dotted lines) or interfacial traps (dashed lines). The fluorescence from specific (*n*, *m*) types is labeled. Reproduced with permission from Ref. [105]. Copyright 2008, American Chemical Society.

offered an alternative pathway for assembling SWNTs and on the other hand also provided a peculiar SWNT-sandwiched W/O interface, which would have potential applications in electron transfer process. Therefore, the contribution of the SWNT-Im to ET at the W/O interface was investigated by scanning electrochemical microscopy, by preparing water phases containing electroactive species (Ru$(NH_3)_6{}^{3+}$). Without SWNT-Im at the W/O interface, the tip current decreased when the tip approached the interface (negative feedback). This means that the initial redox species ($Ru(NH_3)_6{}^{3+}$) in water could not be regenerated and the diffusion of the original redox species ($Ru(NH_3)_6{}^{3+}$) to the tip was blocked. However, at the SWNT-Im sandwiched interface, the tip current increased when the tip approached the interface. This facet indicated that, with the presence of SWNT-Im, the initial form of the redox species could be regenerated at the W/O interface.

Dordick and coworkers observed that an aqueous dispersion of purified SWNTs when contacted with an equal volume of hexane (or isooctane, $CHCl_3$, or CH_2Cl_2) containing a 2 mM concentration of the anionic surfactant Aerosol-OT (AOT, 1,4-dioctoxy-1,4-dioxobutane-2-sulfonic acid, Figure 12.10) led to the transfer of SWNTs from the aqueous phase to the interface [107]. The interfacial assembly was also observed when a dispersion of SWNTs in hexane was poured in solutions of the neutral surfactant Tween (polyoxyethylene (20) sorbitan monolaurate), the cationic surfactant dodecyltrimethylammonium bromide (DTAB), or the cationic lipid 1,2-dimyristoyl-3-trimethylammonium propane (DMTAP).

On the basis of this finding, the authors reasoned that the nanotubes could be exploited as carriers for the transportation of proteins to an aqueous–organic interface, enabling the possibility to enhance the rate of interfacial (bio)transformations.

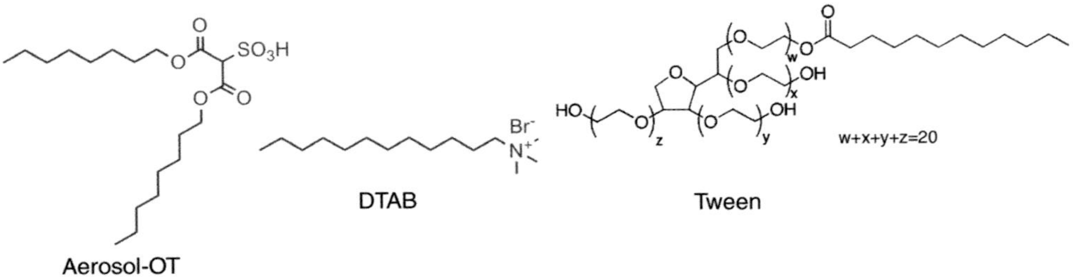

Figure 12.10 Structure of some surfactants employed for the SWNT transfer at the aqueous/hexane interface.

As an example, they adsorbed on SWNTs the protein soybean peroxidase (SBP), which requires a hydrophobic phenol and the hydrophilic H_2O_2, in order to test interfacial catalysis [108]. In the presence of AOT, the SWNT–SBP conjugates assembled at the hexane–water interface. By using UV–vis spectroscopy, no protein or SWNTs were detected in either of the bulk phases. The catalytic activity of the interfacial SWNT–SBP was determined using *p*-cresol, which is a model peroxidase substrate. As a result, the product was detectable only in the organic phase, with an initial reaction rate for interfacial SWNT–SBP of ~75 nM/s. This rate is three orders of magnitude higher than what is observed with identical enzyme concentrations for either native SBP or SWNT–SBP in the aqueous phase of a biphasic system in the absence of AOT, or for native SBP in the aqueous phase in the presence of AOT.

Valcárcel and coworkers employed MWCNTs as extractant for benzene, toluene, and ethylbenzene, the three xylenes isomers (*ortho*, *meta*, and *para*) and styrene, all referred to as BTEX-S [109]. The detection was carried out through headspace/gas chromatographic/mass spectrometric (HS/GC/MS) determination of different olive oil samples. BTEX-S are molecules that are emitted from a variety of sources into the environment, including aerosols, combustion products of fuels and wood, adhesives, industrial paints, and degreasing agents [110]. Their lipophilic nature and common presence in the environment (air, water, and soil) leads to their accumulation in foodstuffs, as fats and edible oils. In this work, the authors used surfactant-coated MWCNTs as additive in liquid–liquid extraction (LLE) for the determination of single-ring aromatic compounds in olive oil samples. After samples treatment, the aqueous extracts collected by this way were analyzed using HS/GC/MS, permitting the quantification of BTEX-S within ~15 min. Each step of the reported LLE/HS/GC/MS setup determines a selectivity enhancement, avoiding the interference of other compounds of the sample matrix. The detection limits resulted in the range of 0.25 ng/ml (obtained for ethylbenzene) and 0.43 ng/ml (for benzene).

12.3.2 Carbon Nanotubes at the Solid–Liquid Interface

A wide range of analytes adsorb irreversibly to the surfaces of single-walled carbon nanotube electronic networks typically used as thin-film transistors or sensors [111–120], although the mechanism is not yet understood. Using thionyl chloride as a model electron-poor adsorbate, Lee and Strano demonstrated that reversible adsorption sites can be created on the nanotube array via noncovalent functionalization with amine-terminated molecules of $pK_a < 8.8$ [121]. A nanotube network composed of single, mostly unbundled nanotubes, near the electronic percolation threshold was necessary for the effective conversion to a reversibly binding array. By testing several types of amine-containing molecules, such as pyridine, aniline, hydrazine, and ethylenediamine, the authors showed that analyte adsorption is largely affected by the basicity (pK_b) of the surface groups. The analyte's binding energy was apparently reduced by its adsorption on the surface chemical groups instead of directly on the SWNT array itself. This adsorption mechanism is supported by X-ray photoelectron spectroscopy (XPS) and molecular potential calculations. By creating a higher adsorption site density with an amino-polymer like polyethyleneimine (PEI), the reversible detection at the parts-per-trillion level was then for the first time demonstrated. On the other hand, some amines failed to produce reversible binding at all. Benzylamine, diethylenetriamine, dimethylamine, and triethylamine are among those that were not effective, meaning that the sensor response remained irreversible as a consequence of strong device–analyte interactions. Another example of the employment of CNT derivatives at the solid–liquid interface is related to their general property to act as sorbents. CNTs present large and biocompatible surface areas, which afford not only huge potential for biological applications but also promising scenario in separation science. A survey of the published literature provides a range of applications of CNTs serving as solid-phase extraction media [122] for the separation/isolation of organic species [123–148] and trace metal [149–151]. In fact, Wang and coworkers used MWCNTs for the isolation of basic proteins from other protein species in biological sample matrices by solid-phase extraction (SPE) [152]. After appropriate pretreatment into a sequential injection system, a microcolumn packed with MWCNTs was incorporated. This device facilitated the online selective sorption of basic protein species such as hemoglobin and cytochrome *c*. In fact, oxidized MWCNTs are negatively charged at pH 6 whereas hemoglobin (isoelectric point, $pI = 7$) and cytochrome *c* ($pI = 10$) are both positively charged. As a consequence, electrostatic interactions occurred between the MWCNT surfaces and the basic proteins, and so their adsorption onto the MWCNT surfaces took place. The retained protein species were subsequently isolated from each other by sequential elution from the microcolumn, by the use of appropriate eluents. Phosphate buffer solution (0.025 mol/l at pH 8.0) allowed an efficient collection of hemoglobin, while a NaCl solution (0.5 mol/l) afforded the quantitative recovery of the retained cytochrome *c*. By using a sample loading volume of 2.0 ml, enrichment factors of 11 and 15 were obtained for hemoglobin and cytochrome *c*, with retention efficiencies of 100% for both species and recovery rates of 98 and 90%, respectively. The practical applicability of this system was demonstrated by processing human blood for isolation of hemoglobin, and satisfactory results were obtained from assay with SDS-PAGE.

Many aromatic molecules, such as porphyrin [153–158] and pyrene [159–161] and their derivatives, interact with the outer sidewalls of SWNTs through π–π stacking

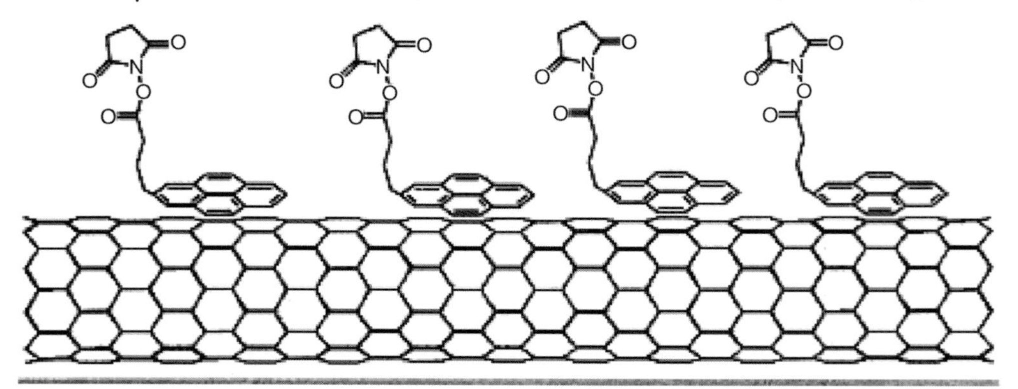

Figure 12.11 Representation of the irreversible adsorption of *N*-succinimidyl-1-pyrenebutanoate on SWNT outer walls. Adapted with permission from Ref. [165]. Copyright 2001, American Chemical Society.

interactions, thus allowing for their noncovalent functionalization [162–164]. Dai and coworkers have reported a broad and smart protocol for the noncovalent functionalization of SWNTs, which allows the subsequent conjugation of (bio)molecules onto SWNTs [165, 166]. Starting from either *N,N*-dimethylformamide (DMF) or MeOH, the bifunctional molecule *N*-succinimidyl-1-pyrenebutanoate used in their studies can be permanently adsorbed on the hydrophobic sidewalls of SWNTs (Figure 12.11). The *N*-succinimidyl-1-pyrenebutanoate molecules anchored at the surfaces of the SWNTs are highly resistant to desorption in aqueous solution, allowing additional functionalization of SWNTs through the succinimidyl ester groups, which are reactive to nucleophilic substitution by amino groups of some proteins, such as ferritin and streptavidin, and biotinyl-3,6-dioxaoctanediamine [165].

This procedure has paved the way for the immobilization of an ample range of molecules on the SWNT sidewalls with high specificity, for the development of SWNT/FET and biosensors. In general, SWNT/FET devices have been found to be responsive to many analytes at the solid–liquid interface [167, 168]. After noncovalent functionalization of SWNTs, these analytes can change the SWNT conductivity for two different reasons. First, there may be a charge/electron transfer between the analytes and the SWNTs, changing the concentration of the carrier (n). Second, the adsorbed analytes can operate as randomly distributed scattering potentials, therefore changing the mobility (μ) of the charge carrier. The conductivity is generally defined by $G = ne\mu$, where e is the electron charge; therefore, the transistor measurements of the transfer characteristics in the SWNT/FET devices can be distinguished between (1) a change in the concentration of the carrier and (2) a change in the mobility of the electrons. Extensive characterizations have established that [167] in the SWNT/FET devices changes in n determine shifts of the threshold voltage (the voltage at which the device turns on for the first time). The binding of electron accepting molecules such as NO_2 to the SWNTs usually leads to a threshold voltage shift toward positive gate voltages, while the binding of electron donating molecules

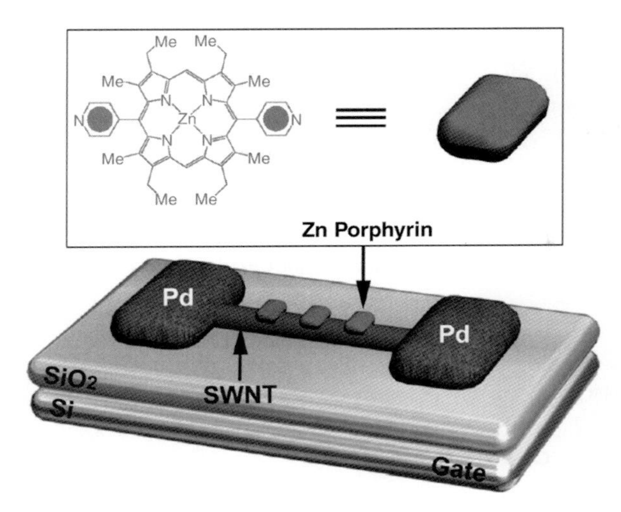

Figure 12.12 Schematic representation of the SWNT/FET device based on Zn–porphyrin/SWNT hybrid. Reproduced with permission from Ref. [169]. Copyright 2006, American Chemical Society.

such as NH_3 leads to a shift toward negative gate voltages. On the other hand, a change in the mobility μ results in a change in the tilt (device transconductance), which is defined by the ratio of the slope of the I_{sd}–V_g curve to its initial slope, where both slopes are measured at zero gate voltage. A decrease in the device mobility can be caused by geometric deformations introduced along the nanotube by the analyte, through change occurring at the intertube interface or by randomly charged scattering centers.

Recently, Stoddart and coworkers prepared SWNT/FET devices to elucidate the electron/charge transfer within the donor–acceptor SWNT hybrids. For example, a SWNT/FET device, noncovalently functionalized with a zinc porphyrin derivative [169], was employed to directly assess a photoinduced electron transfer within the device (Figure 12.12).

The porphyrin molecules act as the electron acceptors whereas the SWNTs act as the electron donors. The photoresponse of the zinc porphyrin-coated SWNT/FET was investigated by its illumination at 420 nm, a value closer to the maximum absorption wavelength (416 nm) of the Soret band characteristic of the zinc porphyrin. The photoresponse of the device causes a shift of the threshold voltage toward positive values, suggesting hole doping of the SWNTs. The direction of the threshold voltage shift indicates that the photoresponse occurs through the electron transfer from the SWNTs to the zinc porphyrin, a fact that is unexpected since porphyrins are usually considered to be electron donors [170]. A possible explanation for this electron transfer process is that, after photoexcitation of the zinc porphyrin, some of the electrons that had been transferred to the SWNT (ground state) are transferred back to the zinc porphyrin molecule in the excited state. The intensity of the photoinduced electron transfer was found to be a function of both the intensity and the wavelength

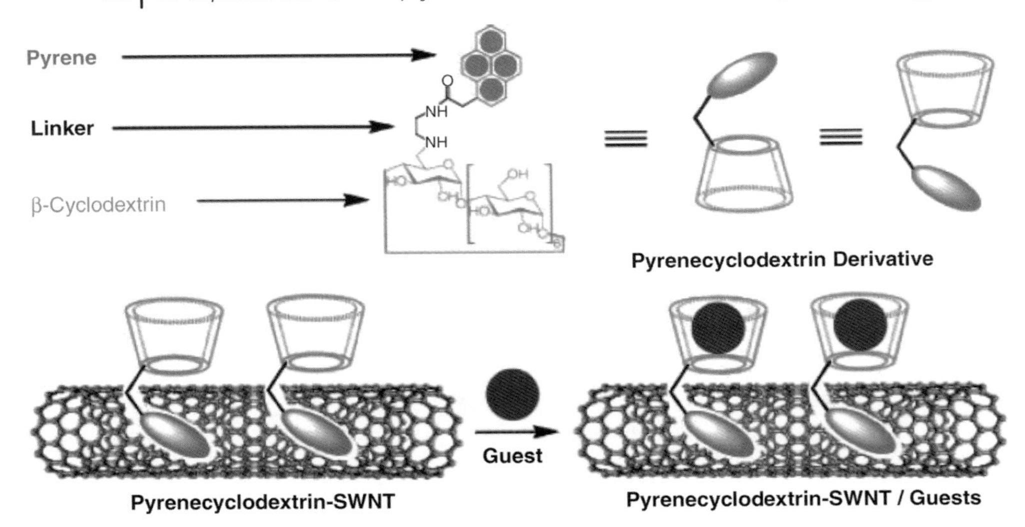

Figure 12.13 Schematic drawing of the pyrenecyclodextrin-decorated SWNT hybrids and how they interact with guest molecules when they are being sensed in an FET device. Reproduced with permission from Ref. [171]. Copyright 2008, Wiley-VCH Verlag GmbH.

of the applied light, with a maximum of 0.37 electrons per zinc porphyrin for the light of 100 W/m^2 at 420 nm [169]. In order to perform quantitative investigations on the chemical sensors based on noncovalently functionalized SWNT/FET devices, Gruner and coworkers prepared a pyrene-modified cyclodextrin (pyrenecyclodextrin) derivative [171]. They used such hybrid to fabricate pyrenecyclodextrin-decorated SWNT/FET devices, which can serve as chemical sensors to selectively detect specific organic molecules, on the basis of their molecular recognition by the cyclodextrin inner cavity. In such SWNT/FET devices, the pyrenecyclodextrin derivatives adsorbed on the SWNTs act as the sensing host (Figure 12.13).

Some organic molecules, as the guests being sensed, can be immobilized by their recognition and assembly in the cavity of the pyrenecyclodextrin derivative in aqueous solution. In the presence of certain organic molecules, the transistor characteristics of the SWNT/FET device shifted toward negative gate voltage, with the following sequence of sensing properties for the guests: 1-adamantanol > 2-adamantanol > 1-adamantanecarboxylic acid > sodium deoxycholate > sodium cholate. These findings indicate that the electrical conductance of the device is highly sensitive to determined organic molecules and moreover it changes appreciably with variations in the surface adsorption of the aforementioned molecules. Interestingly, in the presence of the organic molecules the level of the transistor characteristic movements in the SWNT/FET devices depends linearly on the strength of the complex formation constants (K_S) exhibited by the pyrenecyclodextrin device with these molecules. Therefore, the pyrenecyclodextrin-decorated SWNT/FET devices can serve as chemical sensors to detect organic molecules in aqueous solution, both selectively and quantitatively.

12.3.3 Carbon Nanotubes at the Gas–Solid Interface

The simple and tough architecture of microelectronic devices based on carbon nanotubes, in combination with their environmental sensitivity, places them among the most important candidates for inclusion into ultraportable or wearable chemical analysis devices. Star and coworkers, described the spectroscopic and electrical behavior of simple chemiresistor devices composed of SWNT networks decorated with an oxygen-sensitive Eu^{3+}-containing dendrimer complex [172]. Complexes that contain lanthanides show solution-phase sensitivity toward oxygen, a characteristic that the authors exploited by immobilizing a Eu_8-G3-PAMAM-(1,8-naphthalimide) dendrimer on highly conductive and optically transparent SWNT-based devices (Figure 12.14).

In the Eu_8 structure, the 1,8-naphthalimide groups act as sensitizing agents. Specifically, photoexcited electrons in the excited naphthalimide singlet state undergo intersystem crossing (ISC) into a triplet state (T3) and subsequent energy transfer into the accepting levels of the Eu^{3+} ions produces the sharp Eu^{3+}-centered emission bands in solution. The reversible and reproducible quenching effect of oxygen on the solution-phase Eu_8 emission intensity is in accordance with its predicted behavior. However, they found that the Eu^{3+}-centered emission bands show more sensitivity to oxygen compared to the 1,8-naphthalimide band. The relative emission intensity of solid-state Eu_8 (drop cast onto a quartz substrate) was constant when cycled between atmospheres of pure oxygen and argon, which

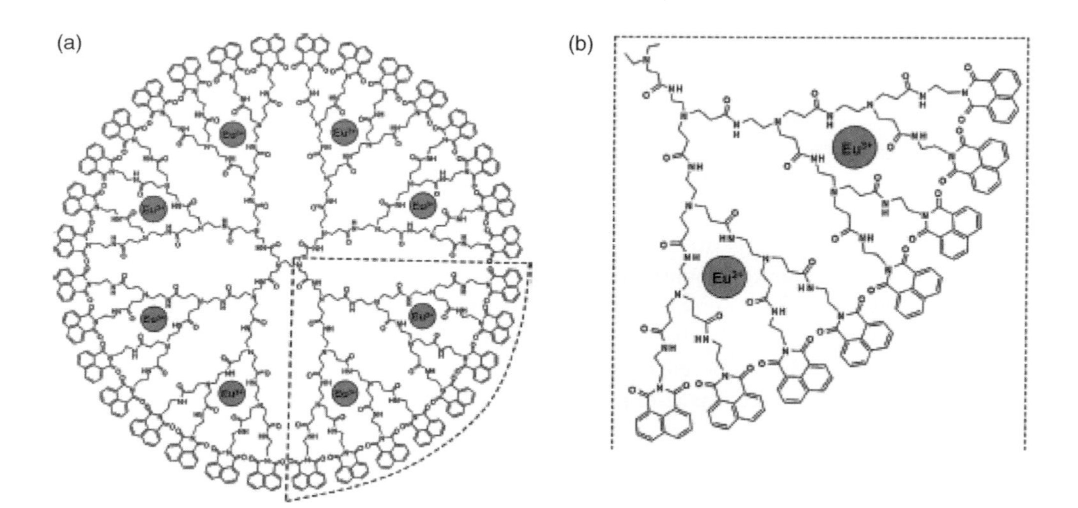

Figure 12.14 (a) Chemical structure of the Eu_8 complex, which contains eight Eu^{3+} cations (dark gray circles) coordinated within a 1,8-naphthalimide-terminated, G3-PAMAM dendrimer core. (b) Expanded view of the Eu_8 structure that illustrates the coordination of the Eu^{3+} ions. Reproduced with permission from Ref. [172]. Copyright 2009, Nature Publishing Group.

highlighted a fundamental difference between the behavior of solid-state and solution-phase samples. However, after illuminating the sample with 365 nm light for 30 min (in-flowing argon), the emission profile of the solid-state Eu_8 developed sensitivity toward oxygen such that the intensity decreased after illumination and partially restored under flowing oxygen. Using simultaneous ultraviolet–visible–near-infrared (UV–vis–NIR) absorbance spectroscopy and network conductance measurements on Eu_8-SWNT devices, it was found that the underlying SWNT network was able to transduce changes in the electronic properties of the Eu_8 layer during illumination with 365 nm light and exposure to pure oxygen gas. After an illumination period of 30 min, the device experienced a decrease in the first semiconducting SWNT absorption band. In addition, illumination triggered an increase in the network conductance. By combining the optical and electrical resistance measurements, the Eu_8-decorated SWNT (Eu_8-SWNT) devices demonstrate a linear sensitivity toward oxygen gas in the environmentally relevant concentration range of 5–27% when they operate at room temperature and ambient pressure, which represents an important step in the development of small-scale and low-power detection platforms for oxygen.

Ha and coworkers investigated the effects of molecular adsorption on the electrical properties of carbon nanotube gas sensors over a wide range of gas concentrations [173]. Gas sensors were fabricated by depositing one–two drops of a solution of purified SWCNTs and dichloroethane on Au electrodes prepatterned on a SiO_2 500 nm/Si100 substrate. Functionalization of the SWCNT networks with PEI or nafion was carried out following previously reported conditions [174]. The *p*-type characteristic was maintained after functionalization of the device with PEI, contrary to the case of individual SWCNTs in which the PEI coating changes the behavior from *p*-type to *n*-type due to electron donation of the amine groups in PEI. In the linear I–V_g region, dI/dV_g is about 1.1×10^{-7} A/V for both raw ($V_{SD} = 50$ mV) and PEI-coated ($V_{SD} = 200$ mV) networks. From the relation $dI/dV_g = \mu_h (C/L^2) V_{SD}$, where C is the capacitance and L is the distance between source and drain, the hole mobility μ_h decreased by about four times after PEI coating, suggesting that the local deformations of the electronic structure at the PEI adsorption sites acted as scattering centers for conducting carriers. Their results suggest that the PEI coating decreases both hole concentration and hole mobility in CNTs because the conductance decreases by about 10 times after PEI coating. The nafion coating, however, did not cause any noticeable change in the I–V_g characteristics. The conductance of PEI-coated devices increases rapidly when exposed to NO_2, but the conductance of the nafion-coated devices decreases upon exposure to NH_3 molecules. Neither device exhibits any noticeable change upon exposure to pure N_2, O_2, Ar, or H_2, the major components of the air.

Strano and coworkers reported the ability to engineer molecular reversibility of the gas adsorption on polymer-functionalized SWCNT sensor array to enable the development of new types of nanoelectronic devices for analyte detection using microelectromechanical system (MEMS)-based microgas chromatography (μGC) [175]. An interdigitated electrode design was chosen to maximize the nanotube surface area for analyte adsorption, as predicted from their model. A SWNT network was formed across the electrodes through AC dielectrophoresis, then polypyrrole

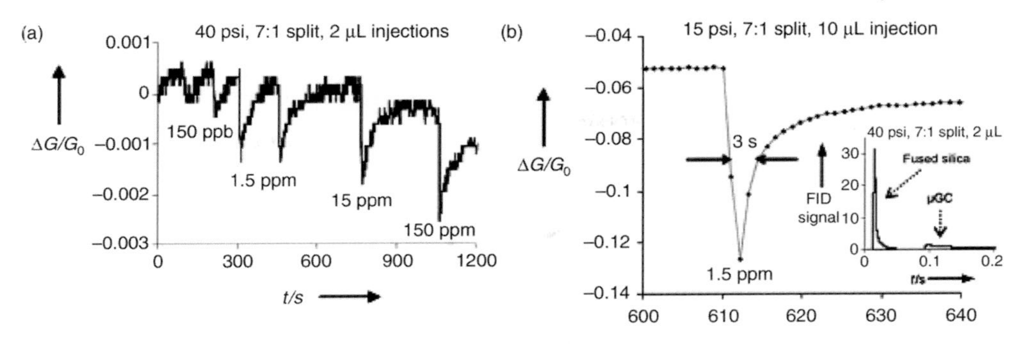

Figure 12.15 Integration of a reversible SWNT sensor with a μGC column. (a) DMMP pulses through the μGC column detected by a PPy-functionalized sensor. A 150 ppb pulse corresponds to 10^9 DMMP molecules. (b) Response of a PPy-functionalized sensor to a DMMP pulse through a conventional fused silica column. The peak is much sharper (FWHM ~3 s) than that from the μGC column (FWHM ~26 s). This is because of peak broadening in the μGC column, as confirmed by the FID signal (*inset*). Reproduced with permission from Ref. [175]. Copyright 2008, Wiley-VCH Verlag GmbH.

(PPy), an amine of $pK_b = 5.4$, was selected as a functionalization material for binding dimethylmethylphosphonate (DMMP), a nerve agent simulant, at the end of a μGC column. Control experiments with only PPy on the electrode gap showed no response to DMMP, whereas a typical change in sensor conductance, upon device exposure to 1 ml pulses of DMMP vapor, was observed. The responses were within the ppb range and were completely reversible, with full-width at half-maximum (FWHM) of only about 4 s. The decrease in conductance was attributed to electron donating DMMP molecules adsorbed on *p*-type semiconducting SWNTs. Also, the response from unfunctionalized sensors was reversible, but only after PPy functionalization did the sensitivity increase by three orders of magnitude, with the reversibility retained. With H_2 carrier gas flowing at 40 psi, a DMMP headspace (2 ml) was manually injected (< 0.3 s) at a 7: 1 split and the conductance was monitored (Figure 12.15). The injector and column temperature were 250 and 30 °C, respectively. The sensor response was reversible and negative, with FWHM of 26 s. The 150 ppb pulse corresponds to approximately 10^9 DMMP molecules, a number that was confirmed by the downstream flame ionization detector (FID). The ability of their experimental approach and detection limits, as low as 10^9 molecules, in this manner was not demonstrated before with any analytical platform.

12.3.4 Carbon Nanotubes at the Biological Interface

Some CNT properties, such as the optical adsorption in the visible–near-infrared (Vis–NIR) region [176], the electrical conductivity [177, 178], and the wide chemical processability [162, 163, 174, 179], make these carbon allotropes very attractive for biomedical applications [161, 166, 180–182]. The most important examples of CNT at the biological interface are related to their use as drug delivery and anticancer

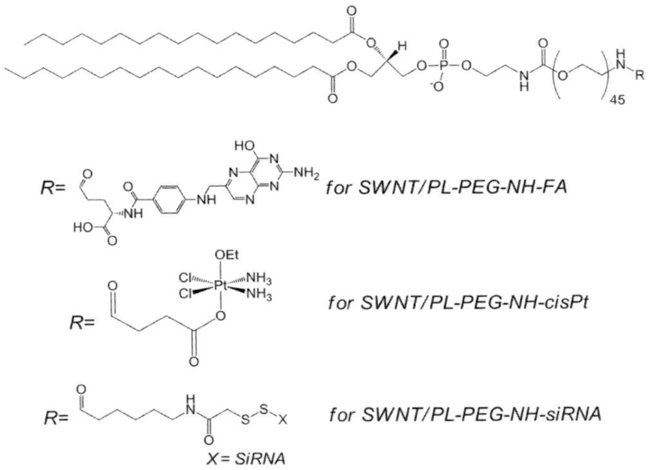

R= for SWNT/PL-PEG-NH-FA

R= for SWNT/PL-PEG-NH-cisPt

R= for SWNT/PL-PEG-NH-siRNA

X = SiRNA

Figure 12.16 Structural representation of the organic moieties used by Dai and coworkers for the supramolecular SWNT functionalization toward different bioapplications.

platforms [183–187], as gene transfection or silencing mediators [188–193], or as substrate for cellular growth [194–202]. Some elegant examples of noncovalent CNT derivatization to elicit important biological properties were reported by Dai and coworkers. They found that the molecule 2-distearoyl-*sn*-glycero-3-phosphoethanolamine-*N*-amino(PEG)$_{2000}$ (PL-PEG-NH_2) irreversibly adsorbs on the outer sidewalls of SWNTs, by the interaction of the aliphatic portion of the phospholipidic moiety. This molecule is, therefore, very useful because the polyethylenic core imparts water solubility and moreover offers a terminal amino group suitable for the further attachment of interesting biomolecules (Figure 12.16).

For instance, PL-PEG-NH_2 was functionalized by carbodiimide chemistry with the folic acid (FA), a naturally occurring product that can be easily internalized in several cell lines that express the membrane-bound folate receptors (FRs) [203]. FRs are overexpressed in several kinds of cancerous cells and for this reason the SWNT/PL-PEG-NH-FA bioconjugate (Figure 12.16) was used to target and selectively internalize SWNTs into FR overexpressing cancerous cells. The aim of the study was to exploit the SWNT property of adsorbing light in the visible and near-infrared regions and to convert it into heat in order to selectively kill the cancerous cells through a hyperthermic approach. To exploit this system, FR-positive HeLa cells (FR + cells) with overexpressed FR on the cell surfaces were obtained by culturing cells in FA-depleted cell medium. Both FR + cells and normal cells without abundant FRs were exposed to SWNT/PL-PEG-NH-FA for 12–18 h, washed, and then radiated with 808 nm laser (1.4 W/cm^2) continuously for 2 min. After the NIR radiation, the authors observed extensive cell death for the FR + cells, evidenced by drastic cell morphology changes, whereas the normal cells remained intact and exhibited normal proliferation behavior over 2 weeks. In another attempt, the authors anchored the cisplatinum derivative [$Pt(NH_3)_2Cl_2(OEt)(O_2CCH_2CH_2CO_2H)$] to the PL-PEG-$NH_2$ moiety adsorbed on SWNTs to use them as longboat prodrug delivery systems for platinum (IV) (Figure 12.16) [204]. Also in this case, as a way to anchor this carboxyl group bearing cisplatinum derivative to the PL-PEG-NH_2 moiety carbodiimide chemistry was employed, allowing to obtain a SWNT longboat (SWNT/PL-PEG-NH-cisPt) with an average loading of 65 platinum (IV) centers per nanotube, as evaluated by atomic absorption spectroscopy (AAS). In order to evaluate the feasibility of this method for delivering a toxic dose of platinum, the authors conducted cytotoxicity assays of SWNT/PL-PEG-NH-cisPt using the testicular carcinoma cell line NTera-2. MTT assay was employed to assess cell viability following treatment for four days with either the free platinum (IV) complex or the SWNT/PL-PEG-NH-cisPt. Compared to *cis*-[Pt $(NH_3)_2Cl_2$] (IC_{50} 0.05 μM), the cytotoxicity of the asymmetrically substituted [Pt $(NH_3)_2Cl_2(OEt)(O_2CCH_2CH_2CO_2H)$] was one order of magnitude less effective ($IC_{50} = 0.5$ μM) and also the amine-functionalized SWNTs/PL-PEG-NH_2 showed negligible cytotoxicity. The SWNT/PL-PEG-NH-cisPt conjugate, however, showed a substantial increase in toxic character with respect to that of the free complex ($IC_{50} = 0.02$ μM) and surpasses that of *cis*-[Pt-$(NH_3)_2Cl_2$] if compared on a per platinum basis. In another study, by employing the heterobifunctional cross-linker sulfosuccinimidyl 6-(3′-[2-pyridyldithio]propionamido)hexanoate (Figure 12.16), the same group obtained SWNT/PL-PEG-NH linker able to anchor thyolated DNA and short interfering RNA (siRNA). By immobilizing these derivatives onto pristine SWNTs, they were able to obtain bioconjugates that showed enhanced activity as carriers of the nucleic acid derivatives inside cancerous HeLa cells [190] and human T lymphocytes [191]. In the former study with the HeLa cells, the ability to release the nucleic acids inside the cells, through the lysosomal-confined cleavage of the disulfidic bonds was ascertained [190]. They used lamin A/C (a protein present inside the nuclear lamina of cells) gene silencing mediated by a particular siRNA sequence, and compared the results obtained with SWNTs acting as cargos with existing transfection agents. HeLa cells were incubated with SWNT/PL-PEG-NH-siRNA for up to 24 h, fixed 48–72 h later, and stained with antilamin and a fluorescently labeled secondary antibody. For comparison, they also employed a commercial transfecting agent lipofectamine for siRNA delivery. Confocal imaging revealed significant reduction in lamin A/C protein expression by SWNT/PL-PEG-NH-siRNA compared to untreated control cells. Furthermore, flow cytometry data showed that for a given siRNA concentration, the percentage of silencing with SWNT/PL-PEG-NH-siRNA was greater than the one obtained with lipofectamine-siRNA. After interfacing SWNT/PL-PEG-NH-siRNA with lymphatic T cells, they investigated SWNT delivery of siRNA against CXCR4, a cell surface coreceptor required for HIV entry into human T cells and infection. About 50–60% knockdown of CXCR4 receptors on H9, Sup-T1, and CEM cells incubated in a solution of SWNT/PL-PEG-NH-siRNA ([siRNA] 50 nm) for 24 h and approximately 90% silencing efficiency was observed upon incubation for 3 days [191]. Also Prato and coworkers reported elegant supramolecular strategies to functionalize CNTs with nucleic acids to alter the expression of selected genes. In fact, after 1,3 dipolar cycloaddition of azomethyne ylides, it was possible to introduce to the CNTs a significant amount of amino functionalities, which are positively charged in the biological context

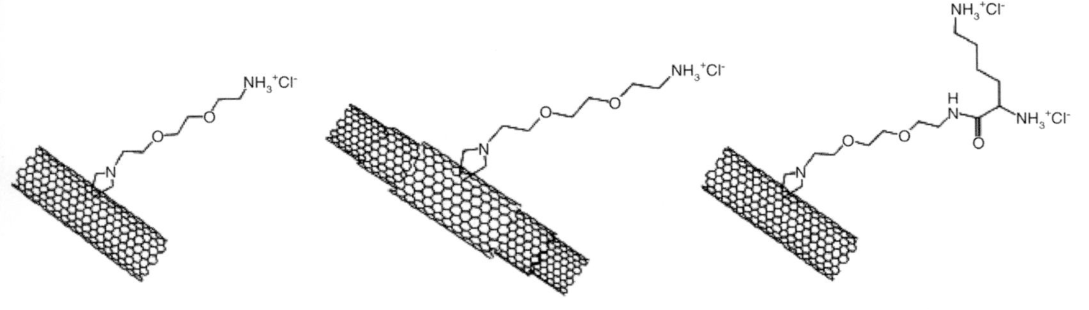

Figure 12.17 Structural representation of ammonium-functionalized CNTs used for the supramolecular functionalization with plasmidic DNA. Reproduced with permission from Ref. [189]. Copyright 2005, American Chemical Society.

[180, 183–186, 205]. Since nucleic acids possess the negative charges arising from the phosphate backbone, electrostatic interaction between the ammonium-CNTs and the plasmidic DNA were at first studied [189, 206].

The ammonium-functionalized CNTs employed in this study were SWNT-NH_3^+, MWNT-NH_3^+, and lysine-functionalized SWNT (SWNT-Lys-NH_3^+) (Figure 12.17). As a plasmidic DNA, the authors employed pCMV-Bgal, a 7.2 kb eukaryotic expression vector that brings the information for the synthesis of β-galactoxidase. CNT–DNA complexes were analyzed by several analytical techniques, such as scanning electron microscopy, surface plasmon resonance, PicoGreen dye exclusion, and agarose gel shift assay. The results indicate that all three types of ammonium-CNTs were able to complex the plasmidic DNA to different degrees, and suggested that both CNT charge density and surface area are important parameters that determine the interaction between ammonium-CNTs with DNA. The authors then focused on the gene transfer efficiency of the various CNT:DNA complexes by transfecting A549 cells. SWNT-NH_3^+ appear to be most efficient in gene transfer when complexed to DNA at an 8 : 1 charge ratio, while SWNT-Lys-NH_3^+ appear most efficient at a 1 : 1 charge ratio. Interestingly, the PicoGreen data indicate that at these charge ratios, approximately 30% of DNA is free to interact with the dye in the case of SWNT-NH_3^+, and a noticeably similar 25% of DNA is free in the case of SWNT-Lys-NH_3^+. In the case of MWNT-NH_3^+, since even at very low charge ratios DNA is fully condensed, there was not much difference in their transfection efficiency across all charge ratios and DNA dose studied.

The same research group has also demonstrated recently that siRNA payloads can be efficiently delivered to mammalian cells by using a dendron-functionalized MWNT series, where the dendron is an alkylated poly-amidoamine (PAMAM) species bearing 2 (for generation 1, G1-MWNTs) or 4 (for generation 2, G2-MWNTs) net positive charges per dendron [192]. Human cervical carcinoma (HeLa) cells were then incubated with the dendron-MWNTs complexed with a noncoding siRNA sequence, which was fluorescently labeled at the 3′ end with Atto 655 that emits light in the far-red region (λ_{em} 690 nm). Each conjugate of the synthesized dendron-

MWNT series was then complexed with the fluorescently labeled siRNA, by using previously optimized conditions (16 : 1 mass ratio) [193]. By this way, the cellular uptake of the different dendron-MWNT/siRNA was observed using confocal microscopy under identical optical conditions. The confocal imaging results showed that the dendron-MWNTs with increased degrees of branching (G2-MWNTs) exhibited more effective intracellular siRNA delivery than did zeroth generation MWNTs and their alkylated derivative, while almost no uptake of siRNA alone was observed. Both the siRNA and the dendron-MWNTs were internalized within the cell cytoplasm, but it was not clear whether the siRNA was detached from the dendron-MWNTs intracellularly. The fluorescent signal from the Atto 655-labeled siRNA was found to be diffused throughout the cytoplasm of cells treated with the dendron-MWNT:siRNA complexes, without any indication of localization in intracellular vesicles (usually evidenced by highly fluorescent pockets within the cell). The cellular morphology did not seem to be affected by the dendron-MWNTs used in these studies, indicating that the siRNA uptake was not a consequence of cellular damage but most probably due to translocation of the dendron-MWNTs through the plasma membrane.

As substrates for cellular growth, CNTs and CNT derivatives were successfully employed for the neuronal cells growth and differentiation [194–198, 200, 201]. The rationale consisted in the consideration of both the electrical conductivity and the high aspect ratio structure of the CNTs, which allows to compare them with the naturally occurring dendrites and axons. In fact, Prato and coworkers demonstrated the possibility of using CNTs as potential devices able to improve neural signal transfer while supporting dendrite elongation and cell adhesion [197]. At first, they performed 1,3 dipolar cycloaddition on MWCNTs, in order to impart them organic solvent solubility, allowing for a product purification [207]. The MWCNTs were then deposited on glass coverlips, which were subsequently heated at 350 °C in order to defunctionalize the MWCNTs and leave a purified and nonfunctionalized material on the glass surface. To address the issue of neuronal and CNT integration within a functional brain network, they used hippocampal neurons isolated and seeded directly on glass coverslips, taken as control, or on a film of purified CNTs, directly layered on glass as described above (Figure 12.18).

To investigate neural network activity when neurons grew integrated into a CNT substrate, they used single-cell patch-clamp recordings. They monitored the occurrence of spontaneous postsynaptic currents (PSCs) since the appearance of PSCs provides clear evidence of functional synapse formation. By this way, neurons deposited on CNTs displayed on average a sixfold increase in the frequency of spontaneous PSCs (6.67 ± 1.04, $n = 15$ cultures, 5 culture series). Then, they further investigated the possibility that CNT substrates affected the balance between inhibition and excitation, by estimating the percentage of excitatory PSCs to the total number of PSCs. This value was similar to that of the control ($64 \pm 8\%$, $n = 4$ cells) and CNT ($64 \pm 3\%$, $n = 4$ cells) sister cultured neurons. The increase in the efficacy of neural signal transmission might be related to the specific properties of CNT materials, such as the high electrical conductivity. Because the transmembrane voltage is a quantity of interest that will affect voltage-dependent membrane processes [208], nanotubes could, in principle, provide a pathway allowing direct electronic current transfer,

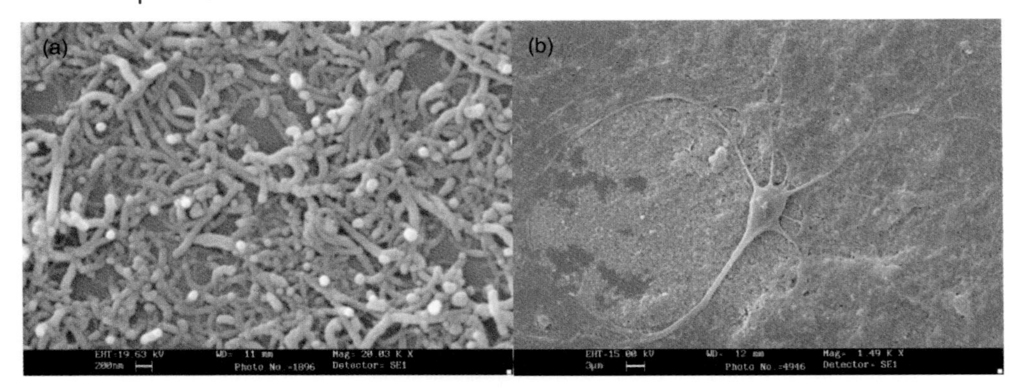

Figure 12.18 (a) Micrographs taken by the scanning electron microscope showing the retention on glass of MWNT films after an 8 day test under culture conditions. (b) Neonatal hippocampal neuron growing on dispersed MWNT after 8 days in culture. The surface structure, composed of films of MWNT and peptide-free glass, allows neuron adhesion. Reproduced with permission from Ref. [197]. Copyright 2005, American Chemical Society.

causing a redistribution of charge along the surface of the membrane. This could account for a reinforcement of a direct electrical coupling between neurons. In a subsequent work, the authors showed, by using single-cell electrophysiology techniques, electron microscopy analysis, and theoretical modeling, that nanotubes improve the responsiveness of neurons by forming tight contacts with the cell membranes that might favor electrical shortcuts between the proximal and the distal compartments of the neuron [209]. They proposed the "electrotonic hypothesis" to explain the physical interactions between the cell and the nanotube, and the mechanisms of how carbon nanotubes might affect the collective electrical activity of cultured neuronal networks. These considerations offer a perspective that would allow to predict or engineer interactions between neurons and carbon nanotubes.

Jan and Kotov have recently reported an interesting example of differentiation of mouse neural stem cells (NSCs) on layer-by-layer (LBL) assembled SWCNT composite [200]. SWNTs were dispersed in a poly(sodium 4-styrene-sulfonate) (PSS, 1 wt%) solution and then LBL assembled with the polyelectrolyte polyethyleneimine. The resulting thin film of coating was referred to as $(PEI/SWNT)_6$. To explore the actions of NSCs on LBL-assembled SWNTcomposite films, mouse embryonic 14-day neurospheres, which are spherical clonal structures of NSCs, from the cortex were seeded and induced to differentiate on the top of round glass coverslips coated with $(PEI/SWNT)_6$. As a control, neurospheres were also seeded and differentiated on coverslips coated with poly-L-ornithine (PLO), a standard substratum commonly used for NSC cultures and studies. In this study, neurospheres attached to both types of substrates and developed neural processes away from the edge of the neurospheres as early as day 1 in culture. The lengths of processes from differentiated NSCs constantly increased over the 7 day culture time. During this time, neural processes developed on PLO-coated substrates remained longer than those developed on (PEI/

SWNT)$_6$-coated substrates; however, the differences were not very significant. These cells extended long neurites that developed into many secondary processes, branching in many directions on the surface. The original shape of the NG105-15 cells is round and fairly small; therefore, such morphology was a clear evidence of the occurred successful differentiation of the cells. The authors then attempted to stimulate the cells by passing current through the nanotube film. The shape of the recorded transients resembles that of the whole-cell stimulation curves obtained for intrinsic excitation with an inserted pipette electrode. These representative currents were evoked by 100 ms, 1–2 μA pulses across the nanotube film at 1 Hz. After an initial capacitive charging artifact, the cells showed strong inward Na^+ currents, which were similar in magnitude and duration to those produced by step changes in membrane potential. All the cells that reported inward currents with step voltage changes could also be stimulated via extrinsic stimulation through the (PEI/SWNT)$_6$-coated substrates. These findings provide evidence of the natural electrical response of the cells to excitation through the laterally applied voltage to the nanotube film and suggest that SWNT-LBL films can be employed to electrically determine significant ion conductance in neuronal cells.

An interesting example that involved the interfacing of CNTs with cell membranes was reported by Bertozzi and coworkers. They coated CNTs with a biomimetic polymer designed to mimic cell surface mucin glycoproteins [210]. The polymers comprises a poly(methylvinylketone) (poly(MVK)) backbone decorated with *R-N*-acetylgalactosamine (R-GalNAc) residues. These sugar residues are reminiscent of the O-linked glycans that decorate mucin glycoproteins. The C_{18} lipid tail provided a hydrophobic anchor for CNT surface assembly (Figure 12.19a).

The coated CNTs were stable in aqueous solution for several months without desorption of the polymer coating. They termed CNTs coated with C_{18}-terminated GalNAc-conjugated polymers as "C_{18}-MMCNTs," where "MM" denotes "mucin mimic." To interface these functionalized CNTs with cells, they took advantage of the *Helix pomatia* agglutinin (HPA), a hexavalent lectin that is specific for R-GalNAc residues and is capable of cross-linking cells and glycoproteins. They reasoned that the complex of HPA with C_{18}-MM-coated CNTs would possess sufficient available HPA binding sites for further complex formation with cell surface glycoproteins. Alternatively, HPA bound to cell surface glycans would present binding sites for GalNAc residues on C_{18}-MM-coated CNTs. In either situation, binding of HPA to GalNAc residues will allow specific interaction of the cells and CNTs. They then complexed C_{18}-MM-coated CNTs with HPA-FITC, and the protein-modified CNTs were then incubated with Chinese hamster ovary (CHO) cells. The labeling observed by fluorescence microscopy and flow cytometry analysis suggested the formation of GalNAc-HPA complexes at both the CNT and the cell surfaces. To evaluate pathway II (Figure 12.19b), the authors required a method for the direct detection of C_{18}-MM-coated CNTs independent of HPA. Thus, they synthesized a C_{18}-MM polymer in which 3% of the GalNAc residues were substituted with the fluorescent dye Texas Red (C_{18}-MM/TR). CHO cells were incubated with unmodified HPA to introduce GalNAc receptors on the cell surface. The cells were then subjected to different concentrations of C_{18}-MM/TR-coated CNTs and finally analyzed by flow cytometry.

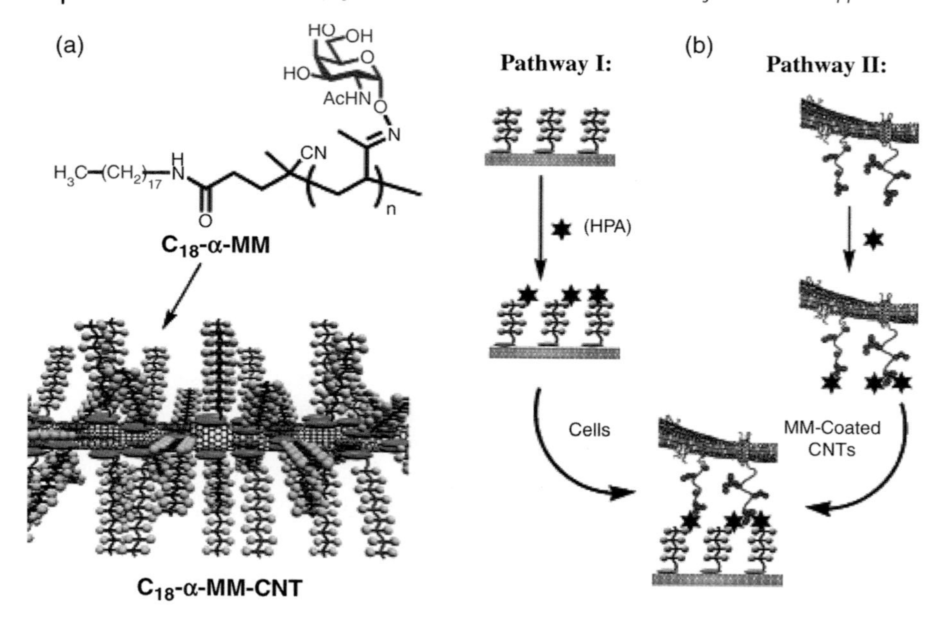

Figure 12.19 (a) Structure of C_{18}-terminated, R-GalNAc-conjugated mucin mimic (C_{18}-R-MM). The mucin mimic polymers assemble on CNT surface in aqueous media through hydrophobic interaction between the C_{18} lipid tails and the CNT surface. The resulting coated CNTs (C_{18}-R-MM-CNTs) were soluble in water. (b) Schematic of interfacing CNTs on cell surfaces via carbohydrate–receptor binding. In pathway I, C18-R-MM-coated CNTs were first bound to HPA, a hexavalent R-GalNAc binding lectin. The complex was then bound to cell surface glycoconjugates using available HPA binding sites presented on CNTs. In pathway II, HPA was first bound to cell surface glycoconjugates. The available HPA binding sites on cell surface were then bound to R-GalNAc residues on C_{18}-R-MM-coated CNTs. Reproduced with permission from Ref. [210]. Copyright 2006, American Chemical Society.

Dose-dependent labeling was observed and the labeling was dependent upon precomplexation of the cells with HPA. Therefore, the glycopolymer coating imparts biocompatibility to the CNTs and simultaneously provides a means for specific cell surface binding. Other examples of carbohydrate-modified CNTs have been recently reviewed by Zhao and Stoddart [164].

12.4 Conclusions

In recent years, a great effort has been made toward the production of fullerene and CNT supramolecular conjugates with peculiar properties at the physical and biological interfaces. Most of the reported examples represent proof of principle of the enormous technological impact that these novel carbon allotropes could have in the coming years. Nevertheless, even though the production of pristine fullerenes and CNTs has reached the multigram scale, most of the aforementioned examples suffer from the feasibility of scale-up conditions, mostly due to the low yields and expensive experimental procedures. For CNT derivatives, the sample inhomogeneity, due to the dissimilar material properties as a consequence of different production methods, further limits their current employment in real-life applications. However, the supramolecular approach has proven to hold great promises in the production and development of new functional materials with defined organization of these novel carbonaceous nanostructures.

References

1 Ruhle, M., Recnik, A., and Ceh, M. (1997) Chemistry and structure of internal interfaces in inorganic materials, in *Solid-State Chemistry of Inorganic Materials. Symposium* (eds P.K. Davies, A.J. Jacobson, C.C. Torardi, and T.A. Vanderah), Materials Research Society, pp. 673–684.

2 Filler, M.A. and Bent, S.F. (2003) The surface as molecular reagent: organic chemistry at the semiconductor interface. *Prog. Surf. Sci.*, **73** (1–3), 1–56.

3 Lehn, J.M. (1988) Supramolecular chemistry: scope and perspectives. Molecules, supermolecules, and molecular devices. *Angew. Chem. Int. Ed.*, **27** (1), 89–112.

4 Bonifazi, D., Mohnani, S., and Llanes-Pallas, A. (2009) Supramolecular chemistry at interfaces: molecular recognition on nanopatterned porous surfaces. *Chem. Eur. J.*, **15** (29), 7004–7025.

5 Ulman, A. (1991) *An Introduction to Ultrathin Organic Films: From Langmuir–Blodgett to Self-Assembly*, Academic Press, San Diego, CA.

6 Schreiber, F. (2000) Structure and growth of self-assembling monolayers. *Prog. Surf. Sci.*, **65** (5–8), 151–256.

7 Diederich, F. and Gomez-Lopez, M. (1999) Supramolecular fullerene chemistry. *Chem. Soc. Rev.*, **28** (5), 263–277.

8 Nierengarten, J.F. (2000) Fullerodendrimers: a new class of compounds for supramolecular chemistry and materials science applications. *Chem. Eur. J.*, **6** (20), 3667–3670.

9 Nierengarten, J.F. (2003) Dendritic encapsulation of active core molecules. *C. R. Chim.*, **6** (8–10), 725–733.

10 Nierengarten, J.F. (2004) Chemical modification of C-60 for materials science applications. *New J. Chem.*, **28** (10), 1177–1191.

11 Guldi, D.M. and Prato, M. (2004) Electrostatic interactions by design. Versatile methodology towards multifunctional assemblies/nanostructured electrodes. *Chem. Commun.*, 2517–2525.

12 Valli, L. and Guldi, D.M. (2002) Langmuir Blodgett films of C_{60} and C_{60}-materials, in *Fullerenes: From Synthesis to Optoelectronic Properties*, Kluwer Academic Publ., Dordrecht, vol. 4, pp. 327–385.

13 Mateo-Alonso, A., Sooambar, C., and Prato, M. (2006) Synthesis and applications of amphiphilic fulleropyrrolidine derivatives. *Org. Biomol. Chem.*, **4** (9), 1629–1637.

14 Nierengarten, J.F. (2003) Fullerodendrimers: fullerene-containing macromolecules with intriguing properties, in *Dendrimers V: Functional and Hyperbranched Building Blocks, Photophysical Properties, Applications in Materials and Life Sciences*, Springer, Berlin, vol. 228, pp. 87–110.

15 Guillon, D., Nierengarten, J.F., Gallani, J.L., Eckert, J.F., Rio, Y., Carreon, M.D., Dardel, B., and

Deschenaux, R. (2003) Amphiphilic and mesomorphic fullerene-based dendrimers. *Macromol. Symp.*, **192**, 63–73.

16 Avdeev, M.V., Khokhryakov, A.A., Tropin, T.V., Andrievsky, G.V., Klochkov, V.K., Derevyanchenko, L.I., Rosta, L., Garamus, V.M., Priezzhev, V.B., Korobov, M.V., and Aksenov, V.L. (2004) Structural features of molecular-colloidal solutions of C_{60} fullerenes in water by small-angle neutron scattering. *Langmuir*, **20** (11), 4363–4368.

17 Wakahara, T., Sathish, M., Miyazawa, K., Hu, C., Tateyama, Y., Nemoto, Y., Sasaki, T., and Ito, O. (2009) Preparation and optical properties of fullerene/ferrocene hybrid hexagonal nanosheets and large-scale production of fullerene hexagonal nanosheets. *J. Am. Chem. Soc.*, **131** (29), 9940–9944.

18 Campidelli, S., Vázquez, E., Milic, D., Prato, M., Barberá, J., Guldi, D.M., Marcaccio, M., Paolucci, D., Paolucci, F., and Deschenaux, R. (2004) Liquid–crystalline fullerene–ferrocene dyads. *J. Mater. Chem.*, **14** (8), 1266–1272.

19 Matsuo, Y., Muramatsu, A., Kamikawa, Y., Kato, T., and Nakamura, E. (2006) Synthesis and structural, electrochemical, and stacking properties of conical molecules possessing buckyferrocene on the apex. *J. Am. Chem. Soc.*, **128** (30), 9586–9587.

20 Nakanishi, T., Schmitt, W., Michinobu, T., Kurth, D.G., and Ariga, K. (2005) Hierarchical supramolecular fullerene architectures with controlled dimensionality. *Chem. Commun.*, 5982–5984.

21 Nakanishi, T., Ariga, K., Michinobu, T., Yoshida, K., Takahashi, H., Teranishi, T., Mohwald, H., and Kurth, D.G. (2007) Flower-shaped supramolecular assemblies: hierarchical organization of a fullerene bearing long aliphatic chains. *Small*, **3** (12), 2019–2023.

22 Michinobu, T., Nakanishi, T., Hill, J.P., Funahashi, M., and Ariga, K. (2006) Room temperature liquid fullerenes: an uncommon morphology of C-60 derivatives. *J. Am. Chem. Soc.*, **128** (32), 10384–10385.

23 Nakanishi, T., Michinobu, T., Yoshida, K., Shirahata, N., Ariga, K., Moehwald, H., and Kurth, D.G. (2008) Nanocarbon superhydrophobic surfaces created from fullerene-based hierarchical supramolecular assemblies. *Adv. Mater.*, **20** (3), 443–446.

24 Asanuma, H., Li, H.G., Nakanishi, T., and Möhwald, H. (2010) Fullerene derivatives that bear aliphatic chains as unusual surfactants: hierarchical self-organization, diverse morphologies, and functions. *Chem. Eur. J.*, **16** (31), 9330–9338.

25 Homma, T., Harano, K., Isobe, H., and Nakamura, E. (2010) Nanometer-sized fluorous fullerene vesicles in water and on solid surfaces. *Angew. Chem. Int. Ed.*, **122** (9), 1709–1712.

26 Nakanishi, T., Morita, M., Murakami, H., Sagara, T., and Nakashima, N. (2002) Structure and electrochemistry of self-organized fullerene–lipid bilayer films. *Chem. Eur. J.*, **8** (7), 1641–1648.

27 Burghardt, S., Hirsch, A., Schade, B., Ludwig, K., and Böttcher, C. (2005) Switchable supramolecular organization of structurally defined micelles based on an amphiphilic fullerene. *Angew. Chem. Int. Ed.*, **44** (19), 2976–2979.

28 Partha, R., Lackey, M., Hirsch, A., Ward Casscells, S., and Conyers, J.L. (2007) Self assembly of amphiphilic C_{60} fullerene derivatives into nanoscale supramolecular structures. *J. Nanobiotechnol.*, **5** (6). DOI: 10.1186/1477-3155-5-6.

29 Wang, X.S., Metanawin, T., Zheng, X.Y., Wang, P.Y., Ali, M., and Vernon, D. (2008) Structure-defined C_{60}/polymer colloids supramolecular nanocomposites in water. *Langmuir*, **24** (17), 9230–9232.

30 Wang, S., He, P., Zhang, J.M., Jiang, H., and Zhu, S.Z. (2005) Novel and efficient synthesis of water-soluble [60] fullerenol by solvent-free reaction. *Synth. Commun.*, **35** (13), 1803–1808.

31 Kokubo, K., Matsubayashi, K., Tategaki, H., Takada, H., and Oshima, T. (2008) Facile synthesis of highly

water-soluble fullerenes more than half-covered by hydroxyl groups. *ACS Nano*, **2** (2), 327–333.

32 Zhou, S., Burger, C., Chu, B., Sawamura, M., Nagahama, N., Toganoh, M., Hackler, U.E., Isobe, H., and Nakamura, E. (2001) Spherical bilayer vesicles of fullerene-based surfactants in water: a laser light scattering study. *Science*, **291** (5510), 1944–1947.

33 Zhang, G., Liu, Y., Liang, D.H., Gan, L.B., and Li, Y. (2010) Facile synthesis of isomerically pure fullerenols and formation of spherical aggregates from $C_{60}(OH)_8$. *Angew. Chem. Int. Ed.*, **49** (31), 5293–5295.

34 Neel, N., Kroger, J., and Berndt, R. (2006) Fullerene nanowires on a vicinal gold surface. *Appl. Phys. Lett.*, **88** (16), 163101–163104.

35 Xiao, W., Ruffieux, P., Aït-Mansour, K., Gröning, O., Palotas, K., Hofer, W.A., Gröning, P., and Fasel, R. (2006) Formation of a regular fullerene nanochain lattice. *J. Phys. Chem. B*, **110** (43), 21394–21398.

36 Ecija, D., Otero, R., Sánchez, L., Gallego, J.M., Wang, Y., Alcamí, M., Martín, F., Martín, N., and Miranda, R. (2007) Crossover site-selectivity in the adsorption of the fullerene derivative PCBM on Au(111). *Angew. Chem. Int. Ed.*, **46** (41), 7874–7877.

37 Hinterstein, M., Torrelles, X., Felici, R., Rius, J., Huang, M., Fabris, S., Fuess, H., and Pedio, M. (2008) Looking underneath fullerenes on Au(110): formation of dimples in the substrate. *Phys. Rev. B*, **77** (15), 153412–153416.

38 Klärner, F.G. and Kahlert, B. (2003) Molecular tweezers and clips as synthetic receptors. Molecular recognition and dynamics in receptor–substrate complexes. *Acc. Chem. Res.*, **36** (12), 919–932.

39 Harmata, M. (2004) Chiral molecular tweezers. *Acc. Chem. Res.*, **37** (11), 862–873.

40 Simonyan, A. and Gitsov, I. (2008) Linear-dendritic supramolecular complexes as nanoscale reaction vessels for "Green" chemistry. Diels–Alder reactions between fullerene C_{60} and polycyclic aromatic hydrocarbons in aqueous medium. *Langmuir*, **24** (20), 11431–11441.

41 Hubble, L.J. and Raston, C.L. (2007) Nanofibers of fullerene C_{60} through interplay of ball-and-socket supermolecules. *Chem. Eur. J.*, **13** (23), 6755–6760.

42 Yoshimoto, S., Honda, Y., Ito, O., and Itaya, K. (2008) Supramolecular pattern of fullerene on 2D bimolecular "chessboard" consisting of bottom-up assembly of porphyrin and phthalocyanine molecules. *J. Am. Chem. Soc.*, **130** (3), 1085–1092.

43 Griessl, S.J.H., Lackinger, M., Jamitzky, F., Markert, T., Hietschold, M., and Heckl, W.M. (2004) Room-temperature scanning tunneling microscopy manipulation of single C_{60} molecules at the liquid–solid interface: playing nanosoccer. *J. Phys. Chem. B*, **108** (31), 11556–11560.

44 Thompson, B.C. and Frechet, J.M. (2008) Organic photovoltaics: polymer–fullerene composite solar cells. *Angew. Chem. Int. Ed.*, **47** (1), 58–77.

45 MacLeod, J.M., Ivasenko, O., Fu, C., Taerum, T., Rosei, F., and Perepichka, D.F. (2009) Supramolecular ordering in oligothiophene–fullerene monolayers. *J. Am. Chem. Soc.*, **131** (46), 16844–16850.

46 Piot, L., Silly, F., Tortech, L., Nicolas, Y., Blanchard, P., Roncali, J., and Fichou, D. (2009) Long-range alignments of single fullerenes by site-selective inclusion into a double-cavity 2D open network. *J. Am. Chem. Soc.*, **131** (36), 12864–12865.

47 Bottari, G., Olea, D., Gómez-Navarro, C., Zamora, F., Gómez-Herrero, J., and Torres, T. (2008) Highly conductive supramolecular nanostructures of a covalently linked phthalocyanine-C_{60} fullerene conjugate. *Angew. Chem. Int. Ed.*, **47** (11), 2026–2031.

48 Li, M., Deng, K., Lei, S.B., Yang, Y.L., Wang, T.S., Shen, Y.T., Wang, C.R., Zeng, Q.D., and Wang, C. (2008) Site-selective fabrication of two-dimensional fullerene arrays by using a supramolecular template at the liquid–solid interface. *Angew. Chem. Int. Ed.*, **47** (35), 6717–6721.

49 Nakanishi, T., Miyashita, N., Michinobu, T., Wakayama, Y., Tsuruoka, T., Ariga, K., and Kurth, D.G. (2006) Perfectly straight nanowires of fullerenes bearing long alkyl chains on graphite. *J. Am. Chem. Soc.*, **128** (19), 6328–6329.

50 Maggini, M., Scorrano, G., and Prato, M. (1993) Addition of azomethine ylides to C_{60}: synthesis, characterization, and functionalization of fullerene pyrrolidines. *J. Am. Chem. Soc.*, **115** (21), 9798–9799.

51 Wang, J., Shen, Y., Kessel, S., Fernandes, P., Yoshida, K., Yagai, S., Kurth, D.G., Möhwald, H., and Nakanishi, T. (2009) Self-assembly made durable: water-repellent materials formed by cross-linking fullerene derivatives. *Angew. Chem. Int. Ed.*, **48** (12), 2166–2170.

52 Barajas-Barraza, R.E. and Guirado-López, R.A. (2002) Clustering of H_2 molecules encapsulated in fullerene structures. *Phys. Rev. B*, **66** (15), 155426–155437.

53 Guirado-López, R.A., Sánchez, M., and Rincón, M.E. (2007) Interaction of acetone molecules with carbon-nanotube-supported TiO_2 nanoparticles: possible applications as room temperature molecular sensitive coatings. *J. Phys. Chem. C*, **111** (1), 57–65.

54 Alldredge, E.S., Bădescu, S.C., Bajwa, N., Perkins, F.K., Snow, E.S., Reinecke, T.L., Passmore, J.L., and Chang, Y.L. (2008) Adsorption of linear chain molecules on carbon nanotubes. *Phys. Rev. B*, **78** (16), 161403–161406.

55 García-Lastra, J.M., Thygesen, K.S., Strange, M., and Rubio, A. (2008) Conductance of sidewall-functionalized carbon nanotubes: universal dependence on adsorption sites. *Phys. Rev. Lett.*, **101** (23), 236806–236810.

56 Arora, G. and Sandler, S.I. (2005) Air separation by single wall carbon nanotubes: thermodynamics and adsorptive selectivity. *J. Chem. Phys.*, **123** (4), 044705–044713.

57 Zhang, Z.Q., Zhang, H.W., Zheng, Y.G., Wang, L., and Wang, J.B. (2008) Gas separation by kinked single-walled carbon nanotubes: molecular dynamics simulations. *Phys. Rev. B*, **78** (3), 035439–035443.

58 Sun, Q., Wang, Q., Jena, P., and Kawazoe, Y. (2005) Clustering of Ti on a C_{60} surface and its effect on hydrogen storage. *J. Am. Chem. Soc.*, **127** (42), 14582–14583.

59 Sun, Q., Jena, P., Wang, Q., and Marquez, M. (2006) First-principles study of hydrogen storage on $Li_{12}C_{60}$. *J. Am. Chem. Soc.*, **128** (30), 9741–9745.

60 Wang, Q., Sun, Q., Jena, P., and Kawazoe, Y. (2009) Theoretical study of hydrogen storage in Ca-coated fullerenes. *J. Chem. Theory Comput.*, **5** (2), 374–379.

61 Thornton, A.W., Nairn, K.M., Hill, J.M., Hill, A.J., and Hill, M.R. (2009) Metal-organic frameworks impregnated with magnesium-decorated fullerenes for methane and hydrogen storage. *J. Am. Chem. Soc.*, **131** (30), 10662–10669.

62 Nagl, S., Baleizão, C., Borisov, S.M., Schäferling, M., Berberan-Santos, M.N., and Wolfbeis, O.S. (2007) Optical sensing and imaging of trace oxygen with record response. *Angew. Chem. Int. Ed.*, **46** (13), 2317–2319.

63 Bonifazi, D., Enger, O., and Diederich, F. (2007) Supramolecular [60]fullerene chemistry on surfaces. *Chem. Soc. Rev.*, **36** (2), 390–414.

64 Bonifazi, D., Spillmann, H., Kiebele, A., de Wild, M., Seiler, P., Cheng, F.Y., Güntherodt, H.J., Jung, T., and Diederich, F. (2004) Supramolecular patterned surfaces driven by cooperative assembly of C_{60} and porphyrins on metal substrates. *Angew. Chem. Int. Ed.*, **116** (36), 4863–4867.

65 Xiao, W., Passerone, D., Ruffieux, P., Aït-Mansour, K., Gröning, O., Tosatti, E., Siegel, J.S., and Fasel, R. (2008) C_{60}/corannulene on Cu(110): a surface-supported bistable buckybowl–buckyball host–guest system. *J. Am. Chem. Soc.*, **130** (14), 4767–4771.

66 Theobald, J.A., Oxtoby, N.S., Phillips, M.A., Champness, N.R., and Beton, P.H. (2003) Controlling molecular deposition and layer structure with supramolecular surface assemblies. *Nature*, **424** (6952), 1029–1031.

67 Schulze, K., Uhrich, C., Schuppel, R., Leo, K., Pfeiffer, M., Brier, E., Reinold, E., and Bauerle, P. (2006) Efficient vacuum-deposited organic solar cells based on a new low-bandgap oligothiophene and fullerene C_{60}. *Adv. Mater.*, **18** (21), 2872–2875.

68 Zhang, H.L., Chen, W., Chen, L., Huang, H., Wang, X.S., Yuhara, J., and Wee, A.T.S. (2007) C_{60} molecular chains on α-sexithiophene nanostripes. *Small*, **3** (12), 2015–2018.

69 Pellarini, F., Pantarotto, D., Da Ros, T., Giangaspero, A., Tossi, D., and Prato, M. (2001) A novel [60]fullerene amino acid for use in solid-phase peptide synthesis. *Org. Lett.*, **3** (12), 1845–1848.

70 Da Ros, T., Bergamin, M., Vázquez, E., Spalluto, G., Baiti, B., Moro, S., Boutorine, A., and Prato, M. (2002) Synthesis and molecular modeling studies of fullerene-5,6,7-trimethoxyindole-oligonucleotide conjugates as possible probes for study of photochemical reactions in DNA triple helices. *Eur. J. Org. Chem.* (3), 405–413.

71 Pantarotto, D., Bianco, A., Pellarini, F., Tossi, A., Giangaspero, A., Zelezetsky, I., Briand, J.P., and Prato, M. (2002) Solid-phase synthesis of fullerene-peptides. *J. Am. Chem. Soc.*, **124** (42), 12543–12549.

72 Bosi, S., Da Ros, T., Spalluto, G., and Prato, M. (2003) Fullerene derivatives: an attractive tool for biological applications. *Eur. J. Med. Chem.*, **38** (11–12), 913–923.

73 Bakry, R., Vallant, R.M., Najam-Ul-Haq, M., Rainer, M., Szabo, Z., Huck, C.W., and Bonn, G.K. (2007) Medicinal applications of fullerenes. *Int. J. Nanomed.*, **2** (4), 639–649.

74 Park, S.J., Kwon, O.H., Lee, K.S., Yamaguchi, K., Jang, D.J., and Hong, J.I. (2008) Dimeric capsules with a nanoscale cavity for [60]fullerene encapsulation. *Chem. Eur. J.*, **14** (17), 5353–5359.

75 Gao, Y.Y., Ou, Z.Z., Yang, G.Q., Liu, L.H., Jin, M.M., Wang, X.S., Zhang, B.W., and Wang, L.X. (2009) Efficient photocleavage of DNA utilizing water soluble riboflavin/naphthaleneacetate substituted fullerene complex. *J. Photochem. Photobiol. A*, **203** (2–3), 105–111.

76 Murthy, C.N. and Geckeler, K.E. (2001) The water-soluble beta-cyclodextrin-[60]fullerene complex. *Chem. Commun.*, 1194–1195.

77 Ikeda, A., Doi, Y., Hashizume, M., Kikuchi, J., and Konishi, T. (2007) An extremely effective DNA photocleavage utilizing functionalized liposomes with a fullerene-enriched lipid bilayer. *J. Am. Chem. Soc.*, **129** (14), 4140–4141.

78 Ikeda, A., Sato, T., Kitamura, K., Nishiguchi, K., Sasaki, Y., Kikuchi, J., Ogawa, T., Yogo, K., and Takeya, T. (2005) Effcient photocleavage of DNA utilising water-soluble lipid membrane-incorporated [60]fullerenes prepared using a [60]fullerene exchange method. *Org. Biomol. Chem.*, **3** (16), 2907–2909.

79 Isobe, H., Nakanishi, W., Tomita, N., Jinno, S., Okayama, H., and Nakamura, E. (2006) Nonviral gene delivery by tetraamino fullerene. *Mol. Pharm.*, **3** (2), 124–134.

80 Klumpp, C., Lacerda, L., Chaloin, O., Da Ros, T., Kostarelos, K., Prato, M., and Bianco, A. (2007) Multifunctionalised cationic fullerene adducts for gene transfer: design, synthesis and DNA complexation. *Chem. Commun.*, 3762–3764.

81 Maeda-Mamiya, R., Noiri, E., Isobe, H., Nakanishi, W., Okamoto, K., Doi, K., Sugaya, T., Izumi, T., Homma, T., and Nakamura, E. (2010) *In vivo* gene delivery by cationic tetraamino fullerene. *Proc. Natl. Acad. Sci. USA.*, **107** (12), 5339–5344.

82 Marchesan, S., Da Ros, T., Spalluto, G., Balzarini, J., and Prato, M. (2005) Anti-HIV properties of cationic fullerene derivatives. *Bioorg. Med. Chem. Lett.*, **15** (15), 3615–3618.

83 Durdagi, S., Mavromoustakos, T., and Papadopoulos, M.G. (2008) 3D QSAR CoMFA/CoMSIA: molecular docking and molecular dynamics studies of fullerene-based HIV-1 PR inhibitors. *Bioorg. Med. Chem. Lett.*, **18** (23), 6283–6289.

84 Wilson, S.R. (2002) Nanomedicine: fullerene and carbon nanotube biology, in *Perspectives of Fullerenes Nanotechnology* (ed. E. Osawa), Springer, pp. 155–163.

85 Tegos, G.P., Demidova, T.N., Arcila-Lopez, D., Lee, H., Wharton, T., Gali, H., and Hamblin, M.R. (2005) Cationic fullerenes are effective and selective antimicrobial photosensitizers. *Chem. Biol.*, **12** (10), 1127–1135.

86 Mroz, P., Pawlak, A., Satti, M., Lee, H., Wharton, T., Gali, H., Sarna, T., and Hamblin, M.R. (2007) Functionalized fullerenes mediate photodynamic killing of cancer cells: type I versus type II photochemical mechanism. *Free Radic. Biol. Med.*, **43** (5), 711–719.

87 Huang, L., Terakawa, M., Zhiyentayev, T., Huang, Y.Y., Sawayama, Y., Jahnke, A., Tegos, G.P., Wharton, T., and Hamblin, M.R. (2010) Innovative cationic fullerenes as broad-spectrum light-activated antimicrobials. *Nanomed. Nanotechnol. Biol. Med.*, **6** (3), 442–452.

88 Guldi, D.M. and Prato, M. (2000) Excited-state properties of C_{60} fullerene derivatives. *Acc. Chem. Res.*, **33** (10), 695–703.

89 Sayes, C.M., Fortner, J.D., Guo, W., Lyon, D., Boyd, A.M., Ausman, K.D., Tao, Y.J., Sitharaman, B., Wilson, L.J., Hughes, J.B., West, J.L., and Colvin, V.L. (2004) The differential cytotoxicity of water-soluble fullerenes. *Nano Lett.*, **4** (10), 1881–1887.

90 Brunet, L., Lyon, D.Y., Hotze, E.M., Alvarez, P.J.J., and Wiesner, M.R. (2009) Comparative photoactivity and antibacterial properties of C_{60} fullerenes and titanium dioxide nanoparticles. *Environ. Sci. Technol.*, **43** (12), 4355–4360.

91 Gharbi, N., Pressac, M., Hadchouel, M., Szwarc, H., Wilson, S.R., and Moussa, F. (2005) [60]Fullerene is a powerful antioxidant *in vivo* with no acute or subacute toxicity. *Nano Lett.*, **5** (12), 2578–2585.

92 Takada, H., Mimura, H., Xiao, L., Islam, R.M., Matsubayashi, K., Ito, S., and Miwa, N. (2006) Innovative anti-oxidant: fullerene (INCI #: 7587) is as "Radical Sponge" on the skin. Its high level of safety, stability and potential as premier anti-aging and whitening cosmetic ingredient. *Fuller. Nanotub. Carbon Nanostruct.*, **14** (2–3), 335–341.

93 Cai, X., Jia, H., Liu, Z., Hou, B., Luo, C., Feng, Z., Li, W., and Liu, J. (2008) Polyhydroxylated fullerene derivative $C_{60}(OH)_{24}$ prevents mitochondrial dysfunction and oxidative damage in an MPP(+)-induced cellular model of Parkinson's disease. *J. Neurosci. Res.*, **86** (16), 3622–3634.

94 Andrievsky, G.V., Bruskov, V.I., Tykhomyrov, A.A., and Gudkov, S.V. (2009) Peculiarities of the antioxidant and radioprotective effects of hydrated C_{60} fullerene nanostuctures *in vitro* and *in vivo*. *Free Radic. Biol. Med.*, **47** (6), 786–793.

95 Osuna, S., Swart, M., and Solà, M. (2010) On the mechanism of action of fullerene derivatives in superoxide dismutation. *Chem. Eur. J.*, **16** (10), 3207–3214.

96 Nakamura, E. and Isobe, H. (2003) Functionalized fullerenes in water. The first 10 years of their chemistry, biology, and nanoscience. *Acc. Chem. Res.*, **36** (11), 807–815.

97 Higashi, N., Inoue, T., and Niwa, M. (1997) Immobilization cleavage of DNA at cationic, self-assembled monolayers containing C-60 on gold. *Chem. Commun.*, 1507–1508.

98 Bernstein, R., Prat, F., and Foote, C.S. (1999) On the mechanism of DNA cleavage by fullerenes investigated in model systems: electron transfer from guanosine and 8-oxo-guanosine derivatives to C-60. *J. Am. Chem. Soc.*, **121** (2), 464–465.

99 Sherigara, B.S., Kutner, W., and D'Souza, F. (2003) Electrocatalytic properties and sensor applications of fullerenes and

carbon nanotubes. *Electroanalysis*, **15** (9), 753–772.

100 Patolsky, F., Tao, G.L., Katz, E., and Willner, I. (1998) C-60-mediated bioelectrocatalyzed oxidation of glucose with glucose oxidase. *J. Electroanal. Chem.*, **454** (1–2), 9–13.

101 Fang, C. and Zhou, X.Y. (2001) The electrochemical characteristics of C-60-glutathione modified Au electrode and the electrocatalytic oxidation of NADH. *Electroanalysis*, **13** (11), 949–954.

102 Ziegler, K.J., Schmidt, D.J., Rauwald, U., Shah, K.N., Flor, E.L., Hauge, R.H., and Smalley, R.E. (2005) Length-dependent extraction of single-walled carbon nanotubes. *Nano Lett.*, **5** (12), 2355–2359.

103 Hudson, J.L., Casavant, M.J., and Tour, J.M. (2004) Water-soluble, exfoliated, nonroping single-wall carbon nanotubes. *J. Am. Chem. Soc.*, **126** (36), 11158–11159.

104 Mayya, K.S. and Caruso, F. (2003) Phase transfer of surface-modified gold nanoparticles by hydrophobization with alkylamines. *Langmuir*, **19** (17), 6987–6993.

105 Wang, R.K., Park, H.O., Chen, W.C., Silvera-Batista, C., Reeves, R.D., Butler, J.E., and Ziegler, K.J. (2008) Improving the effectiveness of interfacial trapping in removing single-walled carbon nanotube bundles. *J. Am. Chem. Soc.*, **130** (44), 14721–14728.

106 Zhang, Y., Shen, Y., Kuehner, D., Wu, S., Su, Z., Ye, S., and Niu, L. (2008) Directing single-walled carbon nanotubes to self-assemble at water/oil interfaces and facilitate electron transfer. *Chem. Commun.*, 4273–4275.

107 Karajanagi, S.S., Vertegel, A.A., Kane, R.S., and Dordick, J.S. (2004) Structure and function of enzymes adsorbed onto single-walled carbon nanotubes. *Langmuir*, **20** (26), 11594–11599.

108 Asuri, P., Karajanagi, S.S., Dordick, J.S., and Kane, R.S. (2006) Directed assembly of carbon nanotubes at liquid–liquid interfaces: nanoscale conveyors for interfacial biocatalysis. *J. Am. Chem. Soc.*, **128** (4), 1046–1047.

109 Carrillo-Carríon, C., Lucena, R., Cárdenas, S., and Valcárcel, M. (2007) Liquid–liquid extraction/headspace/gas chromatographic/mass spectrometric determination of benzene, toluene, ethylbenzene (*o*-, *m*- and *p*-)xylene and styrene in olive oil using surfactant-coated carbon nanotubes as extractant. *J. Chrom. A*, **1171** (1–2), 1–7.

110 Edgerton, S.A., Holdren, M.W., Smith, D.L., and Shah, J.J. (1989) Inter-urban comparison of ambient volatile organic compound concentration in U.S. cities. *J. Air Pollut. Contr.*, **39** (5), 729–732.

111 Collins, P.G., Bradley, K., Ishigami, M., and Zettl, A. (2000) Extreme oxygen sensitivity of electronic properties of carbon nanotubes. *Science*, **287** (5459), 1801–1804.

112 Stan, G., Bojan, M.J., Curtarolo, S., Gatica, S.M., and Cole, M.W. (2000) Uptake of gases in bundles of carbon nanotubes. *Phys. Rev. B*, **62** (3), 2173–2180.

113 Kim, C., Choi, Y.S., Lee, S.M., Park, J.T., Kim, B., and Lee, Y.H. (2002) The effect of gas adsorption on the field emission mechanism of carbon nanotubes. *J. Am. Chem. Soc.*, **124** (33), 9906–9911.

114 Krungleviciute, V., Heroux, L., Talapatra, S., and Migone, A.D. (2004) Gas adsorption on HiPco nanotubes: surface area determinations, and neon second layer data. *Nano Lett.*, **4** (6), 1133–1137.

115 Feng, X., Irle, S., Witek, H., Morokuma, K., Vidic, R., and Borguet, E. (2005) Sensitivity of ammonia interaction with single-walled carbon nanotube bundles to the presence of defect sites and functionalities. *J. Am. Chem. Soc.*, **127** (30), 10533–10538.

116 Saridara, C., Brukh, R., Iqbal, Z., and Mitra, S. (2005) Preconcentration of volatile organics on self-assembled, carbon nanotubes in a microtrap. *Anal. Chem.*, **77** (4), 1183–1187.

117 Snow, E.S., Perkins, F.K., Houser, E.J., Badescu, S.C., and Reinecke, T.L. (2005) Chemical detection with a single-walled carbon nanotube capacitor. *Science*, **307** (5717), 1942–1945.

118 Kingrey, D., Khatib, O., and Collins, P.G. (2006) Electronic fluctuations in nanotube circuits and their sensitivity to gases and liquids. *Nano Lett.*, **6** (7), 1564–1568.

119 Chiashi, S., Watanabe, S., Hanashima, T., and Homma, Y. (2008) Influence of gas adsorption on optical transition energies of single-walled carbon nanotubes. *Nano Lett.*, **8** (10), 3097–3101.

120 Liang, C.W., Sahakalkan, S., and Roth, S. (2008) Electrical characterization of the mutual influences between gas molecules and single-walled carbon nanotubes. *Small*, **4** (4), 432–436.

121 Lee, C.Y. and Strano, M.S. (2008) Amine basicity ($pK_{(b)}$) controls the analyte blinding energy on single walled carbon nanotube electronic sensor arrays. *J. Am. Chem. Soc.*, **130** (5), 1766–1773.

122 Ravelo-Pérez, L.M., Herrera-Herrera, A.V., Hernández-Borges, J., and Rodríguez-Delgado, M.A. (2010) Carbon nanotubes: solid-phase extraction. *J. Chrom. A*, **1217** (16), 2618–2641.

123 Cai, Y., Jiang, G., Liu, J., and Zhou, Q. (2003) Multiwalled carbon nanotubes as a solid-phase extraction adsorbent for the determination of bisphenol A, 4-*n*-nonylphenol, and 4-*tert*-octylphenol. *Anal. Chem.*, **75** (10), 2517–2521.

124 Liu, G., Wang, J., Zhu, Y., and Zhang, X. (2004) Application of multiwalled carbon nanotubes as a solid-phase extraction sorbent for chlorobenzenes. *Anal. Lett.*, **37** (14), 3085–3104.

125 Cai, Y.Q., Cai, Y.E., Mou, S.F., and Lu, Y.Q. (2005) Multi-walled carbon nanotubes as a solid-phase extraction adsorbent for the determination of chlorophenols in environmental water samples. *J. Chrom. A*, **1081** (2), 245–247.

126 Fang, G.Z., He, J.X., and Wang, S. (2006) Multiwalled carbon nanotubes as sorbent for on-line coupling of solid-phase extraction to high-performance liquid chromatography for simultaneous determination of 10 sulfonamides in eggs and pork. *J. Chrom. A*, **1127** (1–2), 12–17.

127 Zhou, Q., Ding, Y., and Xiao, J. (2006) Sensitive determination of thiamethoxam, imidacloprid and acetamiprid in environmental water samples with solid-phase extraction packed with multiwalled carbon nanotubes prior to high-performance liquid chromatography. *Anal. Bioanal. Chem.*, **385** (8), 1520–1525.

128 Zhou, Q., Xiao, J., and Wang, W. (2006) Using multi-walled carbon nanotubes as solid phase extraction adsorbents to determine dichlorodiphenyltrichloroethane and its metabolites at trace level in water samples by high performance liquid chromatography with UV detection. *J. Chrom. A*, **1125** (2), 152–158.

129 El-Sheikh, A.H., Insisi, A.A., and Sweileh, J.A. (2007) Effect of oxidation and dimensions of multi-walled carbon nanotubes on solid phase extraction and enrichment of some pesticides from environmental waters prior to their simultaneous determination by high performance liquid chromatography. *J. Chrom. A*, **1164** (1–2), 25–32.

130 Suádrez, B., Santos, B., Simonet, B.M., Cárdenas, S., and Valcárcel, M. (2007) Solid-phase extraction-capillary electrophoresis-mass spectrometry for the determination of tetracyclines residues in surface water by using carbon nanotubes as sorbent material. *J. Chrom. A*, **1175** (1), 127–132.

131 Wang, S., Zhao, P., Min, G., and Fang, G. (2007) Multi-residue determination of pesticides in water using multi-walled carbon nanotubes solid-phase extraction and gas chromatography-mass spectrometry. *J. Chrom. A*, **1165** (1–2), 166–171.

132 Wang, W.D., Huang, Y.M., Shu, W.Q., and Cao, H. (2007) Multiwalled carbon nanotubes as adsorbents of solid-phase extraction for determination of polycyclic aromatic hydrocarbons in environmental waters coupled with high-performance liquid chromatography. *J. Chrom. A*, **1173** (1–2), 27–36.

133 Yu, J.C., Hrdina, A., Mancini, C., and Lai, E.P. (2007) Molecularly imprinted polypyrrole encapsulated carbon nanotubes in stainless steel frit for micro solid phase extraction of estrogenic compounds. *J. Nanosci. Nanotechnol.*, **7** (9), 3095–3103.

134 Du, D., Wang, M., Zhang, J., Cai, H., Tu, H., and Zhang, A. (2008) Application of multiwalled carbon nanotubes for solid-phase extraction of organophosphate pesticide. *Electrochem. Commun.*, **10** (1), 85–89.

135 Pyrzynska, K. (2008) Carbon nanotubes as a new solid-phase extraction material for removal and enrichment of organic pollutants in water. *Sep. Purif. Rev.*, **37** (4), 372–389.

136 Ravelo-Peréz, L.M., Hernández-Borges, J., and Rodríguez-Delgado, M.A. (2008) Multi-walled carbon nanotubes as efficient solid-phase extraction materials of organophosphorus pesticides from apple, grape, orange and pineapple fruit juices. *J. Chrom. A*, **1211** (1–2), 33–42.

137 Ravelo-Peréz, L.M., Hernández-Borges, J., and Rodríguez-Delgado, M.A. (2008) Multiwalled carbon nanotubes as solid-phase extraction materials for the gas chromatographic determination of organophosphorus pesticides in waters. *J. Sep. Sci.*, **31** (20), 3612–3619.

138 Salam, M.A. and Burk, R. (2008) Novel application of modified multiwalled carbon nanotubes as a solid phase extraction adsorbent for the determination of polyhalogenated organic pollutants in aqueous solution. *Anal. Bioanal. Chem.*, **390** (8), 2159–2170.

139 Asensio-Ramos, M., Hernández-Borges, J., Borges-Miquel, T.M., and Rodríguez-Delgado, M.A. (2009) Evaluation of multi-walled carbon nanotubes as solid-phase extraction adsorbents of pesticides from agricultural, ornamental and forestal soils. *Anal. Chim. Acta*, **647** (2), 167–176.

140 Chen, W., Zeng, J., Chen, J., Huang, X., Jiang, Y., Wang, Y., and Chen, X. (2009) High extraction efficiency for polar aromatic compounds in natural water samples using multiwalled carbon nanotubes/Nafion solid-phase microextraction coating. *J. Chrom. A*, **1216** (52), 9143–9148.

141 Fang, G., Min, G., He, J., Zhang, C., Qian, K., and Wang, S. (2009) Multiwalled carbon nanotubes as matrix solid-phase dispersion extraction absorbents to determine 31 pesticides in agriculture samples by gas chromatography-mass spectrometry. *J. Agric. Food Chem.*, **57** (8), 3040–3045.

142 López-Feria, S., Cárdenas, S., and Valcárcel, M. (2009) One step carbon nanotubes-based solid-phase extraction for the gas chromatographic-mass spectrometric multiclass pesticide control in virgin olive oils. *J. Chrom. A*, **1216** (43), 7346–7350.

143 Salam, M.A. and Burk, R. (2009) Solid phase extraction of polyhalogenated pollutants from freshwater using chemically modified multi-walled carbon nanotubes and their determination by gas chromatography. *J. Sep. Sci.*, **32** (7), 1060–1068.

144 Zhang, W., Sun, Y., Wu, C., Xing, J., and Li, J. (2009) Polymer-functionalized single-walled carbon nanotubes as a novel sol–gel solid-phase micro-extraction coated fiber for determination of polybrominated diphenyl ethers in water samples with gas chromatography-electron capture detection. *Anal. Chem.*, **81** (8), 2912–2920.

145 Hadjmohammadi, M.R., Peyrovi, M., and Biparva, P. (2010) Comparison of C_{18} silica and multi-walled carbon nanotubes as the adsorbents for the solid-phase extraction of Chlorpyrifos and Phosalone in water samples using HPLC. *J. Sep. Sci.*, **33** (8), 1044–1051.

146 Márquez-Sillero, I., Aguilera-Herrador, E., Cárdenas, S., and Valcárcel, M. (2010) Determination of parabens in cosmetic products using multi-walled carbon nanotubes as solid phase extraction sorbent and corona-charged aerosol detection system. *J. Chrom. A*, **1217** (1), 1–6.

147 See, H.H., Sanagi, M., Ibrahim, W.A., and Naim, A.A. (2010) Determination of triazine herbicides using membrane-protected carbon nanotubes solid phase membrane tip extraction prior to micro-liquid chromatography. *J. Chrom. A*, **1217** (11), 1767–1772.

148 Wu, H., Wang, X., Liu, B., Lu, J., Du, B., Zhang, L., Ji, J., Yue, Q., and Han, B. (2010) Flow injection solid-phase extraction using multi-walled carbon

nanotubes packed micro-column for the determination of polycyclic aromatic hydrocarbons in water by gas chromatography-mass spectrometry. *J. Chrom. A*, **1217** (17), 2911–2917.

149 Liang, P., Liu, Y., Guo, L., Zeng, J., and Lu, H. (2004) Multiwalled carbon nanotubes as solid-phase extraction adsorbent for the preconcentration of trace metal ions and their determination by inductively coupled plasma atomic emission spectrometry. *J. Anal. At. Spectrom.*, **19** (11), 1489–1492.

150 Liang, P., Ding, Q., and Song, F. (2005) Application of multiwalled carbon nanotubes as solid phase extraction sorbent for preconcentration of trace copper in water samples. *J. Sep. Sci.*, **28** (17), 2339–2343.

151 Ding, Q., Liang, P., Song, F., and Xiang, A. (2006) Separation and preconcentration of silver ion using multiwalled carbon nanotubes as solid phase extraction sorbent. *Sep. Sci. Technol.*, **41** (12), 2723–2732.

152 Du, Z., Yu, Y.L., Chen, X.W., and Wang, J.H. (2007) The isolation of basic proteins by solid-phase extraction with multiwalled carbon nanotubes. *Chem. Eur. J.*, **13** (34), 9679–9685.

153 Chichak, K.S., Star, A., Altoè, M.V.R., and Stoddart, J.F. (2005) Single-walled carbon nanotubes under the influence of dynamic coordination and supramolecular chemistry. *Small*, **1** (4), 452–461.

154 Guldi, D.M., Rahman, G.M., Jux, N., Balbinot, D., Hartnagel, U., Tagmatarchis, N., and Prato, M. (2005) Functional single-wall carbon nanotube nanohybrids: associating SWNTs with water-soluble enzyme model systems. *J. Am. Chem. Soc.*, **127** (27), 9830–9838.

155 D'Souza, F. and Ito, O. (2009) Supramolecular donor–acceptor hybrids of porphyrins/phthalocyanines with fullerenes/carbon nanotubes: electron transfer, sensing, switching, and catalytic applications. *Chem. Commun.*, 4913–4928.

156 Sgobba, V., Rahman, G.M., Guldi, D.M., Jux, N., Campidelli, S., and Prato, M. (2006) Supramolecular assemblies of different carbon nanotubes for photoconversion processes. *Adv. Mater.*, **18** (17), 2264–2269.

157 Geng, J., Ko, Y.K., Youn, S.C., Kim, Y.H., Kim, S.A., Jung, D.H., and Jung, H.T. (2008) Synthesis of SWNT rings by noncovalent hybridization of porphyrins and single-walled carbon nanotubes. *J. Phys. Chem. C*, **112** (32), 12264–12271.

158 Tu, X. and Zheng, M. (2008) A DNA-based approach to the carbon nanotube sorting problem. *Nano Res.*, **1** (3), 185–194.

159 Ehli, C., Guldi, D.M., Herranz, M.A., Martin, N., Campidelli, S., and Prato, M. (2008) Pyrene-tetrathiafulvalene supramolecular assembly with different types of carbon nanotubes. *J. Mater. Chem.*, **18** (13), 1498–1503.

160 Yang, R., Tang, Z., Yan, J., Kang, H., Kim, Y., Zhu, Z., and Tan, W. (2008) Noncovalent assembly of carbon nanotubes and single-stranded DNA: an effective sensing platform for probing biomolecular interactions. *Anal. Chem.*, **80** (19), 7408–7413.

161 Sudibya, H.G., Ma, J., Dong, X., Ng, S., Li, L.J., Liu, X.W., and Chen, P. (2009) Interfacing glycosylated carbon-nanotube-network devices with living cells to detect dynamic secretion of biomolecules. *Angew. Chem. Int. Ed.*, **48** (15), 2723–2726.

162 Tasis, D., Tagmatarchis, N., Bianco, A., and Prato, M. (2006) Chemistry of carbon nanotubes. *Chem. Rev.*, **106** (3), 1105–1136.

163 Singh, P., Campidelli, S., Giordani, S., Bonifazi, D., Bianco, A., and Prato, M. (2009) Organic functionalisation and characterisation of single-walled carbon nanotubes. *Chem. Soc. Rev.*, **38** (8), 2214–2230.

164 Zhao, Y.L. and Stoddart, J.F. (2009) Noncovalent functionalization of single-walled carbon nanotubes. *Acc. Chem. Res.*, **42** (8), 1161–1171.

165 Chen, R.J., Zhang, Y., Wang, D., and Dai, H. (2001) Noncovalent sidewall functionalization of single-walled carbon nanotubes for protein immobilization. *J. Am. Chem. Soc.*, **123** (16), 3838–3839.

166 Liu, Z., Tabakman, S.M., Chen, Z., and Dai, H. (2009) Preparation of carbon nanotube bioconjugates for biomedical applications. *Nat. Protoc.*, **4** (9), 1372–1382.

167 Kauffman, D.R. and Star, A. (2008) Electronically monitoring biological interactions with carbon nanotube field-effect transistors. *Chem. Soc. Rev.*, **37** (6), 1197–1206.

168 Goldoni, A., Petaccia, L., Lizzit, S., and Larciprete, R. (2010) Sensing gases with carbon nanotubes: a review of the actual situation. *J. Phys. Condes. Matter*, **22** (1), 013001–013008.

169 Hecht, D.S., Ramirez, R.J., Briman, M., Artukovic, E., Chichak, K.S., Stoddart, J.F., and Grüner, G. (2006) Bioinspired detection of light using a porphyrin-sensitized single-wall nanotube field effect transistor. *Nano Lett.*, **6** (9), 2031–2036.

170 Guldi, D.M., Rahman, G.M., Jux, N., Tagmatarchis, N., and Prato, M. (2004) Integrating single-wall carbon nanotubes into donor–acceptor nanohybrids. *Angew. Chem. Int. Ed.*, **43** (41), 5526–5530.

171 Zhao, Y.L., Hu, L.B., Stoddart, J.F., and Gruner, G. (2008) Pyrenecyclodextrin-decorated single-walled carbon nanotube field-effect transistors as chemical sensors. *Adv. Mater.*, **20** (10), 1910–1915.

172 Kauffman, D.R., Shade, C.M., Uh, H., Petoud, S., and Star, A. (2009) Decorated carbon nanotubes with unique oxygen sensitivity. *Nat. Chem.*, **1** (6), 500–506.

173 Kim, S., Lee, H.R., Yun, Y.J., Ji, S., Yoo, K., Yun, W.S., Koo, J.Y., and Ha, D.H. (2007) Effects of polymer coating on the adsorption of gas molecules on carbon nanotube networks. *Appl. Phys. Lett.*, **91** (9). DOI: 093126-093126-3.

174 Pengfei, Q.F., Vermesh, O., Grecu, M., Javey, A., Wang, O., Dai, H.J., Peng, S., and Cho, K.J. (2003) Toward large arrays of multiplex functionalized carbon nanotube sensors for highly sensitive and selective molecular detection. *Nano Lett.*, **3** (3), 347–351.

175 Lee, C.Y., Sharma, R., Radadia, A.D., Masel, R.I., and Strano, M.S. (2008) On-chip micro gas chromatograph enabled by a noncovalently functionalized single-walled carbon nanotube sensor array. *Angew. Chem. Int. Ed.*, **47** (27), 5018–5021.

176 O'Connell, M.J., Bachilo, S.M., Huffman, C.B., Moore, V.C., Strano, M.S., Haroz, E.H., Rialon, K.L., Boul, P.J., Noon, W.H., Kittrell, C., Ma, J., Hauge, R.H., Weisman, R.B., and Smalley, R.E. (2002) Band gap fluorescence from individual single-walled carbon nanotubes. *Science*, **297** (5581), 593–596.

177 Saito, R., Dresselhaus, G., and Dresselhaus, M.S. (1998) *Physical Properties of Carbon Nanotubes*, World Scientific Publishing Company, London.

178 Bryning, M.B., Islam, M.F., Kikkawa, J.M., and Yodh, A.G. (2005) Very low conductivity threshold in bulk isotropic single-walled carbon nanotube–epoxy composites. *Adv. Mater.*, **17** (9), 1186–1191.

179 Banerjee, S., Hemraj-Benny, T., and Wong, S.S. (2005) Covalent surface chemistry of single-walled carbon nanotubes. *Adv. Mater.*, **17** (1), 17–29.

180 Bianco, A., Kostarelos, K., Partidos, C.D., and Prato, M. (2005) Biomedical applications of functionalised carbon nanotubes. *Chem. Commun.*, 571–577.

181 Yang, W.R., Thordarson, P., Gooding, J.J., Ringer, S.P., and Braet, F. (2007) Carbon nanotubes for biological and biomedical applications. *Nanotechnology*, **18** (41), 412001.

182 Liang, F. and Chen, B. (2010) A review on biomedical applications of single-walled carbon nanotubes. *Curr. Med. Chem.*, **17** (1), 10–24.

183 Bianco, A., Hoebeke, J., Partidos, C.D., Kostarelos, K., and Prato, M. (2005) Carbon nanotubes: on the road to deliver. *Curr. Drug Delivery*, **2** (3), 253–259.

184 Pastorin, G., Kostarelos, K., Prato, M., and Bianco, A. (2005) Functionalized carbon nanotubes: towards the delivery of

therapeutic molecules. *J. Biomed. Nanotechnol.*, **1** (2), 133–142.

185 Prato, M., Kostarelos, K., and Bianco, A. (2008) Functionalized carbon nanotubes in drug design and discovery. *Acc. Chem. Res.*, **41** (1), 60–68.

186 Bianco, A., Kostarelos, K., and Prato, M. (2008) Opportunities and challenges of carbon-based nanomaterials for cancer therapy. *Expert Opin. Drug Deliv.*, **5** (3), 331–342.

187 Kostarelos, K., Bianco, A., and Prato, M. (2009) Promises, facts and challenges for carbon nanotubes in imaging and therapeutics. *Nat. Nanotechnol.*, **4** (10), 627–633.

188 Pantarotto, D., Singh, R., McCarthy, D., Erhardt, M., Briand, J.P., Prato, M., Kostarelos, K., and Bianco, A. (2004) Functionalized carbon nanotubes for plasmid DNA gene delivery. *Angew. Chem. Int. Ed.*, **43** (39), 5242–5246.

189 Singh, R., Pantarotto, D., McCarthy, D., Chaloin, O., Hoebeke, J., Partidos, C.D., Briand, J.P., Prato, M., Bianco, A., and Kostarelos, K. (2005) Binding and condensation of plasmid DNA onto functionalized carbon nanotubes: toward the construction of nanotube-based gene delivery vectors. *J. Am. Chem. Soc.*, **127** (12), 4388–4396.

190 Kam, N.W., Liu, Z., and Dai, H. (2005) Functionalization of carbon nanotubes via cleavable disulfide bonds for efficient intracellular delivery of siRNA and potent gene silencing. *J. Am. Chem. Soc.*, **127** (36), 12492–12493.

191 Liu, Z., Winters, M., Holodniy, M., and Dai, H.J. (2007) siRNA delivery into human T cells and primary cells with carbon-nanotube transporters. *Angew. Chem. Int. Ed.*, **46** (12), 2023–2027.

192 Herrero, M.A., Toma, F.M., Al-Jamal, K.T., Kostarelos, K., Bianco, A., Da Ros, T., Bano, F., Casalis, L., Scoles, G., and Prato, M. (2009) Synthesis and characterization of a carbon nanotube-dendron series for efficient siRNA delivery. *J. Am. Chem. Soc.*, **131** (28), 9843–9848.

193 Podesta, J.E., Al-Jamal, K.T., Herrero, M.A., Tian, B.W., Ali-Boucetta, H., Hegde, V., Bianco, A., Prato, M., and Kostarelos, K. (2009) Antitumor activity and prolonged survival by carbon-nanotube-mediated therapeutic siRNA silencing in a human lung xenograft model. *Small*, **5** (10), 1176–1185.

194 Mattson, M.P., Haddon, R.C., and Rao, A.M. (2000) Molecular functionalization of carbon nanotubes and use as substrates for neuronal growth. *J. Mol. Neurosci.*, **14** (3), 175–182.

195 Hu, H., Ni, Y.C., Montana, V., Haddon, R.C., and Parpura, V. (2004) Chemically functionalized carbon nanotubes as substrates for neuronal growth. *Nano Lett.*, **4** (3), 507–511.

196 Hu, H., Ni, Y.C., Mandal, S.K., Montana, V., Zhao, N., Haddon, R.C., and Parpura, V. (2005) Polyethyleneimine functionalized single-walled carbon nanotubes as a substrate for neuronal growth. *J. Phys. Chem. B*, **109** (10), 4285–4289.

197 Lovat, V., Pantarotto, D., Lagostena, L., Spalluto, G., Prato, M., Ballerini, L., Cacciari, B., Grandolfo, M., and Righi, M. (2005) Carbon nanotube substrates boost neuronal electrical signaling. *Nano Lett.*, **5** (6), 1107–1110.

198 Gheith, M.K., Pappas, T.C., Liopo, A.V., Sinani, V.A., Shim, B.S., Motamedi, M., Wicksted, J.R., and Kotov, N.A. (2006) Stimulation of neural cells by lateral layer-by-layer films of single-walled currents in conductive carbon nanotubes. *Adv. Mater.*, **18** (22), 2975–2979.

199 Zanello, L.P., Zhao, B., Hu, H., and Haddon, R.C. (2006) Bone cell proliferation on carbon nanotubes. *Nano Lett.*, **6** (3), 562–567.

200 Jan, E. and Kotov, N.A. (2007) Successful differentiation of mouse neural stem cells on layer-by-layer assembled single-walled carbon nanotube composite. *Nano Lett.*, **7** (5), 1123–1128.

201 Jan, E., Hendricks, J.L., Husaini, V., Richardson-Burns, S.M., Sereno, A., Martin, D.C., and Kotov, N.A. (2009) Layered carbon nanotube-polyelectrolyte electrodes outperform traditional neural interface materials. *Nano Lett.*, **9** (12), 4012–4018.

202 Namgung, S., Kim, T., Baik, K.Y., Lee, M., Nam, J.-M., and Hong, S. (2010;) Fibronectin–carbon-nanotube hybrid nanostructures for controlled cell growth. *Small*, **7** (1), 56–61.

203 Kam, N.W., O'Connell, M., Wisdom, J.A., and Dai, H. (2005) Carbon nanotubes as multifunctional biological transporters and near-infrared agents for selective cancer cell destruction. *Proc. Natl. Acad. Sci. USA.*, **102** (33), 11600–11605.

204 Feazell, R.P., Nakayama-Ratchford, N., Dai, H., and Lippard, S.J. (2007) Soluble single-walled carbon nanotubes as longboat delivery systems for platinum (IV) anticancer drug design. *J. Am. Chem. Soc.*, **129** (27), 8438–8439.

205 Georgakilas, V., Tagmatarchis, N., Pantarotto, D., Bianco, A., Briand, J.P., and Prato, M. (2002) Amino acid functionalisation of water soluble carbon nanotubes. *Chem. Commun.*, 3050–3051.

206 Pantarotto, D., Singh, R., McCarthy, D., Erhardt, M., Briand, J.P., Prato, M., Kostarelos, K., and Bianco, A. (2004) Functionalized carbon nanotubes for plasmid DNA gene delivery. *Angew. Chem. Int. Ed.*, **43** (39), 5242–5246.

207 Georgakilas, V., Kordatos, K., Prato, M., Guldi, D.M., Holzinger, M., and Hirsch, A. (2002) Organic functionalization of carbon nanotubes. *J. Am. Chem. Soc.*, **124** (5), 760–761.

208 Vigmond, E.J., Velazquez, J.L.P., Valiante, T.A., Bardakjian, B.L., and Carlen, P.L. (1997) Mechanisms of electrical coupling between pyramidal cells. *J. Neurophysiol.*, **78** (6), 3107–3116.

209 Cellot, G., Cilia, E., Cipollone, S., Rancic, V., Sucapane, A., Giordani, S., Gambazzi, L., Markram, H., Grandolfo, M., Scaini, D., Gelain, F., Casalis, L., Prato, M., Giugliano, M., and Ballerini, L. (2009) Carbon nanotubes might improve neuronal performance by favouring electrical shortcuts. *Nat. Nanotechnol.*, **4** (2), 126–133.

210 Chen, X., Tam, U.C., Czlapinski, J.L., Lee, G.S., Rabuka, D., Zettl, A., and Bertozzi, C.R. (2006) Interfacing carbon nanotubes with living cells. *J. Am. Chem. Soc.*, **128** (19), 6292–6293.

13
Applications of Supramolecular Ensembles with Fullerenes and CNTs: Solar Cells and Transistors

Hiroshi Imahori and Tomokazu Umeyama

13.1
Introduction

Supramolecular chemistry of fullerenes and carbon nanotubes has been developed rapidly because of their unique structures and electronic properties [1, 2]. For fullerenes, the facile electron accepting ability is one of the most fascinating points. In this regard, intramolecular and intermolecular electron transfer properties of fullerenes and their derivatives as an acceptor have been intensively investigated [1–8]. The investigation has revealed that fullerenes indeed act as an excellent acceptor. The outstanding electron transfer properties can be rationalized by the fact that fullerenes exhibit small reorganization energies in electron transfer [8–11]. Small reorganization energies of donor–acceptor systems allow us to achieve fast photoinduced charge separation and slow charge recombination, which is ideal situation for efficient solar energy conversion. In connection with artificial photosynthesis, fullerenes have been utilized particularly as an acceptor in donor–acceptor-linked systems, yielding a long-lived charge-separated state with a high quantum yield [1–10]. Recently, solar energy conversion has attracted much attention because of increasing demand for the exploitation of renewable energy. In this context, extensive efforts have been made in recent years to develop solar cells and photoelectrochemical devices consisting of donors with fullerenes or their derivatives as an acceptor [12–15]. To increase the device performance, it is crucial to fulfill the following requirements: (i) extensive collection of visible light, (ii) efficient migration of exciton to donor–acceptor interface, (iii) charge separation with a high quantum yield, and (iv) efficient transport of separated electrons and holes into respective electrodes. Considering rather poor light-harvesting properties of fullerenes and their derivatives, it is important to combine them with suitable donors possessing excellent light-harvesting properties in the visible and near-infrared regions. Moreover, the donor–acceptor fabrication in the devices is a vital process for satisfying the requirements. Various supramolecular assemblies by self-assembled monolayers (SAMs), layer-by-layer deposition, electrophoretic deposition,

Supramolecular Chemistry of Fullerenes and Carbon Nanotubes, First Edition. Edited by Nazario Martin and Jean-Francois Nierengarten.

chemical adsorption, and spin coating have been formed on electrodes for fullerene-based solar cells and photoelectrochemical devices.

Analogous concept and strategy can be applied to carbon nanotubes (CNTs) because of similarity in fullerenes and CNTs [2]. Specifically, one-dimensional wire structures of CNTs are appealing for efficient electron and hole transport through the nanowires. Both CNTs and fullerenes are also fascinating materials as building blocks of organic transistors because of their characteristic structures and superior electronic properties.

In this chapter, we focus on fullerenes and CNTs for organic solar cells and transistors. Emphasis lies on supramolecular assemblies of fullerenes and CNTs toward the realization of high device performances. Owing to the limited space and references, we highlight some of the major contributions with the most general applicability and some important aspects.

13.2 Solar Cells

13.2.1 Fullerene-Based Solar Cells

13.2.1.1 Self-Assembled Monolayers

SAMs have recently attracted much attention as a useful methodology for molecular assembly [1]. They enable the molecules of interest to be bound covalently to the surface such as metals, semiconductors, and insulators in a highly organized manner. They also make it possible to arrange functional molecules unidirectionally at the molecular level on electrodes when substituents that would self-assemble covalently on the substrates are attached to a terminal of the molecules.

Imahori *et al.* reported the first multistep electron transfer in SAMs of porphyrin–fullerene dyads [16]. A combination of the porphyrin–fullerene dyads was chosen to exhibit a long-lived charge-separated state with a high quantum yield owing to the small reorganization energies of fullerenes [8–11]. The porphyrin-C_{60} molecules were tilted and nearly parallel to the gold electrode, leading to formation of the loosely packed structures. A maximum adsorbed photon-to-current efficiency (APCE) of 0.5% was obtained in the photoelectrochemical device with the SAM. The short lifetime in the charge-separated state of the zinc porphyrin-C_{60} device led to the poor photocurrent generation, whereas the long lifetime in the exciplex state of the free base porphyrin-C_{60} device resulted in a pronounced increase in the photocurrent [16].

Aso *et al.* designed gold electrodes modified with oligothiophene–fullerene dyads **1** (Figure 13.1) bearing a tripodal rigid anchor, which was a tetraphenylmethane core with three mercaptomethyl arms, allowing such molecules to be well-organized on the electrodes [17]. Maximum APCE values were estimated to be 17% for **1** ($n = 1$) and 35% for **1** ($n = 2$).

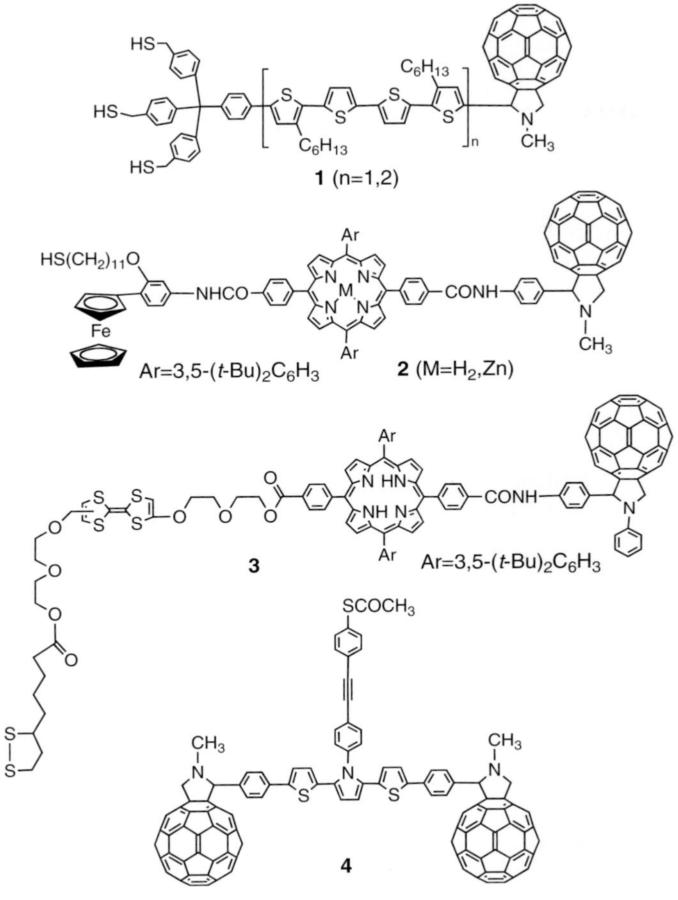

Figure 13.1 Thiol-containing fullerene derivatives 1–4.

To our knowledge, there are a few examples of a SAM of fullerene-containing triad. Imahori *et al.* reported the first alkanethiol-attached triads **2** ($M = H_2$, Zn) with a linear array of ferrocene (Fc), porphyrin (P), and C_{60} (Figure 13.1) [18]. The triad molecules were densely packed with an almost perpendicular orientation on the gold surface. The maximum APCE values were 25% for **2** ($M = H_2$) and 20% for **2** ($M = Zn$) [18]. The use of C_{60} with the small reorganization energy allows the device to accelerate the forward electron transfer and decelerate the undesirable back electron transfer, thus leading to high APCE values.

Imahori *et al.* further combined **2** ($M = H_2$) with boron dipyrrin-alkanethiol as a light-harvesting unit to mimic light harvesting in antenna complexes and multistep electron transfer in reaction centers [19]. The APCE values were increased with increasing the ratio of **2**. A maximum APCE of $50 \pm 8\%$ at 510 nm is one of the highest values ever reported for photocurrent generation at monolayer-modified metal electrodes using donor–acceptor-linked molecules [19].

Zink and coworkers designed a photoactive molecular triad **3** (Figure 13.1) as a nanoscale power supply for a supramolecular machine [20]. A maximum APCE value of 1% was noted. This nanoscale source of electrical energy was utilized to drive the dethreading of a pseudorotaxane comprised of cyclobis(paraquat-*p*-phenylene) cyclophane complexed with 1,5-bis-[(2-hydroxyethoxy)ethoxy]naphthalene (thread). The pseudorotaxane was considered a reasonably stable charge transfer complex between the thread and the cyclophane. Nevertheless, reduction of the viologen moieties in the cyclophane by photoreduction of the electron accepting C_{60} component to the corresponding radical cation weakened the charge transfer interaction, resulting in the decomplexation of the radical cation from thread [20].

Jen and coworkers reported a SAM of C_{60}-tethered 2,5-dithienylpyrrole triad **4** (Figure 13.1) on a gold electrode [21]. The APCE value in the photoelectrochemical device was as high as 51%. The high value may contain significant experimental errors considering that the small absorbance of the adsorbed chromophores on the gold electrode was assumed to be the same as that in solution [21].

Imahori and coworkers prepared ITO electrodes modified chemically with SAMs of porphyrin-C_{60}-linked dyads [22]. IPCE values of the freebase porphyrin-C_{60}-linked dyad and zinc porphyrin-C_{60}-linked dyad devices were 3.4–6.4% and 1.7–8.0%, which were 10–30 times as large as those of the porphyrin reference devices (0.21–0.38%). This is the first example of the remarkable enhancement of photocurrent generation using SAMs of donor–acceptor-linked molecules on ITO electrodes, compared to the corresponding single-chromophore devices [22].

Kim and coworkers designed ITO electrodes covalently modified with C_{60}–metal cluster moiety that was further tethered to zinc porphyrin unit [23]. An APCE value of the device was estimated to be 10.4%. Surprisingly, an addition of 1,4-diazabicyclo [2.2.2]octane (DABCO) into the device resulted in the improvement of the APCE value (19.5%). The authors pointed out that the complexation of DABCO between the two zincporphyrins precluded aggregation with adjacent porphyrins, leading to high performance in photocurrent generation [23]. Boron dipyrrin as a light-harvesting moiety was further tethered to the end of porphyrin to develop the triad on ITO electrode (denoted as ITO/**5**, Figure 13.2). The device exhibited an APCE value of 29% that is the highest value ever reported for ITO electrodes modified with donor–acceptor-linked molecules [24].

Imahori *et al.* reported the first SAMs of ferrocene–porphyrin–fullerene triads on ITO (denoted as ITO/**6**, Figure 13.2) [25]. An APCE value of 11% in the ITO/**6** ($M = Zn$) device was larger than that of 4.5% in the ITO/**6** ($M = H_2$) device. It is worth noting that the APCE values (4.5–11%) for the ITO electrodes are rather lower than the corresponding values (20–25%) for the gold electrodes when similar ferrocene–porphyrin–fullerene triad molecules are attached covalently to the electrodes [25].

The examples given demonstrate that fullerene-based SAMs on gold and ITO electrodes are excellent systems for the realization of efficient photocurrent generation on electrode surfaces. SAMs will pave the way for the development of photoactive molecular devices including solar cells.

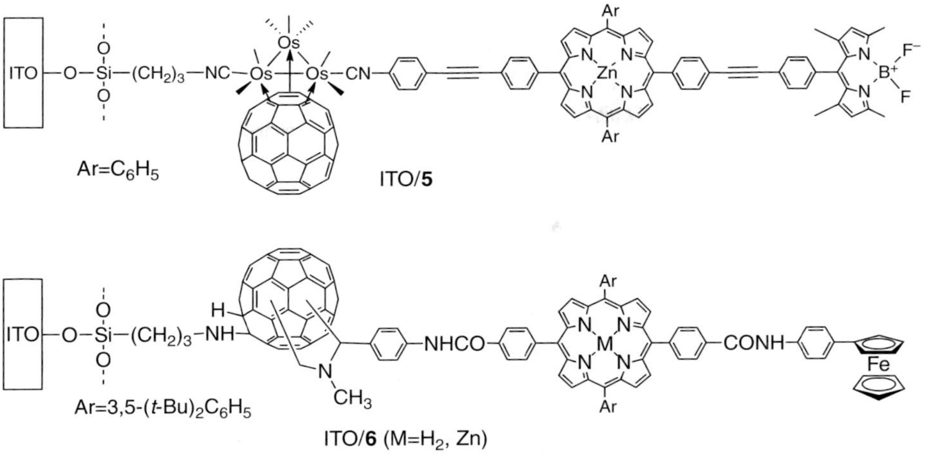

Figure 13.2 Self-assembled monolayers of fullerene-based triads ITO/**5** and ITO/**6**.

13.2.1.2 Layer-by-Layer Deposition

Guldi *et al.* have extensively explored the use of electrostatic and *van der Waals* interactions to fabricate photoactive ITO electrodes deposited with porphyrin–fullerene dyads [26]. At first, a base layer of poly(diallyldimethylammonium) (PDDA) was deposited on the hydrophobic surfaces via simple hydrophobic–hydrophobic interactions. The resulting hydrophilic and positively charged surface promoted the electrostatically driven deposition of poly(sodium 4-styrenesulfonate) (PSS), bearing sulfonic acid functionalities, to yield the PDDA/PSS templates. With the surface sufficiently overlaid with negative charges, the cationic porphyrin–fullerene dyads were deposited. Sequential repetition of the steps for up to 12 times (absorbance = 1.3 at 435 nm) led to the systematic stacking of PSS and the porphyrin–fullerene dyad sandwich layers. The IPCE value of the one-layer device was as low as 0.6% at 440 nm [26]. The flexible cationic tail of the porphyrin–fullerene dyad, leading to the ill-defined arrangement of the porphyrin–fullerene dyad on the ITO electrode, may cause the poor IPCE value.

Guldi tailored the arrangement of photoactive ITO electrodes at the molecular level (Figure 13.3) [27]. C_{60} molecules with negatively charged moieties **7** were adsorbed at the PDDA-modified, positively charged ITO surface (denoted as ITO/PDDA/**7**). Strong van der Waals interactions between C_{60} cores facilitated the layer formation in the ITO/PDDA/**7** device, leaving the anionic dendrimer branches on the surface. In the next step, octacationic porphyrins **8** ($M = H_2$, Zn) were deposited via electrostatic interactions with the anionic dendrimer branches in a monolayer fashion (denoted as ITO/PDDA/**7**/**8**). Subsequent layers were built up analogously utilizing cationic/anionic contacts to yield ITO/PDDA/**7**/**8**/**9** and ITO/PDDA/**7**/**8**/**9**/**10** electrodes, respectively. IPCE values of 1.6, 1.0, 0.08, and 0.01% were noted for the ITO/PDDA/**7**/**8**/**9**/**10**, ITO/PDDA/**7**/**8**/**9**, ITO/PDDA/**7**/**8**, and ITO/PDDA/**7** devices, respectively [27].

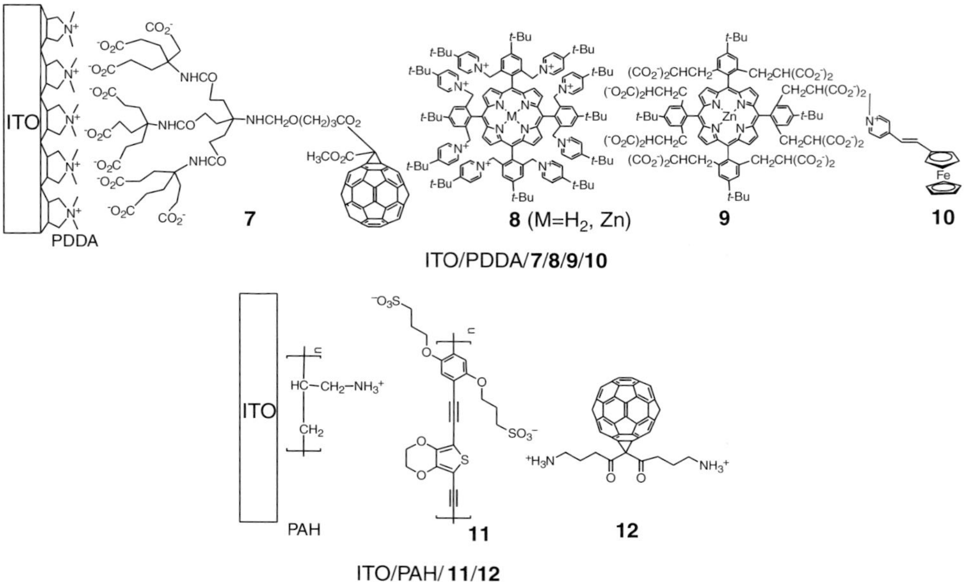

Figure 13.3 ITO/PDDA/**7**/**8**/**9**/**10** and ITO/PAH/**11**/**12** electrodes prepared by layer-by-layer method.

Layer-by-layer fabrication of anionic conjugated polyelectrolyte **11** as a donor and water-soluble dicationic fullerene derivative **12** as an acceptor (Figure 13.3) on a cationic poly(allylamine hydrochloride) (PAH)-modified ITO electrode was conducted to fabricate the active materials for photovoltaic devices (denoted as ITO/(**11**/**12**)$_m$) [28]. An overall power conversion efficiency (η) of 0.04% in this device under AM 1.5 is the highest value ever reported for photovoltaic devices using layer-by-layer technique [28]. Although the cell performance is still low, the better understanding gained through such fundamental systems will ultimately lead to improved performance of polymer-based photovoltaic devices.

13.2.1.3 **Electrochemical Deposition**

Kamat *et al.* reported a novel approach of enhancing the light-harvesting efficiency by electrodeposited C_{60} films from a cluster (aggregate) solution of acetonitrile/toluene (3/1, v/v), which absorbed visible light intensively and revealed much improved photoelectrochemical response [29]. Specifically, a toluene solution of C_{60} molecules was rapidly injected into acetonitrile to form C_{60} clusters due to the lyophobic nature of C_{60} in the mixed solvent. The clusters of C_{60} were deposited as thin films on nanostructured SnO_2 electrodes under the influence of an electric field. A maximum IPCE value of 4% was noted at 420 nm. Direct electron transfer between the excited C_{60} clusters deposited on SnO_2 nanocrystallites and the redox couple (I_3^-/I^-) in solution is responsible for the enhanced photocurrent generation [29].

Kamat *et al.* deposited clusters of C_{60}–aniline-linked dyad as thin films on a nanostructured SnO_2 electrode under the influence of an electric field [30]. A maximum IPCE value of 4% was observed at 450 nm, which is similar to that of similar photoelectrochemical device with pristine C_{60} [30].

Imahori *et al.* applied electrophoretic deposition method to porphyrin–fullerene-linked dyads to assemble the dyad clusters on a nanostructured SnO_2 electrode [31]. Although the light-harvesting property (maximal absorbance = 1.6) was improved remarkably, the maximal IPCE values were found to be still low (0.36–0.42%) [31]. The poor IPCE values may result from macroscopic cancellation of the charge-separated state in the dyad clusters as a result of rather random orientation of the dyad molecules in the clusters.

Imahori, Fukuzumi, and Kamat applied bottom-up self-assembly to porphyrin and fullerene single components and the composites to construct a novel organic solar cell (i.e., dye-sensitized bulk heterojunction solar cell), possessing characteristics of both dye-sensitized and bulk heterojunction devices [32–34]. For instance, porphyrin **13** or **14** makes a complex with C_{60} in nonpolar solvent due to the π–π interaction (Figure 13.4). Then, the supramolecular complex is self-assembled into larger clusters in a mixture of polar and nonpolar solvents because of the lyophobic interaction. Finally, the larger clusters can be further associated with a nanostructured SnO_2 electrode using an electrophoretic deposition technique (denoted as ITO/SnO_2/(**13** + $C_{60})_m$) [32]. A short-circuit photocurrent density (J_{SC}) of 0.18 mA/cm^2, an open-circuit voltage (V_{OC}) of 0.21 V, a fill factor (ff) of 0.35, and an η of 0.012%

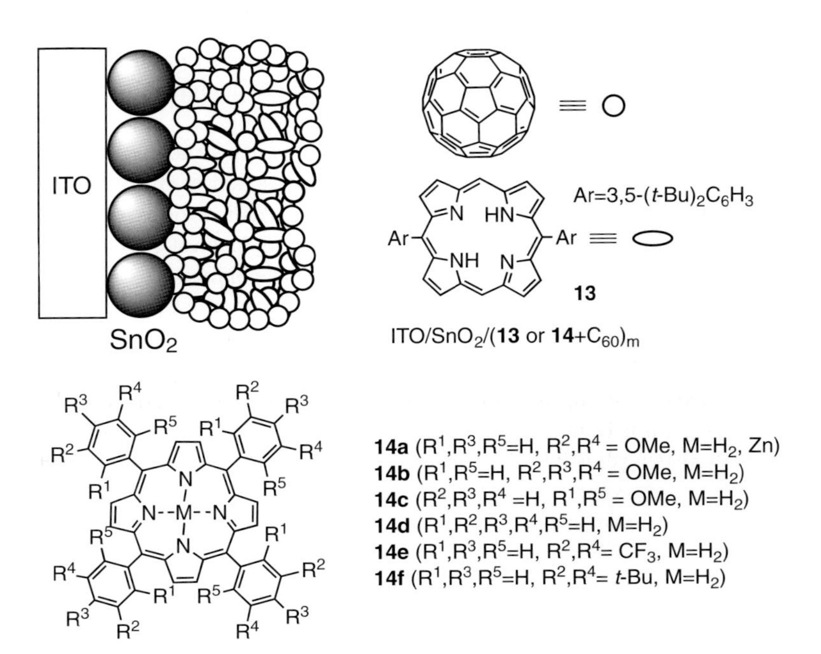

Figure 13.4 Schematic diagram for the ITO/SnO_2/(**13** or **14** + $C_{60})_m$ electrode.

(input power (W_{IN}) = 110 mW/cm^2) were obtained. Photoinduced electron transfer occurs from the porphyrin-excited singlet state ($^1H_2P^*/H_2P^{\bullet+}$ = − 0.7 V versus NHE) to C_{60} ($C_{60}/C_{60}^{\bullet-}$ = − 0.2 V versus NHE). The generated $C_{60}^{\bullet-}$ transfers electrons to a conduction band (CB) of SnO_2 nanocrystallites (E_{CB} = 0 V versus NHE), to yield the current in the circuit. The regeneration of the porphyrin clusters ($H_2P/H_2P^{\bullet+}$ = 1.2 V versus NHE) is achieved by the iodide/triiodide couple (I^-/I_3^- = 0.5 V versus NHE) present in the electrolyte system [32].

Imahori *et al.* further systematically examined the substituent effects of meso-tetraphenylporphyrins **14a–e** on the film structures and their photoelectrochemical properties of the composite clusters of **14a–e** and C_{60} that were electrophoretically deposited on nanostructured SnO_2 electrodes [34]. The photocurrent generation efficiency was found to correlate with the complexation ability of the porphyrin for C_{60}. Basically, the IPCE value was increased with increasing the ratio of the porphyrin to C_{60} in the resulting films, although the amount of C_{60} was much larger than that of the porphyrin (**14** : C_{60} = 1 : > 9). The unique molecular arrangement of the porphyrin **14a** (M = H_2) and C_{60} on SnO_2 electrodes resulted in the largest IPCE value (about 60%) among this type of photoelectrochemical devices [34].

Supramolecular assembly of donor–acceptor molecules is a potential approach to create a desirable phase-separated, interpenetrating network involving molecular-based nanostructured electron and hole highways. However, different complex hierarchies of self-organization going from simple molecules to devices have limited improvement in the device performance. To construct such complex hierarchies comprising donor and acceptor molecules on electrode surfaces, preorganized molecular systems are excellent candidates for achieving the molecular architectures. In particular, they have been three-dimensionally organized using dendrimers [35, 36], oligomers [37], and nanoparticles [33, 38] to combine with fullerenes for organic solar cells.

Dendrimers are well-defined macromolecules with a tree-like structure. Dendrimer structures are, therefore, very attractive for the construction of chromophore arrays because their convergent and/or divergent synthesis enables the assembly of many chromophores in a restricted space and with high topological control. The collaborative efforts of Imahori and coworkers led to the development of solar cells using supramolecular complexes of multiporphyrin dendrimers with C_{60} by clusterization in the mixed solvent and electrophoretic deposition on a SnO_2 electrode [35]. The porphyrin first-generation device exhibited a J_{SC} of 0.29 mA/cm^2, a V_{OC} of 0.22 V, a ff of 0.31, and an η of 0.32% (W_{IN} = 6.2 mW/cm^2). The η value was 10 times as large as that of the ITO/SnO_2/(**14f** + C_{60})$_m$ device (η = 0.035%). The IPCE value was decreased with increasing dendrimer generation [35]. Such an unfavorable trend may originate from the rigid geometry of the porphyrin moieties in the higher generation, which inhibits the efficient incorporation of C_{60} molecules between the porphyrins in the dendrimer.

Imahori and coworkers collaborated to exploit supramolecular photoelectrochemical devices using multifullerene dendrimers **15a–e** (number of fullerenes (n) = 1,2,4,8,16); Figure 13.5) by the similar clusterization and electrophoretic deposition (denoted as ITO/SnO_2/(**15**)$_m$ device) [36]. IPCE value of the ITO/SnO_2/(**15**)$_m$ devices

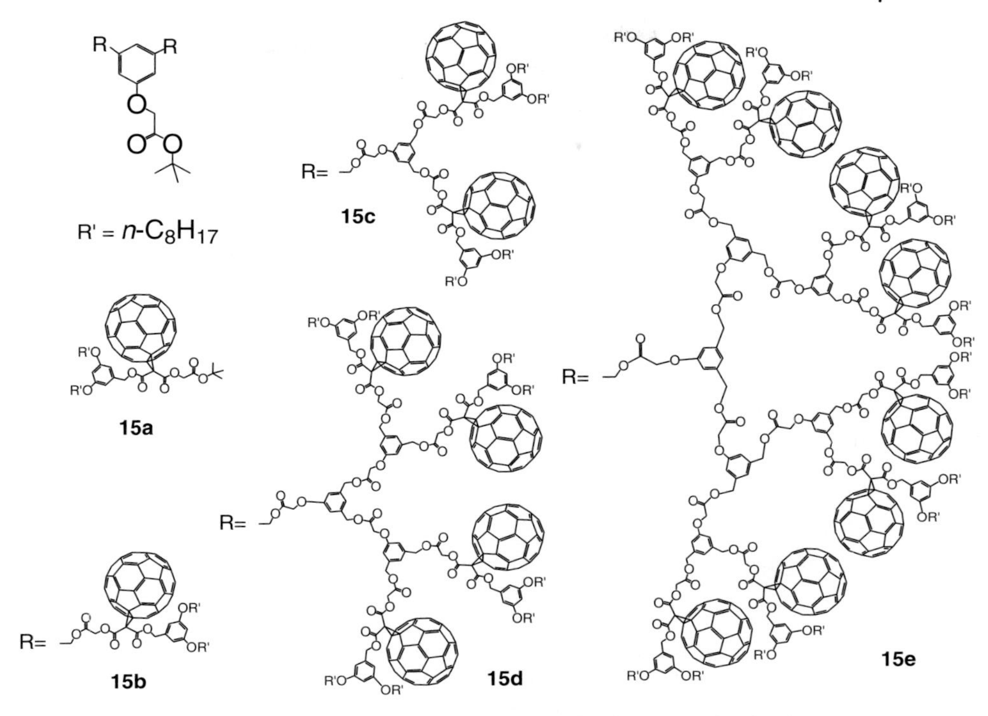

Figure 13.5 Molecular structures of multifullerene dendrimers **15** for supramolecular devices.

under the standard three electrode conditions was increased with increasing the generation number. A maximum IPCE value of 6% in the $SnO_2/(\mathbf{15e})_m$ device was achieved. The structural investigation on the fullerene dendrimers revealed that the higher dendrimer generation led to the formation of densely packed dendrimer clusters with a smaller, compact size. Such structures of fullerene dendrimer clusters on the SnO_2 electrode in the higher generation would make it possible to accelerate the electron injection process from the reduced C_{60} to the CB of the SnO_2 electrode via the more efficient electron hopping through the C_{60} moieties where the average distance between the C_{60} moieties is smaller [36].

Fukuzumi and coworkers presented supramolecular solar cells using intermolecular complexes of porphyrin-peptide oligomers with C_{60} [37]. The SnO_2 electrodes modified with the composite clusters of the porphyrin oligomers with C_{60} were prepared. An η value of 1.6% ($J_{SC} = 0.36\,mA/cm^2$, $V_{OC} = 0.32\,V$, ff $= 0.47$, $W_{IN} = 3.4\,mW/cm^2$) and an IPCE value of 48% at 600 nm were attained in the device with the porphyrin hexadecamer [37].

Imahori *et al.* also focused on nanoparticles as a nanoscaffold [33, 38, 39]. For instance, porphyrin-alkanethiols **16** (number of the methylene groups in the alkanethiol (n) $= 5,11,15$) are three-dimensionally organized on a gold nanoparticle with a diameter of $\sim$2 nm to give multiporphyrin-modified gold nanoparticles **17** ($n = 5,11,15$) with well-defined size ($\sim$10 nm) and spherical shape (step 1,

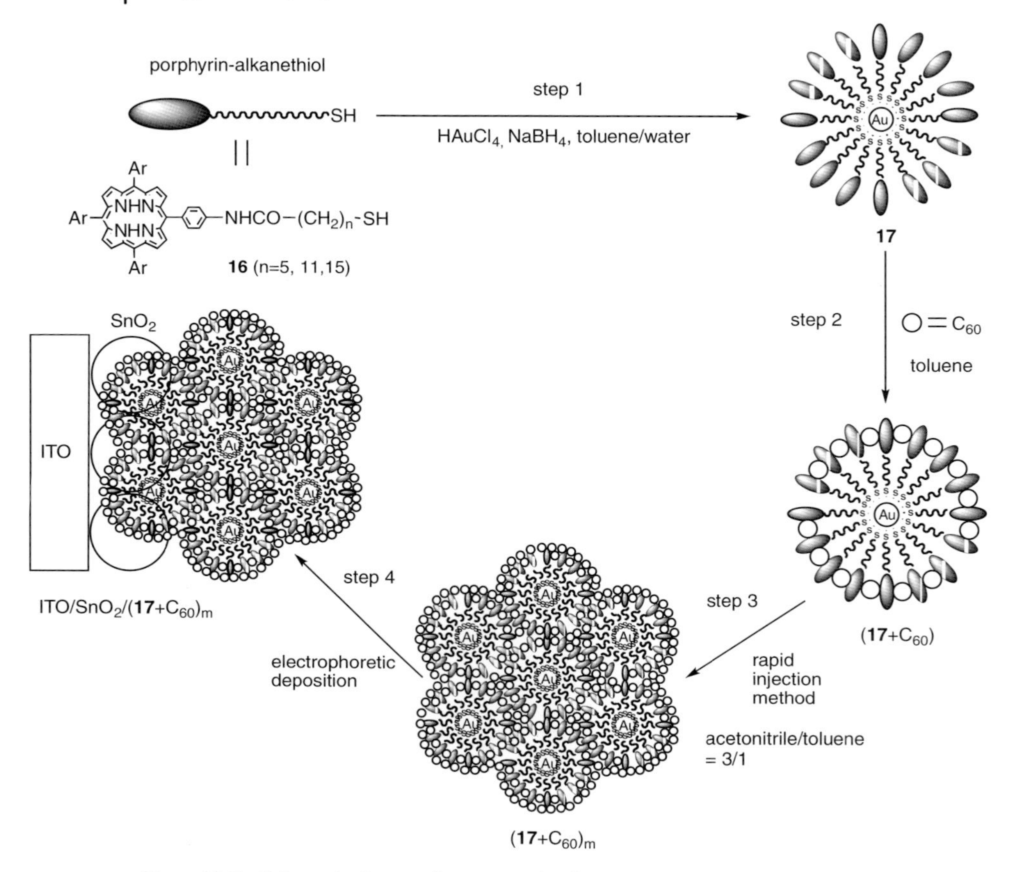

Figure 13.6 Schematic diagram for supramolecular organization of porphyrin alkanethiol **16** with gold nanoparticles to yield porphyrin-modified gold nanoparticles **17** that further assemble with C_{60} on a nanostructured ITO/SnO_2 electrode.

Figure 13.6) [33]. These nanoparticles bear flexible host space between the porphyrins for C_{60}. Therefore, the nanoparticles **17** can be grown into larger clusters $(\mathbf{17} + C_{60})_m$ with a size of ~100 nm in the mixed solvent by associating the nanoparticles with incorporating C_{60} molecules between the porphyrin moieties (steps 2,3). Finally, the large clusters can be deposited electrophoretically on a SnO_2 electrode (step 4) (denoted as ITO/SnO_2/$(\mathbf{17} + C_{60})_m$). IPCE value of the ITO/SnO_2/$(\mathbf{17} + C_{60})_m$ device (up to 54%; $n = 15$) was increased with increasing the chain length ($n = 5,11,15$) between the porphyrin and the gold nanoparticle [33]. The long methylene spacer of **17** allowed suitable space for C_{60} molecules to accommodate them between the neighboring porphyrin rings effectively compared to the nanoparticles with the short methylene spacer, leading to an efficient photocurrent generation. The SnO_2/$(\mathbf{17} + C_{60})_m$ device ($n = 15$) had a $J_{SC} = 1.0\,mA/cm^2$, a $V_{OC} = 0.38\,V$, a ff $= 0.43$, and an $\eta = 1.5\,\%$ ($W_{IN} = 11.2\,mW/cm^2$). The J–V characteristics of the SnO_2/$(\mathbf{29c} + C_{60})_m$

device was also remarkably enhanced by a factor of 45 in comparison with the SnO_2/ (**14f** + C_{60})$_m$ device [33]. These results evidently illustrate that the large improvement in the photoelectrochemical properties arises from three-dimensional interpenetrating network of the porphyrin-C_{60} molecules on the SnO_2 electrode, which facilitates the injection of the separated electrons into the CB.

13.2.1.4 **Solution-Processed Bulk Heterojunction Solar Cells**

For typical solution-processed bulk heterojunction solar cells involving blend films of conjugated polymers and fullerene derivatives, spin-coating method has been used for the fabrication on electrode surfaces [12–14]. To form interpenetrating donor fullerene network in the blend films for efficient photocurrent generation, it is essential to achieve the high solubility of conjugated polymers and fullerene derivatives in spin-coating solvents as well as the percolation of fullerene derivatives between conjugated polymers. Thus, **18** ([60]PCBM) was developed to associate with conjugated polymers such as poly(1,4-phenylenevinylene) (PPV) and poly(3-hexylthiophene) (P3HT) for bulk heterojunction solar cells (Figure 13.7) [40]. The improvement in the light-harvesting property was demonstrated by Janssen and

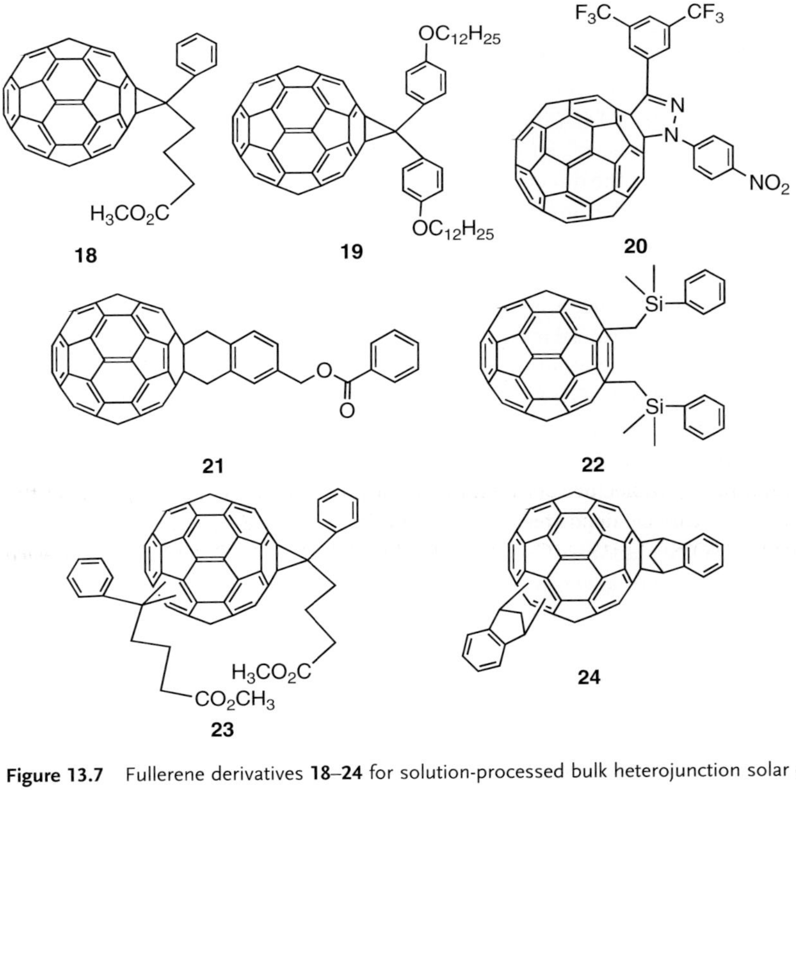

Figure 13.7 Fullerene derivatives **18–24** for solution-processed bulk heterojunction solar cells.

coworkers with the replacement of [60]PCBM by C_{70} analogue [70]PCBM [41]. In recent years, low-bandgap conjugated polymers with low HOMO level have been employed both with [60]PCBM and with [70]PCBM to achieve a high power conversion efficiency of 5–8% in bulk heterojunction solar cells [12–14, 42, 43]. Other fullerene derivatives including **19** (DPM-12) [44], **20** (60BTPF) [45], 70BTPF [46], **21** [47], **22** (SIMEF) [48], and others [49] have been prepared, but [60]PCBM and [70]PCBM are still the best acceptors in terms of overall device performance. It is known that V_{OC} depends on the difference in the HOMO level of donor and the LUMO level of acceptor. Therefore, raising the LUMO level of fullerene derivatives is also a promising strategy for improving the device performance. [60]PCBM bisadduct **23** and trisadduct [50], [60]indene bisadduct **24** (ICBA) [51], and endohedral metallofullerene [52] were synthesized to combine with conjugated polymers for bulk heterojunction solar cells. Double-cable polymer-C_{60} derivatives [1, 2, 12–14] and donor-linked fullerene dyads [53] have been attempted to construct bulk heterojunction solar cells, but the power conversion efficiency remained low because of the difficulty in forming bicontinuous donor–acceptor network in the active layer.

13.2.1.5 Hydrogen Bonding Systems

Hydrogen bonding interaction has been utilized to organize donor and fullerene molecules on electrodes. Imahori *et al.* presented the first mixed films of porphyrin and fullerene with hydrogen bonding on an ITO/SnO_2 electrode to reveal efficient photocurrent generation [54]. The IPCE value was on the order of ITO/SnO_2/**25** + **27** (36%) > ITO/SnO_2/**25** + **28** (28%) > ITO/SnO_2/**25** (26%) > ITO/SnO_2/**26** + **27** (15%) > ITO/SnO_2/**26** + **28** (7%) > ITO/SnO_2/**26** (6%) devices. The IPCE values were increased significantly when **27** rather than **28** was employed. This suggests that not only the direct electron injection takes place but also a competitive electron transfer occurs from the porphyrin excited singlet state to C_{60}, followed by the electron injection from the reduced C_{60} to the CB of the SnO_2 surface. It is worth noting that the IPCE value was improved in the mixed system with hydrogen bonding compared to the reference systems. These results corroborate that photocurrent generation is much enhanced in hydrogen-bonded donor–acceptor system compared to the reference system without hydrogen bonding [54].

Taking into account the fact that the V_{OC} value of dye-sensitized solar cells depends on the difference in the CB edge energies (E_{CB}) of semiconducting electrodes and the redox potential of I^-/I_3^- couple, replacement of the SnO_2 electrode by other semiconducting electrodes (TiO_2, ZnO, etc.) bearing more negative E_{CB} values is an appealing approach to improve the cell performance [55, 56]. For example, the energy level of $C_{60}^{\bullet-}$ (−0.2 V versus NHE) is higher than the CB of SnO_2 electrode (0 V versus NHE), but lower by 0.3 V than the CB of TiO_2 electrode (−0.5 V versus NHE). Thus, no pristine C_{60} can be applied to the solar cell with the TiO_2 electrode instead of the SnO_2 electrode. To realize stepwise electron injection from the porphyrin excited singlet state (−1.0 V versus NHE) to the CB of TiO_2 electrode via C_{60}, we focused on porphyrin carboxylic acid **29** and porphyrin reference **30** and C_{60} hexakisadduct with malonic acid **31** and the C_{60} reference **32** (Figure 13.8) [55]. The hexakisadduct of C_{60} malonic acids **31** was chosen by reason of the suitable

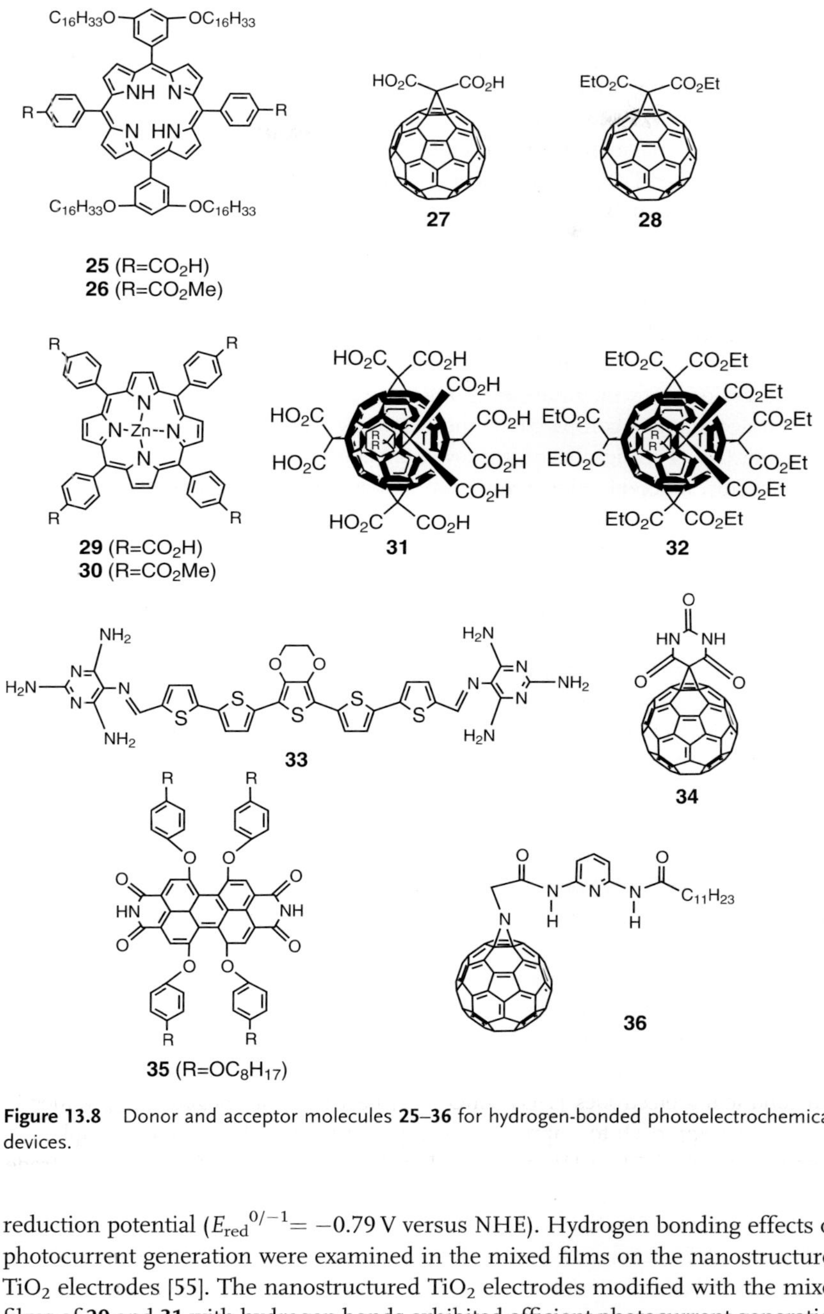

Figure 13.8 Donor and acceptor molecules **25–36** for hydrogen-bonded photoelectrochemical devices.

reduction potential ($E_{red}{}^{0/-1}= -0.79$ V versus NHE). Hydrogen bonding effects on photocurrent generation were examined in the mixed films on the nanostructured TiO_2 electrodes [55]. The nanostructured TiO_2 electrodes modified with the mixed films of **29** and **31** with hydrogen bonds exhibited efficient photocurrent generation (IPCE value of up to 47%) compared to the reference systems without hydrogen bonds. To exemplify an increase in V_{OC} in the present system, we examined the photovoltaic properties of the TiO_2 and SnO_2 electrodes modified with **29** and **31** composite by the spin-coating method: $\eta = 2.1\%$, $J_{SC} = 5.1\ mA/cm^2$, $V_{OC} = 0.58\ V$, and ff = 0.70 for the TiO_2 cell; $\eta = 0.31\%$, $J_{SC} = 2.3\ mA/cm^2$, $V_{OC} = 0.36\ V$, and ff = 0.39 for the SnO_2 cell [55]. These results explicitly exemplify that hydrogen bonding is a potential methodology for the fabrication of donor–acceptor composites on a nanostructured TiO_2 electrode, which exhibits high V_{OC} in comparison with that of the corresponding SnO_2 electrode.

ZnO nanorod electrodes have been applied to solar cells possessing both characteristics of bulk heterojunction and dye-sensitized devices [56]. Namely, first, the ZnO electrodes were covered with **27** to yield the monolayer on the electrodes. Then, **14a** (M = Zn) and **27** were spin coated on the modified surfaces to give the porphyrin–fullerene-modified ZnO electrodes. The porphyrin–fullerene-modified ZnO nanorod devices with the intervening fullerene monolayer exhibited efficient photocurrent generation compared to the reference systems without the fullerene monolayer [56]. The significant improvement of the photocurrent generation efficiency by the fullerene monolayer may be associated with the efficient charge separation in the porphyrin–fullerene composite layer, followed by electron injection to the CB of the ZnO nanorod electrode together with the suppression of charge recombination between the separated charges by the fullerene monolayer.

Bassani and coworkers reported supramolecular photovoltaic devices comprising oligothiophene **33** and fullerene derivative **34** (Figure 13.8) [57]. Codeposition from a solution of symmetric melamine-terminated electron donor oligomers **33** with a complementary barbiturate-labeled electron acceptor fullerene **34** resulted in homogeneous films. Incorporation into photoelectrochemical devices gave a 2.5-fold enhancement in photocurrent generation efficiency compared to the reference device with the nonhydrogen bonding parent C_{60}, although the maximum APCE and IPCE values of the photocurrent generation were moderate (25 and 9.3%, respectively).

Li and coworkers developed a novel hydrogen-bonded supramolecular device of perylene bisimide **35** and fullerene derivative **36** (Figure 13.8) [58]. The poor photocurrent response of the mixed film of **35** and **36** (2 : 1) on an ITO electrode was noted. In addition, the photocurrent generation mechanism was ambiguous.

13.2.1.6 **Coordination Bonding Systems**

Coordination bonding of fullerene derivatives to donor oligomers and conjugated polymers has been applied to solar cells and photoelectrochemical devices [59–61]. Although nanostructures with bicontinuous donor–acceptor arrays have been obtained by self-assembly of donor–acceptor molecules, it is still difficult to achieve desirable vertical arrangement of bicontinuous donor–acceptor arrays on an electrode. Imahori and coworkers reported a novel approach to construct vertical alignment of bicontinuous donor–acceptor arrays on a flat SnO_2 electrode for photoelectrochemical device (Figure 13.9) [60]. A palladium-mediated stepwise self-assembly of zinc porphyrins (ZnP) as a donor ensured the vertical growth of porphyrin chains on the SnO_2 electrode. Pyridylfullerenes (Py-C_{60}) as an acceptor were infiltrated into the porphyrin brush by using the coordination bonding of the

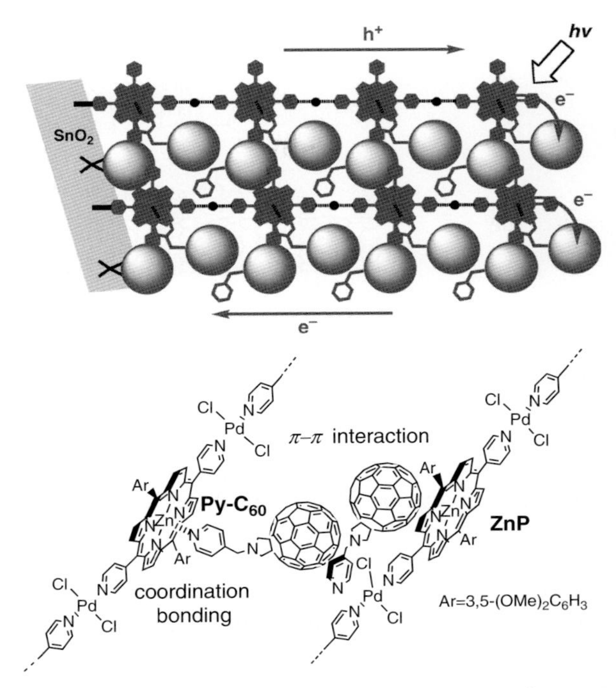

Figure 13.9 Bicontinuous donor–acceptor arrays vertically grown on a SnO_2 electrode for photoelectrochemical devices.

pyridyl moiety to the zinc atom together with π–π interaction between Py-C_{60}. Therefore, it was possible to systematically investigate the relationship between the film structure and the photoelectrochemical properties as a function of the number of porphyrin layer. A maximum IPCE value of 21% is comparable to the highest value (20%) among vertical arrangement of bicontinuous donor–acceptor arrays on electrodes [60]. These results will provide fundamental clue for the molecular design of high-performance organic photovoltaics.

13.2.2 Carbon Nanotubes

Carbon nanotubes (CNTs) are mechanically strong, high modulus graphitic fibers with a diameter of 1–40 nm and a length of micrometers [1, 2, 62, 63]. CNTs are considered to be highly promising for miniaturizing electronics beyond the scale used in electronics at present. Especially, one-dimensional (1D), nanowire-like structure and unique electronic properties of CNTs provide the potential to construct ideal nanohighways for charge carrier on electrodes. Thus, CNTs including single-walled carbon nanotubes (SWNTs) have been used as electron or hole transporting materials in photovoltaic and photoelectrochemical devices [62, 63]. However, the

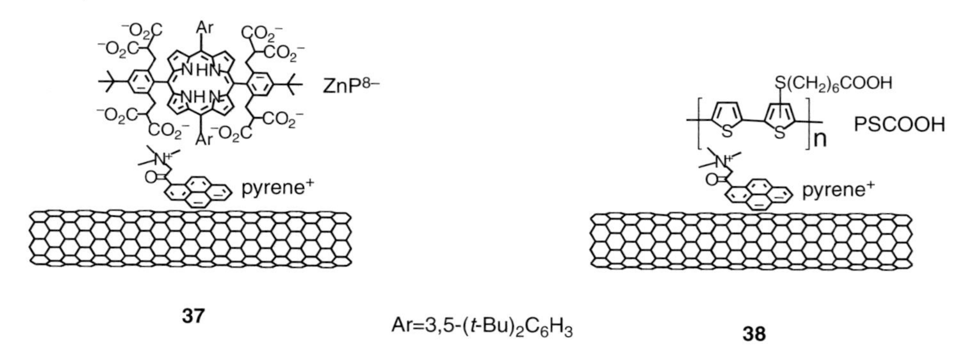

Figure 13.10 Donor SWNT-stacked composites **37** and **38** by van der Waals and electrostatic interactions.

device performance of bulk heterojunction solar cells consisting of CNT with conjugated polymers is much lower than that of the corresponding cells consisting of PCBM instead of CNT [64]. Nevertheless, versatile approaches have been attempted to explore the potential ability as electron or hole transporting materials.

Guldi and Sgobba have extensively examined the use of electrostatic and van der Waals interactions to fabricate photoactive ITO electrodes deposited with CNTs with polymers or small molecules [63]. Namely, the films are formed by depositing alternating layers of oppositely charged materials. For instance, CNT/pyrene$^+$ and negatively charged porphyrin (ZnP^{8-}) **37** (Figure 13.10) were assembled electrostatically on the ITO electrodes [64, 65]. An IPCE value of 10.7% at 425 nm was achieved for the single SWNT/pyrene$^+$/ZnP^{8-} stack device. Similarly, anionic polythiophene (PSCOOH) and SWNT/pyrene$^+$ **38** (Figure 13.10) were integrated together into photoactive ITO electrodes by van der Waals and electrostatic interactions [66]. An IPCE value of 9.3% was attained for the eight-layer SWNT/pyrene$^+$/PSCOOH stack device.

Guldi and coworkers reported the covalent functionalization of SWNTs with 4-(2-trimethylsilyl)ethynylaniline and the subsequent attachment of a zinc phthalocyanine derivative using the reliable Huisgen 1,3-dipolar cycloaddition to yield **39** (Figure 13.11) [67]. An IPCE value of 17.3% was obtained when the composite **39** was fabricated on an ITO electrode.

Imahori and coworkers prepared SWNTs covalently functionalized with alkyl chains or bulky porphyrin units to different degrees [68]. The films of the modified SWNTs were fabricated on ITO/SnO_2 electrodes by electrophoretic deposition. A maximum IPCE value of 4.9% at 400 nm was noted for the porphyrin-linked SWNT **40** (Figure 13.11). It is worth noting that no absorption due to the porphyrin moieties contributed to the photocurrent generation in the photoelectrochemical devices.

Imahori and coworkers developed a novel methodology for the self-organization of fullerene molecules on the sidewall of SWNTs, as illustrated in Figure 13.12 [69]. First, acid treatment cuts pristine SWNT (denoted as p-SWNT) to yield shortened SWNT (denoted as s-SWNT) with carboxylic groups at the open ends and defect sites (step 1). Then, s-SWNT is functionalized with sterically hindered shallow-tailed secondary amine to yield soluble, functionalized SWNT (denoted as f-SWNT)

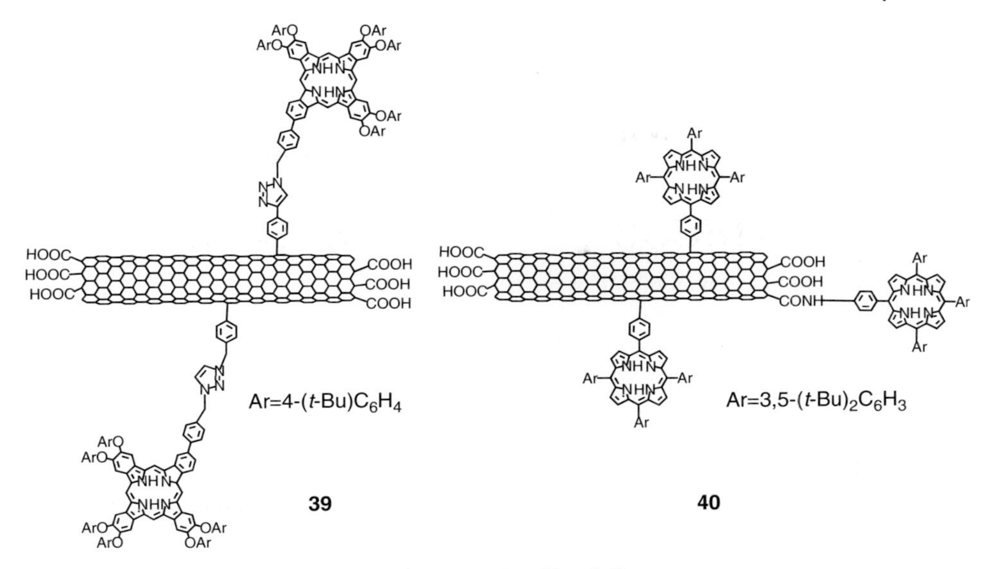

Figure 13.11 Donor-linked carbon nanotube composites **39** and **40**.

in organic solvents (step 2). Finally, poor solvent (i.e., acetonitrile) is rapidly injected into a mixture of C_{60} and f-SWNT dissolved in good solvent (i.e., ODCB), resulting in the formation of the composite clusters of C_{60} and f-SWNT (denoted as $(C_{60} + \text{f-SWNT})_m$, step 3). The SnO_2 electrode modified electrophoretically with the

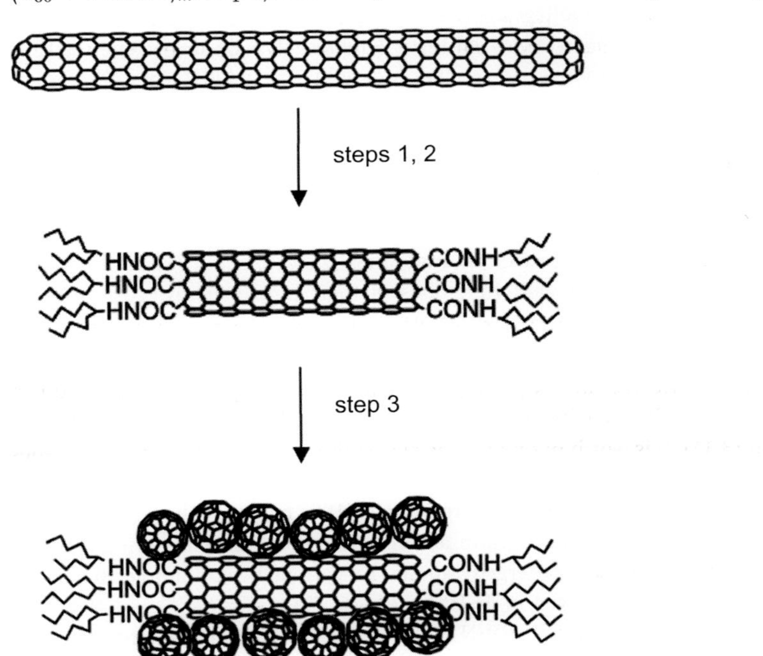

Figure 13.12 Schematic organization of fullerene–carbon nanotube composites.

$(C_{60} + \text{f-SWNT})_m$ exhibited an IPCE value as high as 18% at 400 nm under an applied potential of 0.05 V versus SCE [69]. More importantly, the higher fullerenes (i.e., C_{70} and C_{84}) were found to form single composite clusters exclusively with f-SWNT [69]. These are in marked contrast with the unselective formation of three different clusters in the C_{60}–f-SWNT composites. The C_{70}–f-SWNT photoelectrochemical device exhibited the higher IPCE value (26% at 400 nm) than the C_{60}–f-SWNT device (18%). The higher IPCE value resulted from selective formation of the single composite film, in which the SWNT network was covered with C_{70} molecules, and the high electron mobility (2.4 $cm^2/(V\,s)$) through the C_{70}–SWNT network [69]. Thus, the results obtained here will provide valuable information on the design of molecular devices based on CNTs and fullerenes.

13.3 Transistors

13.3.1 Fullerenes

Conventional silicon-based metal-oxide semiconductor (MOS) field-effect transistor (FET) consists of p–n junctions placed immediately adjacent to the region of the semiconductor controlled by the MOS gate. The carriers enter the structure through the source (S), leave through the drain (D), and are subject to the control by the bias voltage on the gate (G). The voltage applied to the gate is V_G, whereas the drain voltage is V_D. As the gate voltage V_G is increased $V_G > V_T$, where V_T is the depletion inversion transition-point voltage, an inversion containing mobile carrier is formed adjacent to the Si surface, creating a source-to-drain channel. By keeping V_G fixed and varying V_D, the current–voltage characteristic I_D versus V_D of the transistor can be determined for various gate voltages.

Versatile fullerene-based n-channel FETs have been constructed (Figure 13.13) [70–72]. Typically, fullerene layers are formed by vacuum deposition. Fullerenes have exhibited one of the highest electron mobilities up to 6 $cm^2/(V\,s)$ as an n-type material. A major problem with fullerenes is their high sensitivity to ambient conditions, especially oxygen and moisture. Ambipolar FETs consisting of C_{60} derivative and conjugated polymer were also developed [73].

13.3.2 Carbon Nanotubes

The first CNT FETs were independently reported in 1998 by the Dekker group at Delft University [74] and the Avouris group at IBM [75]. After the pioneering work, a number of CNT FETs have emerged rapidly. In particular, supramolecular CNT ensembles have been employed in FET devices [76, 77].

Stoddart and Zhao fabricated SWNT FET devices to examine charge transfer within various donor–acceptor SWNT composites [76]. For instance, a SWNT FET

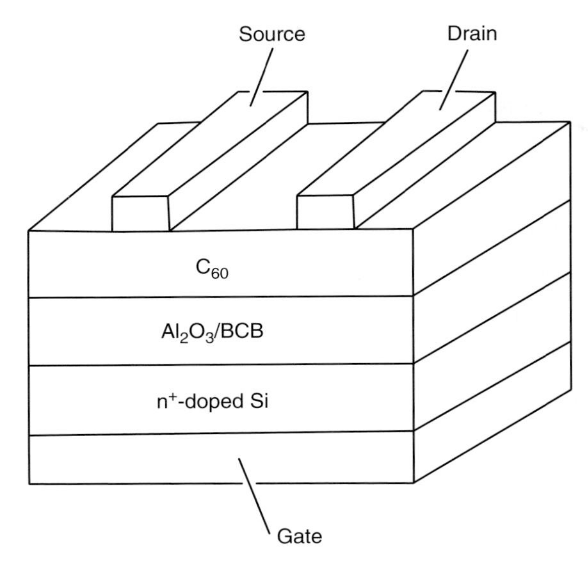

Figure 13.13 Schematics of C_{60} FET with Al_2O_3/BCB (divinyltetramethyldisiloxane bis (benzocyclobutene) dielectric and Al or Ca source/drain contacts [72].

device, functionalized noncovalently with a zinc porphyrin, was utilized to probe photoinduced electron transfer from the SWNT to the porphyrin in the composite **41** (Figure 13.14) [78]. Actually, the response of the device to the light was a shift of the threshold voltage toward positive voltages, suggesting hole doping of the SWNT.

The authors prepared CdSe–SWNT composites **42** (Figure 13.14) by self-assembling the pyrene-linked CdSe nanoparticles (pyrene-CdSe) onto the surfaces of the SWNTs [79]. CdSe-SWNT FET devices have been fabricated using the composites to address the charge transfer from the CdSe nanoparticles to the SWNT. The magnitude of the charge transfer disclosed a strong dependence on the light intensity and wavelength and reached a maximum of 2.2 electrons per pyrene/CdSe nanoparticle.

Stoddart and coworkers have integrated pyrenecyclodextrin-decorated SWNT **43** (Figure 13.14) into FET devices to detect particular organic molecules based on the concept of molecular recognition. In the presence of certain molecules, some of them were incorporated into the cavities of cyclodextrins with moderate binding constants, whereas the others did not have appreciable interactions with the SWNTs [80]. As a result, the transistor characteristics of the pyrenecyclodextrin-decorated SWNT FET device shifted toward negative gate voltage. They used the pyrenecyclodextrin-decorated SWNT FET device as a tunable photosensor in aqueous solution to detect a luminescent ruthenium complex [81].

Eriksson and coworkers have demonstrated the ability to covalently functionalize individual nanotube transistors using an azobenzene-linked anthracene–SWNT

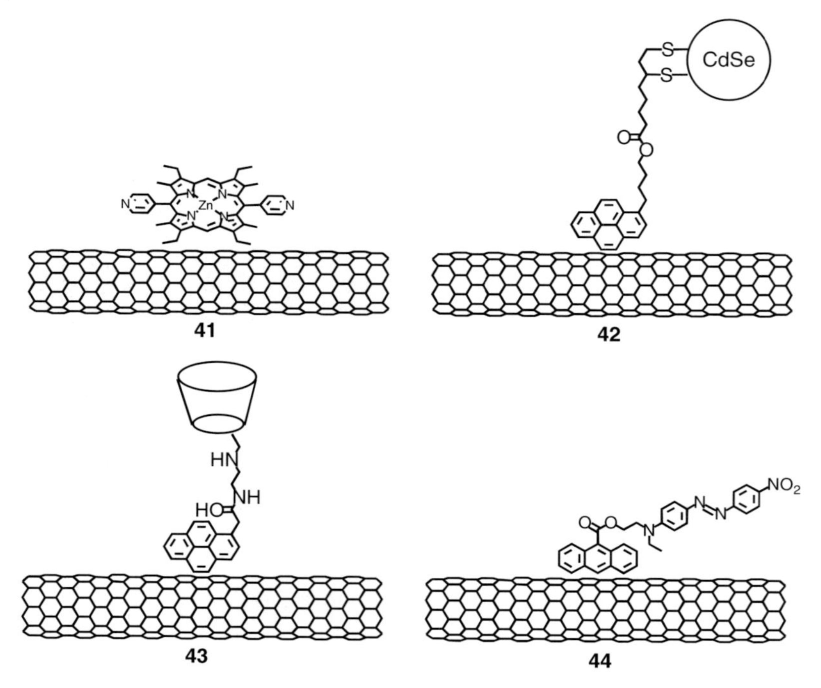

Figure 13.14 Noncovalent functionalization of SWNTs for FETs.

composite **44** (Figure 13.14) [82]. When it photoisomerized, the resulting change in dipole moment modified the local electrostatic potential and modulated the transistor conductance by shifting the threshold voltage.

13.4 Summary

In this chapter, we highlighted some of the recent advances in supramolecular ensembles with fullerenes and CNTs for solar cells and transistors. Fullerenes and their derivatives have been ubiquitously employed as excellent electron accepting and transporting components both in bulk heterojunction solar cells and in transistors. This can be rationalized by small reorganization energies of fullerenes in electron transfer, which allows us to achieve fast charge separation and electron transportation as well as slow charge recombination, compared to conventional acceptors. This is the reason why we have used fullerenes and their derivatives in organic electronics. The methodologies for noncovalent functionalization of SWNTs promise a next generation of SWNT-based integrated multifunctional sensors and devices. The search for novel and better nanocarbon materials including graphenes in addition to the rational design should be continued to develop elaborated supramolecular ensembles with fullerenes and CNTs for solar cells and transistors.

References

1 Langa, F. and Nierengarten, J-.F. (eds) (2007) *Fullerenes*, RSC Publishing, Cambridge.

2 Akasaka, T., Wudl, F., and Nagase, S. (eds) (2010) *Chemistry of Nanocarbons*, John Wiley & Sons Ltd., Chichester.

3 Imahori, H. and Sakata, Y. (1997) Donor-linked fullerenes: photoinduced electron transfer and its potential application. *Adv. Mater.*, **9**, 537–546.

4 Martín, N., Sánchez, L., Illescas, B., and Pérez, I. (1998) C_{60}-based electroactive organofullerenes. *Chem. Rev.*, **98**, 2527–2547.

5 Gust, D., Moore, T.A., and Moore, A.L. (2001) Mimicking photosynthetic solar energy transduction. *Acc. Chem. Res.*, **34**, 40–48.

6 Guldi, D.M. (2002) Fullerene–porphyrin architectures: photosynthetic antenna and reaction center models. *Chem. Soc. Rev.*, **31**, 22–36.

7 Guldi, D.M. (2007) Nanometer scale carbon structures for charge-transfer systems and photovoltaic applications. *Phys. Chem. Chem. Phys.*, **9**, 1400–1420.

8 Imahori, H. (2007) Creation of fullerene-based artificial photosynthetic systems. *Bull. Chem. Soc. Jpn.*, **80**, 621–636.

9 Imahori, H., Hagiwara, K., Akiyama, T., Aoki, M., Taniguchi, S., Okada, T., Shirakawa, M., and Sakata, Y. (1996) The small reorganization energy of C_{60} in electron transfer. *Chem. Phys. Lett.*, **263**, 545–550.

10 Imahori, H., Yamada, H., Guldi, D.M., Endo, Y., Shimomura, A., Kundu, S., Yamada, K., Okada, T., Sakata, Y., and Fukuzumi, S. (2002) Comparison of reorganization energies for intra- and intermolecular electron transfer. *Angew. Chem. Int. Ed.*, **41**, 2344–2347.

11 Fukuzumi, S., Ohkubo, K., Imahori, H., and Guldi, D.M. (2003) Driving force dependence of intermolecular electron-transfer reactions of fullerenes. *Chem. Eur. J.*, **9**, 1585–1593.

12 Günes, S., Neugebauer, H., and Sariciftci, N.S. (2007) Conjugated polymer-based organic solar cells. *Chem. Rev.*, **107**, 1324–1338.

13 Thompson, B.C. and Fréchet, J.M.J. (2008) Polymer-fullerene composite solar cells. *Angew. Chem. Int. Ed.*, **47**, 58–77.

14 Cheng, Y.-J., Yang, S.-H., and Hsu, C.-S. (2009) Synthesis of conjugated polymers for organic solar cell applications. *Chem. Rev.*, **109**, 5868–5923.

15 Imahori, H. and Umeyama, T. (2009) Donor–acceptor nanoarchitecture on semiconducting electrodes for solar energy conversion. *J. Phys. Chem. C*, **113**, 9029–9039.

16 Imahori, H., Ozawa, S., Ushida, K., Takahashi, M., Azuma, T., Ajavakom, A., Akiyama, T., Hasegawa, M., Taniguchi, S., Okada, T., and Sakata, Y. (1999) Organic photoelectrochemical cell mimicking multistep electron transfer in photosynthesis: interfacial structure and photoelectrochemical properties of self-assembled monolayers of porphyrin-linked fullerenes on gold electrodes. *Bull. Chem. Soc. Jpn.*, **72**, 485–502.

17 Hirayama, D., Takimiya, K., Aso, Y., Otsubo, T., Hasobe, T., Yamada, H., Imahori, H., Fukuzumi, S., and Sakata, Y. (2002) Large photocurrent generation of gold electrodes modified with [60] fullerene-linked oligothiophenes bearing a tripodal rigid anchor. *J. Am. Chem. Soc.*, **124**, 532–533.

18 Imahori, H., Yamada, H., Nishimura, Y., Yamazaki, I., and Sakata, Y. (2000) Vectorial multistep electron transfer at the gold electrodes modified with self-assembled monolayers of ferrocene–porphyrin–fullerene triads. *J. Phys. Chem. B*, **104**, 2099–2108.

19 Imahori, H., Norieda, H., Yamada, H., Nishimura, Y., Yamazaki, I., Sakata, Y., and Fukuzumi, S.J. (2001) Light-harvesting and photocurrent generation by gold electrodes modified with mixed self-assembled monolayers of boron–dipyrrin and ferrocene–porphyrin–fullerene triad. *J. Am. Chem. Soc.*, **123**, 100–110.

20 Saha, S., Johansson, E., Flood, A.H., Tseng, H.-R., Zink, J.I., and Stoddart, J.F. (2005) Photoactive molecular triad as a

nanoscale power supply for a supramolecular machine. *Chem. Eur. J.*, **11**, 6846–6858.

21 Kim, K.-S., Kang, M.-S., Ma, H., and Jen, A.K.-Y. (2004) Highly efficient photocurrent generation from a self-assembled monolayer film of a novel C_{60}-tethered 2,5-dithienylpyrrole triad. *Chem. Mater.*, **16**, 5058–5062.

22 Yamada, H., Imahori, H., Nishimura, Y., Yamazaki, I., Ahn, T.K., Kim, S.K., Kim, D., and Fukuzumi, S. (2003) Photovoltaic properties of self-assembled monolayers of porphyrins and porphyrin–fullerene dyads on ITO and gold surfaces. *J. Am. Chem. Soc.*, **125**, 9129–9139.

23 Cho, Y.-J., Ahn, T.K., Song, H., Kim, K.S., Lee, C.Y., Seo, W.S., Lee, K., Kim, S.K., Kim, D., and Park, J.T. (2005) Unusually high performance photovoltaic cell based on a [60]fullerene metal cluster–porphyrin dyad SAM on an ITO electrode. *J. Am. Chem. Soc.*, **127**, 2380–2381.

24 Lee, C.Y., Jang, J.K., Kim, C.H., Jung, J., Park, B.K., Park, J., Choi, W., Han, Y.-K., Joo, T., and Park, J.T. (2010) Remarkably efficient photocurrent generation based on a [60]fullerene–triosmium cluster/Zn–porphyrin/boron–dipyrrin triad SAM. *Chem. Eur. J.*, **16**, 5586–5599.

25 Imahori, H., Kimura, M., Hosomizu, K., Sato, T., Ahn, T.K., Kim, S.K., Kim, D., Nishimura, Y., Yamazaki, I., Araki, Y., Ito, O., and Fukuzumi, S. (2004) Vectorial electron relay at ITO electrodes modified with self-assembled monolayers of ferrocene–porphyrin–fullerene triads and porphyrin–fullerene dyads for molecular photovoltaic devices. *Chem. Eur. J.*, **10**, 5111–5122.

26 Guldi, D.M., Pellarini, F., Prato, M., Granito, C., and Troisi, L. (2002) Layer-by-layer construction of nanostructured porphyrin–fullerene electrodes. *Nano Lett.*, **2**, 965–968.

27 Guldi, D.M., Zilbermann, I., Anderson, G.A., Li, A., Balbinot, D., Jux, N., Hatzimarinaki, M., Hirsch, A., and Prato, M. (2004) Multicomponent redox gradients on photoactive electrode surfaces. *Chem. Commun.*, 726–727.

28 Mwaura, J.K., Mauricio, R.P., Witker, D., Ananthakrishnan, N., Schanze, K.S., and Reynolds, J.R. (2005) Photovoltaic cells based on sequentially adsorbed multilayers of conjugated poly(*p*-phenylene ethynylene)s and a water-soluble fullerene derivative. *Langmuir*, **21**, 10119–10126.

29 Kamat, P.V., Barazzouk, S., Thomas, K.G., and Hotchandani, S. (2000) Electrodeposition of C_{60} cluster aggregates on nanostructured SnO_2 films for enhanced photocurrent generation. *J. Phys. Chem. B*, **104**, 4014–4017.

30 Kamat, P.V., Barazzouk, S., Hotchandani, S., and Thomas, K.G. (2000) Nanostructured thin films of C_{60}–aniline dyad clusters: electrodeposition, charge separation, and photochemistry. *Chem. Eur. J.*, **6**, 3914–3921.

31 Imahori, H., Hasobe, T., Yamada, H., Kamat, P.V., Barazzouk, S., Fujitsuka, M., Ito, O., and Fukuzumi, S. (2001) Spectroscopy and photocurrent generation in nanostructured thin films of porphyrin–fullerene dyad clusters. *Chem. Lett.*, 784–785.

32 Hasobe, T., Imahori, H., Fukuzumi, S., and Kamat, P.V. (2003) Light energy conversion using mixed molecular nanoclusters. Porphyrin and C_{60} cluster films of efficient photocurrent generation. *J. Phys. Chem. B*, **107**, 12105–12112.

33 Hasobe, T., Imahori, H., Kamat, P.V., Ahn, T.K., Kim, S.K., Kim, D., Fujimoto, A., Hirakawa, T., and Fukuzumi, S. (2005) Photovoltaic cells using composite nanoclusters of porphyrins and fullerenes with gold nanoparticles. *J. Am. Chem. Soc.*, **127**, 1216–1228.

34 Imahori, H., Ueda, M., Kang, S., Hayashi, H., Hayashi, S., Kaji, H., Seki, S., Saeki, A., Tagawa, S., Umeyama, T., Matano, Y., Yoshida, K., Isoda, S., Shiro, M., Tkachenko, N.V., and Lemmetyinen, H. (2007) Effects of porphyrin substituents on film structure and photoelectrochemical properties of porphyrin/fullerene composite clusters electrophoretically deposited on nanostructured SnO_2 electrodes. *Chem. Eur. J.*, **13**, 10182–10193.

35 Hasobe, T., Kamat, P.V., Absalom, M.A., Kashiwagi, Y., Sly, J., Crossley, M.J.,

Hosomizu, K., Imahori, H., and Fukuzumi, S. (2004) Supramolecular photovoltaic cells based on composite molecular nanoclusters: dendritic porphyrin and C_{60}, porphyrin dimer and C_{60}, and porphyrin–C_{60} dyad. *J. Phys. Chem. B*, **118**, 12865–12872.

36 Hosomizu, K., Imahori, H., Hahn, U., Nierengarten, J.-F., Listorti, A., Armaroli, N., Nemoto, T., and Isoda, S. (2007) Dendritic effects on structure and photophysical and photoelectrochemical properties of fullerene dendrimers and their nanoclusters. *J. Phys. Chem. C*, **111**, 2777–2786.

37 Hasobe, T., Saito, K., Kamat, P.V., Troaiani, V., Qiu, H., Solladié, N., Kim, K.S., Park, J.K., Kim, D., D'Souza, F., and Fukuzumi, S. (2007) Organic solar cells. Supramolecular composites of porphyrins and fullerenes organized by polypeptide structures as light harvesters. *J. Mater. Chem.*, **17**, 4160–4170.

38 Imahori, H., Fujimoto, A., Kang, S., Hotta, H., Yoshida, K., Umeyama, T., Matano, Y., Isoda, S., Isosomppi, M., Tkachenko, N.V., and Lemmetyinen, H. (2005) Host–guest interactions in the supramolecular incorporation of fullerenes into tailored holes on porphyrin-modified gold nanoparticles in molecular photovoltaics. *Chem. Eur. J.*, **11**, 7265–7275.

39 Imahori, H., Mitamura, K., Shibano, Y., Umeyama, T., Matano, Y., Yoshida, K., Isoda, S., Araki, Y., and Ito, O. (2006) A photoelectrochemical device with a nanostructured SnO_2 electrode modified with composite clusters of porphyrin-modified silica nanoparticle and fullerene. *J. Phys. Chem. B*, **110**, 11399–11405.

40 Yu, G., Gao, J., Hummelen, J.C., Wudl, F., and Heeger, A.J. (1995) Polymer photovoltaic cells: enhanced efficiencies via a network of internal donor–acceptor heterojunctions. *Science*, **270**, 1789–1791.

41 Wienk, M.M., Kroon, J.M., Verhees, W.J.H., Knol, J., Hummelen, J.C., van Hal, P.A., and Janssen, R.A.J. (2003) Efficient methano[70]fullerene/MDMO-PPV bulk heterojunction photovoltaic cells. *Angew. Chem. Ind. Ed.*, **42**, 3371–3375.

42 Kim, J.Y., Lee, K., Coates, N.E., Moses, D., Nguyen, T.-Q., Dante, M., and Heeger, A.J. (2007) Efficient tandem polymer solar cells fabricated by all-solution processing. *Science*, **317**, 222–225.

43 Liang, Y.Y., Xu, Z., Xia, J.B., Tsai, S.T., Wu, Y., Li, G., Ray, C., and Yu, L.P. (2010) For the bright future-bulk heterojunction polymer solar cell with power conversion efficiency of 7.4%. *Adv. Mater.*, **22**, E135–E138.

44 Riedel, I., von Hauff, E., Parisi, J., Martín, N., Giacalone, F., and Dyakonov, V. (2005) Diphenylmethanofullerenes: new and efficient acceptors in bulk-heterojunction solar cells. *Adv. Funct. Mater.*, **15**, 1979–1987.

45 Wang, X., Perzon, E., Delgado, J.L., de la Cruz, P., Zhang, F., Langa, F., Andersson, M., and Inganäs, O. (2004) Infrared photocurrent spectral response from plastic solar cell with low-band-gap polyfluorene and fullerene derivative. *Appl. Phys. Lett.*, **85**, 5081–5083.

46 Wang, X., Perzon, E., Oswald, F., Langa, F., Admassie, S., Andersson, M.R., and Inganäs, O. (2005) Enhanced photocurrent spectral response in low-bandgap polyfluorene and C_{70}-derivative-based solar cells. *Adv. Funct. Mater.*, **15**, 1665–1670.

47 Backer, S.A., Sivula, K., Kavulak, D.F., and Fréchet, J.M.J. (2007) High efficiency organic photovoltaics incorporating a new family of soluble fullerene derivatives. *Chem. Mater.*, **19**, 2927–2929.

48 Matsuo, Y., Sato, Y., Niinomi, T., Soga, I., Tanaka, H., and Nakamura, E. (2009) Columnar structure in bulk heterojunction in solution-processable three-layered p–i–n organic photovoltaic devices using tetrabenzoporphyrin precursor and silylmethyl[60]fullerene. *J. Am. Chem. Soc.*, **131**, 16048–16050.

49 Troshin, P.A., Hoppe, H., Renz, J., Egginger, M., Mayorova, J.Y., Goryachev, A.E., Peregudov, A.S., Lyubovskaya, R.N., Gobsch, G., Sariciftci, N.S., and Razumov, V.F. (2009) Material solubility–photovoltaic performance relationship in the design of novel fullerene derivatives for bulk heterojunction solar cells. *Adv. Funct. Mater.*, **19**, 779–788.

50 Lens, M., Wetzelaer, G.-J.A.H., Kooistra, F.B., Veenstra, S.C., Hummelen, J.C., and Blom, P.W.M. (2008) Fullerene bisadducts for enhanced open-circuit voltages and efficiencies in polymer solar cells. *Adv. Mater.*, **20**, 2116–2229.

51 He, Y., Chen, H.-Y., Hou, J., and Li, Y. (2010) Indene-C_{60} bisadduct: a new acceptor for high-performance polymer solar cells. *J. Am. Chem. Soc.*, **132**, 1377–1382.

52 Ross, R.B., Cardona, C.M., Swain, F.B., Guldi, D.M., Sankaranarayanan, S.G., Keuren, E.V., Holloway, B.C., and Drees, M. (2009) Tuning conversion efficiency in metallo endohedral fullerene-based organic photovoltaic devices. *Adv. Funct. Mater.*, **19**, 2332–2337.

53 Nishizawa, T., Lim, H.K., Tajima, K., and Hashimoto, K. (2009) Efficient dyad-based organic solar cells with a highly crystalline donor group. *Chem. Commun.*, 2469–2471.

54 Imahori, H., Liu, J.-C., Hotta, H., Kira, A., Umeyama, T., Matano, Y., Li, G., Ye, S., Isosomppi, M., Tkachenko, N.V., and Lemmetyinen, H. (2005) Hydrogen bonding effects on the surface structure and photoelectrochemical properties of nanostructured SnO_2 electrodes modified with porphyrin and fullerene composites. *J. Phys. Chem. B*, **109**, 18465–18474.

55 Kira, A., Tanaka, M., Umeyama, T., Matano, Y., Li, G., Ye, S., Isosomppi, M., Tkachenko, N.V., Lemmetyinen, H., and Imahori, H. (2007) Hydrogen bonding effects on film structure and photoelectrochemical properties of nanostructured TiO_2 electrodes modified with porphyrin and fullerene composites. *J. Phys. Chem. C*, **111**, 13618–13626.

56 Hayashi, H., Kira, A., Umeyama, T., Matano, Y., Charoensirithavorn, P., Sagawa, T., Yoshikawa, S., Tkachenko, N.V., Lemmetyinen, H., and Imahori, H. (2009) Effects of electrode structures on photoelectrochemical properties of ZnO electrodes modified with porphyrin–fullerene composite layers with intervening fullerene monolayer. *J. Phys. Chem. C*, **113**, 10798–10806.

57 Huang, C.-H., McCenaghan, N.D., Kuhn, A., Hofstraat, J.W., and Bassani, D.M. (2005) Enhanced photovoltaic response in hydrogen-bonded all-organic devices. *Org. Lett.*, **7**, 3409–3412.

58 Liu, Y., Xiao, S., Li, H., Li, Y., Liu, H., Lu, F., Zhuang, J., and Zhu, D. (2004) Self-assembly and characterization of a novel hydrogen-bonded nanostructure. *J. Phys. Chem. B*, **108**, 6256–6260.

59 Troshin, P.A., Koeppe, R., Peregudov, A.S., Peregusova, S.M., Egginger, M., Lyubovskaya, R.N., and Sariciftci, N.S. (2007) Supramolecular association of pyrrolidinofullerenes bearing chelating pyridyl groups and zinc phthalocyanine for organic solar cells. *Chem. Mater.*, **19**, 5363–5372.

60 Kira, A., Umeyama, T., Matano, Y., Yoshida, K., Isoda, S., Park, J.-K., Kim, D., and Imahori, H. (2009) Supramolecular donor–acceptor heterojunctions by vectorial stepwise assembly of porphyrins and coordination-bonded fullerene arrays for photocurrent generation. *J. Am. Chem. Soc.*, **131**, 3198–3200.

61 Subbaiyan, N.K., Wijesinghe, C.A., and D'Souza, F. (2009) Supramolecular solar cells: surface modification of nanocrystalline TiO_2 with coordinating ligands to immobilize sensitizers and dyads via metal–ligand coordination for enhanced photocurrent generation. *J. Am. Chem. Soc.*, **131**, 14646–14647.

62 Umeyama, T. and Imahori, H. (2008) Carbon-nanotube-modified electrodes for solar energy conversion. *Energy Environ. Sci.*, **1**, 120–133.

63 Sgobba, V. and Guldi, D.M. (2009) Carbon nanotubes: electronic/electrochemical properties and application for nanoelectronics and photonics. *Chem. Soc. Rev.*, **38**, 165–184.

64 Kymakis, E., Koudoumas, E., Franghiadakis, I., and Amaratunga, G.A.J. (2006) Post-fabrication annealing effects in polymer-nanotube photovoltaic cells. *J. Phys. D: Appl. Phys.*, **39**, 1058–1062.

65 Sgobba, V., Rahman, G.M.A., Guldi, D.M., Jux, N., Campidelli, S., and Prato, M. (2006) Supramolecular assemblies of different carbon nanotubes for photoconversion processes. *Adv. Mater.*, **18**, 2264–2269.

66 Rahman, G.M.A., Guldi, D.M., Cagnoli, R., Mucci, A., Schenetti, L., Vaccari, L., and Prato, M. (2005) Combing single wall carbon nanotubes and photoactive polymers for photoconversion. *J. Am. Chem. Soc.*, **127**, 10051–10057.

67 Campidelli, S., Ballesteros, B., Filoramo, A., Díaz, D.D., de la Torres, G., Torres, T., Rahman, G.M.A., Ehli, C., Kiessling, D., Werner, F., Sgobba, V., Guldi, D.M., Cioffi, C., Prato, M., and Bourgoin, J.-P. (2008) Facile decoration of functionalized single-wall carbon nanotubes with phthalocyanines via "click-chemistry." *J. Am. Chem. Soc.*, **130**, 11503–11509.

68 Umeyama, T., Fujita, M., Tezuka, N., Kadota, N., Matano, Y., and Imahori, H. (2007) Electrophoretic deposition of single-walled carbon nanotubes covalently modified with bulky porphyrins on nanostructured SnO_2 electrodes for photoelectrochemical devices. *J. Phys. Chem. C*, **111**, 11484–11493.

69 Tezuka, N., Umeyama, T., Seki, S., Matano, Y., Nishi, M., Hirao, K., Lehtivuori, H., Tkachenko, N.V., Lemmetyinen, H., Nakao, Y., Sakaki, S., and Imahori, H. (2010) Comparison of cluster formation, film structure, microwave conductivity, and photoelectrochemical properties of composites consisting of single-walled carbon nanotubes with C_{60}, C_{70}, and C_{84}. *J. Phys. Chem. C*, **114**, 3235–3247.

70 Haddon, R.C., Perel, A.S., Morrsi, R.C., Palstra, T.T.M., Hebard, A.F., and Fleming, R.M. (1995) C_{60} thin-film transistors. *Appl. Phys. Lett.*, **67**, 121–123.

71 Kobayashi, S., Takenobu, T., Mori, S., Fujiwara, A., and Iwasa, Y. (2003) Fabrication and characterization of C_{60} thin-film transistors with high field-effect mobility. *Appl. Phys. Lett.*, **82**, 4581–4583.

72 Zhang, X.-H. and Kippelen, B. (2008) High-performance C_{60} n-channel organic field-effect transistors through optimization of interfaces. *J. Appl. Phys.*, **104**, 104504.

73 Meijer, E.J., De Leeuw, D.M., Setayesh, S., van Veenendaai, E., Huisman, B.H., Blom, P.W.M., Hummelen, J.C., Scherf, U., and Klapwijk, T.M. (2003) Solution-processed ambipolar organic field-effect transistors and inverters. *Nat. Mater.*, **2**, 678–682.

74 Tans, S.J., Verschueren, A.R.M., and Dekker, C. (1998) Room-temperature transistor based on a single carbon nanotube. *Nature*, **393**, 49–52.

75 Martel, R., Schmidt, T., Shea, H.R., Hertel, T., and Avouris, P. (1998) Single- and multi-wall carbon nanotube field-effect transistors. *Appl. Phys. Lett.*, **73**, 2447–2449.

76 Zhao, Y.-L. and Stoddart, J.F. (2009) Noncovalent functionalization of single-walled carbon nanotubes. *Acc. Chem. Res.*, **42**, 1161–1171.

77 Liu, S., Shen, Q., Cao, Y., Gan, L., Wang, Z., Steigerwald, M.L., and Guo, X. (2010) Chemical functionalization of single-wall carbon nanotube field-effect transistors as switches and sensor. *Coord. Chem. Rev.*, **254**, 1101–1116.

78 Hecht, D.S., Ramirez, R.J.A., Briman, M., Artukovic, E., Chichak, K.S., Stoddart, J.F., and Grüner, G. (2006) Bioinspired detection of light by using a porphyrin-sensitized single-wall nanotube field-effect transistor. *Adv. Mater.*, **6**, 2031–2036.

79 Hu, L., Zhao, Y.-L., Ryu, K., Zhou, C., Stoddart, J.F., and Grüner, G. (2008) Light-induced charge transfer in pyrene/CdSe–SWNT hybrids. *Adv. Mater.*, **20**, 939–946.

80 Zhao, Y.-L., Hu, L., Grüner, G., and Stoddart, J.F. (2008) Pyrenecyclodextrin-decorated single-walled carbon nanotube field-effect transistors as

chemical sensors. *Adv. Mater.*, **20**, 1910–1915.

81 Zhao, Y.-L., Hu, L., Grüner, G., and Stoddart, J.F. (2008) A tunable photosensor. *J. Am. Chem. Soc.*, **130**, 16996–17003.

82 Simmons, J.M., Campbell, V.E., Mark, T.J., Léonard, F., Gopalan, P., and Eriksson, M.A. (2007) Optically modulated conduction in chromophore-functionalized single-wall carbon nanotubes. *Phys. Rev. Lett.*, **98**, 086802.

14
Experimental Determination of Association Constants Involving Fullerenes

Emilio M. Pérez and Nazario Martín

The previous chapters have offered a comprehensive overview of several aspects of the supramolecular chemistry of fullerenes and other carbon nanostructures. Throughout the text, many of the contributors have mentioned association constants and have elaborated discussions on their values. In this chapter, we intend to provide some experimental guidelines for the quantification of association constants involving fullerenes as guests.

Although often considered a straightforward experimental technique – particularly by newcomers to the field – the correct estimation of binding constants is often not as simple as it seems. To arrive at chemically meaningful and quantitatively significant results, we need to carefully address a variety of issues, from planning the titration experiment (choice of spectroscopic method, concentration of host and guest, dilution, number of data points, number of repetitions, etc.) to analyzing the titration data (linear versus nonlinear regression techniques, determination of stoichiometry, choice of equations, estimation of errors, etc.). All these issues have recently been ably reviewed by Thordarson with regard to general host–guest systems [1], besides in often cited textbooks [2]. Thus, here we will go only through some necessary general considerations briefly, to then focus on the peculiarities of fullerenes as guests, and conclude with a practical example. Before moving on, the reader should be aware that the views expressed in this chapter are based on our experience in the determination of binding constants toward the fullerenes, which might not be entirely coincident with that of other experimentalists.

14.1
Planning a Titration Experiment

The equilibrium of association in a host–guest system is defined by

$$m\mathrm{H} + n\mathrm{G} \overset{\beta_{mn}}{\rightleftharpoons} \mathrm{H}_m\mathrm{G}_n, \qquad \text{where } \beta_{mn} = \frac{[\mathrm{H}_m\mathrm{G}_n]}{[\mathrm{H}]^m[\mathrm{G}]^n}$$

Supramolecular Chemistry of Fullerenes and Carbon Nanotubes, First Edition. Edited by Nazario Martín and Jean-Francois Nierengarten.

or for the particular case of $m = n = 1$,[1]

$$H + G \overset{K_a}{\rightleftharpoons} HG, \qquad \text{where } K_a = \frac{[HG]}{[H][G]}$$

In a titration experiment, our objective is to extract K_a from a series of spectroscopic measurements. To achieve this, we will generally prepare a solution of known concentration of host ($[H]_0$) and gradually increase the concentration of host–guest complex by adding to it aliquots of a solution of known concentration of guest ($[G]_0$). Then, we will analyze the variation in a given physical property (e.g., $\Delta\delta$ in the case of NMR or ΔAbs in the case of UV–vis titrations) assuming that it is proportional to the concentration of the host–guest associate (e.g., $\Delta\delta \propto [HG]$ or $\Delta Abs \propto [HG]$). We will then plot that change in physical property against the amount of guest added and choose a mathematical method to analyze the data to estimate K_a.

14.2 Performing a Titration

The first aspect we need to consider before carrying out a titration experiment is the system we wish to study. That includes host, guest, solvent, and temperature. First of all, host, guest, and solvent should be pure. Although obvious, impurities are one of the prime sources of error during titration experiments. The solvent of choice should be one in which all host, guest, and host–guest associate are sufficiently soluble at the working temperature, to prevent precipitation. It should also have a sufficiently high boiling point to minimize errors arising from evaporation. For instance, for a titration at 298 K (25 °C), it is better to use $CHCl_3$ (b.p. = 61.2 °C) than CH_2Cl_2 (b.p. = 39.6 °C). Finally, the temperature needs to be controlled and recorded. Although often found in the literature, particularly in reviews, a binding constant is of little value without the solvent in, and the temperature at, which the titration experiments were carried out.

The expected association constant determines the concentration we should choose for our titration experiment. For an initial estimate of the association constant, a wise starting point is a thorough review of the literature available. If no data on similar host–guest systems are available, chemical intuition (e.g., it is difficult to expect an extremely high association constant from a poorly preorganized receptor) and trial and error with quick preliminary titration experiments will do.

In principle, we should aim to go all the way from a point where all our host is in its free form ($[H] = [H]_0$) to a point as close to saturation as possible, where all the host is in complexed form ($[HG] = [H]_0$). Since complete saturation of the system is often not viable, or at least experimentally sensible, in order to obtain meaningful data we should aim to cover at least 70–80% of the binding isotherm. As a rule of thumb, a good starting point is a concentration of host equal to, or slightly higher than, the

1) Unless explicitly mentioned, in the following we will assume a 1 : 1 binding model for the sake of simplicity. The experimental procedures and mathematical treatment for other stoichiometries can be found in Ref. [1].

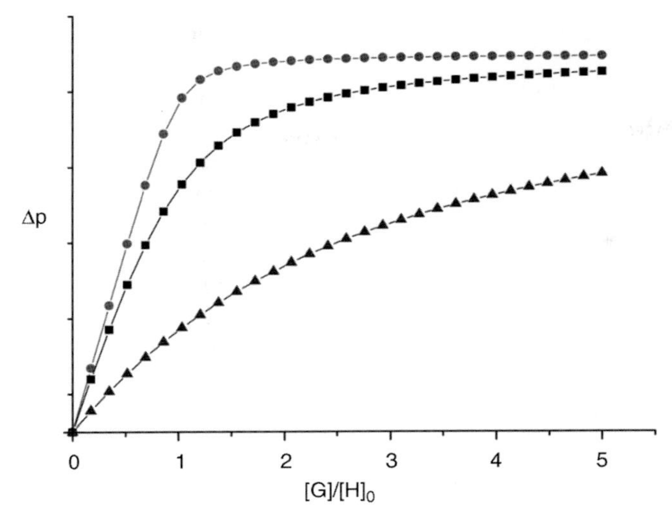

Figure 14.1 Simulated binding isotherms for titration experiments carried out at $[H]_0 = 50 \times K_a^{-1}$ (circles), $[H]_0 = 5 \times K_a^{-1}$ (squares), and $[H]_0 = 0.5 \times K_a^{-1}$ (triangles).

reciprocal of the expected binding constant ($[H]_0 = 5 \times K_a^{-1}$, Figure 14.1, squares). For example, for an expected binding constant of $10^4\ M^{-1}$ we can start with $[H]_0 = 5 \times 10^{-4}$ M. If we work with a much higher concentration than the reciprocal of the binding constant, precise determination of the binding constant will become difficult as the titration curve approaches two straight lines (Figure 14.1, circles). Finally, if we are working at $[H]_0 \ll K_a^{-1}$, it would be difficult to reach saturation, unless we add a huge excess of guest (Figure 14.1, triangles). With regard to the concentration of the guest, $[G]_0$, we will decide considering the number of data points we wish to acquire, and the availability of both host and guest. Generally speaking, we will select a concentration that allows us to reach saturation with the addition of a reasonably small volume of guest solution. That is, we will prepare a solution significantly more concentrated than our solution of host (e.g., $[G]_0 = 5 \times [H]_0$). To prevent dilution factors, we will ensure working at constant concentration of the host by utilizing the solution of host as solvent in the solution of guest.

The volume and number of the aliquots of the solution of guest that we will add during the titration experiment will depend, again, on the estimated binding constant and the number of data points we wish to obtain. For a relatively tight binding and following the guidelines for the concentration of the solutions of host and guest described above, a good working plan is to add 0.1 equiv. aliquots of the guest solution until reaching 3–4 equiv. of guest. Such procedure will yield a sufficient yet manageable number of data points (30–40) and a satisfactory coverage of the binding isotherm. For very weak binding, when we need to work at $[H]_0 \ll K_a^{-1}$ (Figure 14.1, triangles), we will need to add a higher number of molar equivalents of guest, sometimes up to several hundreds, to get sufficient coverage of the binding isotherm. Consequently, even if we can start by adding 0.2–0.5 equiv. aliquots until reaching 1–2 equiv., we should later increase the volume of the aliquots to 1–5 equiv. to avoid an unending experiment.

Before deciding on the number of data points to be collected, we need to take into account that we must repeat each titration experiment several times. If the availability of the host and/or guest advises us to collect fewer data points, we should always consider that it is better to collect three titrations with 15 data points each than one titration with 45 data points.

14.3 Choosing the Spectroscopic Method

Aside from direct methods such as isothermal titration calorimetry (ITC), the most commonly utilized techniques for investigating association equilibria are nuclear magnetic resonance (NMR), electronic absorption spectroscopy (UV–vis), and fluorescence spectroscopy. The choice of one of these spectroscopic methods depends on (1) the chemical nature of host and guest and (2) the concentration at which we are carrying out our titration (and hence on the expected binding constant).

The dependence on the chemical nature of host and guest is immediate; for example, we cannot carry out an NMR titration if either one of them is paramagnetic. On the other hand, the presence of chromophores or fluorophores is a prerequisite for UV–vis and fluorescence titrations, respectively. When considering particular examples, as we shall see with fullerenes, we will need to think about other factors, such as the degree to which the NMR or UV–vis spectra are influenced by complexation.

Aside from this, the concentration of the solutions of host and guest will be the prime factor to consider before choosing a spectroscopic method to analyze the titration.

1) NMR: The sensitivity of ^{1}H NMR on a routine spectrometer will prevent measurements with concentrations much smaller than 10^{-4} M. This implies that NMR is a good method for $K_a < 10^5\ \mathrm{M}^{-1}$. For higher binding constants, it will be difficult to obtain good titration data, as we would need to work at $[\mathrm{H}]_0 \ll 10^{-4}$ M. As a genuine advantage, in the case of systems under slow exchange on the NMR timescale, we will get separate signals for host, guest, and host–guest complex, which will allow us to calculate their concentrations (and thus the binding constant) directly from integration.
2) UV–vis: With a simple set of cuvettes, with path lengths from 0.1 to 1 cm, we can work at a regime where the Beer–Lambert law holds true ($A < 1$) in concentrations ranging approximately from 10^{-3} to 10^{-6} M, although with strongly absorbing chromophores, such as porphyrins, we can work at concentrations below the micromolar. This will allow us to confidently determine binding constants in the range of $10^3 \geq K_a \geq 10^7\ \mathrm{M}^{-1}$.
3) Fluorescence: When it comes to determining binding constants, there is a fundamental difference between fluorescence spectroscopy and NMR/UV–vis, which is often overlooked. Fluorescence methods measure the extent of quenching (or enhancement) of fluorescence of the host by addition of the guest. Since quenching can occur through static (relevant to host–guest association) or

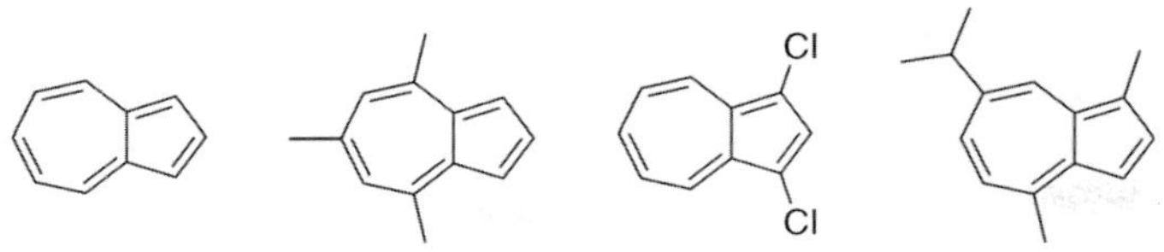

Figure 14.2 Structures of the azulene derivatives that were mistakenly reported to bind C_{60} and C_{70} with log $K_a \sim 5$ through fluorescence titrations [3].

dynamic (collisional) mechanisms, and its intensity depends on other factors aside complexation, such as orientation of the fluorophore and quencher, polarity of the solvent, and so on, it is often the case that the variation in the intensity of fluorescence is not directly proportional to the concentration of host–guest complex. Besides this, fluorescence spectroscopy is more sensitive to impurities (particularly for fluorophores with low fluorescence quantum yields) and can suffer from other detrimental effects, such as second-order transmission, scattering, inner-filter, excited state quenching processes, and photobleaching. If not correctly considered, these factors typically lead to an overestimation of the binding constant. Archetypical of this, and relevant to this chapter, are the binding constants of several derivatives of azulene (Figure 14.2) toward C_{60} and C_{70}, initially overestimated to surprisingly high values of log $K_a \sim 5$ through fluorescence spectroscopy [3] and later reassessed to negligible, if not zero [4]. Although one must always keep an open mind for unexpected results, it is difficult to imagine the relatively small and flat azulenes forming 1 : 1 complexes with the fullerenes of such stability. Moreover, the binding constants initially reported were nearly structure-independent, with very little influence of the derivative of azulene and the fullerene under study [3]. In this regard, it is good to remember that chemical common sense can never be overestimated.

On the other hand, none of these problems arises in systems where fluorescence is enhanced upon formation of the complex. All this said, fluorescence is the most sensitive of the three techniques and will allow us to carry out measurements at submicromolar concentrations, ideal for determination of $K_a \gg 10^6\ M^{-1}$.

14.4 Analyzing the Data

The first approach to analyzing our titration data must be the careful observation of the shape of the binding isotherm. A typical binding isotherm is a hyperbole; different shapes, like sigmoidal, are indicative of phenomena such as cooperativity, and should be analyzed accordingly. For the scope of this chapter, we will consider a standard hyperbolic binding isotherm. The concentration of host and guest and the shape of the isotherm will give us a first (rough!) estimate of the binding constant. For example, if we are working at $[H]_0 = 10^{-4}$ M and our binding isotherm is close to saturation on addition of 1.5–2 equiv. of guest, we can begin to think that K_a will be in the range of 10^4–10^5 M^{-1}. Conversely, if our binding isotherm is close to a straight line, and it is difficult to guesstimate its asymptotic limit, we can conclude that $K_a \ll 10^4\,M^{-1}$, and we should consider repeating the experiment at a higher $[H]_0$.

After this first approximation, we can embark on the mathematical analysis of our titration data. Unless unavoidable, we will refrain from utilizing old-fashioned linear analyses, such as Benessi-Hildebrand, which are based on bold assumptions rarely met experimentally, like $[G]=[G]_0$. Instead, we should perform nonlinear regression analysis on our data. An adequate equation for 1 : 1 equilibria, derived from mass balance is provided as follows:

$$[HG] = \frac{1}{2}\left([G]_0 + [H]_0 + \frac{1}{K_a}\right) - \sqrt{\left([G]_0 + [H]_0 + \frac{1}{K_a}\right)^2 + 4[G]_0[H]_0}$$

$$\text{For NMR titrations: } \Delta\delta = \delta_{\Delta HG}\left(\frac{[HG]}{[G]_0}\right)$$

$$\text{For UV–vis titrations: } \Delta Abs = \varepsilon_{\Delta HG}([HG])$$

Equations for 1 : 2 and 2 : 1 stoichiometry can also be found in Ref. [1]. Most software packages such as Origin®, Matlab®, or Mathematica® are able to solve these equations. If available, we highly recommend employing global analysis software, such as Specfit® for UV–vis titrations, which analyzes all wavelengths of the titration simultaneously and produces calculated spectra for host, guest, and host–guest species, to provide a very good indication of the validity of the fit (see Figure 14.10).

14.5 Determining Stoichiometry

As with planning the titration experiment, our first approach to determine the stoichiometry of our associate should be previous knowledge and chemical common sense. We should consider the structures of host and guest, and any structural information on the host–guest associate, together with the information provided by the titration data. For instance, suppose we have carried out a UV–vis titration experiment, and we observe the formation of an isosbestic point, which is indicative of the presence of just one type of associate. If, together with this, we find that the inflection point of the binding isotherm is located at approximately $[H]_0/[G]_0 = 1$, the data fit satisfactorily to a 1 : 1 binding model, and our host is designed to bind the guest forming 1 : 1 complexes, we can initially conclude that the stoichiometry of our associate is indeed 1 : 1.

Besides this first approximation, the continuous variation plot (Job's plot) method is by far the most widely used protocol for the experimental determination of stoichiometry. It is based on the assumption that the concentration of a complex H_mG_n is at a maximum at $[H]/[G] = m/n$. The experiment consists in analyzing solutions of variable molar fraction of host, from 0 to 1, keeping the $[H] + [G]$ constant, and plotting the variation in a physical property (say, ΔAbs or $\Delta\delta$) versus the molar fraction of host. A maximum at $X = 0.5$ indicates 1 : 1 stoichiometry, whereas a maximum at $X = 0.33$ is indicative of 1 : 2 association. When planning such an experiment, we should be aware that the Job's plot method is most reliable when working at $[H] \gg K_a^{-1}$, that is, in a regime where the system is fully saturated. Working at lower concentrations often results in very broad or multiple maxima. Likewise, it is well known that when there is more than one type of complex present, the Job's method becomes unreliable. If in doubt, we recommend fitting the titration data to all the binding models that make chemical sense and comparing the results.

14.6 Estimating Errors

The experimental determination of binding constants is a delicate exercise. There are many possible sources of error, and sadly enough, many of them are additive (like evaporation of the solvent over time); it requires extreme purity of all the components of the system and adequate measuring equipment (balances, volumetric flasks, pipettes, spectrometers, etc.). As with anything else, practice makes perfect, but it is often the case that calculating a binding constant worth being published takes more time than expected.

Before reporting a binding constant, a key factor is estimating the confidence limits. Let us imagine that after a titration experiment, we have obtained from the fitting software a value of $K_a = (3.02 \pm 0.07) \times 10^5\ M^{-1}$; going to the extremes, should we report a binding constant of $K_a = (3.0 \pm 0.1) \times 10^5\ M^{-1}$ or $K_a = 10^5–10^6\ M^{-1}$? The answer is we should not report either of these, but we shall see which one would be more adequate.

We must *repeat* our experiment to get an idea of the precision and accuracy of our measurement, and we should do it several times (a minimum of three). After repeating the experiment, we can report the binding constant as the mean value of the experiments plus or minus the standard deviation. In this respect, we find it advisable to express binding constants as logarithms, particularly for large association constants. Taking the example above, in our experience and at least for the case of fullerenes as guests, we will find more often that $K_a = (3 \pm 2) \times 10^5\ M^{-1}$ – or $\log K_a = 5.5 \pm 0.5$ – than $K_a = (3.0 \pm 0.1) \times 10^5\ M^{-1}$ – or $\log K_a = 5.48 \pm 0.02$. Besides being an experimental observation, this is easily understood if we consider binding constants in terms of energy. Since $\Delta G = - RT \ln K_a$, we can calculate that going from an association constant of $3 \times 10^5\ M^{-1}$ to $6 \times 10^5\ M^{-1}$ corresponds to an increase in ΔG below 5%.

14.7 Fullerenes as Guests: Spectroscopic Properties

In the following few paragraphs, we will analyze the spectroscopic properties of the fullerenes and their response upon complexation to the techniques generally used in titration experiments: NMR, UV–vis, and fluorescence.

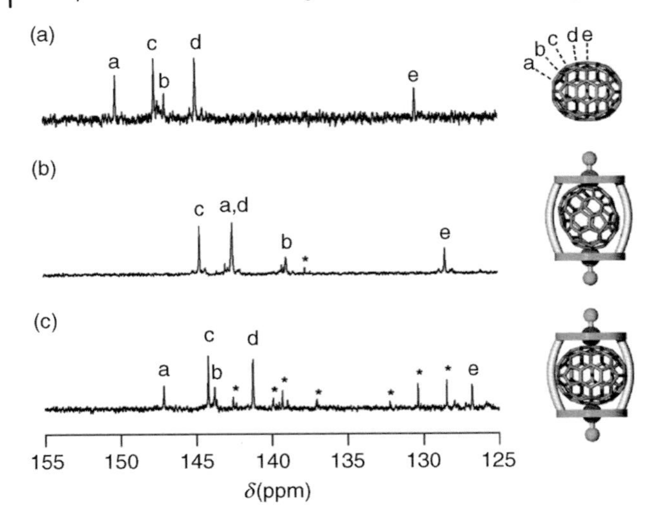

Figure 14.3 Partial ^{13}C NMR spectra ($CDCl_3$/CS_2, 25 °C) of (a) C_{70}, (b) C_{70} bound to an Ir(III) bisporphyrin macrocyclic host, and (c) C_{70} bound to a Rh(III) bisporphyrin macrocyclic host. The inclusion complexes were prepared by mixing ^{13}C-enriched C_{70} with 2.0 equiv. of the corresponding host. Asterisked signals originate from the host. Reproduced with permission of the American Chemical Society from Ref. [5c].

1) NMR: Fullerenes are polyenes, made solely of carbon atoms – except endohedral fullerenes, which are not considered here. C_{60} is a truncated icosahedron, with a carbon atom at the vertices of each polygon and a bond along each polygon edge. Due to its symmetry, all carbon atoms in C_{60} are equivalent, so its ^{13}C NMR presents a single signal at $\delta = 143.3$ ppm in d_5-PhCl at room temperature. The ^{13}C signal of C_{60} can be shifted upfield by up to 2–3 ppm upon complexation by hosts containing recognizing moieties with strong shielding abilities, such as metalloporphyrins [5]. Such shifts can be utilized to determine association constants, although very often the use of ^{13}C-enriched samples is required. Obviously, ^{13}C NMR measurements become more complicated and time consuming with less symmetric fullerenes, such as C_{70}, which presents five different ^{13}C NMR signals. On the other hand, structural information on the host–guest associate can be extracted from the different extent to which each of the signals is affected by complexation (Figure 14.3) [5c].

Hosts featuring purely organic recognizing units produce a noticeable but small shielding effect in the ^{13}C NMR of the fullerene. For instance, phenylene-type groups have been reported to produce upfield shifts ≤ 0.1 ppm at room temperature and ~ 0.3 at -60 °C upon complexation [6]. Association with hosts based on π-extended derivatives of tetrathiafulvalenes results in shifts of similar magnitude, sometimes accompanied with significant broadening (Figure 14.4) [7].

As for the effect of complex formation on the ^{1}H NMR of the host, the shifts are typically very weak <0.05 ppm [6, 7]. Again, larger shifts, up to 0.2 ppm, are

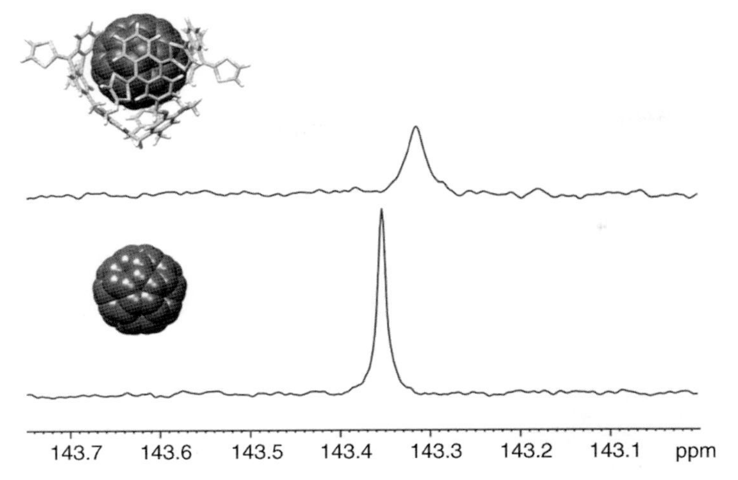

Figure 14.4 ^{13}C NMR (d_5-chlorobenzene, 125 MHz, 25 °C) of C_{60} (*bottom*) and its complex with an exTTF-CTV host (*top*, depicted with carbon atoms in green, sulfurs in yellow, oxygens in red, and hydrogens in white).

observed for metalloporphyrin-based hosts [8]. Besides being quantitatively small, the ^{1}H NMR signals of the host can be shifted both up- and downfield upon complexation [8]. Considering all this, ^{1}H NMR titrations are not often utilized for the determination of binding constants toward fullerenes [9].

2) UV–vis: In hexane solution, C_{60} absorbs all the way from 190 to 620 nm. The most energetic part of the spectrum, 190–410 nm, corresponds to nine allowed $^1T_{1u}-^1T_g$ transitions, while forbidden singlet–singlet transitions are responsible for the less intense absorptions between 410 and 620 nm [10]. The UV–vis spectrum of C_{60} and C_{70} are known to show solvatochromic behavior [11], that is, they are sensitive to the environment and consequently to complexation. Moreover, the electron accepting nature of the fullerenes often results in the appearance of charge transfer bands and/or in a significant modification of the electronic absorption spectrum of the host too. Finally, UV–vis titrations often yield information on the possible structure/stoichiometry of the complex, through the appearance (or not) of isosbestic points (Figure 14.5) [12]. All these facts, together with the range of binding constants typically found for fullerene hosts ($\log K_a = 3$–7), often make UV–vis titrations the method of choice for determination of binding constants toward fullerenes.

3) Fluorescence: Mainly due to the rapid intersystem crossing leading to the formation of the excited triplet state, both C_{60} and C_{70} are very weak fluorophores, with fluoresce quantum yields of 3.2×10^{-4} and 5.7×10^{-4} in toluene at room temperature, respectively.

It has long been known that the fluorescence of the fullerenes is effectively quenched even by simple electron-rich aromatics, such as *N,N*-dialkylanilines, through dynamic quenching mechanisms [13]. It is also often the case that

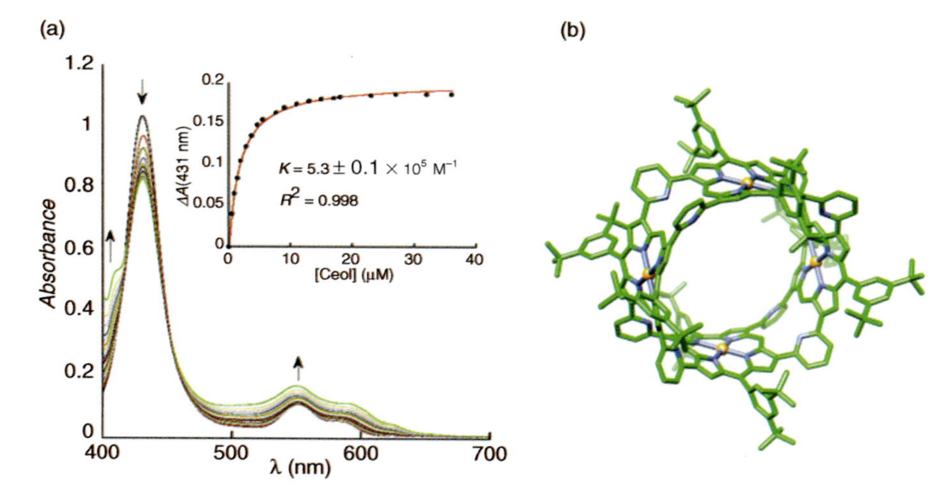

Figure 14.5 (a) UV–vis absorption spectra as recorded during the titration of a tetraporphyrin "nanobarrel" (2.0 μM; (b) with carbons in green, nitrogens in blue, and nickel in orange, hydrogens are omitted for clarity) in toluene in the presence of various amounts of C_{60} ($0 < [C_{60}] < 36\,\mu M$) at 25 °C. Arrows indicate the changes in absorption with increasing $[C_{60}]$. (*Inset*) Plot of $\Delta A_{431\,nm}$ versus $[C_{60}]$. Reproduced with permission of the American Chemical Society from Ref. [12].

fullerene hosts quench the fluorescence of their host upon association in the ground state (i.e., static quenching) so that monitoring the degree of quenching as the concentration of host increases can be used to estimate the association constant. Conversely, the electron acceptor character of the fullerenes makes them good quenchers for electron-rich fluorophores, through energy or electron transfer mechanisms, allowing for fluorescence titration experiments based on addition of the fullerene guest to a solution of the host.

As already mentioned, particular care should be taken if our method of choice for the determination of binding constants is fluorescence quenching. In particular, we should always keep in mind that we can determine association constants precisely only when monitoring purely static quenching processes.[2)] Observing changes in the traits of the UV–vis spectrum upon formation of the host–guest complex is a good indication of static quenching. For example, the binding abilities of a cyclic porphyrin trimer toward C_{60} were recently investigated by Anderson and coworkers by both UV–vis and fluorescence titration experiments to yield a log $K_a = 6.2$ in toluene at room temperature [14]. The changes in the UV–vis spectra of the host upon addition of C_{60} (Figure 14.6a) indicate that a host–guest associate is formed in the ground state, supporting that fluorescence quenching (Figure 14.6b) occurs via a static mechanism.

2) For a more detailed analysis of the possible scenarios, and the adequate equations to utilize in each of them, when determining association constants through fluorescence titrations, see Ref. [1].

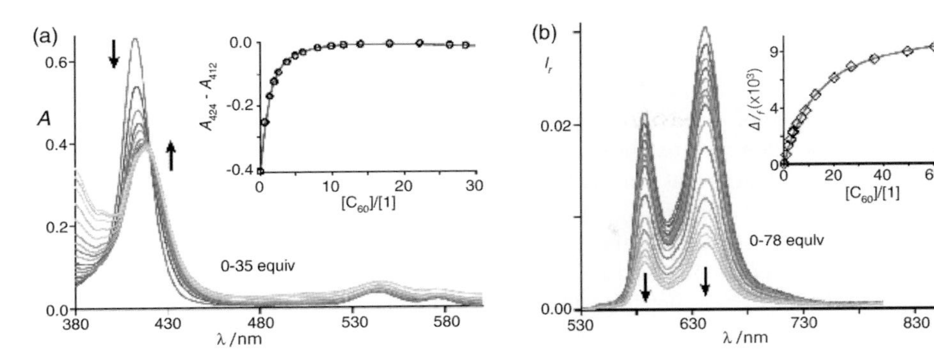

Figure 14.6 (a) Absorption spectral change in the cyclic porphyrin trimer host recently reported by Anderson and coworkers[16] (0.77 μM) in toluene at 298 K upon titration with C_{60} (0 − 27 μM). (b) Fluorescence spectra during the titration of the same host (0.093 μM) with C_{60} (0 − 7.2 μM) in toluene at 298 K (λ_{ex} = 413 nm). Reproduced with permission of the American Chemical Society from Ref. [14].

14.8 Determination of the Binding Constant of an exTTF-based Host toward C_{60}: A Practical Example

We will conclude this chapter by illustrating all the considerations above with a practical example of the determination of the binding constants toward C_{60} of one of our exTTF-based receptors. In particular, we will describe the case of macrocycle 1 (Figure 14.7) [7b,15].

1) Planning the titration experiment.

   We will consider the chemical structure of the host we wish to analyze, and all previous information available to get a first guesstimate of the expected binding constant. The design of **1** is based on previously reported exTTF tweezers-like hosts, which feature two exTTF units linked through a terephthalate or isophthalate ester linker [16]. Such tweezers associate C_{60} with binding

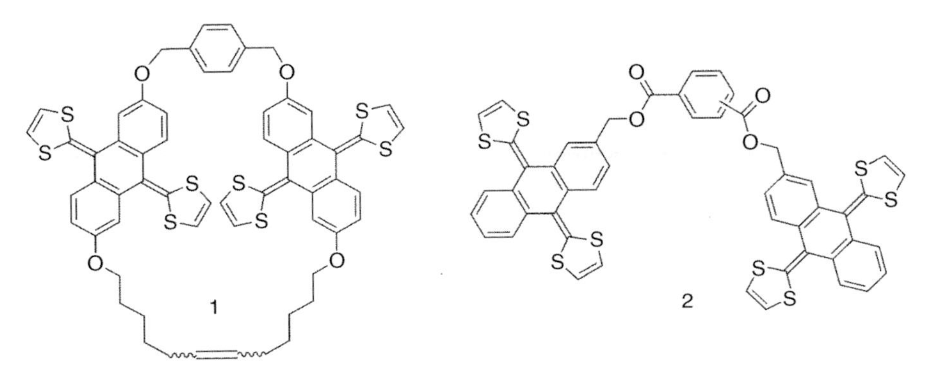

Figure 14.7 Chemical structures of macrocyclic host **1** and the parent tweezers **2**.

constants in the range of log $K_a = 3$–4 in aromatic solvents, despite their lack of preorganization.

In **1** all the basic structural features of **2** are maintained, and a much higher degree of preorganization toward binding of C_{60} is anticipated, given its macrocyclic structure. The length of the alkenyl linker was selected through molecular modeling so that **1** presents a flexible cavity of diameter 11–13 Å, suitable for the encapsulation of C_{60}. Thus, we expected **1** to show significantly larger binding constants than **2**. At the time we carried out this study, the best hosts for C_{60} had been reported by Aida and Tashiro (log $K_a = 5$–8), utilizing macrocyclic structures featuring two metalloporphyrins as recognition units [17].

Considering all this, we anticipated $4 < \log K_a < 8$. Consequently, we decided to employ an initial concentration of host $[\mathbf{1}] = 1.0 \times 10^{-5}$ M and $[C_{60}] = 5.0 \times 10^{-5}$. Taking into account these concentrations, and the fact that we had already proven that the interaction between exTTF and C_{60} produces significant and characteristic changes in the UV–vis spectrum of the host, namely, a decrease in the absorption of the exTTF chromophore at $\lambda \approx 430$ nm accompanied by the appearance of a charge transfer band at $\lambda \approx 475$ nm [16], we decided to utilize UV–vis as the spectroscopic technique to follow the titration.

2) The titration.

   For the sake of comparison with previous results [16], and taking into account the solubility of C_{60}, we decided to utilize PhCl as solvent in the titration experiment. To prepare the 1.0×10^{-5} M stock solution of host, we employed an analytical balance, to weigh an analytically pure sample of **1**, and volumetric glassware to measure the solvent. Prior to titration, the concentration of this solution was double-checked through UV–vis spectroscopy, considering the absorbance of the solution at $\lambda = 425$ nm, and the molar absorptivity of **1** at the same wavelength. This solution of **1** was used as a solvent in the solution of the [60]fullerene guest, to ensure working at a constant concentration of the host during the titration experiment.

   In a typical experiment, we measure 1.0 ml of the host solution and place it in a UV–vis quartz cuvette of 1 cm optical path. This leaves enough free volume to carry out the additions of C_{60} inside the same cuvette, but is sufficient to measure the UV-vis spectrum correctly. Taking into account this volume and the initial concentration of **1**, we perform calculations to determine the volume necessary to add 30×0.1 equiv. aliquots of C_{60}. This can be easily automated in any spreadsheet software. Finally, we perform the additions utilizing Microman® Gilson positive displacement pipettes, which are suitable to measure organic solvents, and record the UV–vis spectra after each addition, to obtain the spectra shown in Figure 14.8.

3) Data analysis.

   Directly from the observation of the UV–vis data, we can extract the following conclusions:

   i) The decrease in absorption at $\lambda = 425$ nm, corresponding to the exTTF chromophore, unambiguously confirms the formation of a complex, keeping in mind that we work at constant concentration of **1**.

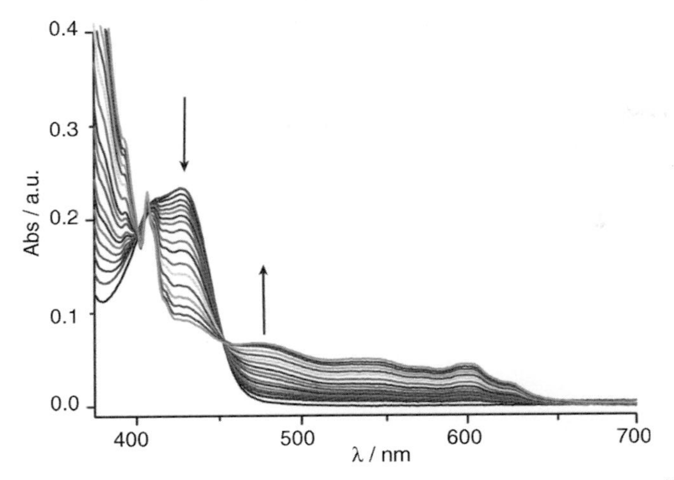

Figure 14.8 UV–vis spectra as obtained during the titration of **1** (1.0×10^{-5} M, PhCl, 298 K) versus C_{60} (5.0×10^{-5} M), recorded at a constant concentration of **1**. Arrows indicate the progress of the titration. Only 18 representative spectra are shown to avoid cluttering.

ii) The degree to which that band is depleted indicates that the relative concentration of the complex is significant. This, together with the working concentration and the number of equiv. of C_{60} added, suggest that log $K_a > 5$.
iii) The formation of a tight isosbestic point at $\lambda = 450$ nm points to the formation of one type of associate only. Considering the precedents, and the structure of 1, the experimental data suggest that a **1**·C_{60} complex of 1 : 1 stoichiometry is formed.

With these first considerations, we can move on to the exact determination of the binding constant. To this end, we performed analysis by two different methods: nonlinear least squares regression of the binding isotherm, obtained from the plot of ΔAbs_{478} versus [C_{60}], utilizing the equation for a 1 : 1 model described above and Microcal Origin® software; and direct analysis of the spectra utilizing Specfit® global analysis software.

For this particular experiment, fitting of the binding isotherm depicted in Figure 14.9 afforded a binding constant of log $K_a = 6.04$, with an $R^2 = 0.996$. After repeating the experiment twice more, we obtained log $K_a = 6.28$ and log $K_a = 5.95$, which provides a mean value of log $K_a = 6.09$ and a standard deviation of 0.17, thus resulting in a value for the binding constant of log $K_a = 6.1 \pm 0.2$.

Analysis with the global analysis software offers remarkable advantages. It is faster, does not require any pretreatment of the data, and all the data in the UV–vis spectra are analyzed simultaneously, instead of the ΔAbs at a single wavelength, which minimizes errors arising from spectral overlap. For example, in our particular case, analyzing the decrease in absorption at 425 nm, instead of the increase in the charge transfer band at $\lambda = 478$ nm, systematically yields overestimated binding constants. Second, the software generates simulated spectra for host, guest, and

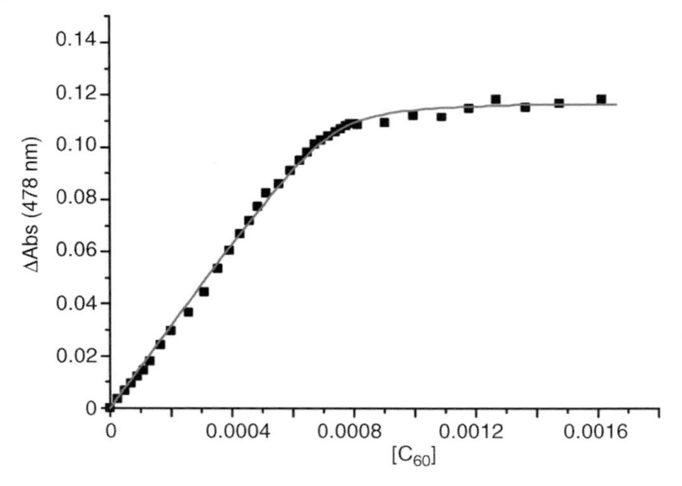

Figure 14.9 Binding isotherm for the titration experiment shown in Figure 14.8. Black squares are experimental data and solid gray line corresponds to the fit.

host–guest species, which are a valuable indication of the quality of the fitting process. For instance, all fits that result in spectra with negative or clearly excessive molar absorptivities can be disregarded, while sensible simulated spectra (Figure 14.10b) are indicative of a reliable fit. Moreover, experimental spectra for pure host and guest can be introduced as data, which facilitates the fitting process. Finally, the program handles more complex stoichiometries in a very straightforward manner.

Analysis with Specfit® yielded a log $K_a = 6.5 \pm 0.5$, identical to the value obtained with Origin® software within error limits.

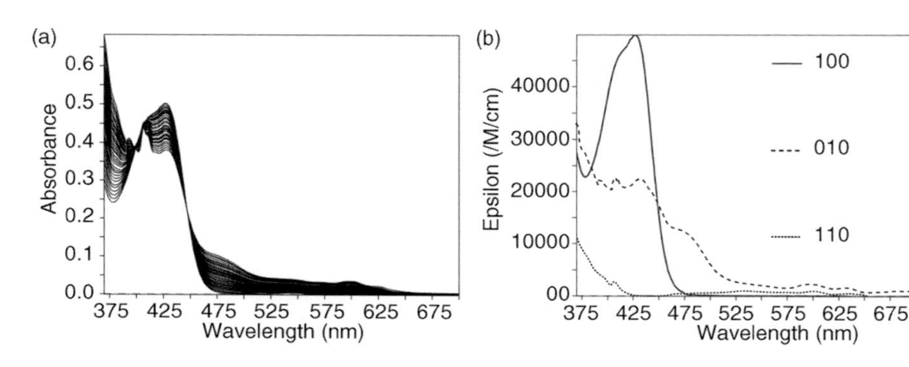

Figure 14.10 (a) Titration data for **1** versus C_{60} and (b) simulated spectra of host (100), guest (010), and host–guest complex (110) for a log $K_a = 6.5$. Note that the simulated spectra for the 1 : 1 associate reproduces very well the trends observed during the titration: depleted absorbance at $\lambda = 425$ nm, isosbestic point with the spectrum of **1** at $\lambda = 450$ nm, charge transfer band at $\lambda = 478$ nm.

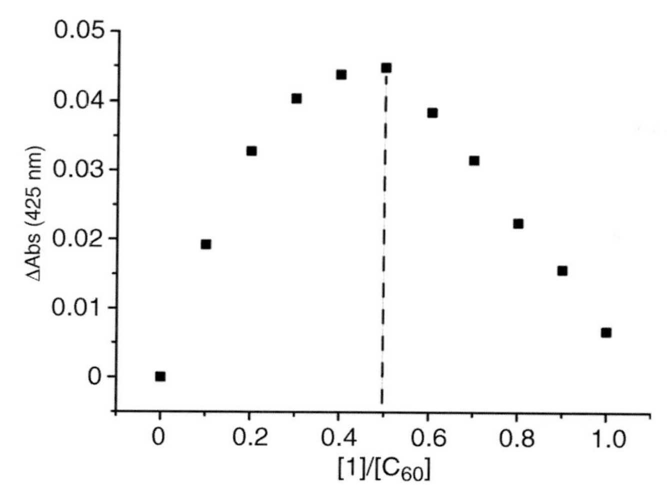

Figure 14.11 Job's plot for **1** versus C_{60}.

4) Determination of stoichiometry.
   At this point, taking into account the precedents, the structural information about **1**, and the very good fits to 1 : 1 binding models, the fact that the $\mathbf{1}{\cdot}C_{60}$ associate presents a 1 : 1 stoichiometry is a fair assumption. Nevertheless, we carried out continuous variation analysis (Job's plots) to confirm the stoichiometry. To do this, it is important to keep in mind that the Job's plot technique works best for systems where there is only one type of complex present, and concentrations at which the system is fully saturated (i.e., at a total concentration significantly higher than K_a^{-1}) [1]. Thus, in our case, we decided to work at $[\mathbf{1}] + [C_{60}] = 10^{-4}$ M. We recorded UV–vis spectra of mixtures of molar fraction of C_{60} from 0 to 1, at 0.1 intervals, and plotted the difference between the observed experimental absorbance at a given wavelength and the expected theoretical absorbance assuming there is no complex formation, calculated as $A_{theor} = \varepsilon_1[\mathbf{1}] + \varepsilon_{C60}[C_{60}]$ for a cuvette of 1 cm optical path length. The resulting graph for $\lambda = 425$ nm is shown in Figure 14.11. The maximum at 0.5 molar fraction confirms the 1 : 1 stoichiometry.

14.9 Conclusions

The correct determination of association constants is far from a trivial experimental technique. Each host–guest system has its own peculiarities, which should be taken into account when deciding the concentration, the spectroscopic technique, and the method of analysis for our titration experiments.

The design of receptors for fullerenes is attracting a great deal of attention, and many groups are starting research lines in this topic. In this chapter, we have presented an overview of what we consider good practice with regard to the experimental determination of binding constants involving fullerenes (C_{60} and C_{70}) as guests, from a very practical, hands-on perspective based on our own experience. We hope these brief guidelines will be useful for those interested in developing their own research lines within the field.

References

1 Thordarson, P. (2011) *Chem. Soc. Rev.*, **40**, 1305.
2 Connors, K.A. (1987) *Binding Constants. The Measurement of Molecular Complex Stability*, John Wiley & Sons, Inc., New York.
3 Rahman, A.F.M.M., Bhattacharya, S., Peng, X., Kimura, T., and Komatsu, N. (2008) *Chem. Commun.*, 1196.
4 Stella, L., Capodilupo, A.L., and Bietti, M. (2008) *Chem. Commun.*, 4744.
5 See, for example, (a) Sun, D., Tham, F.S., Reed, C.A., Chaker, L., Burgess, M., and Boyd, P.D.W. (2000) *J. Am. Chem. Soc.*, **122**, 10704; (b) Wu, Z.-Q., Shao, X.-B., Li, C., Hou, J.-L., Wang, K., Jiang, X.-K., and Li, Z.-T. (2005) *J. Am. Chem. Soc.*, **127**, 17460; (c) Yanagisawa, M., Tashiro, K., Yamasaki, M., and Aida, T. (2007) *J. Am. Chem. Soc.*, **129**, 11912.
6 (a) Eckert, J.-F., Byrne, D., Nicoud, J.-F., Oswald, L., Nierengarten, J.-F., Numata, M., Ikeda, A., Shinkai, S., and Armaroli, N. (2000) *New J. Chem.*, **24**, 749; (b) Lijanova, I.V., Flores Maturano, J., Dominguez Chavez, J.G., Sánchez Montes, K.E., Hernandez Ortega, S., Klimova, T., and Martínez-Garcia, M. (2009) *Supramol. Chem.*, **21**, 24.
7 (a) Huerta, E., Isla, H., Pérez, E.M., Bo, C., Martín, N., and de Mendoza, J. (2010) *J. Am. Chem. Soc.*, **132**, 5351; (b) Isla, H., Gallego, M., Pérez, E.M., Viruela, R., Ortí, E., and Martín, N. (2010) *J. Am. Chem. Soc.*, **132**, 1772.
8 Tong, L.H., Wietor, J.-L., Clegg, W., Raithby, P.R., Pascu, S.I., and Sanders, J.K.M. (2008) *Chem. Eur. J.*, **14**, 3035.
9 For an example where we resorted to ^{1}H NMR titrations for the determination of binding constants toward C_{60}, see Pérez, E.M., Capodilupo, A.L., Fernández, G., Sánchez, L., Viruela, P.M., Viruela, R., Ortí, E., Bietti, M., and Martín, N. (2008) *Chem. Commun.*, 4567.
10 Leach, S., Vervloet, M., Despres, A., Breheret, E., Hare, J.P., Dennis, T.J., Kroto, H.W., Taylor, R., and Walton, D.R.M. (1992) *Chem. Phys.*, **160**, 451.
11 (a) Sun, Y.P. and Bunker, C.E. (1993) *Nature*, **365**, 398; (b) Gallagher, S.H., Armstrong, R.S., Lay, P.A., and Reed, C.A. (1995) *J. Phys. Chem.*, **99**, 5817.
12 Song, J., Aratani, N., Shinokubo, H., and Osuka, A. (2010) *J. Am. Chem. Soc.*, **132**, 16356.
13 Sun, Y.-P., Bunker, C.E., and Ma, B. (1994) *J. Am. Chem. Soc.*, **116**, 9692.
14 Gil-Ramírez, G., Karlen, S.D., Shundo, A., Porfyrakis, K., Ito, Y., Briggs, G.A.D., Morton, J.J.L., and Anderson, H.L. (2010) *Org. Lett.*, **12**, 3544.
15 Canevet, D., Gallego, M., Isla, H., de Juan, A., Pérez, E.M., and Martín, N. (2011) *J. Am. Chem. Soc.* doi: 10.1021/ja111072j
16 (a) Pérez, E.M., Sánchez, L., Fernández, G., and Martín, N. (2006) *J. Am. Chem. Soc.*, **128**, 7172; (b) Gayathri, S.S., Wielopolski, M., Pérez, E.M., Fernández, G., Sánchez, L., Viruela, R., Ortí, E., Guldi, D.M., and Martín, N. (2009) *Angew. Chem., Int. Ed.*, **48**, 815.
17 Tashiro, K. and Aida, T. (2007) *Chem. Soc. Rev.*, **36**, 189.

Index

Supramolecular Chemistry of Fullerenes and Carbon Nanotubes, First Edition. Edited by Nazario Martin and Jean-Francois Nierengarten.

c

d

e

f

o

p

r